Environmental Science and Engineering

The ultimate goal of this series is to contribute to the protection of our environment, which calls for both profound research and the ongoing development of solutions and measurements by experts in the field. Accordingly, the series promotes not only a deeper understanding of environmental processes and the evaluation of management strategies, but also design and technology aimed at improving environmental quality. Books focusing on the former are published in the subseries Environmental Science, those focusing on the latter in the subseries Environmental Engineering.

Long Quoc Nguyen · Luyen Khac Bui ·
Xuan-Nam Bui · Ha Thanh Tran
Editors

Advances in Geospatial Technology in Mining and Earth Sciences

Selected Papers of the 2nd International Conference on Geo-spatial Technologies and Earth Resources 2022

Springer

Editors
Long Quoc Nguyen
Department of Mine Surveying
Hanoi University of Mining and Geology
Hanoi, Vietnam

Luyen Khac Bui
Department of Geodesy
Hanoi University of Mining and Geology
Hanoi, Vietnam

Xuan-Nam Bui
Department of Surface Mining
Hanoi University of Mining and Geology
Hanoi, Vietnam

Ha Thanh Tran
Department of Geology
Hanoi University of Mining and Geology
Hanoi, Vietnam

ISSN 1863-5520 ISSN 1863-5539 (electronic)
Environmental Science and Engineering
ISBN 978-3-031-20462-3 ISBN 978-3-031-20463-0 (eBook)
https://doi.org/10.1007/978-3-031-20463-0

This Springer imprint is published by the registered company Springer Nature Switzerland AG
The registered company address is: Gewerbestrasse 11, 6330 Cham, Switzerland

Preface

This book comprises a series of selected high-quality peer-reviewed papers delivered at the 2nd International Conference on Geospatial Technologies and Earth Resources (GTER 2022). The event will be held during October 14–15, 2022, at Hanoi University of Mining and Geology (HUMG), Hanoi, Vietnam, which is co-organized by HUMG and the International Society for Mine Surveying (ISM) to celebrate the 55th anniversary of the Department of Mine Surveying (HUMG). The conference is financially supported by the Vietnam Mining Science and Technology Association (VMSTA), the Vietnam Association of Geodesy, Cartography and Remote Sensing (VGCR), Vietnam National Coal-Mineral Industries Holding Corporation Limited (VINACOMIN), and Dong Bac Corporation (NECO).

The conference is believed to be an excellent opportunity during the COVID pandemic for the authors and participants to discuss advanced technologies and scientific directions in the fields of geospatial technologies and earth resources. Additionally, via this conference, a chance to exchange new ideas, innovative thinking, and application experiences both virtual and in-person will be provided, by which research or business relations and partner finding for future collaborations would have been established.

Totally, 205 manuscripts have been submitted to the organizing committee. Subsequently, after screening by the selection committee and reviewing by at least two blind reviewers, 34 research and review papers have been selected to be presented at the conference. The selected papers will be delivered over four planned sessions covering different topics of geospatial technologies, earth sciences, water resources, and environmental systems. These 34 papers were also selected for publication in this book with Digital Object Identifier (DOI) references. We believe that this book will provide the readers with an overview of recent advances in the fields of geospatial technologies and earth resources.

We would like to thank all members of the organizing and selection committees, all blind peer reviewers, all chairpersons, and invited speakers for their invaluable contributions. We are also thankful to (i) Mr. Phuong Kim Minh—President of Dong Bac Corporation, (ii) Assoc. Prof. Tran Xuan Truong—President of the HUMG University Council, Assoc. Prof. Trieu Hung Truong—Vice-Rector of HUMG for their administrative work and financial supports. Special thanks are also to Doris Bleier and Viradasarani Natarajan at Springer Nature for supporting book production procedures. Finally, all authors are thankful for their manuscript submissions regardless of whether or not their manuscripts were selected.

Hanoi, Vietnam
June 2022

Long Quoc Nguyen
Luyen Khac Bui
Xuan-Nam Bui
Ha Thanh Tran

Reviewers

Anh Quan Duong, Hanoi University of Mining and Geology, Vietnam
Hoang Bac Bui, Hanoi University of Mining and Geology, Vietnam
Ji Chen, Kyoto University, Japan
Dang Vu Khac, Hanoi National University of Education
Tuyet Minh Dang, Thuy Loi University, Vietnam
Kosta Prodanov Dimitrov, University of Mining and Geology, Sofia, Bulgaria
Mai Dinh Sinh, Le Quy Don Technical University, Vietnam
Cao Dinh Trieu, Institute of Geophysics, Geodynamics Department, Vietnam
Cao Dinh Trong, Institute of Geophysics, Geodynamics Department, Vietnam
Minh Duc Do, Hanoi University of Science, Vietnam
Pham Duc Thang, Vietnam Oil and Gas Group, Vietnam
Nguyen Thanh Duong, Hanoi University of Mining and Geology, Vietnam
Van Hao Duong, Hanoi University of Mining and Geology, Vietnam
Tran Thi Huong Giang, Hanoi University of Mining and Geology, Vietnam
Ropesh Goyal, Indian Institute of Technology Kanpur, India
Thi Hang Ha, Hanoi University of Civil Engineering, Vietnam
Hung Ha, State University of New York at Fredonia, USA
Van Long Hoang, Vietnam Petroleum Institute, Vietnam
Le Hong Anh, Hanoi University of Mining and Geology, Vietnam
Nguyen Huu Hiep, Hanoi University of Mining and Geology, Vietnam
Mo Jialin, National Institute for Environmental Studies, Japan
Fan Jiangtao, Chang'an University, China
Oleg I. Kazanin, Saint-Petersburg Mining University, Russia
Nguyen Khac Long, Hanoi University of Mining and Geology, Vietnam
Hung Khuong The, Hanoi University of Mining and Geology, Vietnam
Duy Thong Kieu, Hanoi University of Mining and Geology, Vietnam
Phu Hien La, Thuy Loi University, Vietnam
Nguyen Lan Chau, University of Transport and Communications, Vietnam
Tomasz Lipecki, AGH University of Science and Technology, Poland
Van Anh Cuong Le, University of Science, Ho Chi Minh City
Tien Dung Le, Hanoi University of Mining and Geology, Vietnam

Valeri Mitkov, University of Mining and Geology, Sofia, Bulgaria
Masoud Monjezi, Tarbiat Modares University, Iran
Thuc Ngo, Mien Tay Construction University, Vietnam
Nguyen Ngoc Thach, Hanoi University of Science, Vietnam
Van Lam Nguyen, Norwegian University of Science and Technology, Norway
Nguyen Ba Duy, Hanoi University of Mining and Geology, Vietnam
Nguyen An, University of Science and Education, Vietnam.
Quoc Dinh Nguyen, Vietnam Institute of Geosciences and Mineral Resources
Thinh Nguyen Van, Hanoi University of Mining and Geology, Vietnam
Pham Nhu Sang, Hanoi University of Mining and Geology, Vietnam
Shishkov Peter, University of Mining and Geology, Sofia, Bulgaria
Nguyen Quang Minh, Hanoi University of Mining and Geology, Vietnam
Vo Thi Hanh, Hanoi University of Mining and Geology, Vietnam
Doan Thi Ngoc Hien, Hanoi University of Science and Technology, Vietnam
Bui Tien Dieu, University College of Southeast Norway, Norway
Nguyen Tien Thanh, Hanoi University of Natural Resources and Environment, Vietnam
Hong Hanh Tran, Hanoi University of Mining and Geology, Vietnam
Dinh Trong Tran, Hanoi University of Civil Engineering, Vietnam
Van Anh Tran, Hanoi University of Mining and Geology, Vietnam
Quoc Cuong Tran, Institute of Geological Sciences, Vietnam
Le Hung Trinh, Le Quy Don Technical University, Vietnam
Stanislav Tsvetkov, University of Mining and Geology, Sofia, Bulgaria
Nguyen Van Duc, Korea Institute of Geosciences and Mineral Resources, South Korea
Nguyen Van Sang, Hanoi University of Mining and Geology, Vietnam
Vu Dinh Toan, Université de Toulouse, Toulouse, France

Contents

About the Editors

Long Quoc Nguyen, Ph.D. is currently Head of the Mine Surveying Department and Senior Lecturer at the Faculty of Geomatics and Land Administration, Hanoi University of Mining and Geology, Hanoi, Vietnam. His current research interests have been focused on the use of new technology equipment for mapping; the use of remote sensing and geographical information system for environmental issues; and applying artificial neural networks for predicting surface displacement. He has been a coordinator of many projects sponsored by the Ministry of Education and Training, the Ministry of Natural Resources and Environment, and industrial companies in Vietnam. He has published three books and more than 60 articles in national and international reputation journals including more than 30 articles in SCIE and Scopus-indexed journals, such as *Scientific Reports*, *Earth Systems and Environment*, *Remote Sensing*, *Natural Resources Research*, *Applied Sciences*, *Inzenia Mineralna*, *Coal Science and Technology*, *Archives of Mining Sciences*, and *World of Mining*. In addition, he has reviewed several manuscripts for Mining Science and Technology, *European Journal of Remote Sensing*, *Inzenia Mineralna*, and *Journal of Mining and Earth Sciences*. e-mail: nguyenquoclong@humg.edu.vn

Luyen Khac Bui, Ph.D. is Lecturer with the Faculty of Geomatics and Land Administration, Hanoi University of Mining and Geology, Hanoi, Vietnam. Currently, he is also holding Post-Doctoral Fellow position with the National Center for Airborne Laser Mapping (NCALM) and the Civil and Environmental Engineering Department, University of Houston, Houston, Texas, USA. His research interests include applying interferometric synthetic aperture radar (InSAR) in natural hazards monitoring, light detection and ranging (LiDAR) and digital elevation model (DEM) uncertainty propagation, using satellite radar altimetry to monitor sea level rise (SLR), geoid modeling, and environmental geodesy. He has published many articles in high-quality scientific journals including *Remote Sensing of Environment*, *IEEE Transactions on Geoscience and Remote Sensing*, *IEEE Geoscience and Remote Sensing Letters*, *Geophysical Journal International*, *Earth, Planets and Space*, *GIScience and Remote Sensing*, among others. He is also contributing to those journals as Reviewer. e-mail: buikhacluyen@humg.edu.vn

Xuan-Nam Bui is currently Full Professor at the Surface Mining Department, Faculty of Mining, Hanoi University of Mining and Geology (HUMG), Vietnam. He received the B.Eng. and M.Eng. degrees in Mining Engineering from HUMG in 1996 and 2001, and the Dr. Eng. degree in Mining Engineering from TU Bergakademie Freiberg, Germany, in 2005. He has been working as Lectuerer at HUMG since 1996. His research interests are environment-friendly mining technology and engineering, occupational safety and health in mining industry, and applications of artificial intelligence and machine learning in geoengineering, mining and environmental issues such as ground vibration, over-pressure, fly rock, air pollution, slope stability, etc. He is Editor-in-Chief of the Journal of Mining and Earth Sciences, HUMG, Potential Reviewer of some SCIE-indexed journals, and Editorial Board Member of several international scientific journals. He is Author and Co-author of more than 11¥0 articles published in SCI, SCIE, Scopus, and international journals, more than other 70 national journal papers, 21 books, and two patents. Currenly, he is Member of the Society of Mining Professors (since 2015) and Vice-President of Vietnam Association of Mining Science and Technology (since 2015). e-mail: buixuannam@humg.edu.vn

Ha Thanh Tran is currently Professor of Geology and Head of the Department of Geology, and Rector of Hanoi University of Mining and Geology (HUMG), Vietnam. He received the M.Sc. degree in Geoscience in 1997 and Ph.D. degree in Geology in 2001, both from the University of Regina, Canada. Professor Tran has been working for more than 30 years on regional geology, structural interpretation, tectonic evolution, and their relationship to natural resources and geohazards in Vietnam and other parts of the world. He is Author and Co-author of more than 40 international scientific articles and more than 40 other domestic papers, six books, and the speaker of more than 40 international presentations relating to his scientific interests. He is Member of the editorial board for several domestic and international scientific journals. Asides his official duty at the HUMG, Prof. Ha Thanh Tran has also been Member of several professional organizations in Vietnam and internationally, including Vice-President (2012–2014) and President (2020–present) of the Earth Science Council of the National Foundation for Science and Technology Development of Vietnam (NAFOSTED), Vice-President (2014, 2017), and President (2015–2016) of the Solid Earth Section (SE) of the Asia-Oceana Geological Society (AOGS). e-mail: tranthanhhai@humg.edu.vn

Application of Unmanned Aerial Vehicles for Surveying and Mapping in Mines: A Review

Long Quoc Nguyen, Minh Tuyet Dang, Luyen K. Bui, Quy Bui Ngoc, and Truong Xuan Tran

Abstract The use of unmanned aerial vehicles (UAVs) is increasing in the mining industry because of the obvious economic and environmental benefits as well as reducing the risk to mineworkers. This paper presents a review of recent developments in relation to the applications of UAVs in surveying and mapping of surface, underground, and abandoned mines. Additionally, after detecting the barriers associated with the deployment of UAV technology in mine surveying, the counter methods to overcome these challenges will be discussed. Finally, the prospects for the development of UAVs are also considered. The results indicate that UAVs can be used for constructing surfaces, creating three-dimensional (3D) models, evaluating their accuracy, and conducting topographic surveying of surface mines. Additionally, this system is a useful tool for mapping underground and abandoned mines. This paper provides a technical reference for expanding the knowledge and recognition of UAV applications in surveying and mapping in mine areas.

Keywords UAV · Drone · Mine · Surveying · Mapping · Terrain surveying

L. Q. Nguyen (✉) · L. K. Bui · Q. B. Ngoc · T. X. Tran
Faculty of Geomatics and Land Administration, Hanoi University of Mining and Geology, Hanoi 100000, Vietnam
e-mail: nguyenquoclong@humg.edu.vn

L. K. Bui
e-mail: buikhacluyen@humg.edu.vn

Q. B. Ngoc
e-mail: buingocquy@humg.edu.vn

T. X. Tran
e-mail: tranxuantruong@humg.edu.vn

L. Q. Nguyen
Innovations for Sustainable and Responsible Mining (ISRM) Research Group, Hanoi University of Mining and Geology, 18 Vien Str., Duc Thang Ward, Hanoi 100000, Vietnam

M. T. Dang
Thuyloi University, 175 Tay Son Street, Hanoi 100000, Vietnam
e-mail: dtminh@tlu.edu.vn

L. Q. Nguyen et al. (eds.), *Advances in Geospatial Technology in Mining and Earth Sciences*, Environmental Science and Engineering,
https://doi.org/10.1007/978-3-031-20463-0_1

1 Introduction

Traditional data collection applying a total station is limited by time and safety and thus often leads to a lack of necessary information for monitoring and mapping [1]. Generally, in mine areas, a total station is used for surveying, of which the measurement data are then processed using a computer-aided design (CAD) software. However, this approach is not effective and has many challenges for hard-to-reach areas [2]. The development of 3S technology, i.e., Remote sensing (RS), Global Navigation Satellite Systems (GNSS), and Geographic Information System (GIS), provides a critical technical guarantee for surveying and mapping in mine areas [3]. Although 3S technology can replace traditional ground mapping, RS with low spatial resolution satellite images is susceptible to weather problems, such as cloud cover. Furthermore, the lower temporal resolution makes satellite imagery difficult to depict the characteristics of mine areas undergoing extreme changes [4]. Therefore, this technology cannot meet the requirements of terrain monitoring and surveying in the mine, in particular in a small-scale or newly worked mine. In comparison, an unmanned aerial vehicle (UAV) method is inexpensive and has broader applicability. In other words, one of the main advantages of using UAVs compared to airborne or traditional field surveys is the significantly lower cost of exploration, especially for remote regions with poor infrastructure [5].

The usage of drones in mine areas is seemingly endless because they can assist to acquire data in the field in real time [6]. Most importantly, this tool can provide access to areas that are hard or dangerous to reach, or inaccessible by human workers, such as vertical cliffs or hills, underground mines, and mining regions. Moreover, UAVs are also capable of doing some mining-related tasks faster and at a lower cost, such as terrain surveying and 3D modeling, land damage assessment, and ecological environment monitoring [3]. They are equipped with various sensors, such as spectral imaging sensors, thermal infrared cameras, and gas sensors, which could give necessary data for different monitoring objects in the mining industry [3]. Rathore and Kumar [6] unlocked the potentiality of UAVs in the mining industry and its implications. They proposed the applications that UAV technology can play an important role in shaping the future of mining concepts, including safety and security, productivity, surveying and mapping, and field data collection [6].

In the mining industry, UAV technology is widely used in terrain surveying, 3D modeling, land damage assessment, ecological monitoring, geological hazards, pollution monitoring, and land reclamation and ecological restoration assessment [3]. According to Mukhamediev et al. [5], in the mineral exploration sphere, due to the gradual depletion of the existing field resources, it is necessary to apply new methods that intensify the processes of exploration. Thus, in recent years, the use of UAVs in mining activities is rapidly expanding. The main spheres of UAV application include mapping, 3D modeling, and conducting geophysical research [5]. The primary surface data, such as Digital Surface Model (DSM), Digital Elevation Model (DEM), and orthomosaic images, can be generated by a UAV system equipped with

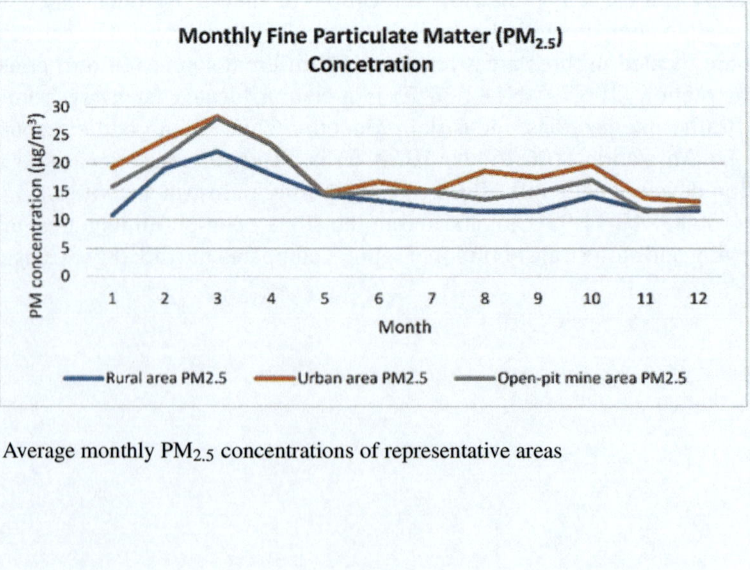

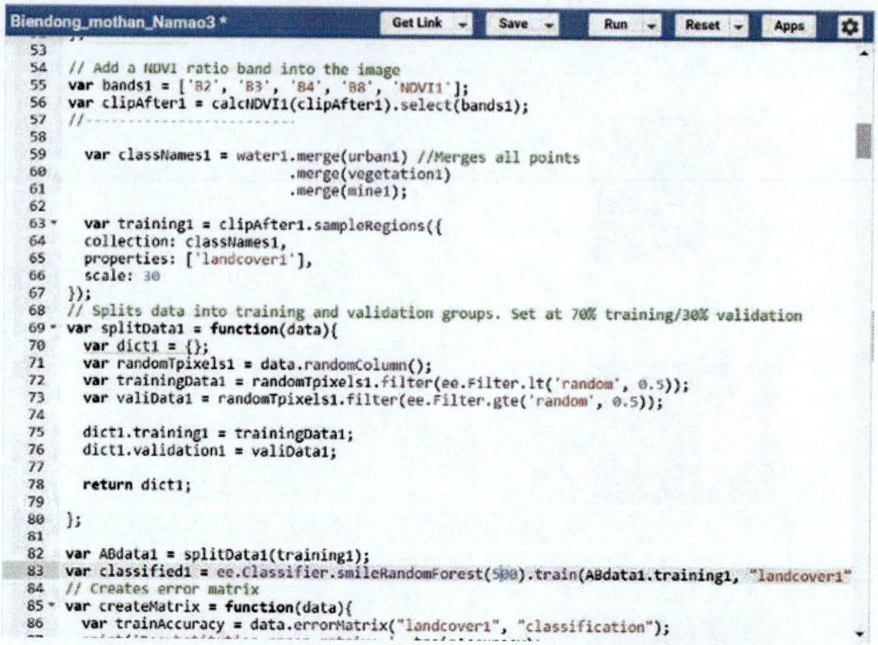


```
// Add a NDVI ratio band into the image
var bands1 = ['B2', 'B3', 'B4', 'B8', 'NDVI1'];
var clipAfter1 = calcNDVI1(clipAfter1).select(bands1);
//-------------------------

  var classNames1 = water1.merge(urban1) //Merges all points
                        .merge(vegetation1)
                        .merge(mine1);

  var training1 = clipAfter1.sampleRegions({
  collection: classNames1,
  properties: ['landcover1'],
  scale: 30
});
// Splits data into training and validation groups. Set at 70% training/30% validation
var splitData1 = function(data){
  var dict1 = {};
  var randomTpixels1 = data.randomColumn();
  var trainingData1 = randomTpixels1.filter(ee.Filter.lt('random', 0.5));
  var valiData1 = randomTpixels1.filter(ee.Filter.gte('random', 0.5));

  dict1.training1 = trainingData1;
  dict1.validation1 = valiData1;

  return dict1;

};

var ABdata1 = splitData1(training1);
var classified1 = ee.Classifier.smileRandomForest(500).train(ABdata1.training1, "landcover1"
// Creates error matrix
var createMatrix = function(data){
  var trainAccuracy = data.errorMatrix("landcover1", "classification");
```

Fig. 6 Code in GEE used to classify the images

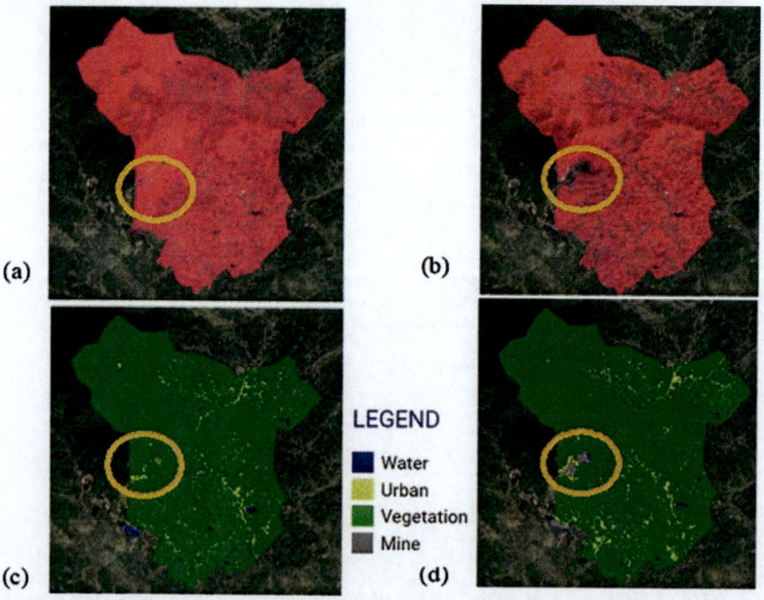

Fig. 7 **a** Sentinel-2 image (2016–2019) median filter, **b** Sentinel-2 images (2019–2021) median filter, **c** RF classification image (2016–2019), and **d** RF classification image (2019–2021)

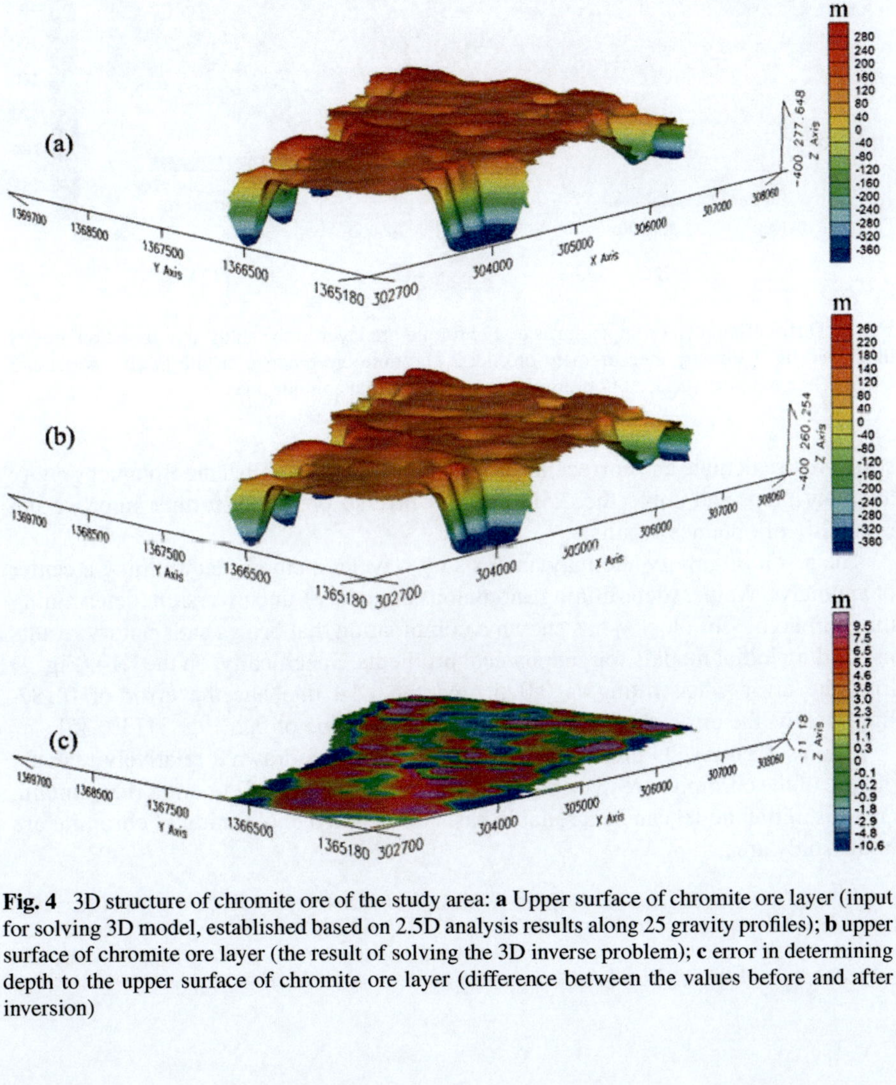

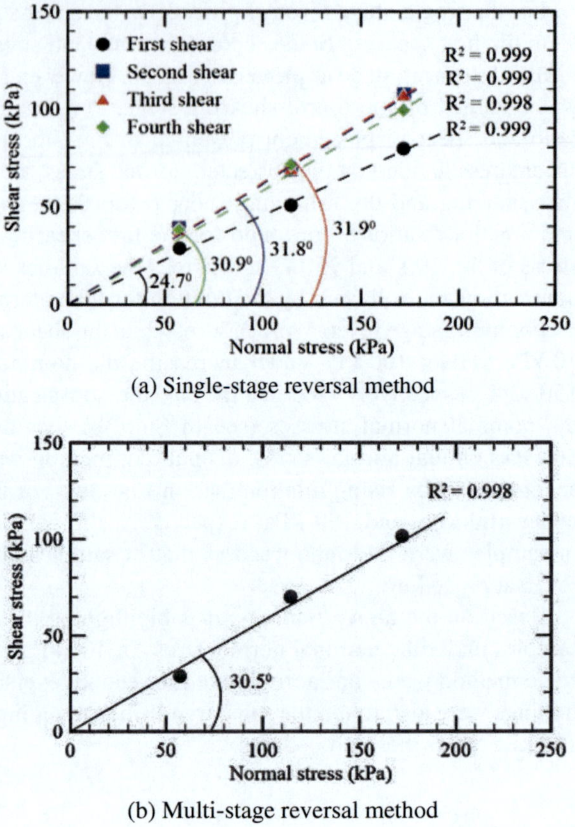

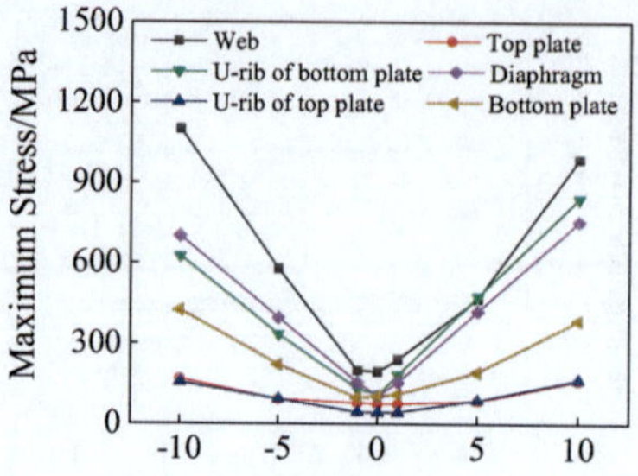

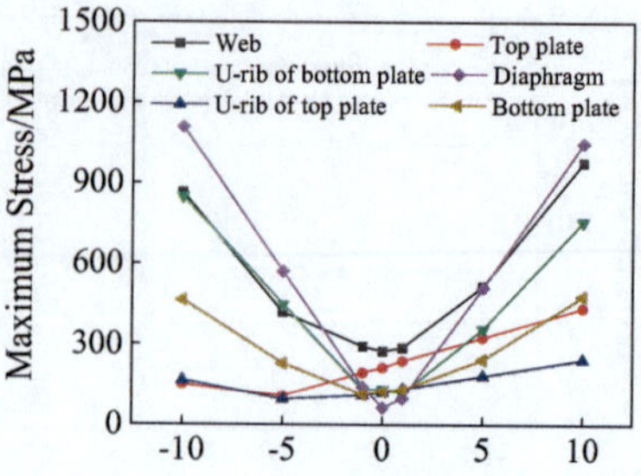

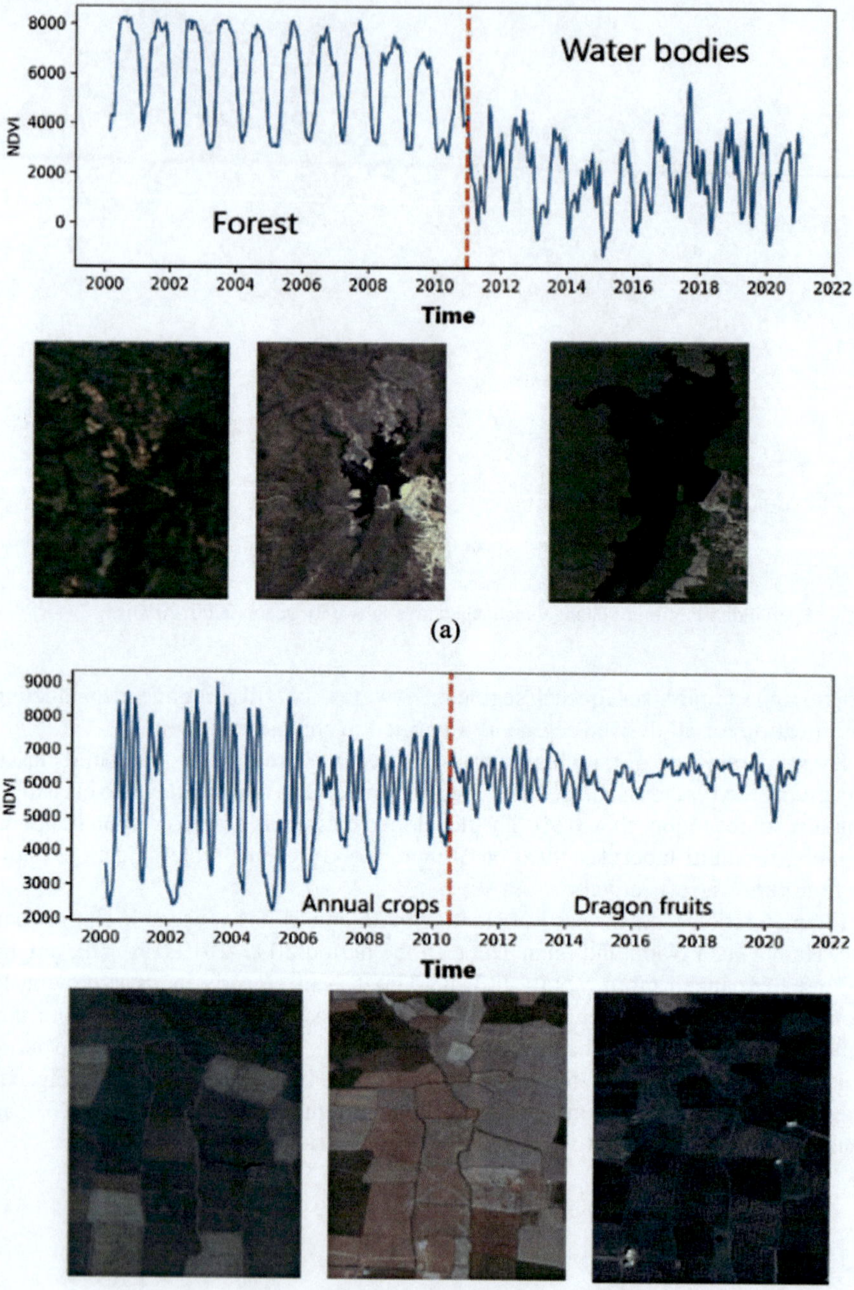

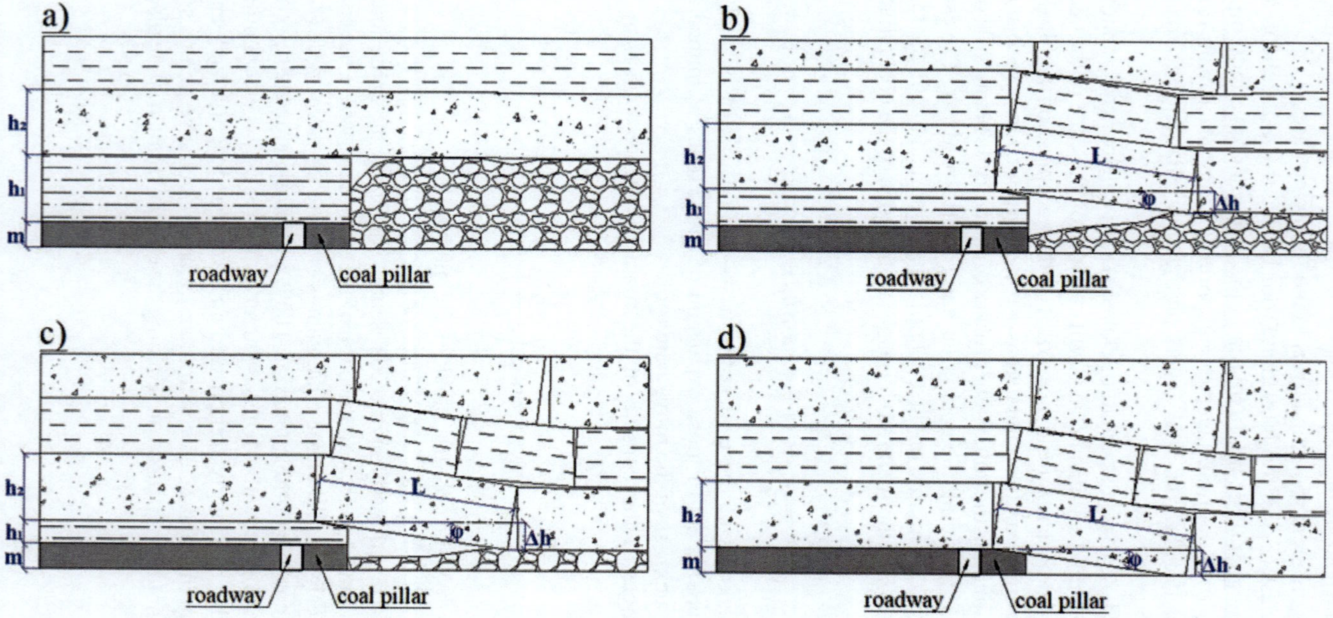

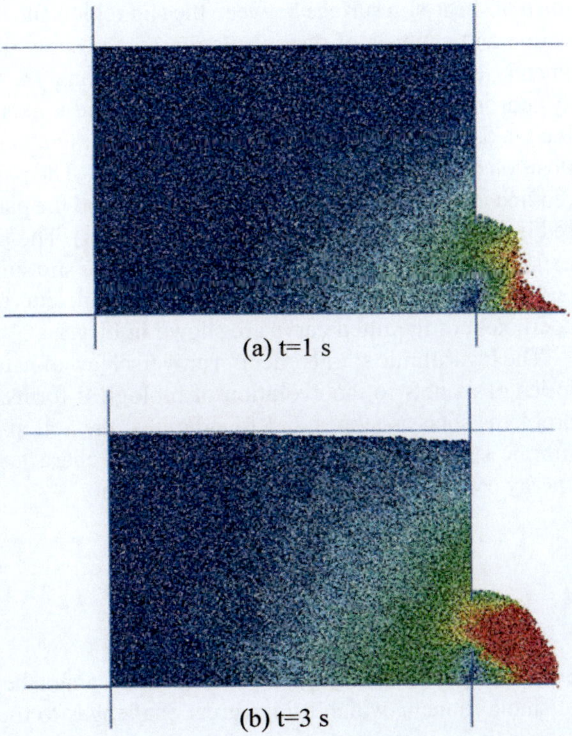

digital cameras [7] with some types of image processing software based on the Structure from Motion (SfM) technique, such as Agisoft Photoscan and Pix4D Mapper. Besides, the obtained surface data not only provides detailed information on the topographical variation but also is important in the calculation of erosion gulley size or slope stability in mine areas [8], estimation of surface extent, and volumetric excavation [7].

Recently, a few scientists have published review papers regarding the UAV application in mining operations. While Park and Choi [9] reviewed the applications of UAVs in mining from Exploration to Reclamation, Shahmoradi et al. [10] performed a comprehensive review to highlight the applications of drone technology in surface, underground, and abandoned mines. Unlike other literature reviews, Lee and Choi presented issues related to UAV technology in mine areas, such as mine surveying, mine operations, drill and blast, mine safety, and construction. Among the applications, using UAVs in terrain surveying and 3D modeling is one of the most popular applications for mining operations. However, no prior reviews have examined to identify and categorize UAV applications related to terrain surveying and mapping mine sites. The goal of this paper is to present a critical review of different applications of UAV in terrain surveying and 3D modeling in mine areas.

2 Data and Methodology

In this study, a systematic review of the extant literature is performed based on a structured analysis of topic specific studies that is related to terrain surveying and 3D modeling in mine areas.

2.1 Search Terms

The first step of a systemic review is to determine the related key works/individual concepts and operationalize them into search terms and syntax. For this study, they are arranged into the following search syntax for study retrieval: ("Unmanned Aerial Vehicles" OR "UAV" OR "Unmanned Aerial System" OR "UAS" OR "Drone"); AND ("Mine" OR "Mining" OR "Surface mine" OR "Open-pit mine" OR "Underground mine" OR "Abandoned mine" OR "Closed mine"); AND ("Surface" OR "DEM" OR "DTM" OR "DSM" OR "Surveying" OR "3D mapping" OR "3D model").

2.2 *Search Procedure*

The review is divided into five sections: defining the research questions, determining relevant studies, choosing studies, charting the data, and collating, summarizing, and reporting the results. A systematic search using the search syntax is performed in the Google Scholar, ScienceDirect, Scopus, and Web of Science databases. The main search language is English. The titles and abstracts of publications are reviewed to determine whether they meet the content of the paper. To be included in this review, the publications are required to be published in peer-reviewed journals or conference papers, issued within the last decade (i.e., 2010 to 2022), and focusing on UAV applications within the mine surveying domain. Some types of studies such as reports or industry trade articles are excluded. In the identification stage, 215 potential related studies are detected. After screening to eliminate duplicate entries, research that was not published within the 2010 to 2022 year range, and non-peer-reviewed papers, 175 documents are carried forward for eligibility analysis. In order to prove the eligibility, the full texts of the papers are reviewed, by which 114 are excluded because they are either not related to the mine surveying domain or do not address UAV applications in mines. A total of 66 articles become the foundation of our systematic literature review.

3 Results and Discussion

3.1 *Application of UAVs for Terrain Surveying and Mapping in Surface Mines*

Terrain surveying or topographic modeling is one of the applications for the mining industry, and it is used primarily for mineral resources and ore reserve estimation, mine planning, and reconciliation. One of the main sources of uncertainty in mining reconciliation is the topographic model updating [11]. Thus, terrain modeling of mine areas in real time is needed. Nevertheless, the traditional methods for mine surveying are expensive and time-consuming, even though it can take months without feedback from the surveying activities. An alternative way to improve the frequency of topography updating is through the use of image-based surfaces acquired by UAVs and processed by specific software [12]. Drones are a cost-effective, quick, and effective data collection tool for the surface generation, such as DEM, DSM, and digital terrain model (DTM). The use of UAVs for visual surveying as well as the creation of 3D models of mine sites has steadily become relevant. UAV technology can acquire high-resolution images, which are then transformed into 3D surface models (DEM, DSM, and DTM), and can be used for producing topographic maps, calculating excavation volume, and showing the mine site in 3D forms. The literature review shows that some scientists have reported the results of constructing DEM for open-pit mines using UAVs [13, 14], and some authors have used other traditional techniques to test the accuracy of DSM obtained using UAVs [12], or analyzing the accuracy of the DEM derived from photogrammetric processing [1, 2, 15, 16]. Additionally, others evaluated the performance of this technology in 3D modeling applications [17–20].

3.1.1 Construction of Surfaces and Accuracy Assessment

Cho et al. [13]. have verified the applicability of UAV photogrammetry to mining engineering. Aerial photos of the test mine area, which were taken by DJI S1000 with 10 cm resolution, were processed with the Agisoft Photoscan software to generate an orthophoto and DEM model [13]. Similarly, Nghia [21] assessed the possibility of developing 3D models for deep open-pit mines from UAV image data. To achieve this goal, the author used the DJI's Inspire 2 device to take photos at the Coc Sau coal mine. The results indicated that the 3D model established from photographic data by Inspire 2 UAV has met the requirements of the accuracy of establishing the mining terrain map at a 1: 1000 scale.

To quantify the uncertainty created by UAV technology, Beretta et al. [12]. compared a DSM formed by photogrammetry using UAVs with those generated through traditional methods. The results showed that the level of detail given by the UAV photogrammetric techniques proved to be more accurate, denser in information, and faster when compared to traditional methods.

According to Forlani et al. [22], the accuracy of photogrammetrically generated DSMs depends on geometric and physical factors, such as the image scale, ground sampling density (GSD), stereo base-length to object distance ratio, camera network geometry, percentages of strip overlap, the accuracy and distribution of ground control points (GCPs), camera calibration, image processing, image matching, point cloud noise, and outlier removal algorithms. The accuracy of DEM depends on flight height also mentioned by Nguyen et al. [23]. Determining DSM quality is therefore a complex task, because the number of variables involved is enormous, and no single experimental study can encompass all of the relevant aspects [22]. Many studies addressed the accuracy of UAV-generated DSM in different environments. Determination of the number of GCPs to ensure the accuracy of mapping and minimize measurements in the field can be found in Nguyen et al. [2]. Similarly, Long et al. [24] used a Light-Weight UAV to choose the number of GCPs for developing precise DSM in the medium-sized open-pit mine.

Kršák et al. [25] also verified the quality of a DSM in mines, which was obtained photogrammetrically using a low-cost UAV. The resultant models demonstrated that the 3D model is multiple times more detailed than the surface formed from the points surveyed by the total station. In addition, UAS-derived DSMs were compared to field measurements of mining pits in the region to assess the accuracy of UAV-derived pit volume measurements [15]. Chirico and DeWitt [15] indicated that UAV imagery and SfM photogrammetric techniques allow DSMs to be produced with a high degree of precision and relative accuracy, but highlighted the difficulties of mapping small artisanal mining pits in remote and data-sparse terrain.

According to Tien Bui et al. [26], the application of UAV and SfM for complex topographic areas, such as open-pit mine areas, is still poorly understood. Therefore, they investigated and verified the potential application of these techniques for building DSM in open-pit coal mine areas and evaluating its accuracy. Because a DSM should only be used after accuracy assessment, in this project, both the horizontal and vertical assessments were carried out by GCPs measured by a Leica total station in terms of Root Mean Square Error (RMSE). The result showed that the DSM model has high accuracy. It can be concluded that small UAVs and SfM are feasible and valid tools for 3D topographic mapping in complex terrains, such as open-pit coal mine areas. Figure 1 that is reused from shows the use of UAV images for generating the DSM of the Nui Beo coal mine, Vietnam.

Like many scientists, Nguyen et al. [1]. believed that the accuracy of UAV-derived DSMs is influenced by topographical factors in the active surface mines. Thus, they performed an experiment to apply the UAV method to three active coal mines, operating at altitudes from −300 m to 300 m. Accordingly, the effects of topographic factors, such as slope, relative elevation, and number of GCPs, on the accuracy of DSMs constructed by the UAV imagery technique were assessed in the experiment. The obtained results revealed that DSMs were generated at a very high horizontal accuracy, i.e., the cm level.

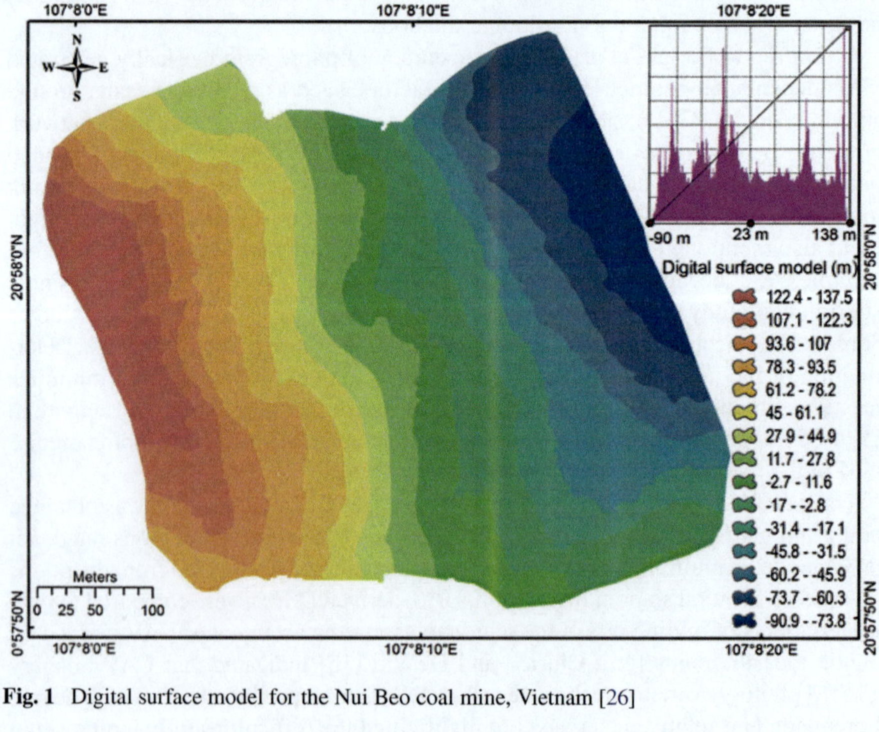

Fig. 1 Digital surface model for the Nui Beo coal mine, Vietnam [26]

In addition, the main topographic information is necessary data for some other purposes. The produced DSM can be used to analyze the progress of the mining process, to estimate ore carrying capacity [27]. Chen et al. [28] examined the characteristics of iron open-pit mines located in Beijing district, China, using high-resolution UAV images. To this end, information regarding DSMs was derived from UAV images and SfM photogrammetry techniques.

Similarly, a spatial query and an analysis of the open-pit mine were performed using the Quantum GIS program and DEMs can be found in the study of Gil and Frąckiewic [29]. In this research, aerial photographs were obtained using a drone by which DTM and DSM were generated. Thereafter, spatial analysis was performed to optimize the location of the observation network points in a surface mine.

Xiang et al. [30] selected an open-pit mine in Beijing, China, as the research area to assess geomorphic changes using a DEM generated using high-resolution UAV images and SfM photogrammetry. Accordingly, the surface of the open-pit mine was analyzed by calculating the difference between the two DEMs.

3.1.2 Creating 3D Models and Evaluating Their Accuracy

3D models are necessary tools for experts in the mining industry because they provide high-quality representations of mining sites. Thus, it is essential to collect accurate data for generating 3D models. Traditional survey methods, which utilize a total station and GNSS receivers to conduct a mine survey, are limited to accurately defining coal stockpiles. Moreover, these approaches increase the risk of injury to the surveyor and demand a high level of safety as well as are often time-consuming, which can lead to additional costs. Alternatively, UAV technology can offer a more cost-effective and safer option [31]. In recent years, UAV photographic measurement technology has played a significant role in 3D modeling in mine areas. Due to UAVs' small size and maneuverability, they can capture data from much lower heights, starting from the ground surface, sweeping through the study areas at various heights and viewpoints, as well as fly-over views above the mine sites [32]. An example of a 3D texture model of the Thuong Tan 3 quarry, Vietnam in 2020 captured by lightweight UAV is shown in Fig. 2.

UAV platforms are increasingly being used as an important source of data for monitoring, surveillance, and 3D modeling of areas influenced by mining activities [17]. They are usually used in combination with digital cameras, and the acquired images are processed using a combination of SfM and Multi-View Stereo (MVS) approaches allowing the extraction of 3D point clouds [33]. The 3D models created from UAV imagery can serve different applications, ranging from natural resource management to civil engineering. Several studies have been carried out in recent years to evaluate the performance of UAVs in 3D modeling applications. Valuable reviews of such studies can be found in [31, 34, 35].

The possibility of 3D modeling in an open-pit limestone mine using a rotary-wing UAV was shown in Kang et al. [36]. The results were used to estimate the amount of mining volume before and after mining of limestone by explosive blasting quickly

Fig. 2 3D texture model of the Thuong Tan 3 quarry, Vietnam in 2020 captured by DJI Phantom 4 Pro

and accurately in a relatively short time. The application of UAVs has been proved as an alternative tool for 3D mapping of open-pit mines in Bui et al. [37]. Based on the UAV images, large-scale 3D topographic maps were successfully modeled. The field test results in this study indicated the applicability of the low-cost UAVs for 3D mapping in large and deep coal pits with relatively high accuracy. This result can be used for optimizing mining operations and also controlling the atmospheric environment.

Similarly, in applying UAVs in building 3D models for large surface mines, Battulwar et al. [18] developed a practical setup to use cost-effective drones for the systematic generation of high-resolution images and 3D models for large open-pit mines. The methods used in this study have been validated using experimental and simulation studies. It can be concluded that the study could generate millimeter resolution 3D models of hazardous inaccessible open-pit slopes without any risks to personnel, who are responsible for the surveys and measurements to obtain multiple parameters of mine slopes.

According to Vassena and Clerici [20], the state-of-the-art 3D surveying technologies, if correctly applied, allow obtaining 3D-colored models of large open-pit mines using different technologies, such as terrestrial laser scanner (TLS) with images combined with UAV-based digital photogrammetry. In the Italian white marble open-pit mine "Botticino", located in northern Italy, a combination of digital photogrammetry by UAV and TLS was proposed. This resulted in an increase in local precision up to ±2 cm [20].

Also, Tong et al. [19] processed and integrated point cloud data created by TLS with UAV imagery and generated a 3D model for mapping and monitoring open-pit mine areas, which achieved the decimeter-level accuracy. Besides using software to create 3D models, such as Agisoft and Pi4D, many studies have shown that integrating UAV data into a GIS environment is an appropriate approach for creating 3D models and post-processing of UAV imagery data. Filipova et al. [38] dealt with UAV data integration into GIS and conducted a spatial analysis to generate a 3D model of an open-pit quarry. UAV digital images as well as contemporary photogrammetric techniques help create accurate geometric 3D models. Through 3D model analysis, more precise information about the current processes and events in the quarry could be gathered, leading to future managerial decisions.

Regarding the UAV's application in building 3D models, Tscharf et al. [39] presented a fully automated end-to-end workflow to obtain precise and geo-accurate reconstructions, especially for complex environments such as open-pit mines. Together with aerial images from a UAV, they were able to enrich 3D models by combining terrestrial images and inside views of an object by joint image processing to generate detailed, accurate, and complete reconstructions.

Ulusoy et al. [40] have used a lightweight drone to collect digital aerial images of an open-pit mine for the ultimate purpose of modeling the terrain using the SfM procedure. They have been able to derive a high-resolution (0.3 m/pixel) DEM and a very high-resolution (0.04 m/pixel) orthorectified aerial photograph. The elevation model dataset has been compared with the regular topographic point measurements of the mine pit, and the accuracy of the aerially derived model has been investigated.

Le Van et al. [41] acquired images in open-pit mines using a post-processed kinematic (PPK) drone and produced a highly accurate DSM. In addition, they experimentally proved the possibility of topographic survey for open-pit mines using drones by analyzing the accuracy improvement based on the increased number of GCPs.

According to Shahbazi et al. [35], UAV-based images have the potential to provide data with unprecedented spatial and temporal resolution for 3D modeling. Thus, they presented theoretical and technical experiments regarding the development, implementation, and evaluation of a UAV-based photogrammetric system for precise 3D modeling in the grave-pit mine. The study was preliminarily assessed for the application of gravel-pit surveying by UAV.

The accuracy of a UAV-based 3D model was mentioned in many studies. Park et al. [31] compared the accuracy of UAV-generated 3D coal stockpile models against traditional field survey techniques. They assessed the effect on the accuracy of the coal stockpile for varying shapes. The results revealed that the UAV-derived 3D models show maximum volume errors of less than 9.0% and minimum volume errors of more than 0.3%. The accuracy of the 3D model reconstructed from UAV images was also assessed by Park et al. [31]. Wang et al. [34] determined the accuracy of 3D geometry from low-attitude UAV images at the Zijin Mine in China. They implemented different algorithms, such as the SfM and the patch-based multiview stereo (PMVS) systems, to create a dense 3D point cloud from the UAV images. They used 17 GCPs to geo-reference a 3D reconstruction point cloud, and the accuracy of the 3D geometry was evaluated by using both the GCPs and the TLS point cloud. The UAV point cloud accuracy was first evaluated at a point level by comparing the absolute coordinates between the UAV point cloud and the GCPs.

In relation to the accuracy assessment of the UAV-derived 3D model, González-Aguilera et al. [16] indicated that even though the image-based modeling workflow requires applying several steps sequentially in order to obtain a real-based 3D model, and thus, error propagation must be mandatory, the level of obtained accuracy is good enough.

3D mapping is a very important aspect of the mining industry. In recent years, the use of UAVs for visual surveying as well as the generation of 3D images of mine sites has steadily become popular. At present, UAV technology has been used widely in the mining industry for terrain surveying. Drones can acquire high-resolution images that are then transformed into 3D surface models. These models are used for topographic mapping or for showing the mine sites in the 3D form. UAVs equipped with low-cost cameras can be considered a great instrument to survey the surface mine. This photogrammetric approach could overcome the low resolution of satellite images and avoid the tiring groundwork of the total station and Global Positioning System (GPS), as well as provide a 3D visualization effect of the study area [3].

Katuruza and Birch [42] used UAV technology in opencast highwall mapping in opencast mines at Isibonelo Colliery of South Africa to produce data for updating geological models and avail the latest information for mine planning to improve the short-term plans. With the UAS, it is possible to obtain digital images of the highwall as well as a multitude of digital terrain points, all in 3D space. The obtained results indicated that the greater the amount of collected high-resolution images, the more detailed the model produced, a dense 3D image of the pit. The generated drone data model was validated against the resource model and actual survey data.

Malpeli and Chirico [43] explored the application of a small UAS for mapping informal diamond mining sites in Africa. They found that this technology provides aerial imagery of unparalleled resolution in a data sparse, difficult to access, and remote terrain. The aerial images were used to develop 10 cm resolution DEMs of the mine site. The authors used ortho-images and DEMs to model the geomorphology of the terrain, and the areas of diamond deposition in the region could be identified.

According to Leo Stalin and Gnanaprakasam [44], a mine map can give information to optimize mining activity. Regular updating of the 3D model and digital mine maps provides an easy way to assess the activity carried out inside the mine. They used Quadrotor UAV to acquire nadir and oblique aerial images. These images are processed in various UAV data processing software to produce high-resolution orthophotograph, DTM, DSM, contour, and 3D virtual reality models. The digital orthophotograph and 3D models generated from this method were used to create the mine map.

Salvini et al. [45] used UAVs to map fractures in a marble quarry and, subsequently, to build 3D discrete fracture network models. Based on the combined use of high-resolution UAV images and engineering geological data on a marble buttress, the construction of a reliable 3D rock mass model can be done. Preliminary results revealed the benefit of modern photogrammetric systems in producing detailed orthophotos and the latter allows accurate mapping in areas difficult to access (one of the main limitations of traditional techniques).

Several literary studies have been conducted to use fixed-wing and rotary-wing UAVs for terrain surveying in surface mines. According to Lee and Choi [46], there are various characteristics between the fixed-wing and rotary-wing UAVs, such as flight height, speed, time, and performance of mounted cameras; thus, they compared the results of topographic surveying at the same site. The fixed wing showed a relatively negligible error when the results of the two types of aerial surveying were compared with ground data. Figure 3 shows orthomosaic images and DSMs of the study area for the two types of UAVs.

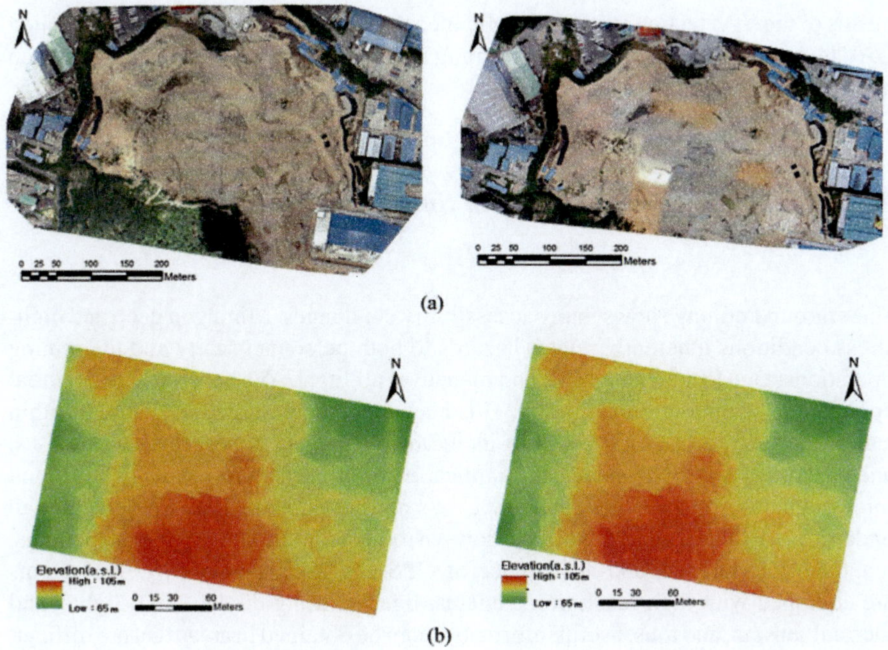

Fig. 3 Results of topographic surveying for the two types of UAVs from [46]: **a** Orthomosaic image, **b** digital surface model

3.2 Application UAV in Terrain Surveying

Besides, some scientists carried out the topographic survey at an open-pit mine using a rotary-wing UAV (DJI Phantom2 Vision+) [47] and a fixed-wing unmanned aerial vchicle (SenseFly eBee) [48]. The obtained results revealed that the fixed-wing UAV has a relatively longer flight time and larger coverage area than rotary-wing UAVs, it can be effectively utilized in large-scale open-pit mines as a topographic surveying tool while rotary-wing UAV is suitable for topographic survey at small-scale open-pit mines [48].

Rossi et al. [49] presented a method to reconstruct the quarry terrain in Bari, Italy, by using nadir and oblique aerial photographs acquired from a UAV and conducted a feasibility analysis after that. It observed that the final position of the point clouds, which show the main geometrical characteristics of the quarry in the topography reconstruction of the study site, can achieve an accuracy of a few centimeters.

In open-pit mines, monitoring of topographic and volumetric changes through time plays an important role in supporting excavation stages and planning rehabilitation strategies. Esposito et al. [7] used UAV photogrammetry to quantify the excavated volume at the Sa Pigada open-pit mine in Sardinia, Italy, and to evaluate the variations in the surface mine extent. They carried out two UAV-based surveys in 2013 and 2015, and 3D dense point clouds and digital orthophotos were obtained by means of the SfM technique. Results obtained in this study suggested that the applied UAV techniques are suitable for performing accurate change detection analysis in an open-pit mine extent.

3.3 Application UAV in Terrain Surveying in Underground Mines

Underground mining shows many accessibility challenges. Mining in deep and high-stress conditions inherently relates hazards to both personnel safety and the mining operations. Due to their small size and maneuver ability, UAVs have various potential applications in underground mining. This may allow them to access locations within a mine that are normally inaccessible, including ore passes, stopes, ventilation raises, and hazardous areas. Nowadays, the application of drone technology in underground mines is in its infancy. There are few UAVs and instruments that are dedicated to underground mines, where we have low-visibility conditions, confined openings, magnetic interference, and an absence of GPS coverage. However, UAV systems are equipped with high-resolution cameras, light emitting diode (LED) lights, and thermal sensors, and thus, useful information can be obtained in areas that are difficult to be accessed by mine workers [9]. In the view of Mitchell and Marshasll [50], the most likely near-term applications for underground UAVs include mine surveying, search, and rescue. In this paper, we only review the studies on UAV application for surveying and mapping of underground mines.

3D map plays an important role in underground mining because it provides accurate data and models of the 3D area of the underground mine. In underground mining, the shape of the underground mine area is dynamic due to a lot of reasons, like the excavation of new tunnels or some natural factors. The accurate and up-to-date 3D model and data of the underground mine environment are, therefore, necessary for an efficient and safe mining process [10]. Li et al. [51] considered the application of UAVs in underground mine mapping and proposed a 3D tunnel system search and mapping algorithm. The tunnel area search and map building are autonomous, and an operator only needs to start or stop the map building in the remote computer. Additionally, a study by Ge et al. [52] presented the results of the work performed by applying UAVs for Tahmoor underground mines in New South Wales (NSW), Australia. They used a UAV to map the underground mine subsidence in these mines. UAV oblique photogrammetry can obtain the three-dimensional (3D) coordinate information of ground features.

Photogrammetry is becoming a more common method for mapping geological and structural features in underground mines. Russell et al. [53] implemented an experiment with photogrammetry conducted from a UAV platform in an underground mine. They assessed the viability of using UAV-based imagery and photogrammetry to model and map rock masses that are inaccessible in underground mines. The obtained results include 3D digital photogrammetry models created from video frame stills and discontinuity map within the digital model.

An interesting example of using an inspection drone in an underground mine for mapping or exploration can be found in the study of Papachristos et al. [54]. They proposed an integrated approach for autonomous navigation and mapping in underground mines using a drone. Stereo cameras have been used for 3D mapping with aerial robots in subterranean tunnels. The research has shown that the usage of long-wave infrared of the electromagnetic spectrum allows the use of thermal cameras in environments with poor visibility and has been included in sensor sets in various underground mapping and localization studies.

According to Turner et al. [55], the advent of inexpensive, open platform UAVs allows to characterize hazardous rock masses by using traditional photogrammetric and forward-looking infrared imagery techniques. In order to prove this, they created a 3D model by thermal imagery using a UAV in an underground mine Barrick Golden Sunlight, Whitehall, Montana, USA, and acquired the geological data from photogrammetry models. The UAV system used in this study included obstacle detection, lighting, thermal imagery, and software. Results concluded that the combination of off-the-shelf technologies with a UAV system can be successfully employed as a geotechnical tool in the underground mining environment.

Similarly, Turner et al. [56] also proved that both thermal and multispectral imaging were successful in the detection and characterization of loose, unstable ground, and adverse discontinuities in the underground mining environment. The datasets, including multiple thermal, multispectral, red, green, and blue (RGB), and Light Detection and Ranging (LiDAR), were acquired in the same study area. They used these data to generate georeferenced 3D point clouds and meshes, and to map discontinuities. Figure 4 shows the DJI Wind 2 in the study of Tuner et al. [56], which could carry a large payload, including a MicaSense RedEdge-M imager, StratusLED ARM lighting, and, most importantly, an Emesent Hovermap SLAM system.

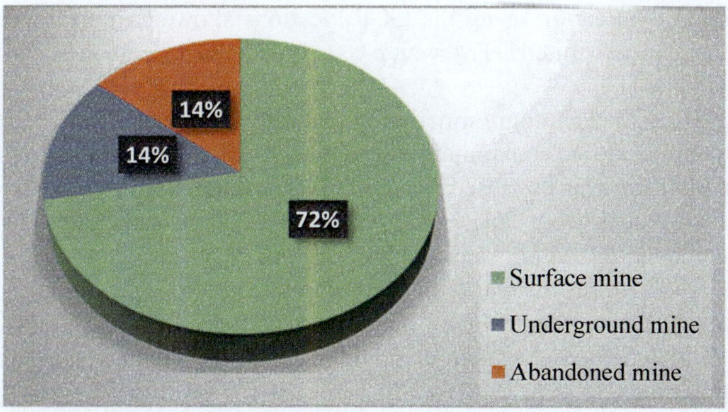

Fig. 4 Percentage distribution and number of reviewed studies of UAV applications in surface, underground, and abandoned mines to surveying and mapping process

3.4 Application of UAVs in Terrain Surveying of Abandoned Mines

Mine survey is an indispensable part of the ecological restoration design of an abandoned open-pit mine because it provides precise survey data for mine management. Furthermore, monitoring and mapping closed mines play an important role to decrease the risk of environmental hazards. However, it is difficult to survey the vast areas with traditional, labor-intensive, expensive monitoring methods. Drone technology, as a financially efficient approach, can be an alternative solution [10]. The most notable recent research effort is by Dai and Xu [57]. In this study, the authors applied UAV photogrammetry technology to 3D modeling and earthwork calculation to solve the problems of high cost, low efficiency, and high labor intensity in traditional manual field mine surveys in the abandoned quarries in Baitu, Suzhou City. 82 ground control points were set up, and elevation data of the ground were acquired using a Dajiang spirit 4 RTK UAV and extracted from a 3D model created with the Context Capture software.

At present, the recultivation of abandoned open-pit mines is a critical environmental task [58]. It is necessary to determine the amount of soil required for the preparation of the investment's expenditure plans. To achieve this, Molnar and Domozi [58] created a 3D surface model based on photogrammetry and UAV photos. As a result, the amount of filling material needed for the recultivation of a closed mine was calculated by 3D models. They matched the volume data computed from geodetic surveys on the mine with UAV-based DSM and the calculation results with the help of the 3D elevation model. The evaluations have indicated that the calculated volume based on photogrammetry has been within the expected accuracy range.

Suh and Choi [59] employed UAV photogrammetry to survey abandoned mine areas. A digital georeferenced orthoimage and DTM with the 5 cm resolution could be obtained by coordinates of pre-installed ground control points (GCPs). Accordingly, contour lines (at the 10 cm contour interval), slope, and curvature were created using the DTM. Validation using the GCP locations showed an error of approximately 14 cm in the generated DTM, which was considered acceptable for subsidence mapping purposes.

Motyka [60] applied photogrammetry for mapping the anthropogenic terrain alterations on exampled closed coal mines in Katowice for land reclamation. In this study, a new UAV platform was proposed for remote laser scanning of terrain from a low altitude, which consisted of an inertial system, a 2D laser scanner, and a system recorder. Moreover, the author presented a mapping of places where some invasive plants exist. Obtained results recommended that, in conditions of difficult geometry of anthropogenic forms requiring constant changes in the UAV flight altitude, it is essential to utilize a parallel system of spatial orientation, such as on-board inertial sensors with GPS.

According to Martin et al. [61], mining of uranium mineral veins in Cornwall, England, has resulted in a significant amount of legacy radiological contamination spread across numerous long-disused mining sites. A newly developed terrain-independent UAV carrying an integrated gamma radiation mapping unit was used for the radiological characterization of a single legacy mining site. It was possible to produce high-spatial-resolution maps, accordingly, determining the radiologically contaminated land areas and rapidly identifying and quantifying the degree of contamination and its isotopic nature. The obtained results showed that the instrument can be considered a viable tool for the characterization of similar sites worldwide.

In order to detect changes in topography linked to anthropogenic and meteorological effects rapidly and precisely and rehabilitate the abandoned mine, Yucel and Turan [62] created 3D terrain models of the mine lakes using high-resolution images from an UAV. 3D modeling of UAV images was performed with the Agisoft software using the most common SfM algorithm. Its workflow, relating to image matching, georeferencing, digital elevation modeling, orthomosaics, 3D point cloud, and 3D textured model creating, was used to create a 3D terrain model for the mine lakes. The study compared the results of two different methods (digitization and classification) within the ArcGIS package. The obtained results proved to be an effective method of visualizing such effects over the short term.

According to Neumann et al. [63], when a coal mine is closed, the methane emission is decreased but does not completely stop. Abandoned mines can release methane at a near-steady rate for an extended period of time. Flooding in the mines can prohibit gas emissions and buildups in empty spaces. This can help to mitigate the dangerous level of working in nearby active underground mines. In their study, based on micro-drone, the author created a 3D virtual mine map from 3D point cloud of optical sensors to calculate the volume capacity for gas storage in abandoned mines.

4 Results and Discussion

In this paper, 69 research publications that encompassed the application of UAVs within surveying and mapping in mine sites were reviewed. Figure 4 shows the percentage distribution of UAV application in three types of mines and the number of studies classified for each type. Table 1 indicates the application in surface mine, underground mine, and abandoned mine and the corresponding number of studies.

As shown in Table 1 and Fig. 4, more than 70% of UAV research were applied to terrain surveying and mapping in surface mines (36/50 research). UAVs were used mainly to generate surfaces, create 3D models, assess their accuracy, and survey terrain in these mines. Of the 36 articles, 13 discussed UAV applications to creating 3D models and evaluating their accuracy, and 14 reported the use of UAVs in the construction of surfaces and assessment of their accuracy. A topographic survey of surface mines based on UAV images can be found in nine studies. Despite advancements in UAV technology, the use of drones in underground mines has been limited. This is reflected in the number of studies applying this technology to survey and

Table 1 UAV application in terrain surveying of surface, underground, and abandoned mines

No	Mine type	Applications	Resources	Numbers
1	Surface mine	Construction of surfaces and assessment their accuracy	[1, 2, 12–15, 22, 23, 25, 26, 28–30, 64]	14
2		Creating 3D models and evaluating their accuracy	[16, 18–20, 31, 34–41]	13
3		Topographic survey	[7, 42–49]	9
4	Underground mine	Mapping 3D in underground mines	[51–56, 64]	7
5	Abandoned mine	Mapping in abandoned mines	[57–63]	7
Sum				50

map the underground mines with only seven research. All these studies focused on generating a 3D model to support an efficient and safe mining process. With the same number, the application of surveying and mapping in closed mines using UAVs was mentioned in seven publications. Of those, six papers conducted research in surface mines and only one studied in underground mines. Notably, while the reviewed research on the application of UAV in surveying and mapping of open-pit mines can apply the rotary-wing or fixed-wing UAV type, all reviewed literatures on UAV applications of surveying in underground mines used UAVs of the rotary-wing type. Thus, it seems that more efforts should be done to improve terrain surveying and mapping in the integration of UAVs with the mine industry.

5 Current Limitations and Future Perspectives in the Use of UAVs for Surveying and Mapping in Mine Sites

5.1 Current Limitations

To our knowledge, this is the first comprehensive scoping review of UAV applications in terrain mapping and surveying of mine sites. Almost all existing review papers have not discussed terrain surveying specifically and are based on a broad search without utilizing a structured approach for literature review, which may result in the omission of related publications. At the moment, it is difficult to perform a systematic review due to the scarcity of experimental studies in the mine areas, in particular, underground and abandoned mines as well as the lack of standardization of mine drones. However, this review followed a structured methodology that includes the application of drones in different types of mines, i.e., surface, underground, and abandoned mines. Therefore, this paper may be the most in-depth review of UAV applications in mine mapping and surveying.

As highlighted in this literature review, UAV technology has been successfully deployed in surveying and mapping mine sites. Although these UAV applications in the mine industry are expanding, there are several obstacles to the more widespread adoption of drone in the mining sector beginning with technical limitations, such as limited battery capacity, flight time, payload, sensor sensitivity, and dependence on climatic conditions. These obstacles affect the different extents of performing the tasks with UAV usage [5]. In surface mines, weather conditions pose a challenge by inducing deviations in the drone's predesignated paths compared to underground mines. In some situations, the weather can affect the UAV system, leading to the failure in their missions [65].

Besides, there are some difficulties in using UAVs in underground mines. In this working environment, mineworkers are always faced with harsh conditions, such as confined space, heat and humidity, dusty concentration, air velocity, poor lighting, lack of wireless communication system, magnetic interference, and an absence of GPS coverage [9, 66]. This leads to an extreme difficulty for an operator to perform a fly by drone in the underground mine. There are few UAVs and instruments that are dedicated to underground environments [9]. Therefore, it is necessary to design an optimized micro-drone that can solve all of these challenges. According to Shahmoradi et al. [10], ideally, the drone should be able to identify and avoid obstacles during its flight in the indoor environment. One of the types of microdrones is multirotor, which allows them to fly in confined spaces, such as underground mines, because they can hover and have high maneuverability. Furthermore, Shahmoradi et al. [10] also designed an autonomous spherical micro-drone for underground mine environments. They proposed a design of optimized multi-rotor UAVs made mainly from carbon fibers, which can reduce the differences for using drones in underground mine environments. In Fig. 5, a schematic view of designed drones for underground mine applications is shown [66, 67]. In addition to these studies, Park and Choi [9] showed that if UAV systems are equipped with high-resolution cameras, LED lights, and thermal sensors, necessary information, such as image (thermal, spectral, etc.), distance, inertial measurement unit, and sound navigation and ranging data, can be acquired in areas where are difficult to be accessed by mine workers.

Another drawback stems from UAV legal restrictions. UAV flights should also adhere to the related legislation and national rules. Moreover, software and algorithmic limitations include restrictions in data processing, while obstruction of software tools and algorithms for UAV control and flight planning are other prohibitive factors [10]. Furthermore, endurance ability has always been a disadvantage point for the UAV industry, in particular for mine applications. Thus, batteries with more endurance should be considered in the future. Also related to this limitation of the UAVs, Tong et al. [19] found that UAVs need more flights to cover a large area due to their low endurance. This is undoubtedly a challenge for remote regions without continuous power supply, such as quarry or abandoned mines.

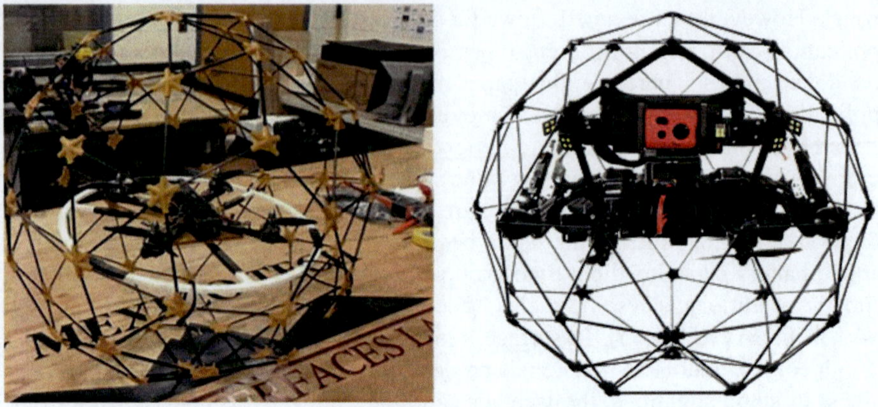

Fig. 5 Designed spherical drone for underground mine applications [66, 67]

5.2 *Future Prospects*

Although there are some obstacles, it can be found that UAVs have been shown to have great prospects for application in surveying and mapping in mine areas. They can replace traditional measurements, perform previously time-consuming tasks rapidly, and acquire periodic data with a much better resolution than satellite images [68]. UAVs in mines have potential applications in surface measurement, such as surface mapping, 3D reconstruction, and terrain surveying. The images from UAVs using laser scanners and high-resolution cameras can be utilized for generating high definition, geographically accurate 3D models, such as DEM, DTM, and DSM. Moreover, these images are converted in point cloud form, which in turn can be conducted to create maps for rock mass stability analysis, calculation of mining subsidence.

Besides, underground mines are one of the most dangerous aspects of the mining industry, and therefore, there are various applications of UAVs to improve safety, including surface roughness mapping, stability analysis, ventilation modeling, hazardous gas and leakage detection, and coal fire detection. In addition, accurate underground terrain mapping could help surveyors master downhole information to provide a plan for further mining activities [10].

The environment of mine areas varies drastically and rapidly. Therefore, in the view of Ren et al. [3], a single data source is usually insufficient to meet the requirements of the actual work. The matching and the integration of multi-source data can achieve complementarity by taking advantage of other methods' strengths. Furthermore, UAV and artificial intelligence (AI) technology have a significant impact on mining activities. Jung and Choi [69] conducted a systematic review to examine the current trends in machine learning research related to the mining industry and analyzed previous studies in the specific subject areas. Thus, it is necessary to integrate AI technology and UAVs in surveying and mapping in mine areas for future research.

6 Conclusion

In this paper, the drone applications in the mining industry for surveying and mapping have been discussed through publications conducted in the past 10 years. The review analyzed 50 papers related to survey and establishing mine topographic maps by UAVs, which were published in academic journals and M.Sc/Ph.D theses. The results revealed that the application of UAVs in mine surveying can be performed in surface, underground, and abandoned mines. As demonstrated in this review, drones are an excellent tool for mapping, surveying, constructing surfaces, and creating a 3D model of mines. Fixed-wing and rotary-wing drones are the most commonly used ones in surveying surface mines, while rotary-wing drones are preferable for underground mines. The obtained results have indicated that UAVs show the benefits of being operable at a low cost and the ability to work in areas difficult to access. Because the UAV system can fly at low altitudes, compared to a manned airborne method or satellite method, the UAV-based approach can capture high-resolution images with dense density.

Despite significant advancements in UAV technology, there are some limitations for the applications of drones in underground and abandoned mines. This is due to many challenges, such as harsh environments, lack of wireless signal, confined spaces, and the concentration of dust and gases. The possible solution for the use of drones in underground mining was mentioned in the review. Encased drones with rotary wing are suggested as a solution to overcome the obstacles in underground mine environments. In the future, the integration of UAV and AI technology promises to bring many useful applications for surveying and mapping in mine areas.

References

1. Nguyen, Q.L., Le Thi, T.H., Tong, S.S., Kim, T.T.H.. UAV photogrammetry-based for open pit coal mine large scale mapping, case studies in Cam Pha City, Vietnam. Sustain. Dev. Mt. Territ. **12**, 501–509 (2020)
2. Nguyen, L.Q.: Accuracy assessment of open-pit mine's digital surface models generated using photos captured by Unmanned Aerial Vehicles in the post-processing kinematic mode. J. Min. Earth Sci. **62**, 38–47 (2021)
3. Ren, H., Zhao, Y., Xiao, W., Hu, Z.: A review of UAV monitoring in mining areas: current status and future perspectives. Int. J. Coal Sci. Technol. **6**, 320–333 (2019)
4. Sayab, M., Aerden, D., Paananen, M., Saarela, P.: Virtual structural analysis of jokisivu open pit using 'structure-from-motion'unmanned aerial vehicles (UAV) photogrammetry: implications for structurally-controlled gold deposits in southwest Finland. Remote Sens. **10**, 1296 (2018)
5. Mukhamediev, R.I., Symagulov, A., Kuchin, Y., Zaitseva, E., Bekbotayeva, A., Yakunin, K., Assanov, I., Levashenko, V., Popova, Y., Akzhalova, A.: Review of some applications of unmanned aerial vehicles technology in the resource-rich country. Appl. Sci. **11**, 10171 (2021)
6. Rathore, I., Kumar, N.P.: Unlocking the potentiality of uavs in mining industry and its implications. Int. J. Innovative Res. Sci., Eng. Technol. **4**, 852–855 (2015)
7. Esposito, G., Mastrorocco, G., Salvini, R., Oliveti, M., Starita, P.: Application of UAV photogrammetry for the multi-temporal estimation of surface extent and volumetric excavation in the Sa Pigada Bianca open-pit mine, Sardinia, Italy. Environ. Earth Sci. **76**, 1–16 (2017)

8. Francioni, M., Salvini, R., Stead, D., Giovannini, R., Riccucci, S., Vanneschi, C., Gullì, D.: An integrated remote sensing-GIS approach for the analysis of an open pit in the Carrara marble district, Italy: slope stability assessment through kinematic and numerical methods. Comput. Geotech. **67**, 46–63 (2015)
9. Park, S., Choi, Y.: Applications of unmanned aerial vehicles in mining from exploration to reclamation: a review. Minerals **10**, 663 (2020)
10. Shahmoradi, J., Talebi, E., Roghanchi, P., Hassanalian, M.: A comprehensive review of applications of drone technology in the mining industry. Drones **4**, 34 (2020)
11. Câmara, T., Fontoura, D., Peroni, R., Capponi, L.: Controlling dilution and mine planning. In: Apcom Symposium Applications of Computers and Operations Research in The Mineral Industry, pp. 1–10 (2013)
12. Beretta, F., Shibata, H., Cordova, R., Peroni, R.d.L., Azambuja, J., Costa, J.F.C.L.: Topographic modelling using UAVs compared with traditional survey methods in mining. REM-Int. Eng. J. **71**, 463–470 (2018)
13. Cho, S.-J., Bang, E.-S., Kang, I.-M.: Construction of precise digital terrain model for nonmetal open-pit mine by using unmanned aerial photograph. Econ. Environ. Geol. **48**, 205–212 (2015)
14. Nguyen, N.V.: Building DEM for deep open-pit coal mines using DJI inspire 2. J. Min. Earth Sci. **61**, 1–10 (2020)
15. Chirico, P.G., DeWitt, J.D.: Mapping informal small-scale mining features in a data-sparse tropical environment with a small UAS. J. Unmanned Veh. Syst. **5**, 69–91 (2017)
16. González-Aguilera, D.: 3D modelling and accuracy assessment of granite quarry using unmmanned aerial vehicle (2012)
17. Moudrý, V., Urban, R., Štroner, M., Komárek, J., Brouček, J., Prošek, J.: Comparison of a commercial and home-assembled fixed-wing UAV for terrain mapping of a post-mining site under leaf-off conditions. Int. J. Remote Sens. **40**, 555–572 (2019)
18. Battulwar, R., Winkelmaier, G., Valencia, J., Naghadehi, M.Z., Peik, B., Abbasi, B., Parvin, B., Sattarvand, J.: A practical methodology for generating high-resolution 3D models of open-pit slopes using UAVs: flight path planning and optimization. Remote Sens. **12**, 2283 (2020)
19. Tong, X., Liu, X., Chen, P., Liu, S., Luan, K., Li, L., Liu, S., Liu, X., Xie, H., Jin, Y.: Integration of UAV-based photogrammetry and terrestrial laser scanning for the three-dimensional mapping and monitoring of open-pit mine areas. Remote Sens. **7**, 6635–6662 (2015)
20. Vassena, G., Clerici, A.: Open pit mine 3D mapping by tls and digital photogrammetry: 3D model update thanks to a slam based approach. Int. Arch. Photogrammetry, Remote Sens. Spat. Inf. Sci. **42**, 1145–1148 (2018)
21. Nghia, N.V.: Building DEM for deep open-pit coal mines using DJI Inspire 2. J. Min. Earth Sci. **61**, 1–10 (2020)
22. Forlani, G., Dall'Asta, E., Diotri, F., Cella, U.M.D., Roncella, R., Santise, M.: Quality assessment of DSMs produced from UAV flights georeferenced with on-board RTK positioning. Remote Sens. **10**, 311 (2018)
23. Nguyen, Q.L., Goyal, R., Le, V.C., Cao, X.C., Pham, V.C., Bui, N.Q., Bui, X.N., Bui, K.L.: Influence of flight height on the accuracy of UAV derived digital elevation model at complex terrain. Inżynieria Mineralna (2020)
24. Long, N.Q., Goyal, R., Bui, L.K., Cuong, C.X., Canh, L.V., Minh, N.Q., Bui, X.N.: Optimal choice of the number of ground control points for developing precise DSM using light-weight UAV in small and medium-sized open-pit mine. Arch. Min. Sci. **66** (2021)
25. Kršák, B., Blišťan, P., Pauliková, A., Puškárová, P., Kovanič, Ľ.M., Palková, J., Zelizňaková, V.: Use of low-cost UAV photogrammetry to analyze the accuracy of a digital elevation model in a case study. Measurement **91**, 276–287 (2016)
26. Tien Bui, D., Long, N.Q., Bui, X.-N., Nguyen, V.-N., Pham, C.V., Le, C.V., Ngo, P.-T.T., Bui, D.T., Kristoffersen, B.: Lightweight unmanned aerial vehicle and structure-from-motion photogrammetry for generating digital surface model for open-pit coal mine area and its accuracy assessment. In: International Conference on Geo-Spatial Technologies and Earth Resources, pp. 17–33. Springer (2017)

27. Lee, S., Choi, Y.: Reviews of unmanned aerial vehicle (drone) technology trends and its applications in the mining industry. Geosys. Eng. **19**, 197–204 (2016)
28. Chen, J., Li, K., Chang, K.-J., Sofia, G., Tarolli, P.: Open-pit mining geomorphic feature characterisation. Int. J. Appl. Earth Obs. Geoinf. **42**, 76–86 (2015)
29. Gil, M., Frąckiewicz, P.: Optimization of the location of observation network points in open-pit mining's. GIS Forum (2019)
30. Xiang, J., Chen, J., Sofia, G., Tian, Y., Tarolli, P.: Open-pit mine geomorphic changes analysis using multi-temporal UAV survey. Environ. Earth Sci. **77**, 1–18 (2018)
31. Park, H.J., Turner, R., Rak Lee, D., One Lee, J.: 3D modelling of coal stockpiles using UAV data in an open cut mine environment. In: Proceedings of the 16th International Congress for Mine Surveying ISM (2016)
32. Tatum, M.C., Liu, J.: Unmanned aircraft system applications in construction. Procedia Eng. **196**, 167–175 (2017)
33. Francesco, N., Fabio, R.: UAV for 3D mapping applications: a review. Appl. Geomatics **6**, 1–15 (2014)
34. Wang, Q., Wu, L., Chen, S., Shu, D., Xu, Z., Li, F., Wang, R.: Accuracy evaluation of 3D geometry from low-attitude uav images: a case study at Zijin Mine. Int. Arch. Photogrammetry, Remote Sens. Spat. Inf. Sci. (4) (2014)
35. Shahbazi, M., Sohn, G., Théau, J., Ménard, P.: UAV-based point cloud generation for open-pit mine modelling. Int. Arch. Photogrammetry, Remote Sens. Spat. Inf. Sci. **40** (2015)
36. Kang, S., Lee, G.-J., Noh, J., Jang, H.D., Kim, S.-M., Ko, C.-S.: 3-dimensional modeling and mining analysis for open-pit limestone mine stope using a rotary-wing unmanned aerial vehicle. J. Eng. Geol. **28**, 701–714 (2018)
37. Bui, X.N., Lee, C., Nguyen, Q.L., Adeel, A., Cao, X.C., Nguyen, V.N., Le, V.C., Nguyen, H., Le, Q.T., Duong, T.H.: Use of unmanned aerial vehicles for 3D topographic mapping and monitoring the air quality of open-pit mines. Inżynieria Mineralna **21** (2019)
38. Filipova, S., Filipov, D., Raeva, P.: Creating 3D model of an open pit quarry by UAV imaging and analysis in GIS. In: International Conference on Cartography and GIS, pp. 652 (2016)
39. Tscharf, A., Rumpler, M., Fraundorfer, F., Mayer, G., Bischof, H.: On the use of UAVs in mining and archaeology-geo-accurate 3d reconstructions using various platforms and terrestrial views. ISPRS Ann. Photogrammetry, Remote Sens. Spat. Inf. Sci. **2**, 15–22 (2015)
40. Ulusoy, İ., Şen, E., Tuncer, A., Sönmez, H., Bayhan, H.: 3D multi-view stereo modelling of an open mine pit using a lightweight UAV. Türkiye Jeoloji Bülteni **60**, 223–242 (2017)
41. van Le, C., Cuong, C.X., Nguyen, Q., Anh, T.T., Xuan-Nam, B.: Experimental investigation on the performance of DJI phantom 4 RTK in the PPK mode for 3D mapping open-pit mines. Inżynieria Mineralna **1**, 65–74–65–74 (2020)
42. Katuruza, M., Birch, C.: The use of unmanned aircraft system technology for highwall mapping at Isibonelo Colliery, South Africa. J. South Afr. Inst. Min. Metall. **119**, 291–295 (2019)
43. Malpeli, K.C., Chirico, P.G.: Testing a small UAS for mapping artisanal diamond mining sites in Africa. Photogramm. Eng. Remote. Sens. **81**, 257–263 (2015)
44. Leo Stalin, J., Gnanaprakasam, R.: Application of unmanned aerial vehicle for mapping and modeling of Indian mines. J. Indian Soc. Remote Sens. **48**, 841–852 (2020)
45. Salvini, R., Mastrorocco, G., Seddaiu, M., Rossi, D., Vanneschi, C.: The use of an unmanned aerial vehicle for fracture mapping within a marble quarry (Carrara, Italy): photogrammetry and discrete fracture network modelling. Geomat. Nat. Haz. Risk **8**, 34–52 (2017)
46. Lee, S., Choi, Y.: Comparison of topographic surveying results using a fixed-wing and a popular rotary-wing unmanned aerial vehicle (drone). Tunnel Undergr. Space **26**, 24–31 (2016)
47. Lee, S., Choi, Y.: Topographic survey at small-scale open-pit mines using a popular rotary-wing unmanned aerial vehicle (drone). Tunnel Undergr. Space **25**, 462–469 (2015)
48. Lee, S., Choi, Y.: On-site demonstration of topographic surveying techniques at open-pit mines using a fixed-wing unmanned aerial vehicle (drone). Tunnel Undergr. Space **25**, 527–533 (2015)
49. Rossi, P., Mancini, F., Dubbini, M., Mazzone, F., Capra, A.: Combining nadir and oblique UAV imagery to reconstruct quarry topography: methodology and feasibility analysis. Eur. J. Remote Sens. **50**, 211–221 (2017)

50. Mitchell, J., Marshall, J.A.: Design of a novel auto-rotating uav platform for underground mine cavity surveying (2017)
51. Li, H., Savkin, A.V., Vucetic, B.: Autonomous area exploration and mapping in underground mine environments by unmanned aerial vehicles. Robotica **38**, 442–456 (2020)
52. Ge, L., Li, X., Ng, A.H.: UAV for mining applications: A case study at an open-cut mine and a longwall mine in New South Wales, Australia. IEEE International Geoscience and Remote Sensing Symposium (IGARSS), pp. 5422–5425 (2016)
53. Russell, E.A., MacLaughlin, M., Turner, R.: UAV-based geotechnical modeling and mapping of an inaccessible underground site. In: 52nd US Rock Mechanics/Geomechanics Symposium. OnePetro (2018)
54. Papachristos, C., Khattak, S., Mascarich, F., Alexis, K.: Autonomous navigation and mapping in underground mines using aerial robots. In: 2019 IEEE Aerospace Conference, pp. 1–8. IEEE (2019)
55. Turner, R., Bhagwat, N., Galayda, L., Knoll, C., Russell, E., MacLaughlin, M.: Geotechnical characterization of underground mine excavations from UAV-captured photogrammetric & thermal imagery. In: 52nd US Rock Mechanics/Geomechanics Symposium. OnePetro (2018)
56. Turner, R.M., MacLaughlin, M.M., Iverson, S.R.: Identifying and mapping potentially adverse discontinuities in underground excavations using thermal and multispectral UAV imagery. Eng. Geol. **266**, 105470 (2020)
57. Dai, H., Xu, J.: Application of UAV photogrammetry on ecological restoration of abandoned open-pit mines, Northern Anhui province, China. Nat. Environ. Pollut. Technol. **21**, 193–199 (2022)
58. Molnar, A., Domozi, Z.: Volume analysis of surface formations on the basis of aerial photographs taken by drones. Int. J. Signal Process., Image Process. Pattern Recognit. **1**, 152–159 (2016)
59. Suh, J., Choi, Y.: Mapping hazardous mining-induced sinkhole subsidence using unmanned aerial vehicle (drone) photogrammetry. Environ. Earth Sci. **76**, 1–12 (2017)
60. Motyka, Z.: Systems for spatial and physico-chemical parameters mapping of anthropogenic landscape forms and plants formations in mining areas with the use of photogrammetry and remote laser sensing from low height. Czasopismo Inżynierii Lądowej, Środowiska i Architektury (2017)
61. Martin, P.G., Payton, O.D., Fardoulis, J.S., Richards, D.A., Scott, T.B.: The use of unmanned aerial systems for the mapping of legacy uranium mines. J. Environ. Radioact. **143**, 135–140 (2015)
62. Yucel, M.A., Turan, R.Y.: Areal change detection and 3D modeling of mine lakes using high-resolution unmanned aerial vehicle images. Arab. J. Sci. Eng. **41**, 4867–4878 (2016)
63. Neumann, P.P., Hernandez Bennetts, V., Lilienthal, A.J., Bartholmai, M., Schiller, J.H.: Gas source localization with a micro-drone using bio-inspired and particle filter-based algorithms. Adv. Robot. **27**, 725–738 (2013)
64. Azhari, F., Kiely, S., Sennersten, C., Lindley, C., Matuszak, M., Hogwood, S.: A comparison of sensors for underground void mapping by unmanned aerial vehicles. In: UMT 2017: Proceedings of the First International Conference on Underground Mining Technology, pp. 419–430. Australian Centre for Geomechanics (2017)
65. Shakhatreh, H., Sawalmeh, A.H., Al-Fuqaha, A., Dou, Z., Almaita, E., Khalil, I., Othman, N.S., Khreishah, A., Guizani, M.: Unmanned aerial vehicles (UAVs): a survey on civil applications and key research challenges. IEEE Access **7**, 48572–48634 (2019)
66. Shahmoradi, J., Roghanchi, P., Hassanalian, M.: Drones in underground mines: challenges and applications. In: 2020 Gulf Southwest Section Conference (2020)
67. https://www.flyability.com/mine-drone. Accessed 10 May 2022
68. Agarwal, R., Singh, D., Chauhan, D., Singh, K.: Detection of coal mine fires in the Jharia coal field using NOAA/AVHRR data. J. Geophys. Eng. **3**, 212–218 (2006)
69. Jung, D., Choi, Y.: Systematic review of machine learning applications in mining: exploration, exploitation, and reclamation. Minerals **11**, 148 (2021)

Mining-Induced Land Subsidence Detected by Persistent Scatterer InSAR: Case Study in Pniówek Coal Mine, Silesian Voivodeship, Poland

Thi Thu Huong Kim, Hong Ha Tran, Tuan Anh Phan, and Tomasz Lipecki

Abstract In Europe, coal energy has recently been limited due to environmental pollution. However, other sources still do not provide enough energy for European countries. Additionally, the limitation in coal exploitation has resulted in reducing labor market in mining workers. Therefore, despite having to pay a fine every day, coal mines are still operated in Poland. Pniówek is a fifty-year coal mine in the south of Poland with over 1000 m depth. It is one of the largest coal reserves in Poland, with a total operative resource 101.9 million tons. This mine provides about 12,200 tons of coal every day. This paper uses 44 images Sentinel-1A satellite data from June 19th 2018 to December 23rd 2019 with InSAR technology to determine subsidence and ongoing deformation on and around this mine. The study results are useful for policymakers, managers, and authorities with a better management system. This is because land subsidence is happening and destroying infrastructure, and threatening people live in the area. The results, which are processed by the Persistent Scatterer InSAR (PSInSAR) method with the Sentinel Application Platform (SNAP) and the Stanford Method for Persistent Scatterers software packages (StaMPS), show that the subsidence has occurred in most of the areas. The maximum line-of-sight (LOS) displacement of up to −40 mm/yr is found in the residential area near the mine in Pawłowice village, while the LOS displacement within the Pniówek mine reaches −36.7 mm/yr.

T. T. H. Kim · H. H. Tran · T. A. Phan
Hanoi University of Mining and Geology, Hanoi, Vietnam
e-mail: tranhongha@humg.edu.vn

T. A. Phan
e-mail: phan@agh.edu.pl

T. T. H. Kim (✉) · T. A. Phan · T. Lipecki
AGH University of Science and Technology, Kraków, Poland
e-mail: huongkim@agh.edu.pl

T. Lipecki
e-mail: lipecki@agh.edu.pl

H. H. Tran
TU Bergakademie, Freiberg, Germany

L. Q. Nguyen et al. (eds.), *Advances in Geospatial Technology in Mining and Earth Sciences*, Environmental Science and Engineering,
https://doi.org/10.1007/978-3-031-20463-0_2

Keywords InSAR · PSInSAR · Sentinel-1 · Pniówek · Upper Silesian coal basin

1 Introduction

Land subsidence is a global environmental geological hazard due to natural and human activities. It has been observed for decades in many places around the world with about 150 countries recorded subsidence [1]. Land subsidence may destroy infrastructures and buildings and affect land surface and subsurface. Additionally, it dangers human safety and animals living on the surface. Therefore, it is necessary to find solutions to minimize its negative effects on both the environment and humans.

Underground mining exploitation is one of the human activities that has been widely known for causing land subsidence. Mining-induced land subsidence results from the displacement of rock masses above the seam and post-extraction void migration to the terrain surface [2]. Some studies about mining-induced land subsidence have been conducted in, e.g., USA [3], Turkey [4], China [5], Poland [6], and Vietnam [7, 8]. Mining-induced land subsidence has been measured for a long time by conventional methods, such as geodetic levelling, total station, or the Global Navigation Satellite System (GNSS) [6]. These methods are disadvantageous in that the measurements are conducted on limited and discrete points on the Earth's surface. Therefore, they are difficult to obtain a complete overall look at the movement over the entire study area. In recent years, some modern measurement methods have been applied to monitor the land subsidence, such as Unmanned Aerial Vehicle (UAV) [9], Interferometric Synthetic Aperture Radar (InSAR) [10], and Laser Scanner [11]. These methods allow for monitoring surfaces in larger areas with a relatively low spatial spacing. These new methods are applied more and more in both science and industry, giving many benefits for monitoring land subsidence.

In recent decades, InSAR has been applied effectively to monitor land subsidence with high accuracy and spatial resolution. Furthermore, it almost does not depend on weather conditions. Differential SAR interferometry (DInSAR) is a technique that can obtain the deformation over a large spatial coverage (>10^4 km^2) and with a high spatial resolution of up to several meters [12]. It estimates the displacement along the line-of-sight (LOS) direction using the phase changes between two overlapping SAR scenes. In DInSAR, the contribution of topography is removed. However, the results of DInSAR are still affected by the various error and noise sources, including geometrical and temporal decorrelation, and atmospheric disturbances. Temporal decorrelation is caused by physical surface changes between the images taken at different acquisition times. These changes affect the scattering properties of the surface, causing the loss of coherence [13]. Geometrical decorrelation is due to a large spatial baseline between two image acquisitions which results in inconsistency [14]. The atmospheric disturbances are due to the noise caused by signal delay. The more images are used, the smaller magnitude of error and noise we likely have. Thus, different multi-temporal InSAR (MTInSAR) methods have been proposed which usually use a big SAR dataset. This technique is capable of detecting small changes at an unprecedented level of spatial details (centimeters to millimeters) in land surface displacement [2, 10]. Small baseline subset (SBAS) [15] and Persistent Scatterer InSAR (PSInSAR) [16] are among the most commonly used MTInSAR methods. The advantage of SBAS is that redundant interferograms are adopted that are useful in reducing noises, especially in areas with low deformation magnitude affected by high noise [17]. On the other hand, in PSInSAR, the use of a single primary (formerly master) image allows reducing its noise as it exists in all interferograms [10].

In Poland, PSInSAR was applied for the first time by Perski et al. in 1998 [18]. Since then, there have been many published studies using InSAR technology, which was applied to different types of mines in Poland. Hejmanowski et al. studied the subsidence caused by nearby mining-induced earthquakes by analyzing the spatio-temporal distribution of ground movements caused by two tremors, and the main scientific aim is in the areas where high-energy tremors occur and ground movement overlap [19]. In 2021, Wojciech Witkowski et al. integrated leveling observations and PSInSAR results for monitoring deformation caused by underground mining [20]. Also in this year, he used InSAR to estimate mining-induced horizontal stain tensor of land surface. In this study, Witkowski filled a gap in the current knowledge for determining horizontal strain tensors based on double-geometry InSAR [2]. Milczarek et al. applied PSInSAR for surface deformations in post-mining area in Walbrzych hard coal basin in 2016, 2017 [21, 22]. Lesniak et al. employed PSInSAR to monitor surface deformation in the Dabrowa Basin of Poland in 2020 [23]. Recently, Kim et al. showed the surface deformation detected by PSInSAR in a salt mine in Wapno, Poland [6]. From 2016 until recently, there have been over three hundred research articles about the application of PSInSAR in Poland (data from google scholar page).

In this study, we apply PSInSAR to monitor the land subsidence in the areas around Pniówek Coal Mine where InSAR-derived deformation has not been studied so far. The study is based on approximately 1.5 year InSAR data, spanning between June 2018 and December 2019. The objective of this study is to show the potential of PSInSAR with Sentinel-1 SAR data to quantify surface deformation around the Pniówek Coal Mine.

The remainder of the paper is structured as follows: Sect. 2 introduces the history and characteristic of the Pniówek coal mine. Section 3 introduces the effects of the Pniówek coal mine exploitation on infrastructures. In Sect. 4, the study area, dataset, and method are provided. In Sect. 5, experimental results are provided, and Sect. 6 concludes the study.

2 History of the Pniówek Coal Mine, Silesian Voivodeship, Poland

Jastrzębska Spółka Węglowa (JSW) Capital Group is the largest producer of high-quality hard coking coal in the European Union and one of the leading producers of coke used for steel smelting. Coking coal, known as metallurgical coal, has been recognized as one of the 30 critical raw material for the European Union, which means strategic economic importance for the European industry. Coking coal is an essential component in the production of steel. Currently, there are no alternative and economically justified technologies for smelting steel without the use of coking coal. The main recipients of the JSW Group's products are customers in Poland, Germany, Austria, the Czech Republic, Slovakia, Italy, and India.

Pniówek Coal Mine is one of the JSW plants, of which the quality of coal is increased with depth. The deposit of the Pniówek mine covers a 28.6 km^2 area, which is located in the Pawłowice commune (87.1%) and the city of Jastrzębie-Zdrój (12.9%). It was built from 1963 to 1974 and is still in operation recently [24]. The mining area of Pniówek Coal Mine is bordered by the Zofiówka coal mine to the west, the Borynia coal mine to the north-west, the Pawłowice–Warszowice–North coal bed to the north, the "Pawłowice I" deposit to the east, and the "Bzie-Dębina" deposit to the south.

Exploration and geological works in the first mine area of Pniówek and its vicinity began in the late nineteenth century. The first holes were drilled in the years 1890–1909 [25]. After World War II, in connection with the expansion of the Rybnicki Okreg coalfield, intensive geological and exploration works through the holes began in 1956. The first mining site in the nowadays Pniówek mine area was established in 1962 as the name "Krzyżowice". Due to reduced investment costs, by the decision of the Minister, from 24 February 1964, the construction of the Pniówek mine was suspended by the theory of the so-called Carbon Twilight. In 1966, the construction was continued with a new concept for the Pniówek mine: a single-stage model for the production of 15,000 tons of hard coal, taking into account a high rate of center extraction. After drilling the hole "Krzyżowice No. 1 $\div$ 27" and creating geological documentation in 1968, in 1970, it was decided to start construction of a mine. In the same year, it was also built a new site, named "Krzyżowice I" and the shaft started running. Trench drilling at the depths of 580 m (ventilation) and 705 (mining exploitation) was started in 1972 [25].

On December 4, 1974, the official Pniówek coal mine was established. The vertical entrance is made of 5 shafts (three shafts are on the main plot and 2 shafts are on the periphery). The main underground transport is a fully automatic conveyor system with coal expansion tanks. The enrichment of all raw coal production takes place in the coal processing factory. The Pniówek coal mine is up to 1100 m deep, with the main coal types of 35.1, 35.2.A, and 35.2.B. The other minor coal types recorded are 34.1, 34.2, and 33 locally.

The main milestones in the history of the Pniówek coal mine are:

- 12th July 1971: embedding the foundation act of building Pniówek coal mine.
- 4th December 1974: the official Pniówek coal mine was established.
- September 1976: operating a mechanical coal processing plant.
- 1st October 1982: connection of Pniówek coal mine with a Coal Mine Associations in Jastrzębie-Zdrój.
- December 1986: reaching the production capacity of 12,000 tons/day.

- 1st February 1990: introducing a work time registration automatic system.
- 1st April 1993: Pniówek coal mine become a part of Jastrębska Spółka Węglowa SA company.
- May 1995: start building at level 1000 m deep.
- May 2000: commissioning of central air conditioning, based on a combined energy and cooling system.
- 13rd December 2006: extraction of 100 million tons.
- 4th December 2007: putting into operation level 1000 m—driving people in shaft II.
- 2008: commencement of works related to the expansion of the "Pawłowice 1".
- 6th July 2011, JSW Group debuted on the Warsaw Stock Exchange.
- 20th April 2022, a methane explosion occurred in the Pniówek coal mine with 15 deaths (miners and mine rescuers, who after the first explosion, went to help the injured) [26].

Pniówek is one of the historical methane deposits in Europe. The methane status of the mine area is controlled by an automated system. The mine uses sensors to control methane and fire hazards with continuous measurement, period of the sensors: 4 min [25].

3 Effects of the Pniówek Coal Mine Exploitation on Infrastructures

Pniówek Coal Mine is one of the mines in Jastrzębska Spółka Węglowa SA company in the Southern part or Upper Silesia Coal Basin. Due to mining activities in the Pniówek mine, roads, buildings, and other infrastructures have been damaged. Observations by geodetic surveying showed that, in the vicinity of the mine, both continuous and discontinuous deformations were found [27]. In the study about the influence of longwall exploitation carried out in the vicinity of a protection pillar on a rectified building, Tadeusz Majcherczyk said that, from 1977 to 2002, the survey of surface deformation was carried out along the road from Dębina to Pniówek, which is located to the northwest from the mine. The line's points close to the mine, measured by levelling method, indicated the subsidence of approximately 500 mm. Over the other points on this line, the subsidence rates appeared to increase over time [27]. The subsidence was found to relate to mining exploitation. Levelling observations since 1994 showed that a building in the southern area and near the Pniówek Coal Mine inclined 3.4‰ southward. Over the same building, the deflection was found unchanged in the northern segment, but increased up to 16.7 ‰ in the south-westward in 2001 [27]. Figure 1 shows the fracture of the building due to mining-induced land subsidence near the Pniówek mine.

On March 22, 2021, the Radio99FM published a newspaper about a crack of a few centimeters wide occurring across the Pałowice bypass (Fig. 2). At that time, the crack is found about 2–3 cm wide. Not only the cracks have been found on the road

Fig. 1 Damaged wall of the house near the Pniówek mine due to coal mine exploitation [27]

surface, but also over all layers of the road structure. This damage was reported to be related to the mine activities; as a result, the mine will be repaired at the owners' expense.

Fig. 2 A few centimeter wide crack across the Pałowice bypass [28]

4 Study Area, Dataset, and Method

4.1 Study Area and Datasets

The study area is located in the west-southern part of Poland, on the south side of the Rybnik Plateau, which is part of Upper Silesian Basin (see Fig. 3). This place is built up of brown and dust soil from dust, dust argillaceous, and loess-like sediment, and from clay crusts. The substance of this area is built of carbon rock with hard coal deposit, on which there are salt deposits, gypsum, and sulfur deposit [29].

The study area covers the Pniówek hard coal mine. It is limited from 49°54′N to 50°2′N and from 18°34′E to 18°46′E (see Fig. 4). This coal mine subsidence affects directly all Pniówek village with 1322 people, Pałowice village with 3391 people, Warszowice village with 1754 people, and Krzyżowice village with 352 people (data in 2015 from https://www.city-facts.com/accessed on March 15th 2022).

The C-band Sentinel-1 data has a 5.6 cm wavelength. Sentinel-1 satellites operate with four acquisition modes: Stripmap (SM), Interferometric Wide Swath (IW), Extra-wide swath (EW), and Wave mode (WV) with different spatial resolutions. Sentinel-1 consists of two platforms currently operating in the same orbit, phased at 180 degrees to each other. They are Sentinel-1A, which was launched in April 2014,

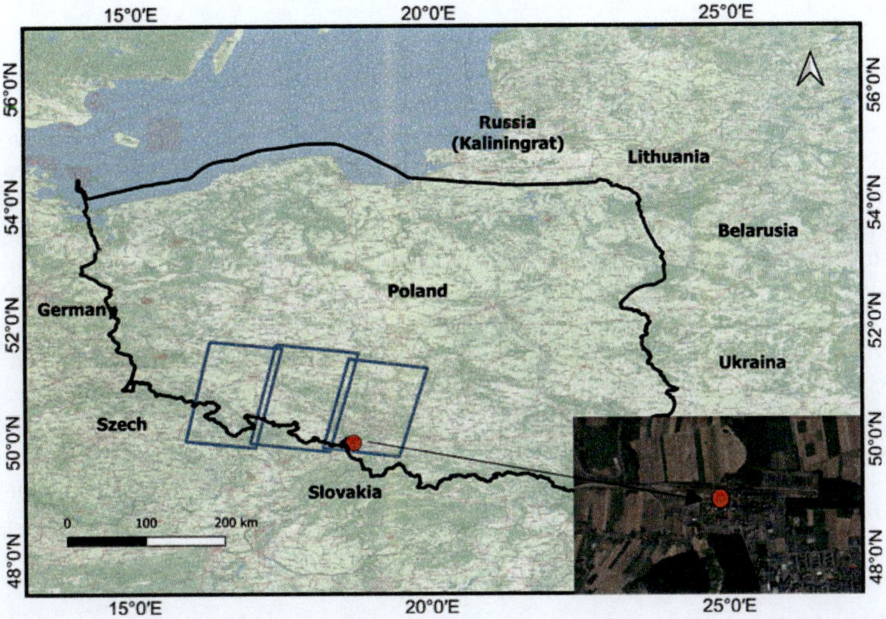

Fig. 3 The study area (green box) in Poland map. The extent of the Sentinel-1A data sub-swaths is indicated by blue boxes. Red dots are the office of the Pniówek coal mine company (KWK Pniówek)

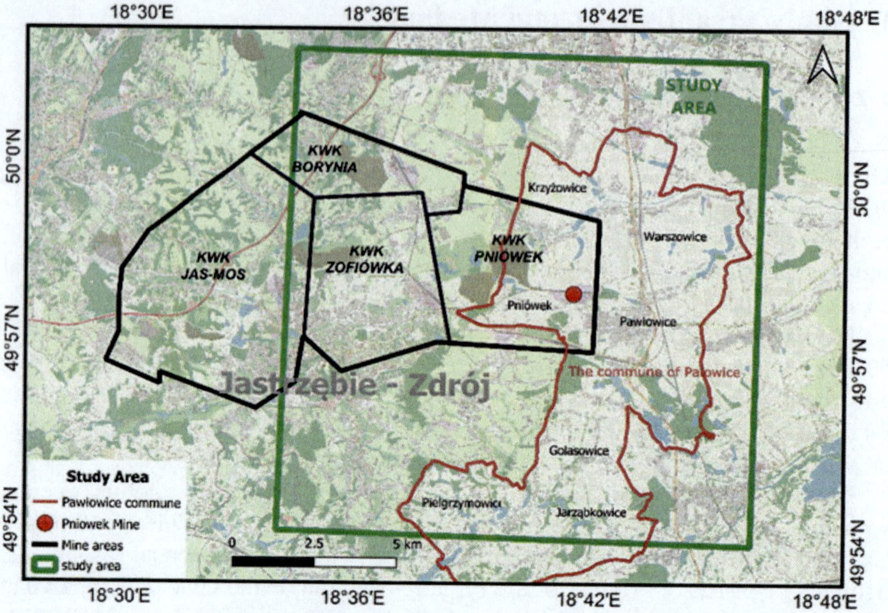

Fig. 4 Study area cover part of Jastrzębie-Zdrój city; in it, underground mine KWK Pniówek and residence areas. Red dots are the office of KWK Pniówek company

and Sentinel-1B, which was launched in April 2016. To study land surface deformation with the PSInSAR method, Sentinel-1 data in the Single Look Complex (SLC) format and IW mode are used. The C-band radar used in PSInSAR has been proven to be a valuable tool for monitoring slow-moving landslides with good accuracy (millimeters per year) [10, 30, 31].

This study uses Sentinel-1A data, which can be freely and easily obtained from https://scihub.copernicus.eu/dhus/#/home. The data is of the Interferometric Wide-swath mode and acquired in the descending orbit number 124. A total of 44 Sentinel-1A images have been downloaded. The dataset spans a 1.5 year period, from June 19, 2018, to December 31, 2019, with a 12-day temporal interval. The image captured on April 15, 2019, is chosen as the primary (formerly master) image. The images were selected according to good weather days and combined with the primary image to create short spatial baselines. Table 1 shows the information associated with the 44 images, including the acquisition dates, perpendicular baselines, and temporal baselines.

4.2 Method

The PSInSAR method uses multi-temporal SAR images, based on the DInSAR method. DInSAR consists of the comparison of the interferometric phase of two

Table 1 List of Sentinel-1A SAR scenes used in this study

No	Date of acquisition	Perpendicular baseline (m)	Temporal baseline (days)
1	19 June 2018	−95.05	300
2	01 July 2018	−23.47	288
3	13 July 2018	23.83	276
4	25 July 2018	−26.88	264
5	6 August 2018	−52.43	252
6	18 August 2018	−94.61	240
7	30 August 2018	11.43	228
8	11 September 2018	−12.13	216
9	05 October 2018	−75.59	192
10	17 October 2018	−31.42	180
11	29 October 2018	14.03	168
12	10 November 2018	73.11	156
13	22 November 2018	49.75	144
14	04 December 2018	58.13	132
15	16 December 2018	−85.23	120
16	28 December 2018	6.58	108
17	09 January 2019	23.66	96
18	21 January 2019	93.20	84
19	02 February 2019	85.30	72
20	14 February 2019	−63.31	60
21	26 February 2019	−28.92	48
22	10 March 2019	33.05	36
23	22 March 2019	65.59	24
24	03 April 2019	64.26	12
25	15 April 2019	0	0
26	27 April 2019	−21.97	−12

(continued)

Table 1 (continued)

No	Date of acquisition	Perpendicular baseline (m)	Temporal baseline (days)
27	09 May 2019	−103.02	−24
28	21 May 2019	52.22	−36
29	02 June 2019	46.39	−48
30	14 June 2019	−69.71	−60
31	26 June 2019	−84.77	−72
32	08 July 2019	−96.59	−84
33	20 July 2019	−7.34	−96
34	01 August 2019	10.84	−108
35	13 August 2019	9.01	−120
36	25 August 2019	−48.29	−132
37	18 September 2019	−85.58	−156
38	12 October 2019	−16.91	−180
39	30 September 2019	81.20	−168
40	24 October 2019	−93.16	−192
41	05 November 2019	−21.57	−204
42	17 November 2019	26.08	−216
43	29 November 2019	83.06	−228
44	23 December 2019	35.77	−252

SAR radar images acquired over the same areas on different dates. Let us assume there is point P located on the ground which is subsiding to the point P1 over the time. There are two images acquired at different times t_0 and t (see Fig. 5). The phase difference between the two images can use to determine the land deformation between two moments of point P:

$$\Delta\varphi_{\text{int}} = \varphi_S - \varphi_M = \frac{\text{SP1} - \text{MP}}{\frac{\lambda}{4\pi}} + \varphi_{\text{scat}_S} - \varphi_{\text{scat\_}M} \quad (1)$$

where $\Delta\varphi_{\text{int}}$ is the phase difference between the phase of primary image φ_M and secondary (formerly slave) image φ_S, M and S are the satellite positions of the primary and secondary images, MP and SP1 are the distance from the sensor of satellite to the point P at time t_0 and t_1, λ is the radar wavelength, and φ_{scat_S}, $\varphi_{\text{scat\_}M}$ are the phase change during the interaction between radar wave and the target.

Fig. 5 DInSAR measurement

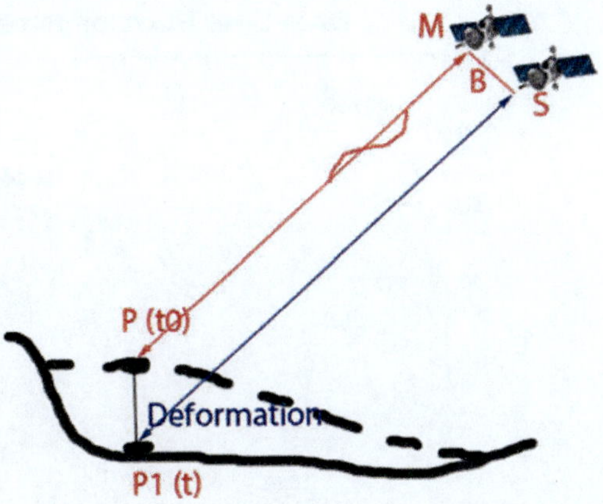

The interferometric phase in Eq. (1) includes the topography component, surface deformation, and various error and noise sources. Thus, the land deformation is determined by removing the topographic component, noises, and errors from the interferometric phase:

$$\Delta\varphi_{D_{\mathrm{int}}} = \Delta\varphi_{\mathrm{int}} - \Delta_{\mathrm{Topo}_{\mathrm{simu}}} = \varphi_{\mathrm{Displ}} + \varphi_{\mathrm{Topo}_{\mathrm{res}}} + \varphi_{\mathrm{Atm}_S} - \varphi_{\mathrm{Atm}_m} + \varphi_{\mathrm{orb}_S} - \varphi_{\mathrm{orb}_m} + \varphi_{\mathrm{noise}} + 2k\pi \quad (2)$$

where $\Delta\varphi_{D_{\mathrm{int}}}$ is the DInSAR signal, $\Delta_{\mathrm{Topo}_{\mathrm{simu}}}$ is the topographic component, φ_{Displ} corresponds to the surface deformation, $\varphi_{\mathrm{Topo}_{\mathrm{res}}}$ is the residual topographic error (RTE), φ_{Atm_S}, φ_{Atm_m} are the atmospheric delay at the secondary and primary images, φ_{orb_S}, φ_{orb_m} are the phase components due to the orbital error, φ_{noise} is the phase noise, and k is an integer as phase ambiguity corresponding to the number of full wavelengths.

DInSAR is usually limited by temporal, geometrical decorrelation, and atmospheric delay. It only works properly in areas where interferograms are characterized by high coherence. In typical surfaces of low coherence, such as forests, snow, and the other vegetable areas, insufficient coherence value of corresponding pixels gives unreliable phase difference value.

Multi-temporal DInSAR (MTInSAR) techniques, known as Advanced-DInSAR, can partly overcome the DInSAR limitations. PSInSAR is one of the two most used methods of MTInSAR. This method uses multiple SAR images acquired in the same area to generate an interferogram network with a single primary image. In this study, the Sentinel-1A data is processed by the PSInSAR method with the primary image chosen on April 15, 2019. The interferogram network is shown in Fig. 6, in which the acquired time and the perpendicular baseline length are described along the x and y axes.

The PSInSAR processing is divided into three main workflows: (i) data pre-processing, (ii) DInSAR, and (iii) the PSInSAR processing using StaMPS. The two first steps were processed by the Sentinel Application Platform software package

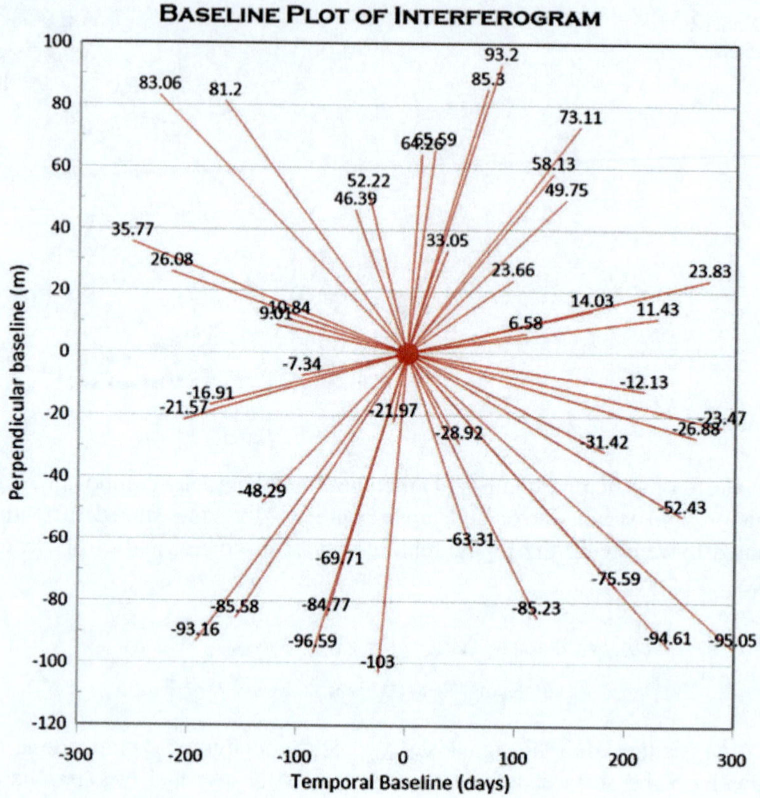

Fig. 6 Baseline plot of Interferogram. The red dot is the primary image, which was captured on April 15, 2019

(SNAP) by European Space Agency (ESA), and then, the third step was analyzed by the Stanford Method for Persistent Scatterers (STaMPS) software package. These steps are described in Fig. 7.

In workflow (i):

- Firstly, the primary image is selected from the dataset in a way that reduces the overall temporal and perpendicular baselines (see Table 1). The Sentinel-1A images are split with the selected sub-swath, bursts, and polarization. Then, the orbit state vectors in the metadata of the products are updated, based on the accurate information of satellite position and velocity.
- The secondary images are made the co-registration with a primary image on the same sub-swath. Then, the steps of removing black path between bursts and making subset of the study area are done for all split Sentinel-1 images.

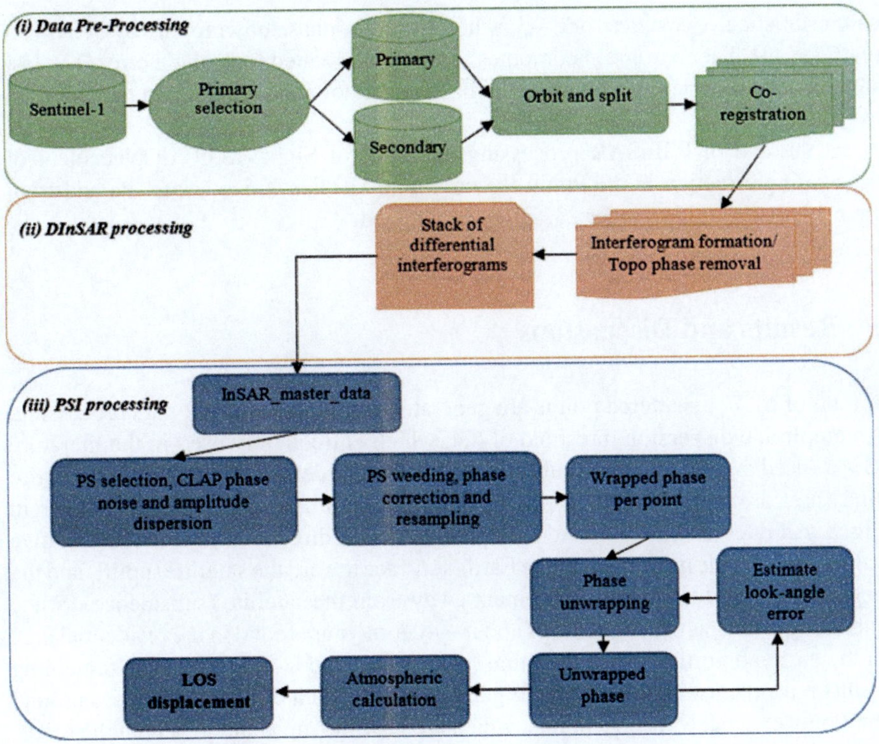

Fig. 7 Workflow of the PSInSAR data processing

In workflow (ii):

- 43 differential interferograms between the primary and secondary images are analyzed. The Shuttle Radar Topography Mission (SRTM) Digital Elevation Model (DEM) is used for removing the topography phase from these interferograms by the SNAP software. In this step, the research area is subsetted.
- Products of workflow (i) and (ii) are exported to be subsequently processed in STaMPS. The exported folders will be created to save all resulted interferograms. They consist of folders used to store single images of SLC views for all SAR scenes, folders contain all interferograms, and folder storing the coordinates of primary image and cropped DEM of the study area.

In workflow (iii):

The data, which exported from SNAP, is imported to STaMPS and saved in the Matlab workspace. The steps in STaMPS are as follows: Phase noise estimation step is the estimation of the phase noise for each pixel in interferogram. Then, qualified PS pixels on the basic of noise characteristics are selected. The third step, PS noise or PS affected by signal contribution from neighboring elements are removed. Phase correction step helps correct the wrapped phase. The interferograms are unwrapped using statistical—cost network—flow algorithm for phase unwrapping by SNAPHU method [32]. Then, program computes spatially correlated look angle error. The last step, in atmospheric filtering, triangle mesh generator and Delaunay triangulator are exploited.

At the end of PSInSAR processing, the Line of Sight (LOS) displacement of descending direction in the Pniówek coal mine, Silesian Voivodeship, Poland from June 19, 2018, to December 23, 2019, is obtained.

5 Results and Discussions

A total of 61,771 scattered points are generated from the PSInSAR data processing. An amplitude dispersion threshold of 0.4, which is frequently used in the literature [33], is used to select an initial subset of PS pixels. The deformation rates for the entire study area are then computed from the deformation time series, which are shown in Fig. 8. All rates are shown in the light of sight (LOS) direction, in which the positive values indicate the movement of the Earth's surface toward the satellite (uplift) and the negative values describe the movement away from the satellite (subsidence). Some subsiding points with a high rate of about −40 mm/yr are found in the residential area in the east and north-west of the mine (positions B and C in Fig. 9). The remaining points have the lower subsidence. Figures 8 and 9 indicate subsidence areas around the Pniówek Coal Mine. PS points, which are near the mine and locate on the roads, also residential area, are the most important points in this study because they directly affect the people living in these areas. Subsidence areas appeared to happen in most residence areas, which are marked as A, B, C, D, and E in Fig. 9.

Figure 9 shows the subsidence around the Pniówek coal mine. Green colors describe the uplift of the Earth's surface. Red, orange, and yellow colors are the subsidence. The areas with large subsidence values are determined in the A, B, and C areas.

Over the northern part of the mine (position A), the largest subsidence values reached the maximum value of −31.4 mm/yr (point P1 in Fig. 10. Figure 9 reveals the subsidence map around the Pniówek mine. The map is created by the QGIS software from subsidence values of PS points from −30 mm/yr to 15 mm/yr by triangulated irregular network (TIN) interpolation [34]. In Fig. 9, the red color is the area with subsidence over −30 mm/yr. The areas in orange indicate subsidence rates ranging from −20 mm/yr to −30 mm/yr. All the residential areas in position A are of subsidence below −31.4 mm/yr.

At position B, the subsidence values are larger than those in the A position. The maximum subsidence value in B is −40 mm/yr (point P8 in Fig. 10). At this position, the average subsidence is about −20 mm/yr.

At the C position, the maximum subsidence is −43.3 mm/yr (point P4 in Fig. 10). In the residential area of this position, most PS subsidence points are from −10 mm/yr to −30 mm/yr. The coordinates and subsidence rates of some high-subsiding points (over −31 mm/yr) in Fig. 10 are shown in Table 2.

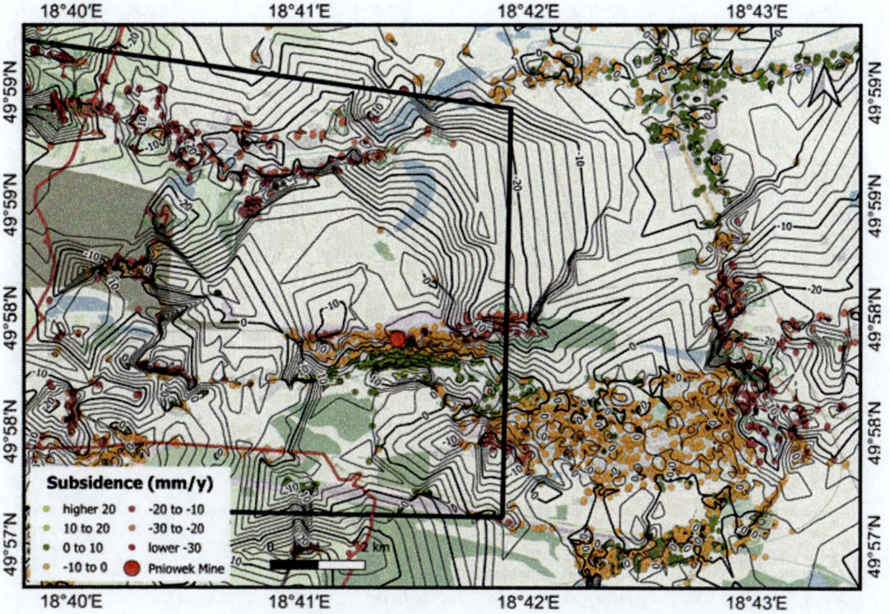

Fig. 8 Deformation rates computed over the study area between June 19th, 2018, and December 23rd, 2019. The big red point is the mine office of the KWK Pniówek company

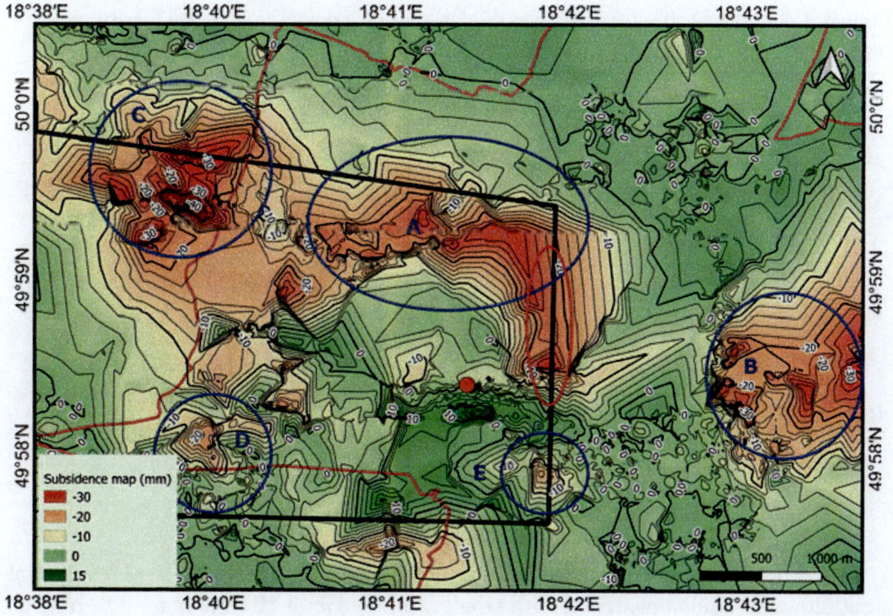

Fig. 9 Subsidence map around the Pniówek mine

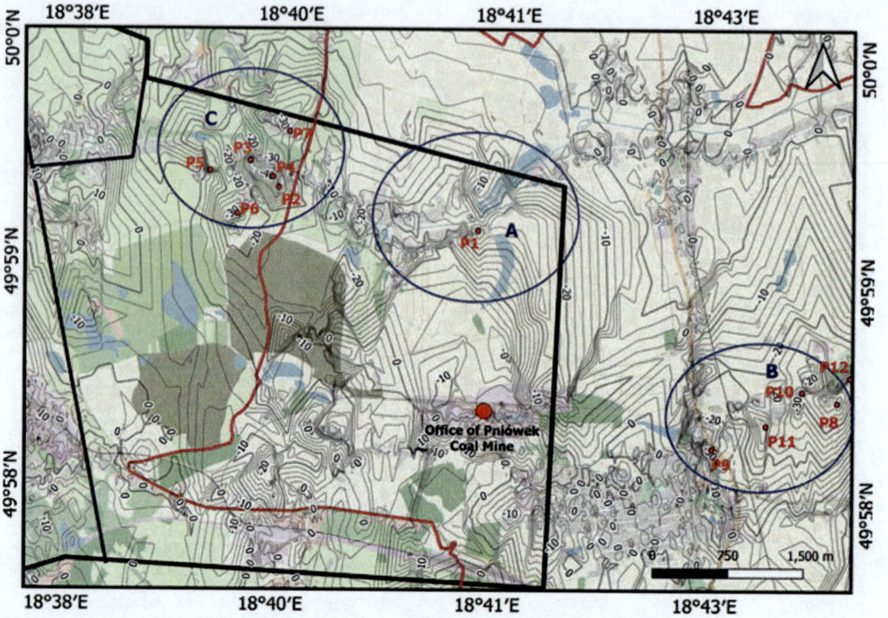

Fig. 10 Points of high subsidence rates extracted in positions A, B, and C

Table 2 Coordinates and subsidence rates of high-subsiding points in A, B, and C positions

No	Points	Latitude (deg)	Longitude (deg)	Subsidence rate (mm/yr)
1	P1	49.98883	18.66058	−31.4
2	P2	49.98822	18.65932	−33.7
3	P3	49.98990	18.65492	−40.3
4	P4	49.98863	18.65809	−43.3
5	P5	49.98916	18.64947	−31.7
6	P6	49.98517	18.65401	−33.8
7	P7	49.99311	18.65894	−37.4
8	P8	49.97028	18.73767	−40.0
9	P9	49.96495	18.72073	−33.4
10	P10	49.97079	18.73247	−33.5
11	P11	49.96923	18.73266	−31.8
12	P12	49.97248	18.73950	−35.7

In addition to the three areas with large subsidence rates, there are also 2 residential areas showing land subsidence with smaller rates from −10 mm/yr to −20 mm/yr. They are named the E and D positions in Fig. 9. Although the subsidence value in D and E are not as high as those in A, B, and C positions, these areas need to focus, because there are people living in their areas. The surface deformation has been

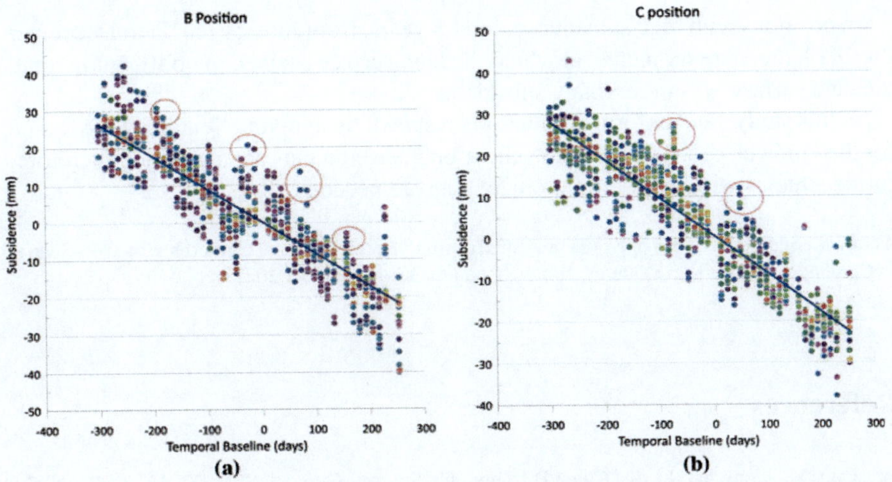

Fig. 11 Behavior of the land surface in the high subsidence positions B and C during time from June 19, 2018, to December 23, 2019

found in most of the residential areas around the Pniówek mine. It also extends from position A to the KWK Pniówek company office (dark-red oval in Fig. 9). However, this area is only field with grass and trees.

In positions with high subsidence B and C, the plots of deformation time series are also created (see Fig. 11a, b). In these positions, there are moments, which have been marked by red circles in Fig. 11, the land surface seems to uplift, but in the total period from June 19, 2018, to December 23, 2019, the land surface is more likely subsided.

6 Conclusions

With 44 Sentinel-1A SAR images between June 19th, 2018, and December 23rd, 2019, the PSInSAR method has been applied with the SNAP and StaMPS software packages in this study. The surface deformation over the Pniówek coal mine and the surrounding areas are found which are summarized as follows:

Sentinel-1A can be used with the PSInSAR method to detect and determine deformation caused by mining operations in the Pniówek coal mine. In this study, there were many subsidence PS points, but the largest subsiding areas were mostly located in the residential areas. There are 3 positions with very high subsidence rates located in the northern and eastern parts of the coal mine. Moreover, there are 2 low-rate subsiding areas located in the residential areas in the southern part of mine. The maximum subsidence detected in residential areas around the Pniówek coal mine reaches −43.3 mm/yr.

In positions with high subsidence, for 1.5 years, from June 19, 2018, to December 23, 2019, there are moments, in which the land surface seemed to uplift, but in total, the land surface was more likely subsided.

In this study, land deformation has been shown by applying Sentinel-1 SAR data for the Pniówek coal mine, but no validation measurements are available. Therefore, further studies with more measurement data are needed.

Acknowledgements The paper was presented during The International Conference on Geo-Spatial Technologies and Earth Resources, Hanoi, Vietnam, 13–14 October 2022.

References

1. Gao, M., Gong, H., Li, X., Chen, B., Zhou, C., Shi, M., Guo, L., Chen, Z., Ni, Z., Duan, G.: Land subsidence and ground fissures in Beijing capital international airport (BCIA): evidence from Quasi-PS InSAR analysis. **11**(12), 1466 (2019). https://doi.org/10.3390/rs11121466
2. Witkowski, W.T., Łukosz, M., Guzy, A., Hejmanowski, R.: Estimation of mining-induced horizontal strain tensor of land surface applying InSAR. **11**(7), 788 (2021). https://doi.org/10.3390/min11070788
3. Grzovic, M., Ghulam, A.: Evaluation of land subsidence from underground coal mining using TimeSAR (SBAS and PSI) in Springfield, Illinois, USA. Nat. Hazards. **79**(3), 1739–1751 (2015). https://doi.org/10.1007/s11069-015-1927-z
4. Abdikan, S., Arıkan, M., Sanli, F.B., Cakir, Z.: Monitoring of coal mining subsidence in peri-urban area of Zonguldak city (NW Turkey) with persistent scatterer interferometry using ALOS-PALSAR. Environ. Earth Sci. **71**(9), 4081–4089 (2014). https://doi.org/10.1007/s12665-013-2793-1
5. Yue, H., Liu, G., Guo, H., Li, X., Kang, Z., Wang, R., Zhong, X.: Coal mining induced land subsidence monitoring using multiband spaceborne differential interferometric synthetic aperture radar data. J. Appl. Remote Sens. **5**(1), 053518 (2011). https://doi.org/10.1117/1.3571038
6. Huong, K.T.T., Ha, T.H., Luyen, B.K., Tomasz, L.: Mining-Induced Land Subsidence Detected by Sentinel-1 SAR Images: An Example from the Historical Tadeusz Kościuszko Salt Mine at Wapno, Greater Poland Voivodeship, Poland, Inzynieria Mineralna (2021). https://doi.org/10.29227/IM-2021-02-04
7. Long, N., Anh, T., Bui, L.: Determination of ground subsidence by Sentinel-1 SAR data (2018–2020) over Binh Duong Quarries, Vietnam, VNU. J. Sci. Earth Environ. Sci. **37** (2021). https://doi.org/10.25073/2588-1094/vnuees.4605
8. Tran, V.A., Bui, X.N., Nguyen, Q.L., Tran, T.A.: Land subsidence detection in Tan My-Thuong Tan open pit mine and surrounding areas by time series of Sentinel-1 images. Inżynieria Mineralna 1 Nr 2 (2020). https://doi.org/10.29227/IM-2020-02-22
9. Pérez, V.M.G., Alegria, O.C., Medellin-Aguilar, G.: On the measure of land subsidence throughout DEM and orthomosaics using GPS and UAV. Investigación y Tecnología **22**(1), 6 (2021)
10. Bui, L.K., Le, P.V.V., Dao, P.D., Long, N.Q., Pham, H.V., Tran, H.H., Xie, L.: Recent land deformation detected by Sentinel-1A InSAR data (2016–2020) over Hanoi, Vietnam, and the relationship with groundwater level change. GIScience Remote Sens. **58**(2), 161–179 (2021). https://doi.org/10.1080/15481603.2020.1868198
11. Lipecki, T.: Non-contact diagnostics of the geometry of a historic wooden building as an element of periodic safety assessment. Sensors **22**, 1301 (2022). https://doi.org/10.3390/s22041301

12. Maghsoudi, Y., van der Meer, F., Hecker, C., Perissin, D., Saepuloh, A.: Using PS-InSAR to detect surface deformation in geothermal areas of West Java in Indonesia. Int. J. Appl. Earth Obs. Geoinformation **64**, 386–396 (2018). https://doi.org/10.1016/j.jag.2017.04.001
13. Sousa, J.J., Ruiz, A.M., Hanssen, R.F., Bastos, L., Gil, A.J., Galindo-Zaldívar, J., de Galdeano, C.S.: PS-InSAR processing methodologies in the detection of field surface deformation—Study of the Granada basin (Central Betic Cordilleras, Southern Spain). J. Geodyn. **49**(3), 181–189 (2010). https://doi.org/10.1016/j.jog.2009.12.002
14. Yu, J., Ng, A.H.-M., Jung, S., Ge, L., Rizos, C.: Urban monitoring using persistent scatterer InSAR and photogrammetry. The international archives of the photogrammetry. Remote Sens. Spat. Inf. Sci. **37**, 257–262 (2008)
15. Berardino, P., Fornaro, G., Lanari, R., Sansosti, E.: A new algorithm for surface deformation monitoring based on small baseline differential SAR interferograms. IEEE Trans. Geosci. Remote Sens. **40**(11), 2375–2383 (2002). https://doi.org/10.1109/TGRS.2002.803792
16. Hooper, A., Segall, P., Zebker, H.: Persistent scatterer interferometric synthetic aperture radar for crustal deformation analysis, with application to Volcán Alcedo, Galápagos. J. Geophys. Res.: Solid Earth. **112**, B7 (2007). https://doi.org/10.1029/2006JB004763
17. Bui, L.K., Featherstone, W., Filmer, M.S.: Disruptive influences of residual noise, network configuration and data gaps on InSAR-derived land motion rates using the SBAS technique. Remote Sens. Environ. **247**, 111941 (2020). https://doi.org/10.1016/j.rse.2020.111941
18. Perski, Z.: Applicability of ERS-1 and ERS-2 InSAR for land subsidence monitoring in the Silesian coal mining region, Poland. Int. Arch. Photogrammetry Remote Sens. **32**, 555–558 (1998)
19. Hejmanowski, R., Malinowska, A.A., Witkowski, W.T., Guzy, A.: An analysis applying InSAR of subsidence caused by nearby mining-induced earthquakes. Geoscience **9**(12), 490 (2019). https://doi.org/10.3390/geosciences9120490
20. Witkowski, W.T., Mrocheń, D., Sopata, P., Stoch, T.: Integration of the levelling observations and PSInSAR results for monitoring deformations caused by underground mining. In: 2021 IEEE International Geoscience and Remote Sensing Symposium IGARSS, pp. 6614–6617. IEEE (2021). https://doi.org/10.1109/IGARSS47720.2021.9553988
21. Milczarek, W., Blachowski, J., Grzempowski, P.: Application of PSInSAR for assessment of surface deformations in post-mining area—case study of the former Walbrzych hard coal basin (SW Poland). Acta. Geodyn. Geomater. **14**(1), 185 (2017). https://doi.org/10.13168/AGG.2016.0026
22. Milczarek, W., Blachowski, J.: Analysis of mining deformations based on PSInSAR technique–case study of the Walbrzych coalmines, Poland. In: Proceedings of the 16 th International Congress for Mine Surveying, Brisbane, Australia, pp. 12–16 (2016)
23. Leśniak, A., Porzycka-Strzelczyk, S.: The pattern of surface deformations in the Dąbrowa Basin (Poland) based on PSInSAR analysis. Geol. Quart. **64**(2), 304–318 (2020). https://doi.org/10.7306/gq.1537
24. https://www.jsw.pl/o-nas/zaklady/pniowek. J.S.W.SA. website of Jastrzębska Spółka Węglowa SA
25. Tobiczyk, S., Szczurek, C., Modrzejewski, D., Knapczyk, A.: Historia Kopalni Węgla Kamiennego'Pniówek'i jej ogólna charakterystyka. J. Wiadomości Górnicze **60**(11), 642–653 (2009)
26. https://www.gazetaprawna.pl/wiadomosci/kraj/artykuly/8411432,koniec-akcji-w-kopalni-pniowek.html. PAP; End of action at the Pniówek mine. Rescuers will return for seven miners after the fire is extinguished (May 2022)
27. Majcherczyk, T., Małkowski, P., Niedbalski, Z.: The influence of longwall exploitation carried out in the vicinity of a protection pillar on a rectified building. Gospodarka Surowcami Mineralnymi **25**, 247–259 (2009)
28. https://www.radio90.pl/z-powodu-eksploatacji-gorniczej-doszlo-do-uszkodzenia-obwodnicy-pawlowic.html. Radio90PM; Z powodu eksploatacji górniczej doszło do uszkodzenia obwodnicy Pawłowic (2021)

29. Pala, P.: Effect on surface deformation resulting from mining activity on value of property valuation. A case study from Jastrzębie-Zdrój region, Contemporary Trends in Geoscience. **7**(1), 127–132 (2018). https://doi.org/10.2478/ctg-2018-0008
30. Ciampalini, A., Bardi, F., Bianchini, S., Frodella, W., Del Ventisette, C., Moretti, S., Casagli, N.: Analysis of building deformation in landslide area using multisensor PSInSAR™ technique. Int. J. Appl. Earth Obs. Geoinformation **33**, 166–180 (2014). https://doi.org/10.1016/j.jag.2014.05.011
31. Ferretti, A., Prati, C., Rocca, F., Casagli, N., Farina, P., Young, B.: Permanent Scatterers technology: a powerful state of the art tool for historic and future monitoring of landslides and other terrain instability phenomena. In: Proceedings of 2005 International Conference on Landslide Risk Management, (Vancouver, Canada, 2005) (2005)
32. Chen, C.W., Zebker, H.A.: Network approaches to two-dimensional phase unwrapping: intractability and two new algorithms. J. Opt. Soc. Am. A **17**(3), 401–414 (2000). https://doi.org/10.1364/JOSAA.17.000401
33. Sousa, J.J., Hooper, A.J., Hanssen, R.F., Bastos, L.C., Ruiz, A.M.: Persistent Scatterer InSAR: a comparison of methodologies based on a model of temporal deformation versus spatial correlation selection criteria. Remote Sens. Environ. **115**(10), 2652–2663 (2011). https://doi.org/10.1016/j.rse.2011.05.021
34. Bui, L.K., Glennie, C.L., Hartzell, P.J.: Rigorous propagation of LiDAR point cloud uncertainties to spatially regular grids by a TIN linear interpolation. IEEE Geosci. Remote Sens. Lett. **19**, 7003105 (2022)

Slope Stability Evaluation of Fenghuangshan Landfill Under Rainfall Condition: A Case Study

Yuru Chen, Jun Kuang, Renmin Zhu, Jianlin Cao, Jun Zhou, and Qiang Tang

Abstract The landfill is usually instable by rainfall. This paper uses GeoStudio software to analyze the stability of the landfill. Four methods, including Morgenstern-Price method, Bishop method, Janbu method, and Spencer method, were used to calculate and evaluate the stability of the slope. The results show that with the duration of rainstorm, the pore water pressure increases gradually, and the slip resistance moment decreases linearly. The factor of safety is highest when the anti-slip pile is set in the middle of the slope with a maximum safety factor of 2.12. As the length of the anti-slip pile increases, the circular damage surface of the sliding bed in front of the pile gradually moves downward. When the pile length is 30 m, the sliding surface runs through the whole sliding area from the top to the foot of the slope; at this point, the sliding resistance on the slope of the central anti-slip pile reached 2921.58 kN. The shear strength of the soil is fully utilized, and the economic efficiency is optimized. Compared to the slip resistance moment under storm conditions, the minimum slip resistance moment is increased by 11,506.23 kN·m.

Keyword Landfill · Slope stability analysis · Pore water pressure · Sliding surface · Anti-slip pile

1 Introduction

As a country with frequent geological disasters in China, both landslides induced by natural conditions and landslides caused by human factors have caused great losses to domestic economic construction and people's lives and property [1, 2]. A large number of landslide accident statistics show that rainfall often plays an important

Y. Chen · Q. Tang (✉)
School of Rail Transportation, Soochow University, Suzhou 215141, China
e-mail: tangqiang@suda.edu.cn

J. Kuang · R. Zhu · J. Cao · J. Zhou
Jiangsu Suzhou Geological Engineering Survey Institute, Suzhou 215011, China

The Fourth Geological Brigade of Jiangsu Geological and Mineral Bureau, Suzhou 215004, China

L. Q. Nguyen et al. (eds.), *Advances in Geospatial Technology in Mining and Earth Sciences*, Environmental Science and Engineering,
https://doi.org/10.1007/978-3-031-20463-0_3

role in slope instability accidents [3, 4]. The occurrence of slope instability is usually accompanied by rainfall, whether it is the natural rock and soil slope formed in the natural environment or the artificial slope formed by filling and digging for various construction needs [5, 6]. Therefore, it is of great significance to study the stability of tunnel abandoned slag during rainstorm infiltration.

Limit equilibrium method is one of the most commonly used methods in slope stability analysis and support design [7]. However, the reliability of the calculation results of slope safety factor is affected due to the simplified calculation process by ignoring some forces between strips. With the continuous development of computer technology and the improvement of calculation level, more strict limit equilibrium methods were put forward [8–10], which can satisfy the balance of overall force and moment of slope considering interstrip force as comprehensively as possible. In this paper, GeoStudio software was used to use representative limit equilibrium methods satisfying both torque and force balance as examples such as Bishop method satisfying torque balance, Janbu method satisfying force balance, Morgenstern-Price method, and Spencer method. The results of different limit equilibrium methods were discussed in terms of stress assumption of strip and global equilibrium equation.

Previous researches mainly focus on the infinite length inclined straight slope, ignoring the fact that the high slope with finite length has steps, and the slope is not straight. So, we take the broken line-shaped slope as the research object as the innovation point of this paper. Fenghuangshan landfill is located in Suzhou, China. The existing slope is steep. Due to the large rainfall in Suzhou, it is more prone to landslides and other safety accident. This paper takes Fenghuangshan landfill as an example. SEEP/W in GeoStudio software is used to calculate the pore water pressure of the engineering slope. Morgenstern-price method, Bishop method, Janbu method, and Spencer method are used for comparative analysis of slope stability coefficient. The variation law of pore water pressure, anti-sliding moment, and sliding moment with rainfall time and the evolution law of stability coefficient with position and depth of anti-slide pile under the working condition of anti-slide pile were investigated.

2 Theoretical Assumption

In terms of the construction of equilibrium equation, the limit equilibrium method can be divided into three categories: satisfying torque balance, satisfying force balance, and satisfying both torque and force balance [11, 12]. Taking simplified Bishop method, strict Janbu method and Morgenstern-Price method as examples, this paper discusses the differences of force modes and equilibrium equation construction of strip with limit equilibrium methods satisfying moment equilibrium, force equilibrium and both moment and force equilibrium.

Simplified Bishop method assumes that the shape of the sliding surface is a strict circular arc so that all the positive pressures at the bottom of different strips pass through the center of the circle, and the torque generated by the positive pressure at the bottom of the strip is 0. In the simplified Bishop method, only, the gravity of the strip and the tangential force at the bottom of the strip produce torques on the center of the circle. The simplified Bishop method satisfies the overall torque balance, but does not satisfy the overall force balance, and its stability coefficient is expressed by the bending moment of each strip to the rotating center [13, 14]. According to the equation of torque balance, the safety factor of slope can indicated as:

$$F_m = \frac{\sum_{i=1}^{n} (c_i b_i + W_i \tan \varphi_i)/m_{\alpha i}}{\sum_{i=1}^{n} W_i \sin \alpha_i} \tag{1}$$

$$m_{\alpha i} = \cos \alpha_i + \sin \alpha_i \tan \varphi_i / F_m \tag{2}$$

where F_m is the safety factor; $m_{\alpha i}$ is the calculated coefficient of the ith block; W_i is the weight of the ith bar; c_i is the cohesion of the ith strip; b_i is the length of the ith bar; φ_i is the effective internal friction angle of the ith strip.

Janbu method is a limit equilibrium analysis method satisfying force equilibrium. In this method, the tangential and normal forces between strips are considered, and the acting positions of the tangential and normal forces between strips are assumed to reduce the variables. At the same time, the tangential force and normal force at the bottom of the strip are considered completely. Since Janbu method does not need to meet the balance of the overall torque, there is no special requirement for the shape of the sliding surface, and it can be applied to the sliding surface of any shape. The Janbu method satisfies the overall force balance, but does not satisfy the overall torque balance, and its stability coefficient is expressed by the sliding driving force of each strip and the anti-sliding force at the bottom of the strip. According to the equation of force balance, the slope safety factor can be indicated as [15, 16]:

$$F_f = \frac{\sum_{i=1}^{n} [c_i b_i + (W_i + H_i) \tan \varphi_i]/m_{\alpha i}}{\sum_{i=1}^{n} (W_i + H_i) \sin \alpha_i} \tag{3}$$

$$m_{\alpha i} = \cos \alpha_i^2 + \sin \alpha_i \cos \alpha_i \tan \varphi_i / F_f \tag{4}$$

where H_i is the height of the ith strip.

Morgenstern-Price method is a limit equilibrium analysis method which satisfies both global force equilibrium and moment equilibrium. The sliding body is split into infinitesimal strips; each strip is subjected to the normal force and tangential force between the strips and at the bottom of the strip. However, because there are many variables involved in this method, it is necessary to construct balance equation of force and moment balance equation to solve the stability coefficient. According to the equation of force balance, the slope safety factor can be indicated as [17]:

$$F_m = \frac{\sum_{i=1}^{n} (c_i b_i / \cos \alpha_i - N_i \tan \varphi_i)/R_i}{\sum_{i=1}^{n} W_i x_i / \cos \alpha_i} \tag{5}$$

where N_i is the force at the bottom against the clod of the ith strip; R_i is the slide radius of the ith strip.

According to the balance of the force, the equation of safety factor can be indicated as:

$$F_f = \frac{\sum_{i=1}^{n} (c_i b_i - N_i \tan \varphi_i \cos \alpha_i)}{\sum_{i=1}^{n} (N_i / \sin \alpha_i)} \tag{6}$$

Spencer method holds that the dip angle of interstrip force is constant, and that the direction of interstrip force is consistent. According to the balance conditions of force and moment, the angle and safety factor Fs are calculated as the basis of slope stability evaluation [18].

$$F_s = \frac{\sum_{i=1}^{n} \left\{ \left[(P_i - P_{i-1}) \sin(\alpha_i - \theta) + W_i \cos \alpha_i \right] \tan \varphi_i + c_i b_i \right\}}{\sum_{i=1}^{n} \left[(P_i - P_{i-1}) \cos(\alpha_i - \theta) + W_i \sin \alpha_i \right]} \tag{7}$$

where P_i is the resultant force between soil strips.

In this paper, GeoStudio software is used to make comparative analysis of slope stability coefficient by using Bishop method satisfying torque balance, Janbu method satisfying force balance, Morgenstern-Price method, and Spencer method satisfying torque and force balance at the same time, so as to obtain more reliable calculation results of slope safety coefficient.

3 Case Analysis

3.1 Project Profile

Fenghuangshan in SuZhou, China, is a north sub-tropical maritime monsoon climate area with abundant rainfall. There are typhoons in late summer and early autumn. The annual average precipitation is 1035.9 mm, mainly concentrated in June to September. The annual average evaporation is 1065.8 mm. The existing landfill is covering an area of about 42,000 m^2. The south yard is 30 ~ 95 m wide from north to south and 130 ~ 165 m long from east to west. The top of the slope covers an area of 9036 m^2, and the elevation of the top is between 51 and 55 m. The total height of the south yard is about 40 m.

According to the exploration, the soil (rock) body of the proposed site can be segmented into four engineering geological layers according to the difference of its engineering characteristics, which are ① miscellaneous fill, ② clay, ③ plain fill, and ④ gravel soil. The section of the South Slope of the dump is shown in Fig. 1a.

In the south side of the slope, due to the action of surface water scour and infiltration in rainy season, a small area of landslide collapse has occurred locally. In addition, due to the influence of tectonic fissures and weathering, there are many

(a) The mountain form of Fenghuangshan

(b) Status of South Slope collapse

Fig. 1 Status of Fenghuangshan

dangerous rock masses in the upper rock cliff left by the excavation of south storage yard, some of which have collapsed and collapsed. Figure 1b shows the current situation of slope collapse of south storage yard.

3.2 *Selection of Calculation Parameters*

As the groundwater in this paper is below the slip surface, according to the study of Yang et al. [16] and Jing-ren et al. [17], groundwater has little influence on slope slip, so its influence is ignored in this simulation.

Through field sampling, rock mechanics test in the laboratory, to determine the soil strength parameters, under the condition of rainfall, mineral rock softening occurs, results in the decrease of strength, calculation, first of all, according to the field investigation situation underground seepage line is determined, the saturation line above the rock mass strength parameters under natural state, the following section for saturated rock mass strength parameters. The final physical properties of soil mass are shown in Table 1.

Table 1 Physical properties of soil

	Gravity (kN/m^3)	Cohesion (kPa)	Friction angle (°)	Permeability coefficient (m/s)	Optimum moisture content (%)	Maximum dry density (g/cm^3)
① Miscellaneous fill	17.2	0.0	28.0	5.0×10^{-7}	5.0	2.27
② Clay soil	19.2	31.0	12.8	6.1×10^{-8}	5.7	2.21
③ Plain fill	19.0	15.0	13.1	5.5×10^{-8}	6.3	2.18
④ Gravel soil	23.0	5.0	35.0	5.0×10^{-5}	7.4	2.12

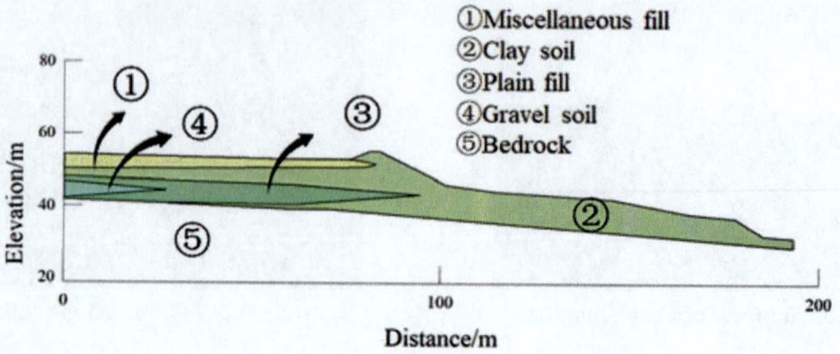

Fig. 2 Slope model drawing

3.3 Model Establishment

GeoStudio is an overall analysis tool for geological structure model software [19, 20], including SLOPE/W, SEEP/W, and other professional analysis software. Using multi-field coupling analysis function, it can also improve the overall analysis ability of SLOPE/W and other software modules [21]. In SEEP/W software, through seepage finite element calculation, the pore water pressure of slope under the condition of uneven saturation and unsaturated can be analyzed, as well as the transient pore water pressure when the slope is stable.

According to the characteristics of the soil filled slope, the South Slope profile was imported into the GeoStudio model module to establish the model, as shown in Fig. 2. Seep/W module was used to conduct seepage simulation of the slope, and the pore water pressure of the soil filled slope under the initial and rainstorm conditions was obtained. Then, Slope/W module is used to simulate slope sliding. The simulation process is as follows: When SLOPE/ W and SEEP/W are coupled, four methods are selected as the analysis types of simulation, respectively. After the material parameters are given, the entry and exit forms of the slip surface are selected. Set the slip surface form and the stability analysis of the slope start.

4 Slope Stability Analysis Under Rainfall Condition

4.1 Seepage Field

For the South Slope filled with soil, the seepage field and stability of the slope are analyzed by using GeoStudio software. The stress/strain boundary conditions were set in the displacement field analysis: Only, the *x*-type displacement was set as 0 on the left side of the model, and the *x*-type and *y*-type displacements were set as 0 on the bottom, and the rest parts were set as free displacements.

The Seep/W module of GeoStudio software was used to conduct steady-state seepage simulation of pore water pressure distribution in the southern slope under annual average rainfall and transient seepage simulation of pore water pressure distribution in the downslope under rainstorm conditions (100 mm/d). The distribution of pore water pressure in the slope is shown in Fig. 3. Table 2 is the parameter of soils in nature and saturated conditions.

Figure 3 shows that the volumetric water content in the slope body increases with the slope depth, and the pore water pressure at the same depth increases accordingly. Compared with the pore water pressure under the initial groundwater level condition, the pore water pressure contour becomes dense and gradually tends to form a ring under the rainstorm condition, indicating that the position of the infiltration peak inside the slope deepens with the increase of the water content of the slope.

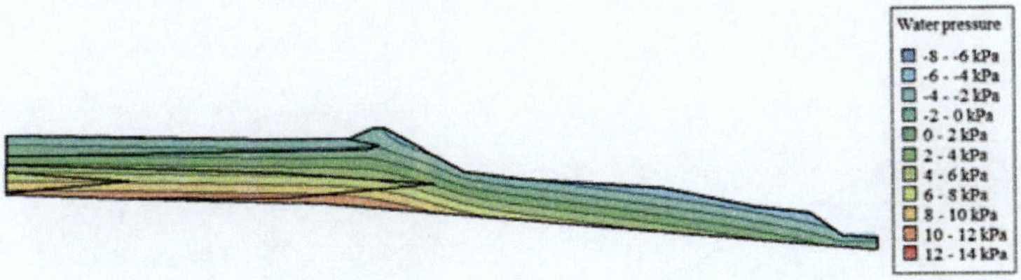

(a) Natural state

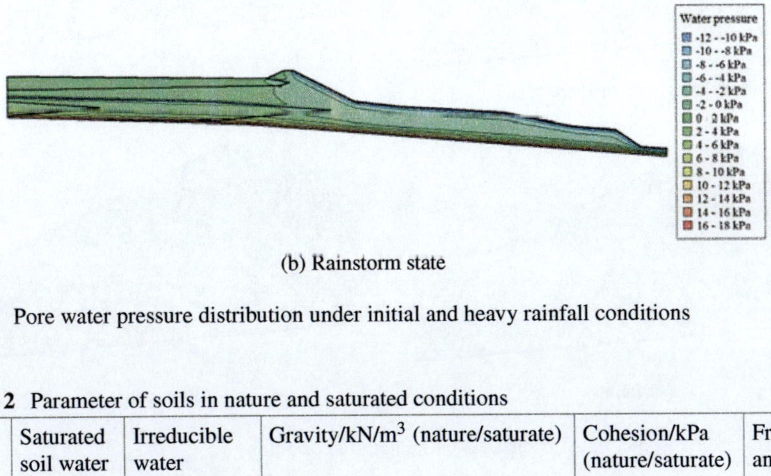

(b) Rainstorm state

Fig. 3 Pore water pressure distribution under initial and heavy rainfall conditions

Table 2 Parameter of soils in nature and saturated conditions

	Saturated soil water content/%	Irreducible water saturation/%	Gravity/kN/m^3 (nature/saturate)	Cohesion/kPa (nature/saturate)	Friction angle/°
Clay soil	20	3	16.5/19.2	31.0/10.8	12.8
Plain fill	23	2	16.1/19.0	15.0/8.1	13.1

4.2 Rainstorm State

The position of potential slip surface of the four methods without precipitation is shown in Fig. 4. Table 3 shows the specific values of anti-sliding moment, sliding moment, and sliding surface radius of slope under the four methods.

It can be seen from the results that the location of potential slip surface calculated by Spencer method is mainly in the middle and lower part of the slope, and the location of potential slip surface calculated by other three methods is relatively similar. The sliding surface radius of the four methods is very close, and the sliding surface radius of Bishop method is the smallest, which is 15.86 m. From the mechanical point of view, anti-sliding moment and sliding moment are two components of slope safety factor. The sliding moment calculated by Janbu method is the smallest, which is 966.6 kPa. The maximum anti-sliding moment calculated by Spencer method is

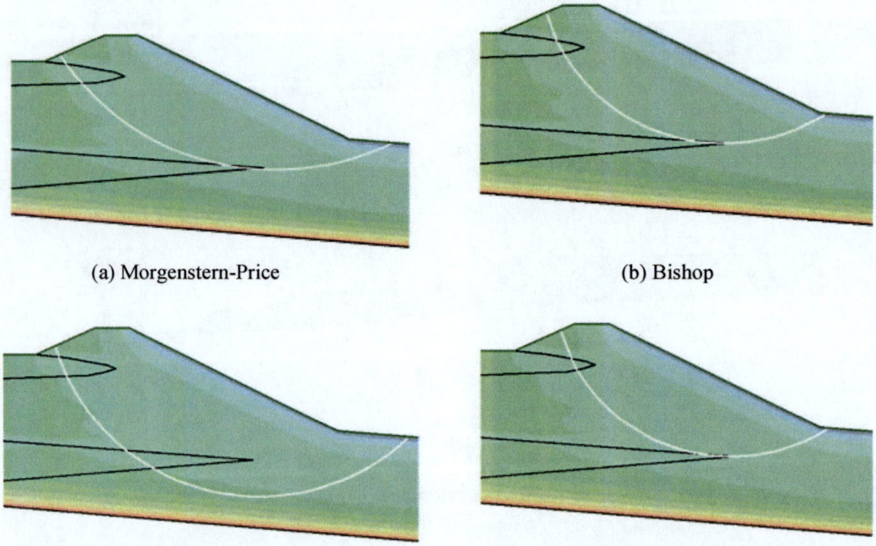

(a) Morgenstern-Price (b) Bishop

(c) Janbu (d) Spencer

Fig. 4 Location of potential slip surfaces under natural conditions

Table 3 Comparison of slip resistance, slip moment, and sliding surface radius

	Morgenstern-Price	Bishop	Janbu	Spencer
Anti-slip moment/kN·m	26,076.23	23,893.84	1611.61	27,540.07
Slipping moment/kN·m	13,610.21	12,874.55	966.630	14,561.75
Sliding radius/m	9.27	8.86	10.11	9.92
Safety factor	1.905	1.851	1.667	1.891

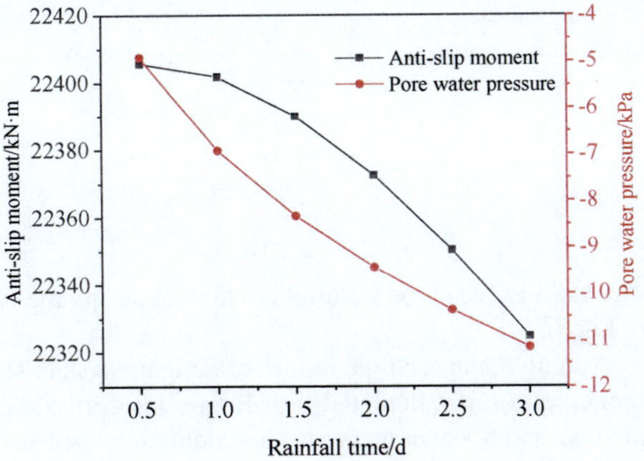

Fig. 5 Pore water pressure and anti-slip moment variation diagram

27540.1 kPa. The safety factor of slope is equal to the ratio of anti-sliding moment to sliding moment, and the lowest safety factor is Janbu method. In order to ensure the safety of the project, more conservative methods with less safety factor are often adopted. Therefore, Janbu method is used to analyze the different working conditions of rainstorm and anti-slide pile.

As can be seen from Fig. 5, in the early rainfall period, the slope surface ground fill is relatively dry, and the soil saturation and material permeability coefficient are low, so the soil at this stage has a strong infiltration capacity. As the rainfall progresses and the rainfall duration continues to increase, the rainwater infiltration rate gradually decreases; the surface pore water pressure accumulates; the surface matric suction diminishes gradually; the unsaturated zone in the slope body decreases; the humid front migrates to the interior, and the internal pore water pressure changes from insignificant to gradually increase. After the rainfall stops, the soil water above the saturated area is replenished downward, and the saturated area gradually becomes long and narrow.

4.3 Stability

In accordance with the engineering geological characteristics of the landfill, its stability coefficient and classification standard of stable state are shown in Table 4 [22].

On the basis of simulating the pore water pressure distribution of the southern slope under the annual average rainfall and the slope under the rainstorm condition, the Seep/W module coupled with the Slope/W module, Morgenstern-price method, Bishop method, Janbu method and Spencer method are used to figure the stability coefficient of slope, respectively. The rainfall in July is concentrated in Suzhou, and the average rainfall days are 3–7 days [23, 24]. When analyzing slope stability under rainfall conditions, the rainfall duration is 72 h, and the daily rainfall is 100 mm/d.

Table 4 Stable state classification criteria

Safety factor	Status
Fs < 1.00	Unstable
1.00 ≤ Fs < 1.05	Less stable
1.05 ≤ Fs < 1.15	Basically stability
1.15 ≤ Fs	Stable

The variation of slope stability coefficient during the continuous rainstorm is shown in Fig. 6.

Aiming at the arc-type failure of Fenghuangshan slope, the stability of the North Slope section is calculated by Morgenstern-Price method, Bishop method, Janbu method, and Spencer method. The calculation results show that, except for the calculation of Janbu method, the other three methods are relatively close to the calculation results of slope stability coefficient. It can be considered that the southern slope of Fenghuangshan is in a stable state or basically stable state under normal working conditions. However, under extreme rainfall conditions, instability failure, namely circular sliding failure, may occur.

The Janbu method has been widely used in slope stability studies because of its simplicity and avoidance of non-convergence problems [25, 26]. However, in view of the analysis of the theoretical assumptions in Chap. 2, the method assumes that the vertical shear between the bars is zero, which can only meet the requirements when the bars are densely divided, and the moment balance on the center point of the bottom surface of the soil bars is not considered in the derivation process, so its safety factor is the lowest. Morgenstern-Price method is a limit equilibrium analysis method which satisfies both global force equilibrium and moment equilibrium. This method simplifies the strip stress and is suitable for sliding surfaces of arbitrary shape, so its safety factor is the highest.

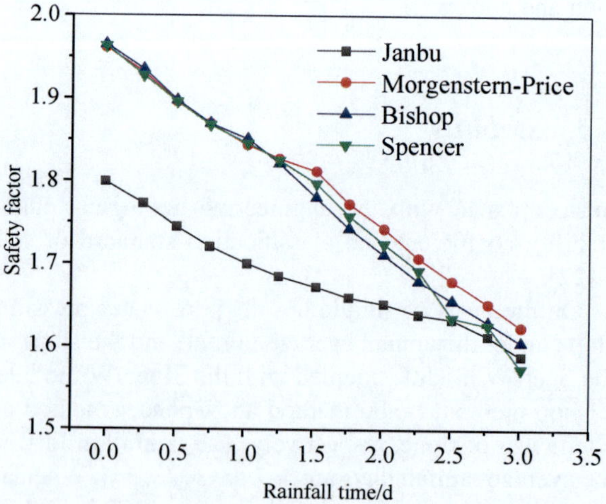

Fig. 6 Safety factor varies with rainfall time under different methods

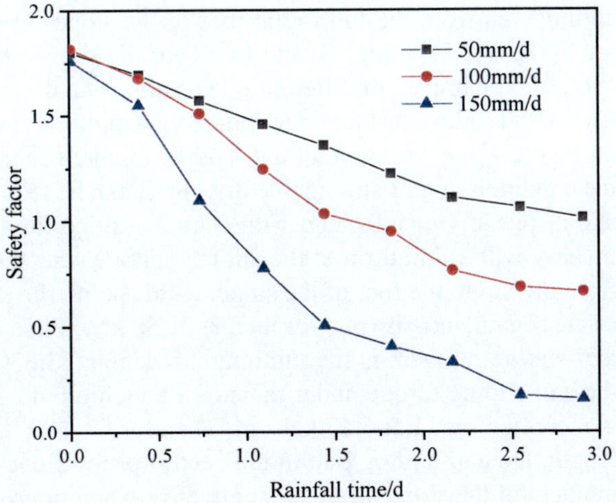

Fig. 7 Safety factor varies with rainfall time under different rainfall intensity

According to Tables 3 and 4, the stability coefficient of the South Slope under normal working conditions is about 1.20, which is in a stable state and meets the safety requirements stipulated in the code. In the case of heavy rain, slope stability coefficient is about 0.97, which is in an unstable state. Corresponding engineering actions should be taken to precaution and control the slope.

As can be seen from Fig. 7, when the rainfall intensity is 50 mm/d, the descend range of the safety factor is the smallest. The reason for this result is that the rainfall intensity is small, which results in less water infiltration into the soil and less influence on the pore water pressure and matric suction inside the soil. Therefore, the influence on the shear strength of the soil is relatively low. So, there was only a small reduction in safety. With the increase of rainfall intensity, the decreasing range of safety factor also increases. When the rainfall intensity is 150 mm/d, the decreasing range of safety factor curve reaches the maximum. In the case of 100 mm/d and 150 mm/d rainstorm, the safety factor decreases to 1 after 48 h, and the slope becomes unstable.

5 Anti-Slide Pile Reinforcement Methods

As an effective slope reinforcement measure, anti-slide pile plays an important role in landslide control. It has the characteristics of flexible layout, simple construction, small disturbance to the slope, large bearing capacity, fast construction speed, etc., and has been widely used at present [27].

The section size of anti-slide pile is $1.2 \times 1.8\ m^2$, which made of C30 concrete. The pile spacing is 4.0 m, and the shear bearing capacity of the anti-slide pile is 3000 kN. It can be seen from Fig. 8 that as the length of the anti-slip pile increases, the circular damage surface of the sliding bed in front of the pile gradually moves downward. When the pile length is 30 m, the sliding surface runs through the whole sliding area from the top to the foot of the slope. The positions of anti-slide piles are at the top A, central B, and foot C of the slope when the length of pile is 30 m (Fig. 9). The impact of different pile positions on the mechanical properties of slope fill and the sliding surface under rainstorm conditions is explored with Janbu method.

Figure 10a is the curve of anti-sliding moment changing with rainfall time under the condition of anti-sliding pile driving. It can be seen from the figure at when anti-sliding pile driving is located in the central of slope, the anti-sliding moment decreases linearly with rainfall time; the sliding surface runs through the whole sliding area from the top to the foot of the slope; at this point, the sliding resistance on the slope of the central anti-slip pile reached 2921.58 kN. When the rainfall time is 3 days, the anti-sliding moment is the minimum, reaching 33,830.96 kN · m. Compared with the anti-sliding torque under rainstorm condition, the minimum anti-sliding torque is increased by 11,506.23 kN · m. Therefore, under the condition of the same pile length, the anti-sliding pile in the central of the slope can increase the anti-sliding moment of the slope filling most effectively and improve the stability of the slope.

Figure 10b is the curve of safety factor changing with rainfall time under the condition of anti-sliding pile driving. The safety factor of the anti-slide pile is the largest when it is located in the central of the slope, and the safety factor of the slope near the foot of the slope is the smallest, which is in a stable state and meets the safety requirements stipulated in the code. When the anti-slide pile is placed in the central and lower part of the slope, a penetrating plastic shear zone is generated inside the slope body behind the pile, and the sliding body passes over the pile top to produce a whole sliding type. When the anti-slide pile is set in the central and upper part of the slope, the plastic shear band is formed inside the front slope body of the pile, and the slope is separated from the anti-slide pile to produce the overall traction sliding. When the anti-slide pile is placed in the central of the slope, the effect of the slope reinforcement is the best.

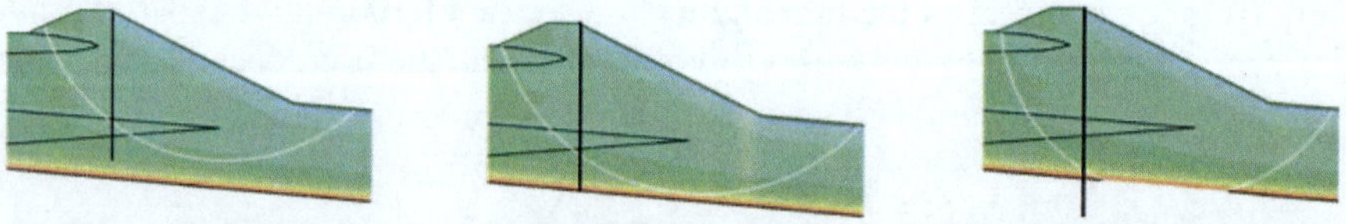

(a) pile length is 20 m (b) pile length is 25 m (c) pile length is 30 m

Fig. 8 Circular damage surface of sliding bed with different pile length

Fig. 9 Diagram of the different locations of the anti-slip piles

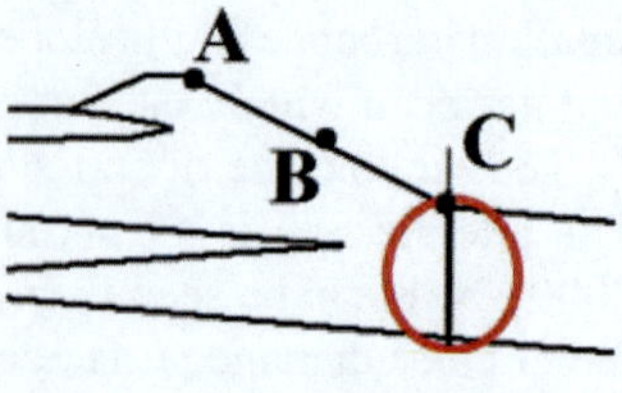

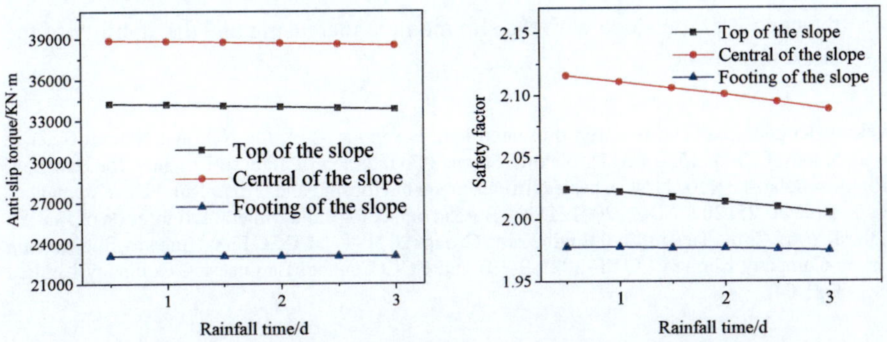

(a) Anti-slip torque (b) Safety factor

Fig. 10 Variation curves of slip moment and safety factor at different locations of the anti-slip pile

6 Conclusion

In this research, GeoStudio software was used to simulate the stability of the slope of the Fenghuangshan landfill under rainfall condition in Suzhou, China, and the following conclusions were obtained.

(1) Compared with the pore water pressure under the initial conditions, the pore water pressure contours on the slope under storm conditions become dense and gradually tend to form a ring, and the location of the infiltration peak inside the slope body deepens as the water content of the slope body increases.
(2) The southern side slope is in a steady state under normal situation, while it is in an unstable state when the slope under the heavy rainfall situation. Four methods, namely the Morgenstern-Price, the Bishop, the Janbu, and the Spencer, were used to figure up the stability of the southern slope. The results of the stability coefficients of the other three methods are close to each other, except for the low results of the individual calculation method (Janbu method). Under heavy rainfall conditions, circular sliding damage may occur on the soil slope of Fenghuangshan.
(3) The radii of the sliding surfaces are very close under the four methods, with the Bishop method having the smallest radius of 8.86 m. The Janbu method has the smallest sliding moment of 966.6 kPa, while the Spencer method has the largest anti-slip moment of 27,540.1 kPa. The method of Morgenstern-Price has the highest factor of safety, so the method is the best method of limit equilibrium for the analysis of the two different working conditions of the storm and the installation of anti-slip piles.
(4) The safety factor is the highest when the pile length is 30 m, and the safety factor is the lowest when the pile is located in the central of the slope, and near the foot of the slope is in a stable state, meeting the safety requirements of the code. The minimum slip moment is increased by 11,506.23 kN · m. Compared with the slip moment under the rainstorm condition, the piles are arranged for

the central of the slope with the slip moment increasing and the stability of the slope improving.

Acknowledgements The research presented here is supported by the National Natural Science Foundation of China (52078317), Natural Science Foundation of Jiangsu Province for Excellent Young Scholars (BK20211597), project from Bureau of Housing and Urban-Rural Development of Suzhou (2021–25; 2021ZD02; 2021ZD30), Bureau of Geology and Mineral Exploration of Jiangsu (2021KY06), China Tiesiju Civil Engineering Group (2021–19), CCCC First Highway Engineering Group Company Limited (KJYF-2021-B-19) and CCCC Tunnel Engineering Company Limited (8gs-2021-04).

References:

1. Yuan, X.k., Zhang, X.: Research on slope stability of quarry based on geostudio and madis/Gts. J. World Sci. Res. J. **7**(12), 96–105 (2021)
2. Aldulaimi, A.A.: Safety evaluation of the chamrga earth dam. a seepage deformation, and stability analysis with GeoStudio. J. IOP Conf. Ser.: Earth Environ. Sci. **664**(01), 1–15 (2021)
3. Harish, K.R.P., Ramya, K.V., Srinivas K.: Design of landfill liners for fine grained soils using CTRAN/W. J. Mater. Today: Proc. **45**(02), 3413–3418 (2021)
4. Wang, J.M., David, C., Yang, X.J.: Investigating the stability and anchor support parameters of slopes subjected to wetting and drying cycles in relation to the Nanfen open-pit mine, China. J. Arab. J. Geosci. **13**(20), 721–746 (2020)
5. Daniele, C., Flavio, P., Paolo, T.: Small-scale slope instability on the submarine flanks of insular volcanoes: the case-study of the Sciara del Fuoco slope (Stromboli). J. Int. J. Earth Sci. 1–16 (2020)
6. Chen, F.F., Sun, D.Y., Guo, Y.M.: Correction: spatiotemporal scan and age-period-cohort analysis of hepatitis C virus in Henan, China. J. PLoS ONE **10**(08), 2005–2012 (2017)
7. Song, J., Fan, Q.Q., Feng, T.G.: A multi-block sliding approach to calculate the permanent seismic displacement of slopes. J. Eng. Geol. **255**, 48–58 (2019)
8. Chia-Chi, C., Meng-Chia, W.: DEM simulation of planar sliding using a particulate interface model considering velocity-dependent friction. J. Comput. Geotech. **112**, 51–59 (2019)
9. Guo, P., Li, W.C.: Development and implementation of Duncan-Chang constitutive model in GeoStudio2007. J. Procedia Eng. 395–402 (2012)
10. Ellen, M.R., George, A.: A unified model for predicting earthquake-induced sliding displacements of rigid and flexible slopes. J. Eng. Geol. **112**(01), 51–60 (2010)
11. Xie, J.R., Taro, U.: Predicting the sliding behavior of rotational landslides based on the tilting measurement of the slope surface. J. Eng. Geol. **269**, 105554–105554 (2020)
12. Tao, Z.G., Shu, Y., Yang, X.J.: Physical model test study on shear strength characteristics of slope sliding surface in Nanfen open-pit mine. J. Int. J. Min. Sci. Technol. **30**(03), 421–429 (2020)
13. Dong, J., Wang-Ru, L., Ming, L.L.: Effects of seismic anisotropy and geological characteristics on the kinematics of the neighboring Jiufengershan and Hungtsaiping landslides during Chi-Chi earthquake. J. Int. J. Tectonophysics **466**(03), 438–457 (2009)
14. Changfu, W.U., Yao, G., Chen, L.: Slope stability analysis for a ionic rare earth mine in-situ leaching based on geo-studio. J. Chin. J. Geol. Hazard Control **27**(02), 72–77 (2016)
15. Yuan, Y.Z.: The application of GEO-slope in a fill slope stability analysis. J. Changchun Inst. Technol. (Nat. Sci. Ed.) **14**(01), 96–99 (2013)
16. Yang, G.L.: Safety evaluation of seepage stability of a reservoir homogeneous dam based on geo-studio. J. Jilin Water Res. (06) 53–56 (2018)

17. Jing-Ren, M.A.: Discussion on the application of Geo-studio software in the stability analysis of landslide. J. World Nonferrous Metals (13), 131–133 (2016)
18. Li, R.F., Lou, X.G., Sun, Y.R.: The evaluation of man-made rock slope stability in city: a case study of Duilujie landslip of urban area in Anshan city. J. China Min. Mag. **20**(05) (2011)
19. Mukhlisin, M., Aziz, N.: Study of horizontal drain effect on slope stability. J. Geol. Soc. India 483–490 (2016)
20. Bao, H., Song, S.: Theoretical analysis of slope stability and support scheme selection based on software geo-studio. J. Geol. Explor. **55**(05), 1329–1335 (2019)
21. Sharma, P., Mouli, B., Jakka, R.S.: Economical design of reinforced slope using geosynthetics. J. Geotech. Geol. Eng. 1631–1637 (2020)
22. Zhao, Y., Zhu, L.B., Hu, W.Y.: Analysis of stability factors and interaction rules of soil slope under heavy rainfall. J. IOP Conf. Ser. Earth Environ. Sci. **546**(03), 1–7 (2020)
23. Zhi, P., Li, T.H., Yang, Z.Y.: Finite element analysis of failure characteristics of structural soil slope under different rainfall patterns, 152–158 (2019)
24. Luo, Z., Bui, X.N., Nguyen, H.: A novel artificial intelligence technique for analyzing slope stability using PSO-CA model. J. Eng. With Comput. 1–12 (2019)
25. Wang, L., Wu, C., Li, Y.: Probabilistic risk assessment of unsaturated slope failure considering spatial variability of hydraulic parameters. J. KSCE. J. Civil Eng. 5032–5040 (2019)
26. Wei, L.A., Wang, Z.J., Zhang, H.: Study on influence ofrainfall type on seepage stability of soil slope. J. Hydro Sci. Cold Zone Eng. **1**(09), 5–11 (2018)
27. Zou, T., Zhang, B.Q. Zuo, H.Y.: Dynamic of seepage stability of rockfill dam based on geo-studio software. J. Water Conservancy Sci. Technol. Econ. **24**(10), 12–17 (2018)

Forecasting PM10 Concentration from Blasting Operations in Open-Pit Mines Using Unmanned Aerial Vehicles and Adaptive Neuro-Fuzzy Inference System

Xuan-Nam Bui, Chang Woo Lee, and Hoang Nguyen

Abstract In this paper, a state-of-the-art technology for modeling and controlling dust concentration from blasting operations in open-pit mines was introduced. Accordingly, a variety of smart sensors were mounted on an unmanned aerial vehicle to measure dust concentration (i.e., PM10) from blasting operations at the Thuong Tan IV quarry (Binh Duong). The meteorological conditions were also considered related to the air quality in open-pit mines. The dataset was then used to develop an artificial intelligence model for forecasting PM10 concentration in the spatial of the quarry, namely adaptive neuro-fuzzy inference system (ANFIS). The results indicated that PM10 induced by blasting operations in the quarry exceeds the allowable limit many times, and the ANFIS model can forecast PM10 concentration in the quarry with a high acceptable accuracy (~90%). It can be used to evaluate and control the air quality in the entire quarry. The paper also provided the evidence to develop better machine learning/artificial intelligence models for forecasting PM10 concentration induced by blasting operations, as well as other parameters in the air quality controlling in open-pit mines.

Keywords Air quality in open-pit mine · Unmanned aerial vehicle · Smart mine · Blasting · Adaptive neuro-fuzzy inference system

---

X.-N. Bui (✉) · H. Nguyen
Department of Surface Mining, Mining Faculty, Hanoi University of Mining and Geology, Hanoi 100000, Vietnam
e-mail: buixuannam@humg.edu.vn

H. Nguyen
e-mail: nguyenhoang@humg.edu.vn

Innovations for Sustainable and Responsible Mining (ISRM) Research Group, Hanoi University of Mining and Geology, Hanoi 100000, Vietnam

C. W. Lee
Department of Energy and Mineral Resources, College of Engineering, Dong-A University, Busan 49315, Korea
e-mail: cwlee@dau.ac.kr

L. Q. Nguyen et al. (eds.), *Advances in Geospatial Technology in Mining and Earth Sciences*, Environmental Science and Engineering,
https://doi.org/10.1007/978-3-031-20463-0_4

1 Introduction

Mining is well-known as one of the operations having adverse effects on environment, such as land cover; air, soil, and water pollution; ground vibration, to name a few [1–5]. Of those, air pollution is one of the most visible negative effects. In the air pollution, atmospheric particulate matter (PM) induced by operations in open-pit mines is a considerable concern since its adverse effects on the environment and human health [6–10]. Among the operations in open-pit mines, blasting is one of the operations that generate a large amount of dust with dust clouds containing high distribution. Effective control of such dust clouds presents a major challenge for mining and environmental engineers.

To mitigate the adverse effects of dust caused by blasting operations, Khazins and Shuvalov [11] numerical modeled dust cloud to calculate the dust dispersion in the wind stream with a very interesting result. It can provide an overview of dust dispersion for each blast with a specific capacity. Nevertheless, the prediction of dust concentration induced by blasting operations has not been solved in this study. For this aim, several researchers studied and applied artificial intelligence (AI) techniques to forecast dust emission/concentration not only from blasting operations but also from other sources in open-pit mines. For example, Bakhtavar et al. [12] developed an AI-based model for forecasting dust emission using the fuzzy number and Monte Carlo methods. The results are pretty good with an RMSE of 7.018 and R^2 of 0.875 in practical engineering. In another study, Bakhtavar et al. [13] applied the fuzzy cognitive map and neural network to predict vertical and horizontal dust distributions with a promising result, as well. An artificial neural network (ANN) integrated with dimensional analysis (DA) was also discovered to forecast dust emission by Hosseini and Monjezi [14]. They showed that the ANN-DA can forecast dust emission with high accuracy.

Although several studies regarding the applications of AI-based methods in predicting dust caused by blasting operations in open-pit mines have been introduced; however, all of them presented the dust emission problem. Meanwhile, dust concentrations caused by blasting operations in open-pit mines are much more important due to their distribution and concentrations per unit volume is the basis for assessing air quality according to standards. This study, therefore, presented an application of AI in forecasting PM10 concentrations caused by blasting operations in a quarry (in Vietnam). The adaptive neuro-fuzzy inference system (ANFIS) model was considered and configured for this aim. The results would be important basis for developing other models intend to forecast different PM concentrations, as well as controlling air quality, in open-pit mines. The details of methodology as well as the obtained results are presented and discussed in the next sections.

2 Materials and Method

2.1 Materials

To carry out this study, an air quality monitoring system was designed by the Innovations for Sustainable and Responsible Mining (ISRM) Research Group at the Hanoi University of Mining and Geology, Hanoi, Vietnam, which is mounted on a UAV to monitor different factors of air quality in the mine, including PM10, as shown in Fig. 1.

For forecasting PM10, various input parameters were collected, including blasting parameters, meteorological conditions, and monitoring distance. In this study, the blasting parameters, such as explosive charge (W), powder factor (P), stemming (ST), spacing (S), were considered and collected from the blast patterns. In addition, the air humidity (AH), moisture content (MC), air pressure (AP), wind speed (WS), wind direction (WD), and the distance from the blast site to the position of UAV, were collected for the purposes of this study. Finally, a total of 142 observations were collected as a dataset and it is summarized in Table 1. It is worth noting that the WD parameter was converted from the labels (e.g., North, South, West, East, etc.) to numeric to develop the regression predictive model.

Fig. 1 Data collection at the mine using dust sensors and UAV

Table 1 Summary of the dust dispersion dataset used in this study

Statistics	W	P	ST	S	AH	–
Min	12,248	5.100	2.800	4.000	69.00	–
1st Qua	14,598	5.200	3.800	5.800	75.00	–
Median	15,404	5.500	4.300	6.400	80.50	–
Mean	15,625	5.484	4.369	6.365	79.86	–
3rd Qua	16,941	5.700	4.800	7.000	84.00	–
Max	19,688	5.900	6.300	8.700	95.00	–
Statistics	MC	AP	WS	WD	D	PM10
Min	1.740	102.0	0.020	1.000	29.9	20.00
1st Qua	2.513	104.0	1.725	5.000	116.4	67.25
Median	3.455	105.0	2.300	9.000	215.1	83.00
Mean	3.327	105.3	2.245	8.563	203.8	81.27
3rd Qua	3.980	106.0	3.013	12.750	263.1	96.75
Max	5.540	109.0	3.970	15.000	386.2	135.00

Note "–" is not a number

2.2 *Method*

To forecast PM10 concentration induced by blasting operations in open-pit mines, this study used the ANFIS model, which is one of the artificial neural networks designed based on the fuzzy systems. Such fuzzy systems allow the ANFIS model that can simulate nonlinear relationships of variables with the adaptability and rapid learning capabilities [15]. To do this, ANFIS requires a number of membership functions and IF–THEN fuzzy rules. The relationship between the inputs and output can be presented through five layers excluding the input and output layers, as shown in Fig. 2.

In ANFIS, these five hidden layers act as the membership functions, fuzzification, rule-base, normalization, and defuzzification, and their mission is to calculate the weights of the network. Finally, the calculated weights are summarized and used to compute the output.

For training the network, ANFIS can use the gradient descend-based algorithms, back propagation, to name a few. These algorithms can provide matrix multiplications with high accuracy to find the optimal weights of the network. The details of ANFIS can be referred to the literature [16–19].

3 Results and Discussion

To forecast PM10 concentration induced by blasting operations, 142 blasting events were measured, as described above. Subsequently, the dataset was randomly divided

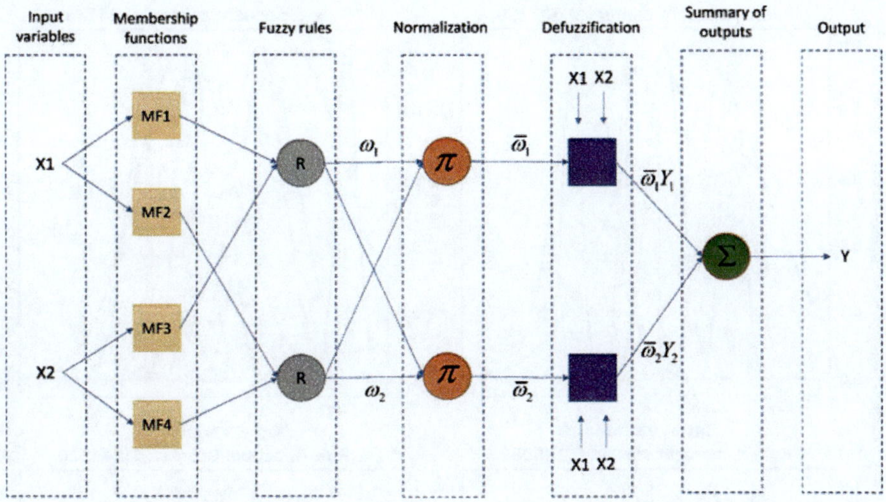

Fig. 2 ANFIS structure

into two parts: 70% for training the ANFIS model, and the remaining 30% for testing the trained ANFIS model. Before training the model, the dataset was normalized to reduce the error of the model during training the model. Furthermore, a fivefold cross-validation technique was also used to avoid overfitting of the model when training and evaluating the model. This technique was repeated 5 times to ensure the balance of the model.

Regarding the parameters of the ANFIS model, 40 membership functions (MFs) were selected based on our experience. The Huber loss function was used to evaluate the model during training and testing the model under 1000 epochs. To train the ANFIS model in this study, the Adam optimization algorithm was applied with the learning rate which was selected as 0.001. The training performance of the ANFIS model for forecasting PM10 concentration is shown in Fig. 3, and the optimization of MFs is shown in Fig. 4.

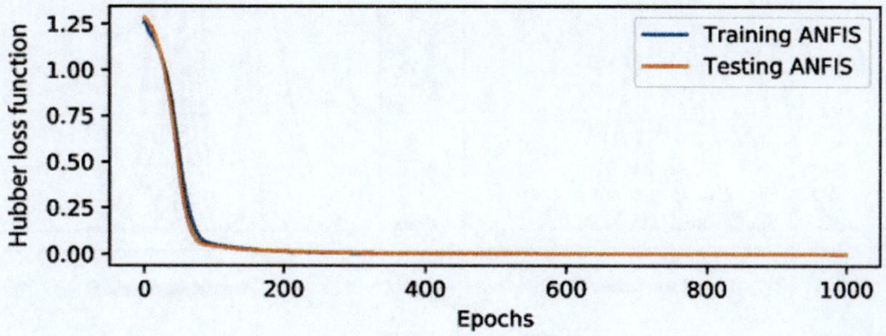

Fig. 3 Training and testing curves of the ANFIS model with the Adam optimization and the Huber loss function

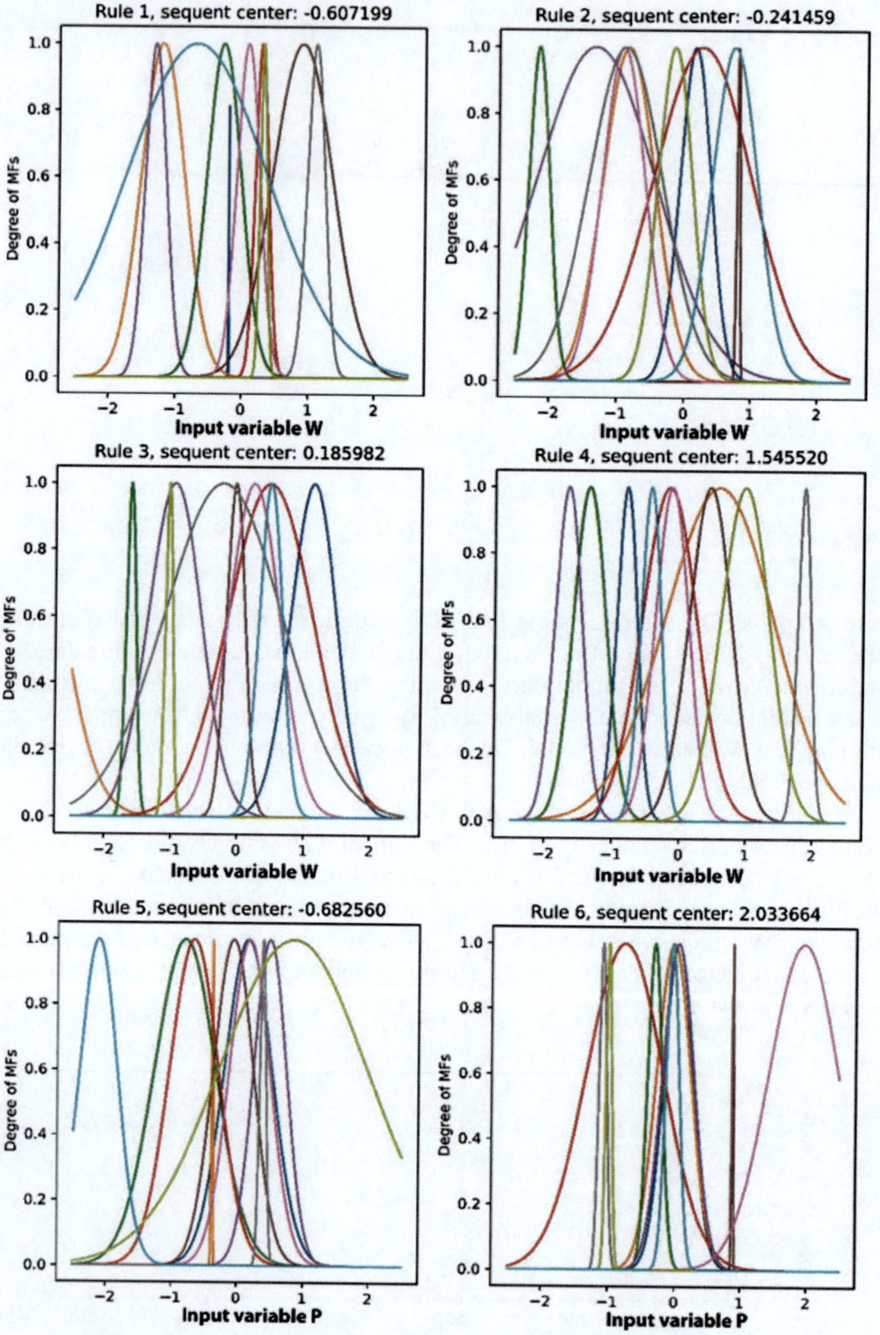

Fig. 4 Distribution of MFs of the ANFIS model with the Adam optimization

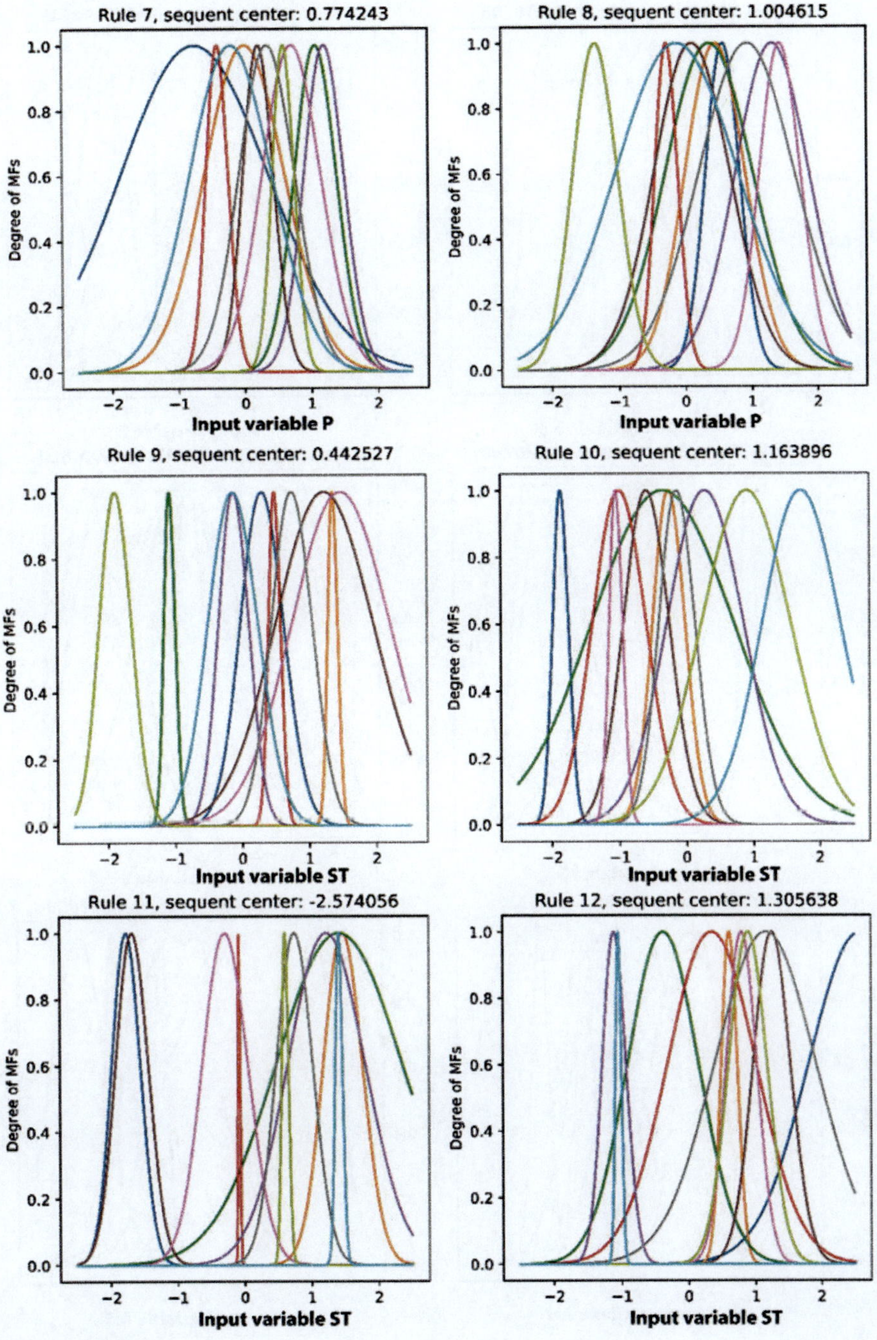

Fig. 4 (continued)

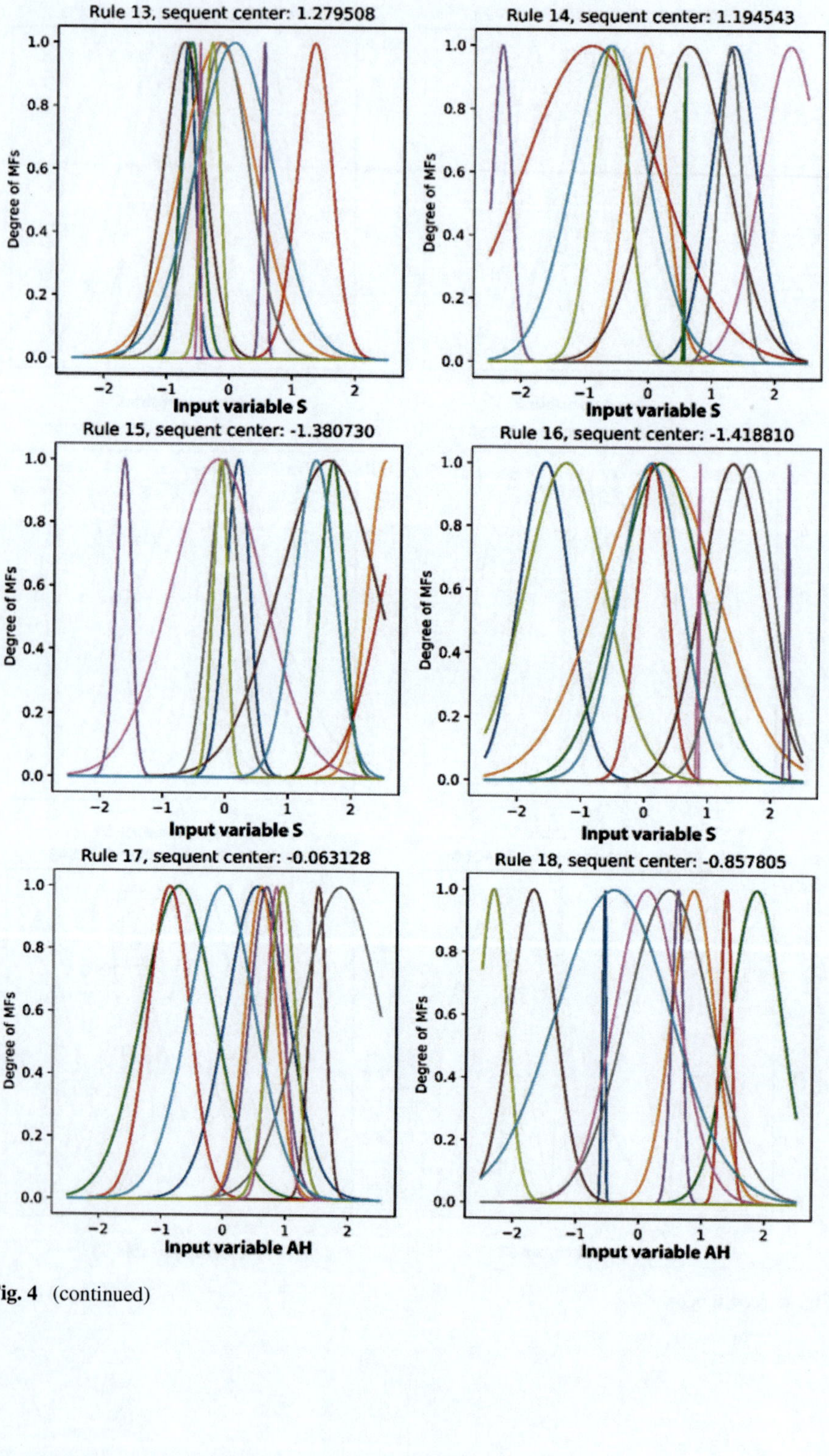

Fig. 4 (continued)

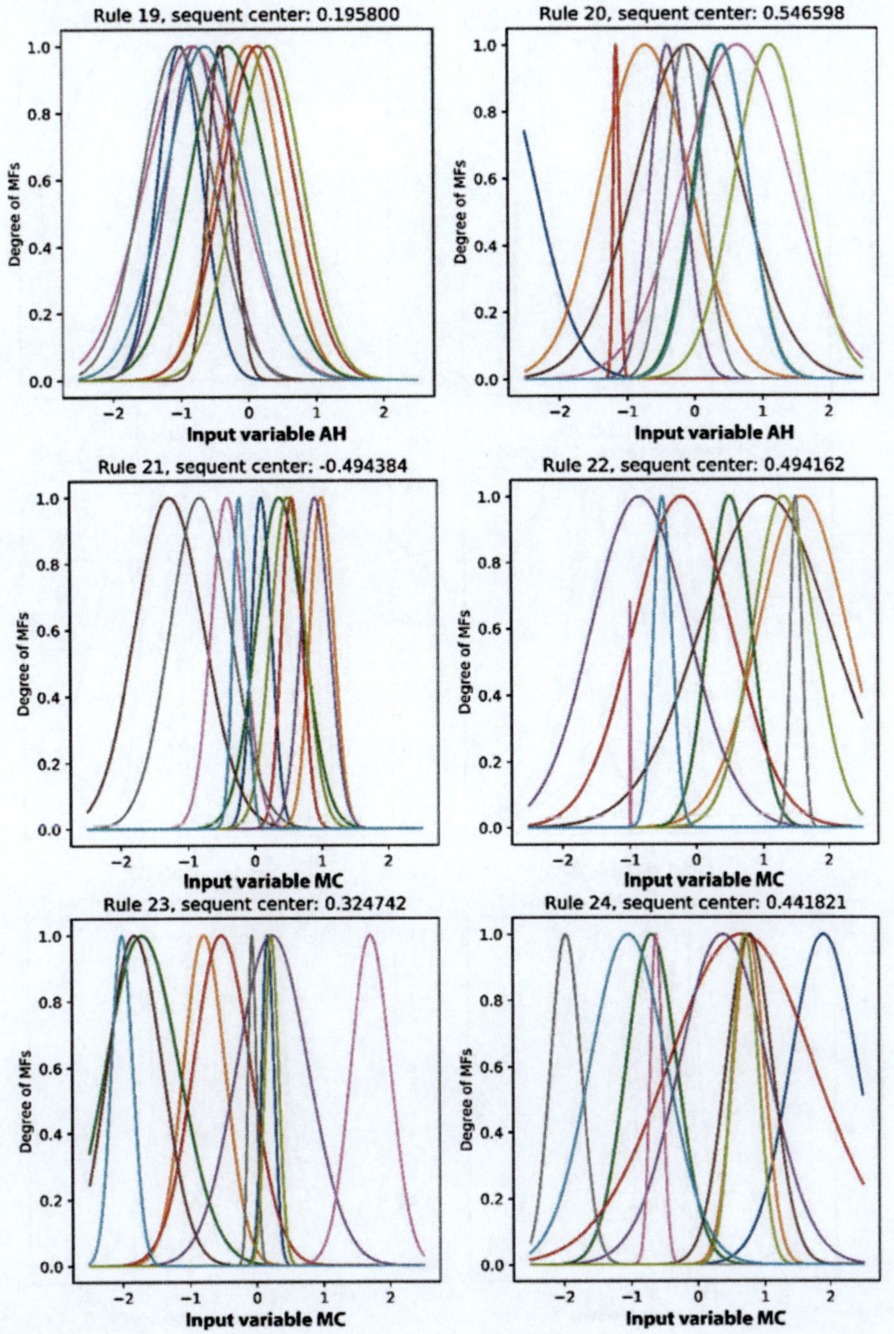

Fig. 4 (continued)

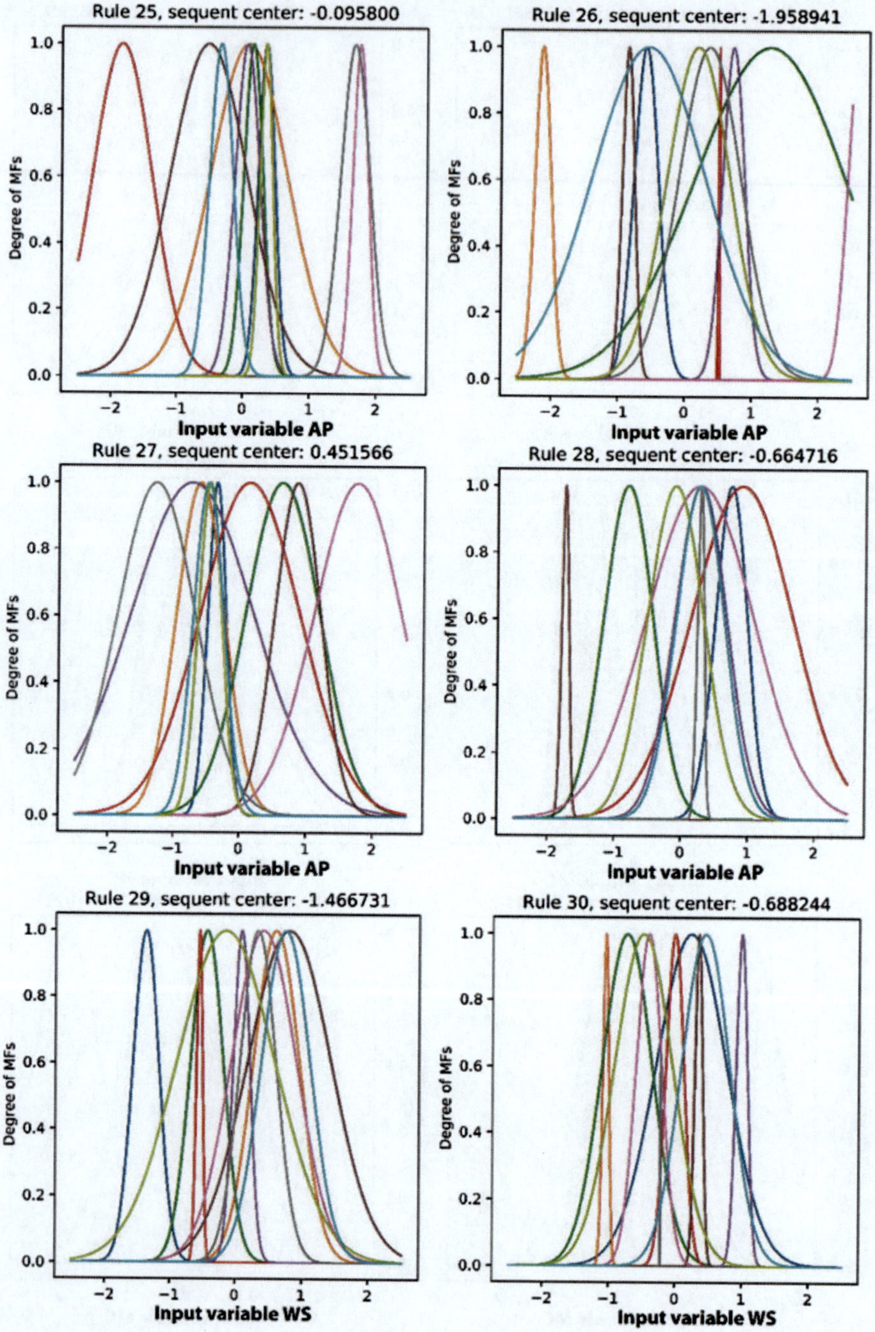

Fig. 4 (continued)

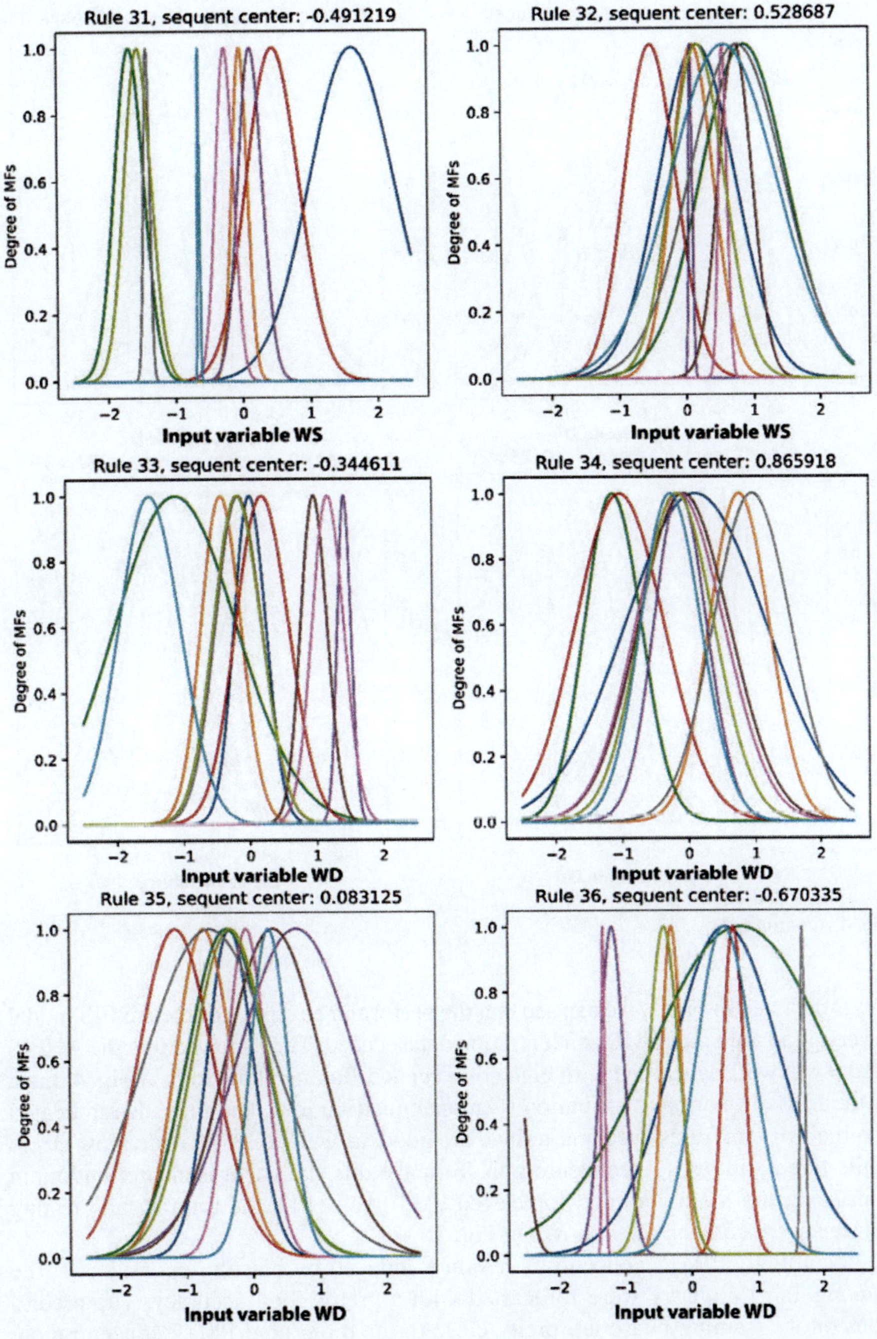

Fig. 4 (continued)

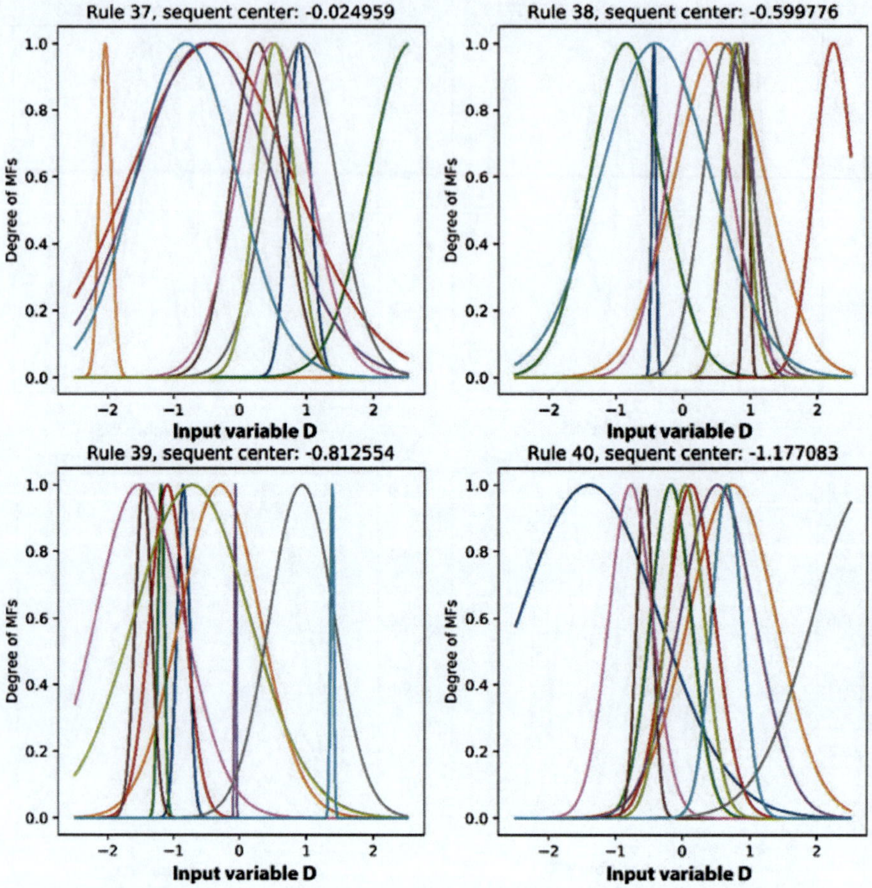

Fig. 4 (continued)

As depicted in Fig. 3, we can see that the performance curves of the ANFIS model is very good with the PM10 concentration dataset used. They showed that the ANFIS model was well-developed with high convergence. Taking along look at Fig. 4, most of the MFs are normal distribution (Gaussian distribution), and they also indicated that the MFs and rules used and trained are good, as well. They provided low errors while forecasting PM10 concentration. To make this statement is clearer and more consistent, the monitored and forecasted PM10 values on the training and testing datasets were compared, as shown in Fig. 5.

Accordingly, PM10 concentration values induced by blasting operations at the Thuong Tan IV quarry were forecasted with a pretty good accuracy. The predictions on the training phase are pretty close to the monitored PM10 concentration. Meanwhile, it is slightly lower on the testing dataset. Overall, the outcome predictions provided by the ANFIS model are pretty good and the model has no overfitting

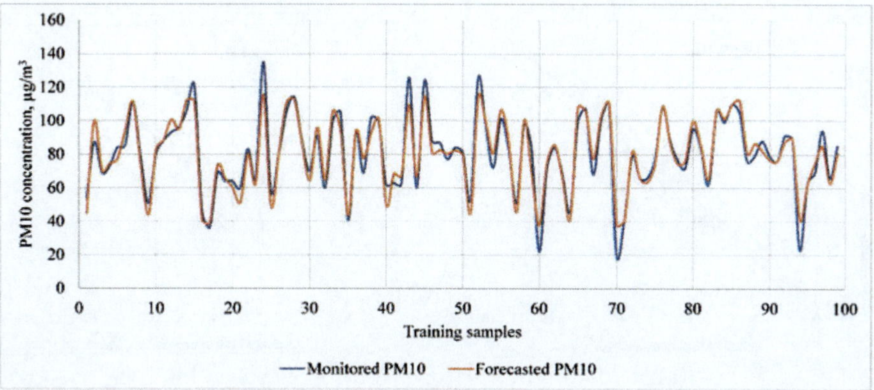

(a)

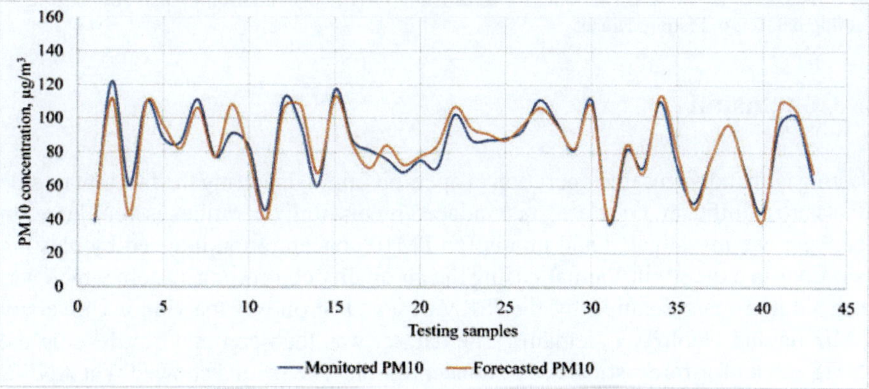

(b)

Fig. 5 Comparison of monitored and forecasted PM10 by the optimized ANFIS model, **a** Training phase; **b** Testing phase

problem. The performance metrics (i.e., RMSE, MAE, R^2) indicated that the accuracy of the ANFIS model in forecasting PM10 concentration in practical engineering is acceptable with an RMSE of 6.743 and 6.702; MAE of 5.142 and 5.240; R^2 of 0.912 and 0.900, for the training and testing datasets, respectively. The correlation between actual PM10 and forecasted PM10 is also evaluated in Fig. 6.

As can be seen in Fig. 6, the convergence of the ANFIS model is pretty high with the 80% of confidence level when forecasting PM10 concentration induced by blasting operations at the Thuong Tan IV quarry. Taking a closer look at Fig. 6, it is conspicuous that only one data point is outside 80% confidence level for both training and testing datasets. In other words, the reliability of the ANFIS model in the 80% confidence level is acceptable and the stable of the model is conspicuously enough. This can make the model stable in practical engineering upon forecasting PM10 concentration induced by blasting operations.

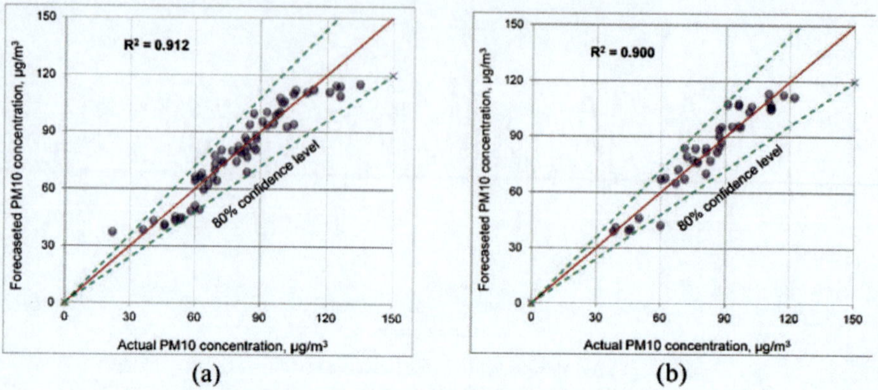

Fig. 6 Accuracy of the ANFIS model for forecasting PM10 concentration induced by blasting, **a** Training dataset; **b** Testing dataset

4 Conclusion

Blasting is an indispensable operation in open-pit mines, inspiring the fact that its side effects are significant. Of those, dust induced by blasting operations is very large. In this paper, we investigated and monitored PM10 concentration induced by blasting operations in a quarry in Vietnam using the air quality controlling system which was designed and manufactured by the ISRM research group of the Hanoi University of Mining and Geology (Vietnam). The dataset was then compiled to develop the ANFIS model for forecasting PM10 concentration. The results showed that ANFIS is a potential model with an accuracy of 90% in forecasting PM10 concentration. Also, they indicated that the developed ANFIS model is very stable, and it can be considered to use in forecasting PM10 concentration in practical engineering.

Acknowledgements The authors would like to thank Dr. Nguyen Van Duc at the Dong-A university (Korea) and colleagues to support the air quality controlling system.

References

1. Monjezi, M., et al.: Environmental impact assessment of open pit mining in Iran. Environ. Geol. **58**(1), 205–216 (2009)
2. Lu, X., et al.: Prediction into the future: a novel intelligent approach for $PM_{2.5}$ forecasting in the ambient air of open-pit mining. Atmos. Pollut. Res. **12**(6), 101084 (2021)
3. Kia, S., et al.: Atmospheric transport over open-pit mines: the effects of thermal stability and mine depth. J. Wind Eng. Ind. Aerodyn. **214**, 104677 (2021)
4. Florea, R., et al.: Water pollution in gold mining industry: a case study in Roşia Montană district, Romania. Environ. Geol. **48**(8), 1132–1136 (2005)
5. Artiola, J., et al.: Soil and land pollution. In: Environmental and Pollution Science, pp. 219–235. Elsevier (2019)

6. Laney, A.S., Weissman, D.N.: Respiratory diseases caused by coal mine dust. J. Occup. Environ. Med./Am. Coll. Occup. Environ. Med. **56**(10), S18 (2014)
7. Bao, Q., et al.: Microscopic characterization and mesoscopic simulation of the interaction between chemically grafted copolymer and coal dust in an open-pit coal mining environment. Sustain. Chem. Pharm. **22**, 100470 (2021)
8. Petsonk, E.L., Rose, C., Cohen, R.: Coal mine dust lung disease. New lessons from an old exposure. Am. J. Respir. Crit. Care Med. **187**(11), 1178–1185 (2013)
9. Kasap, Y., Subaşı, E.: Risk assessment of occupational groups working in open pit mining: analytic hierarchy process. J. Sustain. Min. **16**(2), 38–46 (2017)
10. Lilic, N., et al.: Dust and noise environmental impact assessment and control in Serbian mining practice. Minerals **8**(2), 34 (2018)
11. Khazins, V.M., Shuvalov, V.V., Soloviev, S.P.: Numerical modeling of formation and rise of gas and dust cloud from large scale commercial blasting. Atmosphere **11**(10), 1112 (2020)
12. Bakhtavar, E., et al.: Air pollution risk assessment using a hybrid fuzzy intelligent probability-based approach: mine blasting dust impacts. Nat. Resour. Res. **30**(3), 2607–2627 (2021)
13. Bakhtavar, E., et al.: Green blasting policy: simultaneous forecast of vertical and horizontal distribution of dust emissions using artificial causality-weighted neural network. J. Clean. Prod. **283**, 124562 (2021)
14. Hosseini, S., et al.: Prediction of dust emission due to open pit mine blasting using a hybrid artificial neural network. Nat. Resour. Res. **30**(6), 4773–4788 (2021)
15. Ghenai, C., et al.: Short-term building electrical load forecasting using adaptive neuro-fuzzy inference system (ANFIS). J. Build. Eng. **52**, 104323 (2022)
16. Walia, N., Singh, H., Sharma, A.: ANFIS: adaptive neuro-fuzzy inference system-a survey. Int. J. Comput. Appl. **123**(13) (2015)
17. Zhou, J., et al.: Performance evaluation of hybrid FFA-ANFIS and GA-ANFIS models to predict particle size distribution of a muck-pile after blasting. Eng. Comput. 1–10 (2019)
18. Jang, J.-S.: ANFIS: adaptive-network-based fuzzy inference system. IEEE Trans. Syst. Man Cybern. **23**(3), 665–685 (1993)
19. Armaghani, D.J., Asteris, P.G.: A comparative study of ANN and ANFIS models for the prediction of cement-based mortar materials compressive strength. Neural Comput. Appl. 1–32 (2020)

Assessing the Effect of Open-Pit Mining Activities and Urbanization on Fine Particulate Matter Concentration by Using Remote Sensing Imagery: A Case Study in Binh Duong Province, Vietnam

Thanh Dong Khuc, Long Quoc Nguyen, Dinh Trong Tran, Van Anh Tran, Quynh Nga Nguyen, Xuan Quang Truong, and Hien Quang Pham

Abstract Fine particulate matter ($PM_{2.5}$) leads to respiratory and cardiovascular diseases. Nowadays, $PM_{2.5}$ monitoring on a large scale is a significant concern. The study aims to estimate the $PM_{2.5}$ concentrations and assess the effect of open-pit mining and urbanization on $PM_{2.5}$ concentrations by using satellite observations and ground-based stations over Binh Duong province, Vietnam. Aerosol Optical Depth (AOD) from Moderate Resolution Imaging Spectroradiometer (MODIS) and hourly data of $PM_{2.5}$ concentrations, temperature, and relative humidity at ground-based

T. D. Khuc · D. T. Tran (✉) · H. Q. Pham
Hanoi University of Civil Engineering, Hanoi 100000, Vietnam
e-mail: trongtd@huce.edu.vn

T. D. Khuc
e-mail: dongkt@huce.edu.vn

H. Q. Pham
e-mail: hien66763@nuce.edu.vn

L. Q. Nguyen · D. T. Tran · D. T. Tran · V. Anh Tran
Hanoi University of Mining and Geology, Hanoi 100000, Vietnam
e-mail: nguyenquoclong@humg.edu.vn

V. Anh Tran
e-mail: tranvananh@humg.edu.vn

Q. N. Nguyen
National Academy of Public Administration, Hanoi 100000, Vietnam
e-mail: nqn1@napa.edu.vn

X. Q. Truong
Hanoi University of Natural Resources and Environment, Hanoi 100000, Vietnam
e-mail: txquang@hunre.edu.vn

L. Q. Nguyen · V. Anh Tran
Innovations for Sustainable and Responsible Mining (ISRM) Research Group, Hanoi University of Mining and Geology, Hanoi 100000, Vietnam

L. Q. Nguyen et al. (eds.), *Advances in Geospatial Technology in Mining and Earth Sciences*, Environmental Science and Engineering,
https://doi.org/10.1007/978-3-031-20463-0_5

stations in 2020 (May–December) are matched to establish the relationship between AOD and $PM_{2.5}$. Linear regression and multiple linear regression methods show a good correlation with $R = 0.806$ and $R = 0.817$, respectively. The study applies both regression models to estimate the spatial distribution of monthly $PM_{2.5}$ concentrations over Binh Duong province and validate with ground-based stations data ($R = 0.878$; RMSE = 6.62 $\mu g/m^3$). The results show that the annual $PM_{2.5}$ concentrations are high in the southern districts where it is densely populated and tend to decrease in the northern districts of which the highest are found in Di An (18.76 $\mu g/m^3$) and Thuan An (18.76 $\mu g/m^3$), the lowest is in Phu Giao (14.20 $\mu g/m^3$). The study demonstrates that the development of urban built-up density and population density increases the contribution of fine particles. At the same time, open-pit mine areas are the main contribution of coarse particles.

Keywords $PM_{2.5}$ · AOD · Urbanization · Open-pit mining

1 Introduction

Over the last two decades, air pollution has gotten a lot of attention because of its negative impacts on the global climate, declining air quality, and dramatically affecting human health [1–4]. Aerosols (also known as particulate matter) have significant direct and indirect radiative forcing effects due to their scattering and absorption properties, which are affected by changing atmospheric stability [5, 6]. Besides, several studies have found that $PM_{2.5}$ (particulate matter with a diameter of less than 2.5 μm) exposure might result in serious health consequences such as cardiovascular disease, respiratory infections, morbidity, and mortality [7–10]. Air quality monitoring and regulation in industrial and fast-growing cities become increasingly important [11].

Ground-based $PM_{2.5}$ is monitored by air quality monitoring stations, which give accurate and reliable $PM_{2.5}$ data. On the other hand, ground-based stations are appropriate for assessing air pollution in the near vicinity, but they are often spatially constrained due to the high cost of the equipment and their operation [12, 13]. Furthermore, significant spatial and temporal variation of $PM_{2.5}$ could not be adequately represented by just using point-level data to address the regional air pollution problem. Advanced satellite remote sensing technology may provide a viable answer for ground-level observations. Satellite sensors are more beneficial than ground stations when extending point-level observations across a larger scale, allowing researchers to more efficiently estimate and monitor ground-level $PM_{2.5}$ [14–16]. For example, the Moderate Resolution Imaging Spectroradiometer (MODIS) onboard the National Aeronautics and Space Administration's (NASA) Earth Observing System (EOS), a polar-orbiting satellite, can measure daily aerosol optical depth (AOD) with extraordinary precision for aerosol investigations [17, 18]. Previous research utilized MODIS AOD to link ground-level PM concentrations for air quality [19–22].

In recent years, a lot of research has applied regression analysis to estimate ground-level $PM_{2.5}$ concentrations from satellite-derived AOD [20, 22–26]. In general, these research show a high correlation between AOD and $PM_{2.5}$ [24–26]. However, the correlation between them is primarily influenced by other factors such as meteorological and aerosol properties [21]. These studies are based on observations of $PM_{2.5}$ and related parameters that can establish estimated functions [14, 21, 27]. Several methods have been used, such as: simple linear regression between AOD and $PM_{2.5}$, multiple linear regression with more parameters (e.g., AOD, relative humidity, temperature...), artificial neural network (ANN) algorithms [19, 21, 25, 26, 28].

Vietnam is a developing country with high population growth especially in urban areas [29–31]. The development of industrial zones increases the emission of dust and pollutants into the surrounding environment. The development of urban increases the emission of dust and pollutants into the surrounding environment through activities such as cooking, biomass burning, and traffic activities. The process of building infrastructure is the driving force behind the exploitation of construction materials and is the cause of the increase in the amount of fine particulate matter released into the environment through mining and transportation [32, 33]. Currently, many provinces of Vietnam have open-pit mines, such as Quang Ninh, Binh Duong, Ha Nam. In particular, some mines are located near residential areas because of the convenience of the transportation system. The growth of open-pit mining and urbanization increases human exposure to fine particulate matter. The study effects of urbanization and open-pit mining on $PM_{2.5}$ concentration will be a reliable document to develop policies in controlling emission activities and reducing air pollution in urban areas [34].

This study aims to evaluate the impact of urbanization and open-pit mining on $PM_{2.5}$ concentrations using earth observation and ground-level data. Fine particulate concentrations are assessed on a large scale through spatial distribution maps. Monthly and yearly average maps of $PM_{2.5}$ are created using regression models of $PM_{2.5}$, AOD values, and meteorological factors. $PM_{2.5}$ concentrations over different representing regions (i.e., urban, rural, open-pit mining) are extracted to evaluate the $PM_{2.5}$ variation of each area over time. In addition, the population density and building density are also connected to $PM_{2.5}$ concentration data for influence analysis.

2 Materials and Methods

2.1 Study Areas

Binh Duong is a province in the Southeast region of Vietnam with geographical coordinates 10°51′46″–11°30′ North latitude and 106°20′–106°58′ East longitude (Fig. 1). Binh Duong is located in the tropical monsoon climate [35]. The rainy season runs from May to November, while the dry season runs from December to April the next year. This is a developing province and tends to promote the development of industrial parks and attracted a large number of migrant workers. By 2019, the total

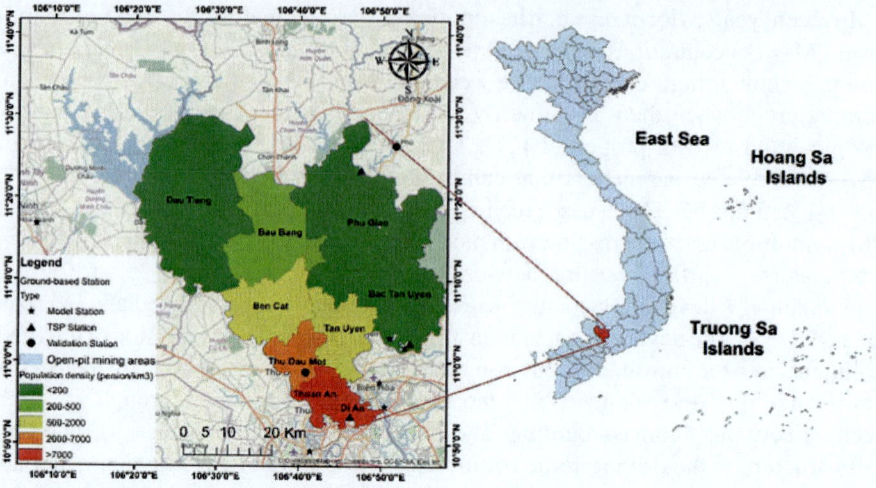

Fig. 1 Location map of study region in Binh Duong province

population of Binh Duong province is nearly 2.5 million people [36]. The population is concentrated in the southern districts and tends to decrease in the North. Similar to other provinces of Vietnam with mining activity, Binh Duong has several large stone open-pit mining sites such as Thuong Tan, Tan My, Tan Dong Hiep. These mines are located near residential areas leading to air pollution from mining and transportation activities (Fig. 1).

2.2 *Data Description*

This study uses Corrected Optical Depth Land products from Terra (MOD04) and Aqua (MYD04) of MODIS Level-2 AOD products at a spatial resolution of 3 km. MODIS instruments onboard EOS Terra and Aqua are in a sun-synchronous polar orbit at an altitude of about 705 km, passing over the equator around 10:30 a.m. and 1:30 p.m. local time, respectively. Standard data products were downloaded in hierarchal data format (HDF) from the NASA website (https://ladsweb.nascom.nasa.gov/). There is a good agreement between Terra and Aqua MODIS AOD and AERONET observations in global validation studies, with a correlation coefficient of about 0.9 and little bias.

The study uses hourly $PM_{2.5}$, Relative Humidity, and Temperature data at ground-based stations to establish regression models and validate $PM_{2.5}$ estimation results from satellite images (Table 1). The selected stations are evenly distributed inside and surrounding the study area (Fig. 1). The data from May to December 2020 of stations Tan Phong B, Thanh Thai, Nguyen Van Nghi, An Binh, Le Hong Phong, and Tan Lien are measured by optical sensor PAS-OA318 developed by PAMair (the air

Table 1 Description of data in the study

Data	Station	Location	Period
AOD MODIS product		Binh Duong province and surrounding	January 2020 to December 2020
Model station ($PM_{2.5}$, temperature, relative humidity)	Tan Phong B	(10.96° N; 106.83°E)	May 2020 to December 2020
	Nguyen Van Nghi	(10.82° N, 106.69°E)	May 2020 to December 2020
	Thanh Thai	(11.28° N, 106.13°E)	May 2020 to December 2020
	An Binh	(10.91° N, 106.84°E)	May 2020 to December 2020
	Long Son	(11.03° N, 106.88°E)	July 2020
	Tan My	(11.05° N, 106.85°E)	July 2020
$PM_{2.5}$ validation station	Le Hong Phong	(10.98° N; 106.68°E)	May 2020 to December 2020
	Tan Lien	(11.44° N; 106.86°E)	May 2020 to December 2020
Total suspended particles station (TSP)	Phu Giao	(11.39° N;106.79°E)	2020
	Di An Center	(10.89° N; 106.77°E)	2020
	Tan Uyen Stone Mine	(11.04° N; 106.89°E)	2020
Population density data		Binh Duong province	2019
NDBI map		Binh Duong province	2020

quality network operating over 400 monitoring points across 63 cities and provinces of Vietnam) (Fig. 2a) [37]. Besides, the PMS5003 optical sensor mounted on the ground-based instrument was used to measure data at Tan My and Long Son stations (Fig. 2b). Both of these instruments use the light-scattering method to measure the $PM_{2.5}$ concentrations.

Total suspended dust (TSP) refers to the sum of particles suspended in the air with an aerodynamic diameter of less than or equal to 100 microns [38]. The Binh Duong Center of Natural Resources and Environmental Technical–Monitoring under the Department of Natural Resources and Environment of Binh Duong province provides TSP ground-based station data including monthly average data of three stations representing three areas of different emission sources. The Phu Giao station is located in a sparsely populated area, far from industrial zones representing rural areas. Second, The Di An Center station represents an urban area located in the urban center area with a high population density, many vehicles of transport, and is influenced by several surrounding industrial activities. Finally, Tan Uyen Stone Mine station is located in the open-pit mining area, influenced by the process of mining and transporting stone representing the open-pit mining area.

The population density data provided by the General Statistics Office of Vietnam is the population density data of the districts of Binh Duong province in 2019 [39]. The

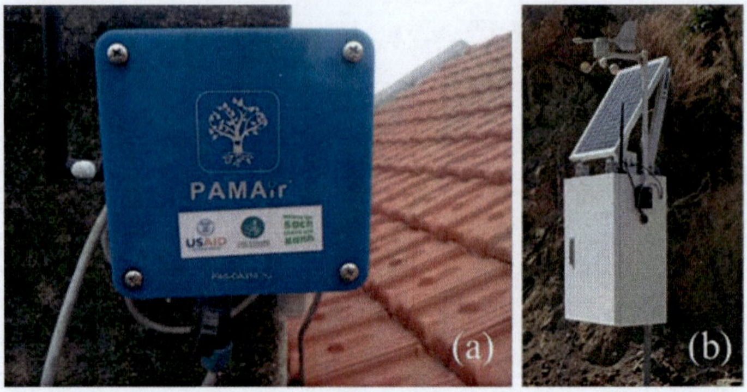

Fig. 2 Ground-based station used in the study: **a** sensor PAS-OA318 developed by PAMair; **b** Ground-based station use PMS5003 sensor

Normalized Difference Built-Up Index (NDBI) indicates the built-up area density of the study area [40]. This index is calculated through satellite images Landsat 8-OLI, operated by the US Geological Survey with a spatial resolution of the electromagnetic spectrum channel (0.43–2.3 μm) is 30 m, the spatial resolution with the thermal infrared channel (10.6–12.5 μm) is 100 m. Satellite images of Binh Duong province are downloaded from the website https://earthexplorer.usgs.gov/.

2.3 *Regression Model*

Simple Linear Regression

In this study, a simple linear regression model is established between the independent variable MODIS AOD_{550} nm of the ground station and the dependent variable $PM_{2.5}$ to understand that $PM_{2.5}$ and AOD are stable in the range of certain space [17, 41]. This simple regression model is used for the months when there is no data to measure meteorological parameters in the study, namely from January 2020 to April 2020. The time the satellite cross over the position of the ground station $PM_{2.5}$ is determined to coincide with a delay of ±30 min and the used AOD value is the average of 3 km around the $PM_{2.5}$ station location [41, 42]. Therefore, the regression function can be given as follows [17, 40, 43, 44]:

$$PM_{2.5} = \alpha + \beta \times AOD \quad (1)$$

where $PM_{2.5}$ refers to fine particulate concentration (μg/m^3); AOD is the aerosol optical depth; β is the slope and α is the intercept.

Multiple Linear Regression

Some studies have shown that meteorological variables influence the relationship between $PM_{2.5}$ concentration and AOD [14, 16, 21]. Multiple Linear Regression explains a constant relationship between dependent and two or more independent variables [14]. The observed meteorological data are the important factors affecting fine particulate concentration. The hourly data of $PM_{2.5}$, Relative Humidity, and Temperature parameters are matching with the AOD data with the same method as in the simple linear regression model. The study used $PM_{2.5}$ data at six ground-based stations to match with the AOD satellite for regression model building. Then use two stations such as Le Hong Phong and Tan Lien to validate (Table 1). The used parameters in the multiple regression function are shown in the regression formula below [14, 39]:

$$\mathrm{PM}_{2.5} = (\alpha + \varepsilon_1) + (\beta_1 + \varepsilon_2) \times \mathrm{AOD} + (\beta_2 + \varepsilon_3) \times \mathrm{Temp} + (\beta_3 + \varepsilon_4) \times \mathrm{RH} \tag{2}$$

where Temp is the temperature (°C); RH is the relative humidity (%); α and β are the fixed coefficients; and ε is the random error.

2.4 $PM_{2.5}$ Concentration Mapping

The study uses AOD, Relative Humidity, and Temperature maps as input values of the regression model to build a daily average $PM_{2.5}$ map. First, the daily average AOD map is created after removing pixels affected by clouds. Next, average hourly relative humidity and temperature data at monitoring stations are converted to daily average data. This study uses the Kriging interpolation method to fill up empty pixels and interpolate mean daily temperatures and relative humidity factors. Finally, the monthly average $PM_{2.5}$ and annual average $PM_{2.5}$ maps are created by averaging the daily and monthly average $PM_{2.5}$ maps, respectively.

2.5 Evaluating Indicator

This paper estimates the value of $PM_{2.5}$ concentration using a regression algorithm. It concerns whether the predicted $PM_{2.5}$ is the same as the observed $PM_{2.5}$. Correlation coefficient (*R*) and root mean square error (*RMSE*) are chosen to evaluate the estimate accuracy. The values of *R* and *RMSE* are calculated based on the following formula:

$$RMSE = \sqrt{\frac{1}{n}\sum_{i=1}^{n}(y_i - y_i')^2} \tag{3}$$

where y_i is the ith estimated $PM_{2.5}$ value; y'_i is the ith observed $PM_{2.5}$ value; n is the sample size.

$$R = \frac{\sum_{i=1}^{n} (x_i - \overline{x})(y_i - \overline{y})}{\sqrt{\sum_{i=1}^{n} (x_i - \overline{x})^2 \sum_{i=1}^{n} (y_i - \overline{y})^2}} \tag{4}$$

where x_i and y_i are the ith sample points; $\bar{x}$ and $\bar{y}$ are the sample means; n is the sample size.

3 Results and Discussion

3.1 Regression Model of $PM_{2.5}$ Estimation

Figure 3 illustrates the scatter plot of the simple linear relationship between MODIS AOD and $PM_{2.5}$ over Binh Duong province in 2020. The values of $PM_{2.5}$ used in the model range from 1.67 to 46.45 ($\mu g/m^3$), while the AOD range from 0.18 to 1.62. There is a high correlation between MODIS AOD and fine particulate matter concentration with a correlation coefficient (R) equal to 0.806 when applying a simple linear regression model. Table 2 shows the regression functions and performance evaluation of both models. The multiple linear regression model that considers the influence of meteorological factors including relative humidity and temperature with 33 points model showed slightly better performance ($R = 0.817$) when compared to the simple regression model. The linear regression and multiple linear regression functions were $PM_{2.5} = 22.306AOD_{550} + 8.390$ and $PM_{2.5} = 97.484 + 21.439AOD_{550} -$ 0.564RH-1.646 Temp, respectively.

The correlation coefficient between estimated and measured $PM_{2.5}$ is 0.878, and the root means squared error (RMSE) is just 6.62 g/m^3 (Fig. 4). The least-squares method is used to establish the linear trend with the highest fitness value among the distributed points in this study. This validation chart demonstrates that there is a good performance of estimating $PM_{2.5}$ concentrations model. In the scatter plot of observed and predicted values, the points were dispersed around a 1:1 line. The fitting line of the measured–estimated $PM_{2.5}$ scatter plot has a slope and intercept of 1.05 and 0.499, respectively. The results can be improved when overcoming the lack of data on meteorological variables such as wind speed, planetary boundary layer height, etc., or satellite AOD calibration with ground-based AOD before putting it into the model.

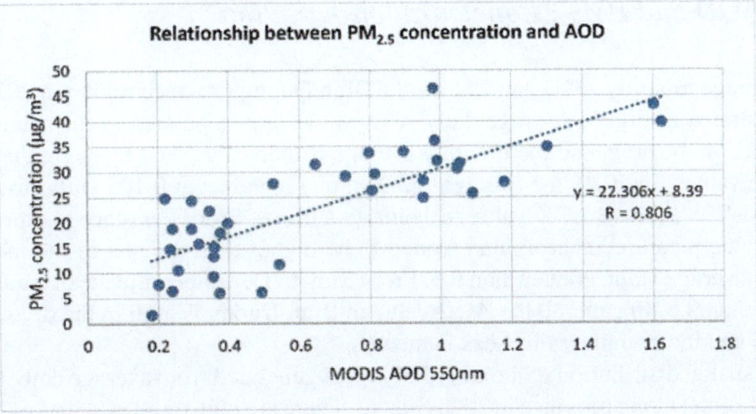

Fig. 3 Linear regression of the MODIS AOD and $PM_{2.5}$

Table 2 Description of simple linear regression and multiple linear regression

Model	Regression functions	Evaluation
Linear regression	$PM_{2.5} = 22.306AOD_{550} + 8.390$	$R = 0.806$
Multiple linear regression	$PM_{2.5} = 97.484 +$ $21.439AOD_{550}-0.564RH\text{-}1.646Temp$	$R = 0.817$ p value = 0.000

* Remark: $p < 0.01$

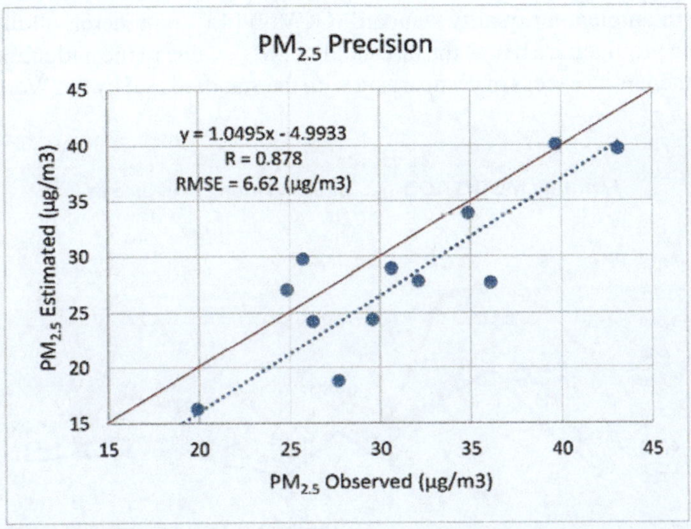

Fig. 4 Scatter plot for models fitting result

3.2 AOD and $PM_{2.5}$ Concentration Variation

The average monthly AOD values over the Binh Duong province region in 2020 are demonstrated in Fig. 5. Average daily AOD values are calculated as the average of all pixels on the area, and the monthly average is then calculated by using the daily average values. In 2020, the average AOD value ranged from 0.105 (July) to 0.860 (March). The average AOD value in January is 0.118. From February to April, the AOD value increased sharply and tended to be higher than the rest of the months, with an average value greater than 0.5. From May to December, the mean values are all less than 0.5. In general, the AOD value in Binh Duong is high in the dry season months and lower in the rainy season months.

The spatial distribution of monthly $PM_{2.5}$ was generated from average daily $PM_{2.5}$ maps of Binh Duong province in 2020 (Fig. 6). There is a high trend in values of $PM_{2.5}$ in southern districts such as Di An, Thuan An, Tan Uyen, Thu Dau Mot. In addition, there is a slight increase in the Northern and Northwest counties in May. The average $PM_{2.5}$ concentration of Binh Duong province shows a markedly high increase from February to April, with the highest value in March (23.655 $\mu g/m^3$). In contrast, the months with low mean concentrations belong to November (11.850 $\mu g/m^3$) and December (12.027 $\mu g/m^3$). The results demonstrate a clear difference in the value of $PM_{2.5}$ concentrations when comparing the rainy season and the dry season.

The average annual value of the districts in Binh Duong province in 2020 is shown in Fig. 7. These values were compared with the average $PM_{2.5}$ value with the 2005 standards of the World Health Organization (WHO) and National technical regulation in ambient air quality standard (QCVN) [45]. In general, all districts of Binh Duong province are below the threshold of $PM_{2.5}$ value in the national standard. However, there is an excess of comparison with the standards set by the World Health

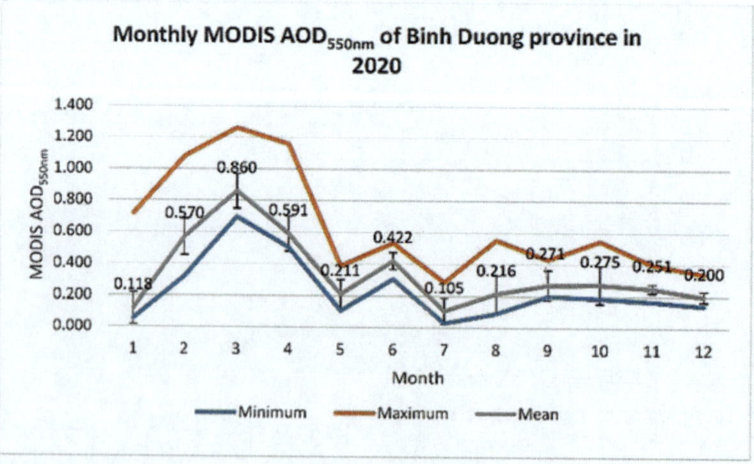

Fig. 5 Average monthly MODIS AOD of Binh Duong province in 2020

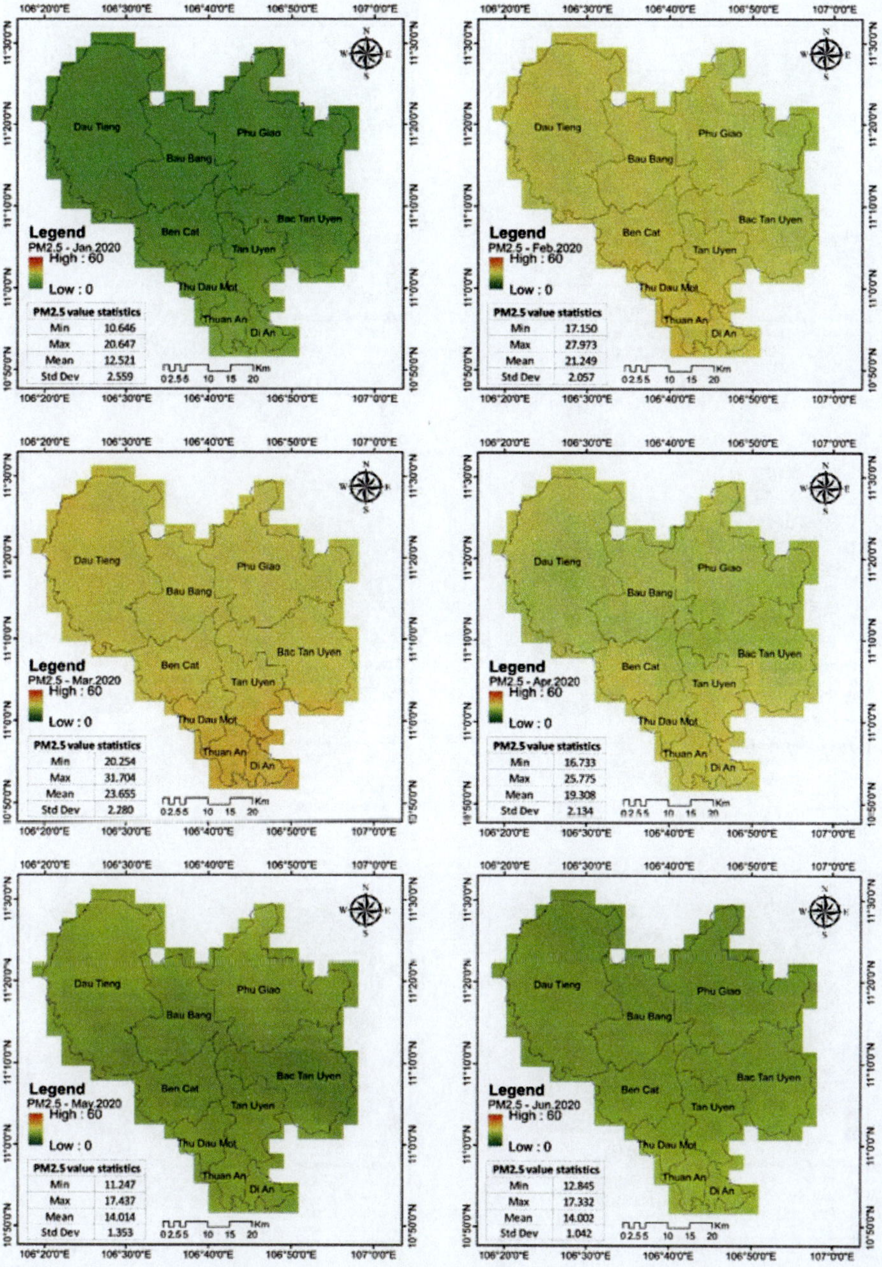

Fig. 6 Spatial distribution of monthly average $PM_{2.5}$ concentrations of Binh Duong province in 2020

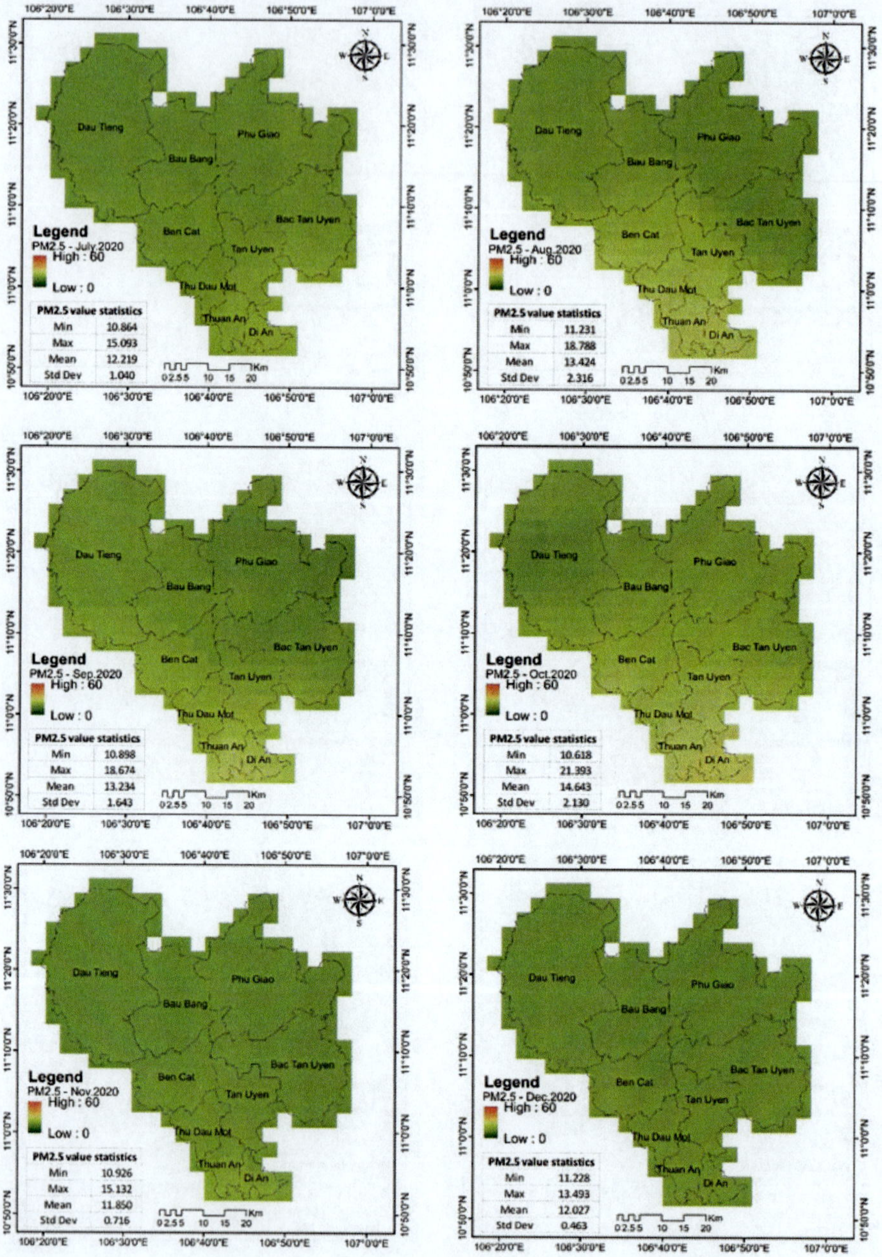

Fig. 6 (continued)

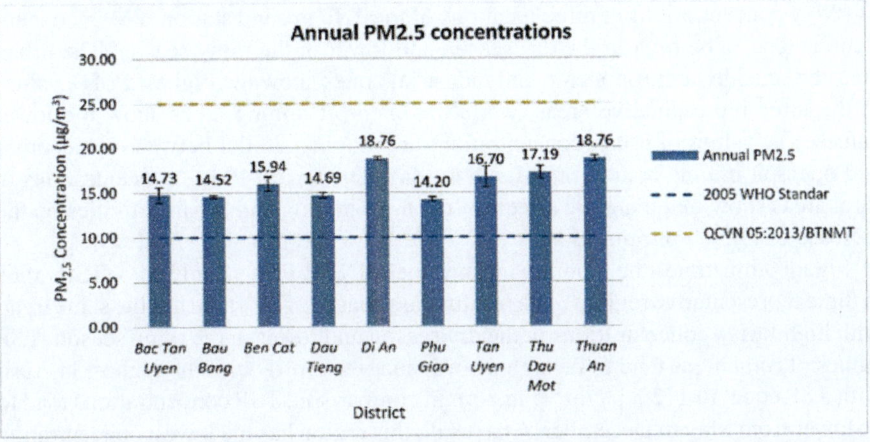

Fig. 7 Average annual $PM_{2.5}$ concentration of districts of Binh Duong province in 2020

Organization. There is a marked increase when comparing the average values of the southern districts such as Di An, Thuan An, Thu Dau Mot, Tan Uyen with the rest. These areas are mainly urban areas with a high population, industrial zones, and open-pit mines. Figure 7 shows that the highest mean annual $PM_{2.5}$ value belongs to Di An and Thuan An districts (18.76 $\mu g/m^3$), while the lowest is Phu Giao district (14.20 $\mu g/m^3$).

3.3 *Evaluation of the Variation of TSP, $PM_{2.5}$ Concentrations in Representative Areas*

The monthly average $PM_{2.5}$ values at the locations of three TSP ground-based stations are extracted to compare with the monthly mean of TSP observed. The selected TSP stations are located in three areas representing different sources of dust emissions. Phu Giao station (106.79° N; 11.30°E) is a station located far away from urban centers, traffic intersections, industrial production zones with good environmental quality. Di An station (106.76° N; 10.88°E) is an urban center area with a high population density and is influenced by neighboring industrial activities. Tan Uyen Station (106.89° N; 11.03°E) is located in the stone open-pit mining area, affected by quarrying and stone transportation. Figure 8 compares the monthly average value of $PM_{2.5}$ concentrations at three locations of the TSP ground station. $PM_{2.5}$ concentrations tend to be high in the dry season and lower in the rainy season. The urban areas affected by anthropogenic and industrial zones show the highest $PM_{2.5}$ values in the three representative areas, while the open-pit mining areas show the lower values. The values of $PM_{2.5}$ concentrations are roughly similar between urban areas and open-pit mining areas from March to May. The lowest $PM_{2.5}$ concentrations in rural areas show that there are effects of open-pit mining and urban activities on the increase of $PM_{2.5}$ concentrations.

Figure 9 illustrates the monthly mean values of TSP, $PM_{2.5}$, and $PM_{2.5}$/TSP ratios in three representative regions. The results show that the TSP value has the same trend with high $PM_{2.5}$ concentrations in the dry season and lower in the rainy season. TSP values of rural areas tend to be high from January to April, with the highest in April with TSP equal to 112.8 $\mu g/m^3$ (Fig. 9a). In comparison, TSP concentrations tend to be lower from May to December. As a result, this region has the lowest concentration of TSP compared with the other two representative areas. This reason leads to some months of the year with the $PM_{2.5}$/TSP ratio being as high as May (56.00%), July (74.23%), October (73.05%). The urban area reported high TSP concentrations from January to May, with the highest concentration in March (161.3 g/m^3). Then the concentration decreased between June and September with a threshold of less than 50.0 $\mu g/m^3$. Finally, there was a sudden increase in October (84.8 $\mu g/m^3$) and a gradual decrease in the last two months of the year (Fig. 9b). $PM_{2.5}$/TSP ratio in this area tends to be lower than in rural areas. The concentration of TSP in the open-pit mining area is superior to high in urban and rural areas. In the first five months of the year, the TSP level of this region increased sharply, with all values exceeding 200 $\mu g/m^3$, of which the highest was in March (614 $\mu g/m^3$). In this period, the $PM_{2.5}$/TSP ratio has a low value in this region, with all months being less than 10% (Fig. 9c). Overall, the $PM_{2.5}$/TSP ratio in the open-pit mining areas has the lowest value among the three regions. This percentage increases gradually in the last

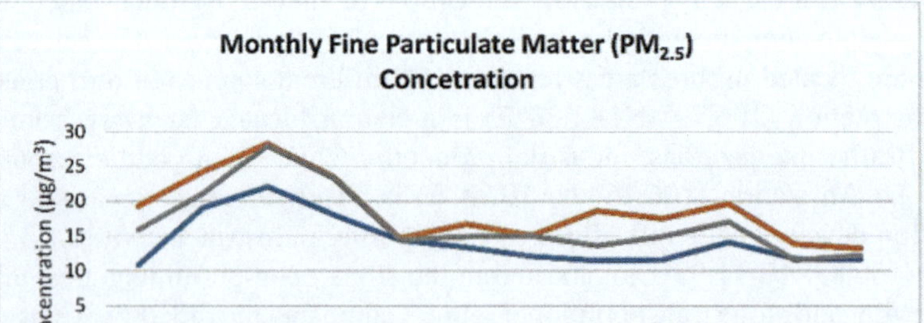

Fig. 8 Average monthly $PM_{2.5}$ concentrations of representative areas

months of the year and reaches the highest level in December (40.59%). The results demonstrated that the primary emission of open-pit mining is coarse particles.

3.4 *Effect of Built-Up Density and Population Density on $PM_{2.5}$ Concentration*

In Vietnam, urbanization tends to cause an increase in built-up density and population density. Figure 10 illustrates the scatter plot of the relationship between $PM_{2.5}$ concentrations with NDBI index and population density in the districts of Binh Duong city in 2020. There is a good correlation between $PM_{2.5}$ and NDBI index and population density in Binh Duong province, with $R = 0.875$ and $R = 0.950$, respectively. The results demonstrate that the increase in built-up density and population density contributes to the increase in $PM_{2.5}$ concentrations.

4 Conclusions

The study established a model to estimate the concentration of $PM_{2.5}$ in a case study in Binh Duong province by using satellite image data and ground-based data. The results show that the simple linear regression model and the multiple linear regression model show good evaluation results with $R = 0.806$ and $R = 0.817$, respectively. The daily average AOD, Relative Humidity, and Temperature maps were used to calculate the daily average $PM_{2.5}$ spatial distribution using both regression models. The monthly average $PM_{2.5}$ concentration results showed a high increase in $PM_{2.5}$ concentrations in the dry season and low in the rainy season months in Binh Duong province. The spatial distribution of $PM_{2.5}$ concentration values in the southern districts tends to be higher than in the rest of the regions caused by urban development, industrial development, and mining activities. The study also demonstrated that open-pit mining areas provide a large proportion of coarse-particle composition while urban areas have a high proportion of fine particles. The study also demonstrated that increasing building density and increasing population density cause an increase in $PM_{2.5}$ concentrations. The study has the potential to be applied to areas with open-pit mining operations and high population density. However, there will be limits to areas with few ground $PM_{2.5}$ stations.

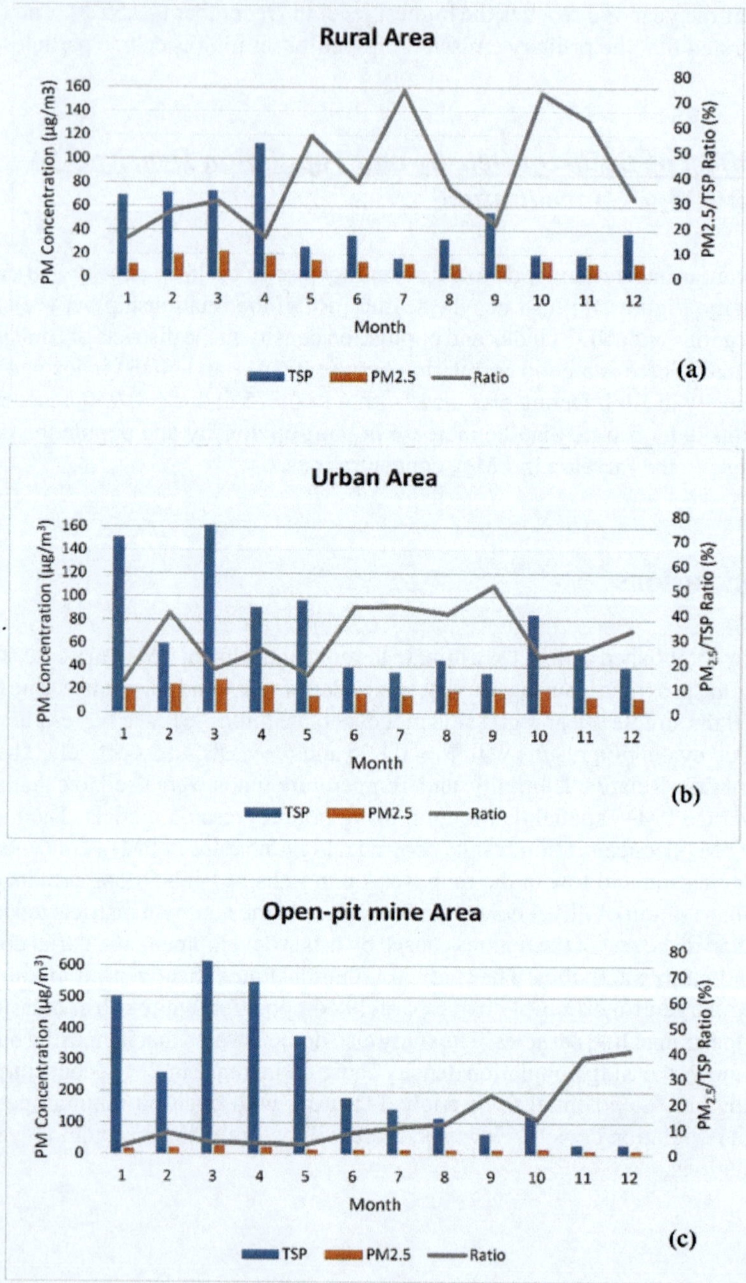

Fig. 9 Monthly mean values of TSP, $PM_{2.5}$, and $PM_{2.5}$/TSP ratios of representative **a** rural areas; **b** urban areas; **c** open-pit mining areas

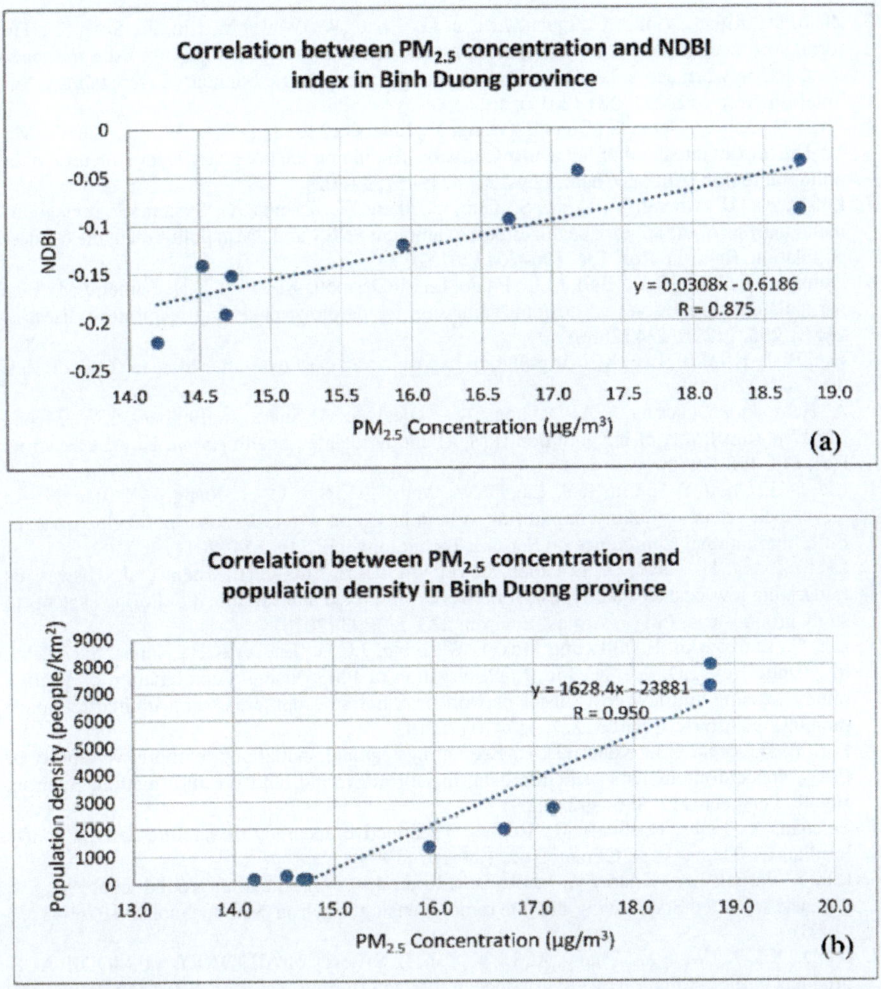

Fig. 10 Correlation between $PM_{2.5}$ concentrations and **a** NDBI index; **b** population density in the districts of Binh Duong city

References

1. Franchini, M., Mannucci, P.M.: Impact on human health of climate changes. Eur. J. Int. Med. **26**, 1–5 (2015)
2. Kampa, M., Castanas, E.: Human health effects of air pollution. Environ. Pollut. **151**, 362–367 (2008)
3. Tagaris, E., Liao, K.-J., DeLucia, A.J., Deck, L., Amar, P.: Russell AG: potential impact of climate change on air pollution-related human health effects. Environ. Sci. Technol. **43**, 4979–4988 (2009)
4. Gu, H., Cao, Y., Elahi, E., Jha, S.K.: Human health damages related to air pollution in China. Environ. Sci. Pollut. Res. **26**, 13115–13125 (2019)

5. Zhao, D., Xin, J., Gong, C., Quan, J., Liu, G., Zhao, W., Wang, Y., Liu, Z., Song, T.: The formation mechanism of air pollution episodes in Beijing city: insights into the measured feedback between aerosol radiative forcing and the atmospheric boundary layer stability. Sci. Total Environ. **692**, 371–381 (2019)
6. Zhao, D., Xin, J., Gong, C., Quan, J., Wang, Y., Tang, G., Ma, Y., Dai, L., Wu, X., Liu, G., Ma, Y.: The impact threshold of the aerosol radiative forcing on the boundary layer structure in the pollution region. Atmos. Chem. Phys. **21**, 5739–5753 (2021)
7. Lu, F., Xu, D., Cheng, Y., Dong, S., Guo, C., Jiang, X., Zheng, X.: Systematic review and meta-analysis of the adverse health effects of ambient $PM_{2.5}$ and PM_{10} pollution in the Chinese population. Environ. Res. **136**, 196–204 (2015)
8. Dominici, F., Peng, R.D., Bell, M.L., Pham, L., McDermott, A., Zeger, S.L., Samet, J.M.: Fine particulate air pollution and hospital admission for cardiovascular and respiratory diseases. JAMA **295**, 1127–1134 (2006)
9. Lee, B.-J., Kim, B., Lee, K.: Air pollution exposure and cardiovascular disease. Toxicol. Res. **30**, 71–75 (2014)
10. Al-Hemoud, A., Gasana, J., Al-Dabbous, A., Alajeel, A., Al-Shatti, A., Behbehani, W., Malak, M.: Exposure levels of air pollution ($PM_{2.5}$) and associated health risk in Kuwait. Environ. Res. **179**, 108730 (2019)
11. Liu, J.-H., Chen, Y.-F., Lin, T.-S., Lai, D.-W., Wen, T.-H., Sun, C.-H., Juang, J.-Y., Jiang, J.-A.: Developed urban air quality monitoring system based on wireless sensor networks. In: 2011 Fifth International Conference on Sensing Technology, pp. 549–554 (2011)
12. Li, J., Zhang, H., Chao, C.-Y., Chien, C.-H., Wu, C.-Y., Luo, C.H., Chen, L.-J., Biswas, P.: Integrating low-cost air quality sensor networks with fixed and satellite monitoring systems to study ground-level $PM_{2.5}$. Atmos. Environ. **223**, 117293 (2020)
13. Lin, C., Labzovskii, L.D., Leung Mak, H.W., Fung, J.C.H., Lau, A.K.H., Kenea, S.T., Bilal, M., Vande Hey, J.D., Lu, X., Ma, J.: Observation of $PM_{2.5}$ using a combination of satellite remote sensing and low-cost sensor network in Siberian urban areas with limited reference monitoring. Atmos. Environ. **227**, 117410 (2020)
14. Liu, Y., Paciorek, C.J., Koutrakis, P.: Estimating regional spatial and temporal variability of $PM_{2.5}$ concentrations using satellite data, meteorology, and land use information. Environ. Health. Perspect. **117**, 886–892 (2009)
15. Zelenka, A., Perez, R., Seals, R., Renné, D.: Effective accuracy of satellite-derived hourly irradiances. Theor. Appl. Climatol. **62**, 199–207 (1999)
16. Liu, Y., Sarnat, J.A., Kilaru, V., Jacob, D.J., Koutrakis, P.: Estimating ground-level $PM_{2.5}$ in the eastern United States using satellite remote sensing. Environ. Sci. Technol. **39**, 3269–3278 (2005)
17. Fu, D., Xia, X., Wang, J., Zhang, X., Li, X., Liu, J.: Synergy of AERONET and MODIS AOD products in the estimation of $PM_{2.5}$ concentrations in Beijing. Sci. Rep. **8**,10174 (2018)
18. Wang, W., Mao, F., Pan, Z., Du, L., Gong, W.: Validation of VIIRS AOD through a comparison with a sun photometer and MODIS AODs over Wuhan. Remote. Sens. **9**, 403 (2017)
19. Wang, X., Sun, W., Zheng, K., Ren, X., Han, P.: Estimating hourly $PM_{2.5}$ concentrations using MODIS 3 km AOD and an improved spatiotemporal model over Beijing-Tianjin-Hebei, China. Atmos. Environ. **222**, 117089 (2020)
20. Hu, Z.: Spatial analysis of MODIS aerosol optical depth, $PM_{2.5}$, and chronic coronary heart disease. Int. J. Health. Geogr. **8**, 27 (2009)
21. Ni, X., Cao. C., Zhou, Y., Cui, X., Singh, P.R.: Spatio-temporal pattern estimation of $PM_{2.5}$ in Beijing-Tianjin-Hebei region based on MODIS AOD and meteorological data using the back propagation neural network. Atmos. **9**, 105 (2018)
22. He, Q., Huang, B.: Satellite-based mapping of daily high-resolution ground $PM_{2.5}$ in China via space-time regression modeling. Remote. Sens. Environ. **206**, 72–83 (2018)
23. Mao, L., Qiu, Y., Kusano, C., Xu, X.: Predicting regional space–time variation of $PM_{2.5}$ with land-use regression model and MODIS data. Environ. Sci. Pollut. Res. **19**, 128–138 (2012)
24. Hu, H., Hu, Z., Zhong, K., Xu, J., Zhang, F., Zhao, Y., Wu, P.: Satellite-based high-resolution mapping of ground-level $PM_{2.5}$ concentrations over East China using a spatiotemporal regression kriging model. Sci. Total. Environ. **672**, 479–490 (2019)

25. Lee, H.J., Chatfield, R.B., Strawa, A.W.: Enhancing the applicability of satellite remote sensing for $PM_{2.5}$ estimation using MODIS deep blue AOD and land use regression in California, United States. Environ. Sci. Technol. **50**, 6546–6555 (2016)
26. You, W., Zang, Z., Zhang, L., Li, Y., Wang, W.: Estimating national-scale ground-level PM25 concentration in China using geographically weighted regression based on MODIS and MISR AOD. Environ. Sci. Pollut. Res. **23**, 8327–8338 (2016)
27. Đông, N., Phong, D.: Mối quan hệ thực nghiệm giữa $PM_{2.5}$ và độ sâu quang học aerosol AOD ở khu vực nội thành Hà Nội. Vietnam J. Hydrometeorol. **718**, 22–31 (2020)
28. Trà, T., Hiếu, N., Nguyen, D.: Đánh giá sự phân bố nồng độ bụi $PM_{2.5}$ tại khu vực TP. HCM bằng công nghệ viễn thám–một số kết quả ban đầu. Vietnam J. Hydrometeorol. **721**, 80–91 (2021)
29. Saksena, S., Fox, J., Spencer, J., Castrence, M., DiGregorio, M., Epprecht, M., Sultana, N., Finucane, M., Nguyen, L., Vien, T.D.: Classifying and mapping the urban transition in Vietnam. Appl. Geogr. **50**, 80–89 (2014)
30. Pham, H.M., Yamaguchi, Y.: Urban growth and change analysis using remote sensing and spatial metrics from 1975 to 2003 for Hanoi. Vietnam. Int. J. Remote. Sens. **32**, 1901–1915 (2011)
31. Nam, N.H.: Đánh giá ảnh hưởng của cơ cấu dân số và tình hình lao động đến thực hiện pháp luật của người dân trên địa bàn tỉnh Đồng Tháp (2021)
32. Miatto, A., Dawson, D., Nguyen, P.D., Kanaoka, K.S., Tanikawa, H.: The urbanisation-environment conflict: Insights from material stock and productivity of transport infrastructure in Hanoi. Vietnam. J. Environ. Manage. **294**, 113007 (2021)
33. Schiappacasse, P., Müller, B., Linh, L.T.: Towards responsible aggregate mining in Vietnam. Resour. **8**, 138 (2019)
34. Li, G., Fang, C., Wang, S., Sun, S.: The effect of economic growth, urbanization, and industrialization on fine particulate matter ($PM_{2.5}$) concentrations in China. Environ. Sci. Technol. **50**, 11452–11459 (2016)
35. Thuy, V.T., Huyen, N.T., Tu, L.H., Loi, N.K.: Status of bamboos in Binh Duong province, Vietnam: distribution, species diversity, conservation and utilization. Trees, For. People. **6**, 100137 (2021)
36. Long, N.V.: Strategic planning vision and some key directions for economic breakthroughs during 2021–2025 period: a study of binh duong-vietnam and smart city program. J. Archit. Plan. **20**, 63–78 (2019)
37. Vuong, Q., Choi, S.-D., Bac, V., Thang, H., Hue, N., Lan, T., Duong, H., Tuyen, T., Thang, P.: Spatial and temporal variations of the $PM_{2.5}$ concentrations in Hanoi metropolitan area, Vietnam, during the COVID-19 lockdown. Int. J. Environ. Anal. Chem. 1–13 (2021)
38. Alias, M., Hamzah, Z., Kenn, L.: PM10 and total suspended particulates (TSP) measurement in various power stations. Malays. J. Anal. Sci. **11**(1), 255–261 (2007)
39. General Statistics Office of Vietnam. https://www.gso.gov.vn/
40. Zhang, Y., Odeh, I.O.A., Han, C.: Bi-temporal characterization of land surface temperature in relation to impervious surface area, NDVI and NDBI, using a sub-pixel image analysis. Int. J. Appl. Earth. Obs. Geoinf. **11**, 256–264 (2009)
41. Liaw, J.-J., Huang, Y.-F., Hsieh, C.-H., Lin, D.-C., Luo, C.-H.: $PM_{2.5}$ concentration estimation based on image processing schemes and simple linear regression. Sensors. **20**, 2423 (2020)
42. Yuan, S., Li, Y., Gao, J., Bao. F.: A new coupling method for $PM_{2.5}$ concentration estimation by the satellite-based semiempirical model and numerical model. Remote Sens. **14**, 2360 (2022)
43. Lennartson, E.M., Wang, J., Gu, J., Castro Garcia, L., Ge, C., Gao, M., Choi, M., Saide, P.E., Carmichael, G.R., Kim, J., Janz, S.J.: Diurnal variation of aerosol optical depth and $PM_{2.5}$ in South Korea: a synthesis from AERONET, satellite (GOCI), KORUS-AQ observation, and the WRF-chem model. Atmos. Chem. Phys. **18**, 15125–15144 (2018)
44. Jeong, J.I., Park, R.J., Yeh, S.-W., Roh, J.-W.: Statistical predictability of wintertime $PM_{2.5}$ concentrations over East Asia using simple linear regression. Sci. Total. Environ. **776**, 146059 (2021)

45. World Health Organization: WHO global air quality guidelines: particulate matter ($PM_{2.5}$ and PM_{10}), ozone, nitrogen dioxide, sulfur dioxide and carbon monoxide (2005)
46. Yao, F., Wu, J., Li, W., Peng, J.: A spatially structured adaptive two-stage model for retrieving ground-level $PM_{2.5}$ concentrations from VIIRS AOD in China. ISPRS J. Photogramm. Remote Sens. **151**, 263–276 (2019)

Effect of Loading Frequency on the Liquefaction Resistance of Poorly Graded Sand

Sung- Sik Park, Dong- Kiem -Lam Tran, Tan-No Nguyen, Seung-Wook Woo, and Hee -Young Sung

Abstract Cyclic simple shear tests were carried out to study the effect of frequency (f) on the poorly graded sand's liquefaction resistance. The samples are prepared in the medium state ($D_r = 60\%$) by the deposition method, and a wide range of f (f = 0.03, 0.05, 0.1, 0.2, and 0.5 Hz) is considered. The results show that the liquefaction resistance is uninfluenced by low f (0.03 and 0.05 Hz). When the f is higher than the optimum frequency ($f_{\text{opt}} = 0.1$ Hz), the liquefaction resistance increases with increasing f.

Keywords Cyclic simple shear test · Liquefaction resistance · Loading frequency · Poorly graded sand

1 Introduction

It has been extensively studied how loading frequency (f) affects liquefaction resistance of sand; however, the issue is highly controversial. In several studies, liquefaction resistance of sandy soil was found to be small or unaffected by loading frequency. Peacock et al. [1] applied several cyclic simple shear tests (CSS) with the range from 1/6 to 4 Hz of f and noted that the frequency effects were small. The

S.-. S. Park · D.-. K. .-L. Tran (✉) · T.-N. Nguyen · S.-W. Woo · H. .-Y. Sung
Department of Civil Engineering, Kyungpook National University, Daegu 41566, Republic of Korea
e-mail: lam.trandongkiem@uah.edu.vn

S.-. S. Park
e-mail: sungpark@knu.ac.kr

T.-N. Nguyen
e-mail: nonguyen@knu.ac.kr

S.-W. Woo
e-mail: geowsw@knu.ac.kr

H. .-Y. Sung
e-mail: bryanfriend@naver.com

L. Q. Nguyen et al. (eds.), *Advances in Geospatial Technology in Mining and Earth Sciences*, Environmental Science and Engineering,
https://doi.org/10.1007/978-3-031-20463-0_6

same conclusions have been reached by other studies [2, 3]. More recently, Zhu et al. [4] performed cyclic triaxial tests to highlight the effect of f on the sand liquefaction response. Medium dense and dense samples were subjected to a various f (in both stress-controlled mode and strain-controlled mode). It was found that the effect of the f was insignificant for dense sand in stress-controlled mode.

Moreover, numerous researchers [5, 6] reported that the sand exhibited a higher cyclic strength at the higher loading frequencies. A series of CSSs was conducted on Nakdong sand with a wide range of f (0.05–1 Hz) by Nong et al. [6] to examine the f effects on the cyclic strength. It showed that an increase in the f resulted in increasing liquefaction resistance, regardless of the vertical effective stress or relative density.

In contrast, other researchers [7, 8] found that varying the frequency resulted in a decrease in cyclic strength. Dash et al. [8] observed that cyclic liquefaction resistance decreased as f increased in range of 0.1–0.5 Hz, due to an increase in the excess pore water pressure.

Therefore, the relationship between the loading frequency and liquefaction resistance of sand remains unclear and requires more investigation. In this study, CSS tests on poorly graded medium density sand samples are conducted with various f (0.03, 0.05, 0.1, 0.2, and 0.5 Hz) to evaluate the influence of f on the liquefaction resistance of sand.

2 CDSS Tests

2.1 *Material*

The characteristic and particle size analysis of the silica sand are shown in Table 1 and Fig. 1, respectively. The specific gravity (G_S) is 2.60, the maximum void ratio (e_{max}) and minimum void ratio (e_{min}) are 1.02 and 0.63, respectively. Based on the Unified Soil Classification System, the sand is classified as poorly graded sand (SP).

2.2 *Testing Condition and Sample Preparation*

The CSS device used in this study is described in Fig. 2. The system includes a shear box, a lateral load cell, a vertical load cell, and a horizontal and vertical LVDT. In the CSS tests, typical cylindrical samples have initial diameter of 63.5 mm and height of 25 mm, respectively. The dry funnel deposition technique is used to prepare the sample in this study. All dry sand samples are conducted with a relative density of 60%, consolidated under the initial vertical effective stress (σ') of 100 kPa. The cyclic stress ratio (CSR) is defined as the ratio between the cyclic shear stress (τ_{cyc}) and the initial vertical effective stress (σ'). Upon the consolidation stage, the cyclic stage

Table 1 Physical properties of testing samples

Index property	Value
Specific gravity (G_S)	2.60
D_{10} (mm)	0.128
D_{30} (mm)	0.28
D_{50} (mm)	0.64
D_{60} (mm)	0.68
C_u	5.31
C_c	0.87
e_{max}	1.02
e_{min}	0.63
Unified soil classification system classification	SP

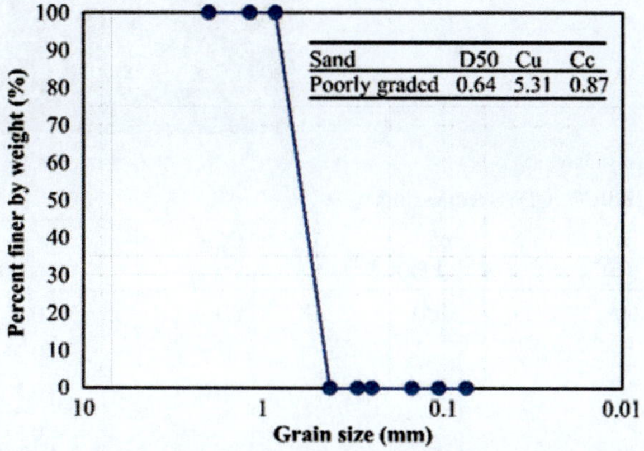

Fig. 1 Grain size distribution curve

is conducted with CSR values (0.1, 0.12, 0.15, and 0.18), as well as f (0.03, 0.05, 0.1, 0.2, and 0.5 Hz).

3 Result and Discussion

3.1 Test Results

The details of CSS test results, including CSR, f and average number of cycles to liquefaction (N_{cyc}), are shown in Table 2. Figure 3 presents the typical result for the case CSR = 0.1 and $f = 0.1$ Hz. To determine the cyclic resistance of silica sand, at least three CSS distinct test cases with various CSRs are conducted for a given f.

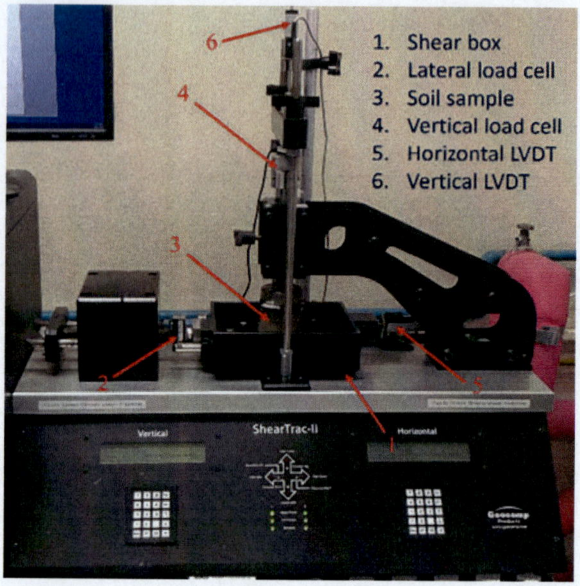

Fig. 2 CDSS device

Table 2 CDSS test results

Dr (%)	σ′ (kPa)	CSR	*f* (Hz)	N_{cyc}
60	100	0.1	0.03	69
			0.05	69
			0.1	69
			0.2	77
			0.5	111
		0.12	0.03	35
			0.05	35
			0.1	35
			0.2	53
			0.5	67
		0.15	0.03	11
			0.05	11
			0.1	11
			0.2	16
			0.5	18
		0.18	0.2	5
			0.5	6

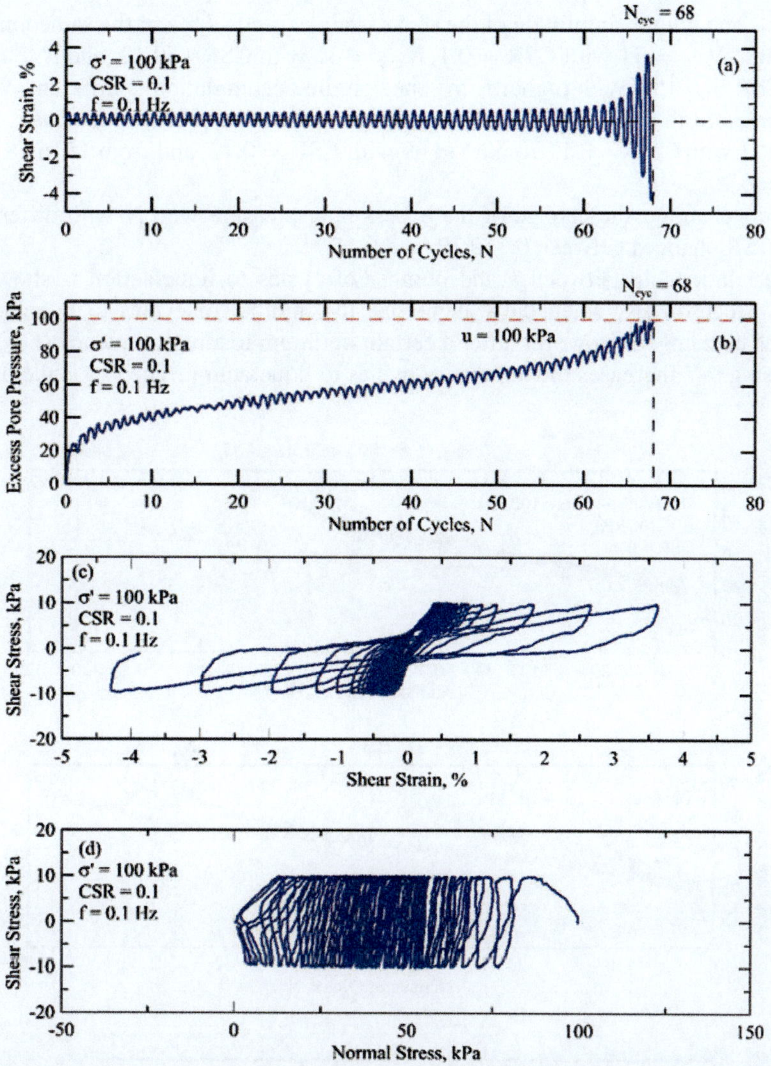

Fig. 3 CDSS test results for the case CSR = 0.1, $f = 0.1$ Hz, **a** shear strain and N, **b** excess pore pressure and N, **c** shear stress versus shear strain, and **d** shear stress versus normal stress

3.2 Liquefaction Resistance of Poorly Graded Sand

Effect of f on cyclic behavior of poorly graded sand with various CSRs is shown in Fig. 4. The figure describes the undrained cyclic simple shear response of poorly graded silica sand with various f_s ($f = 0.03$, 0.05, 0.1, 0.2, and 0.5 Hz) at different CSRs ((a) CSR = 0.1, (b) CSR = 0.12, and (c) CSR = 0.15). The shear strain is accumulated continuously and nearly the same with low f ($f = 0.03$, 0.05, and

0.1 Hz). The double amplitude of the shear strain exceeds 7.5% at the same number of cycles ($N_{cyc} = 71$ with CSR = 0.1, $N_{cyc} = 31$ with CSR = 0.12, and $N_{cyc} = 15$ with CSR = 0.15). With higher f, the shear strain accumulation is different. When f increases from 0.1 to 0.5 Hz, the number of cycles to liquefaction increases from 71 to 111 with CSR = 0.1, from 31 to 69 with CSR = 0.12, and from 15 to 18 with CSR = 0.15.

Figure 5 shows the change of the excess pore pressure with N with different f when CSR changes between 0.1, 0.12, and 0.15.

The relationship between f and number of cycles to liquefaction is shown in Fig. 6. Accordingly, when the f increases, the number of cycles to liquefaction remains unchanged. However, after a certain optimum loading frequency (f_{opt}), the increasing in f increases the number of cycles to liquefaction. The f_{opt} is the one at

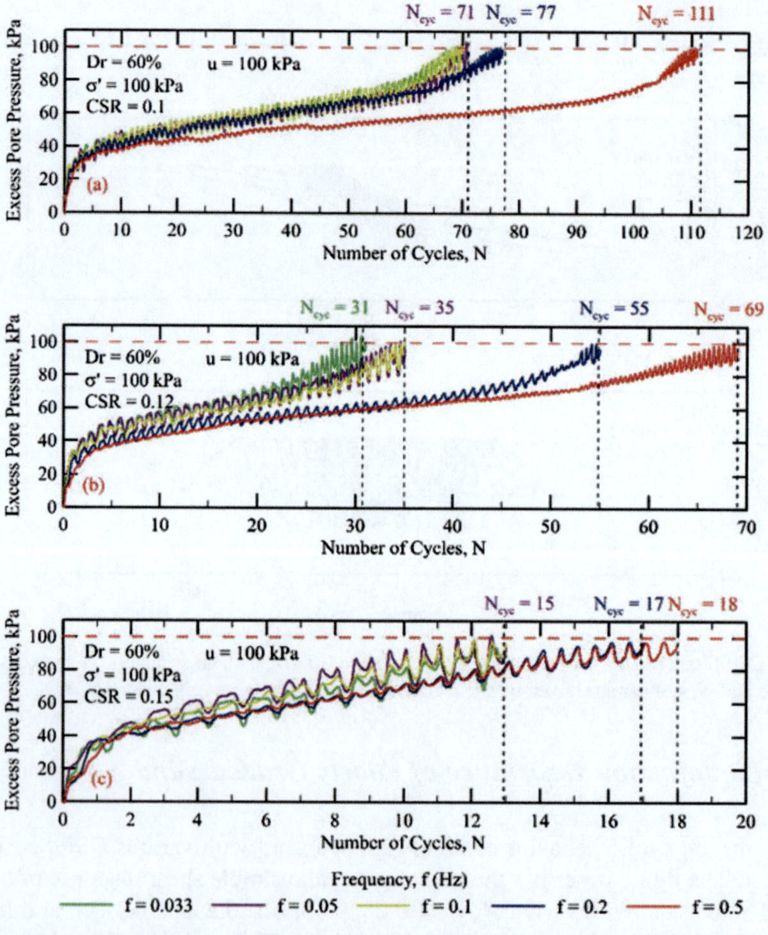

Fig. 4 Effect of f on the shear strain accumulation (medium sand and $\sigma' = 100$ kPa): **a** CSR = 0.1, **b** CSR = 0.12, and **c** CSR = 0.15

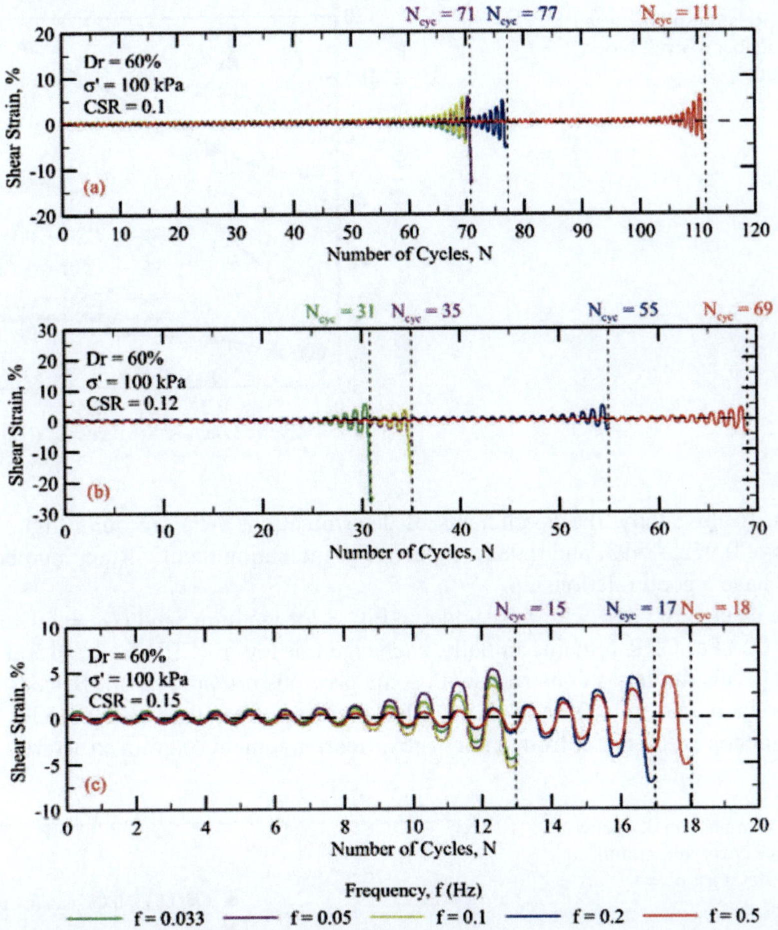

Fig. 5 Effect of f on the excess pore pressure accumulation (medium sand and $\sigma' = 100$ kPa): **a** CSR = 0.1, **b** CSR = 0.12, and **c** CSR = 0.15

which the liquefaction resistance remains unchanged with decreasing f and increases with increasing f. When f is less than 0.1 Hz, the number of cycles to liquefaction is not influenced by the f at a constant CSR ($N_{cyc} = 15$ at CSR = 0.15, $N_{cyc} = 35$ at CSR = 0.12, and $N_{cyc} = 71$ at CSR = 0.1). The increasing in f corresponds to the increasing trend of the liquefaction resistance. N_{cyc} increases from 71 to 111 at CSR = 0.1, from 31 to 69 at CSR = 0.12, and from 15 to 18 at CSR = 0.15 when f increases from 0.1 to 0.5 Hz. Moreover, the f_{opt} ($f_{opt} = 0.1$ Hz) is the same for three cases of CSRs. Therefore, it can be stated that the f_{opt} does not depend on the CSR.

Figure 7 shows the cyclic resistance curves of poorly graded sand for various f. The trend curves shown in Fig. 7 define the relationship between CSR and N_{cyc}. The liquefaction resistance of sand, expressed in terms of the cyclic resistance ratio (CRR), is defined as the cyclic stress ratio at 15 cycles based on the cyclic resistance

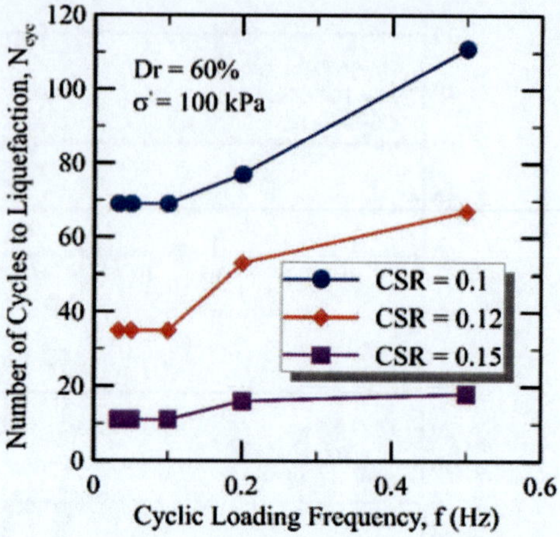

Fig. 6 Relationship between *f* and number of cycles to liquefaction

curves. In this study, the coefficients of determinations (R^2) are quite high, with values of 0.972, 0.993, and 0.985, respectively, indicating that CSR and number of cycles have a good relationship.

The change of CRR with *f* is shown in Fig. 8 for medium sand ($D_r = 60\%$) and $\sigma' = 100$ kPa. CRR remains initially unchanged at low *f* (CRR = 0.1413 at $f \leq 0.1$ Hz). This finding is consistent with some previous research [1–3]. Besides that, when *f* increases from 0.1 to 0.5 Hz, CRR increases from 0.1413 to 0.1532. This phenomenon could result from a slippage or rearrangement of grain structure.

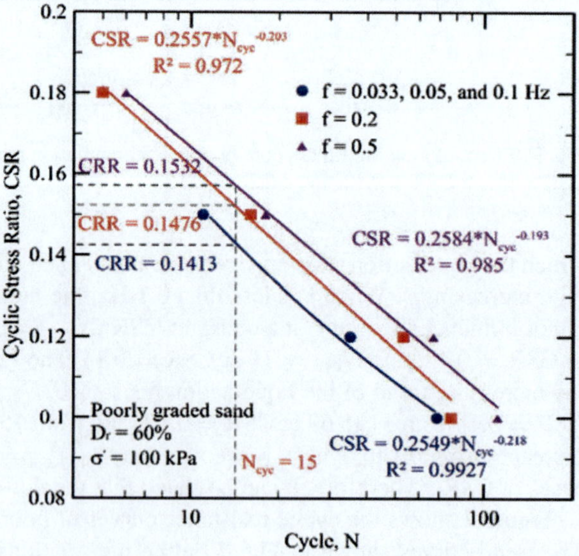

Fig. 7 Comparison of cyclic resistance curve at medium density states for σ' = 100 kPa

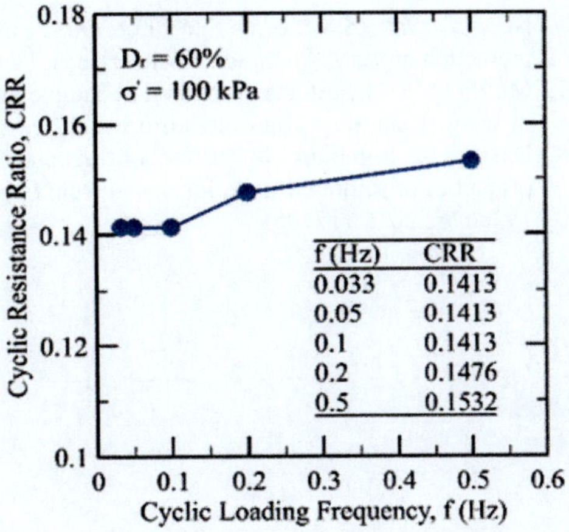

Fig. 8 Effect of loading frequency on cyclic resistance

4 Conclusion

In this study, the influence of f on cyclic liquefaction resistance of poorly graded sand has been investigated. Numerous CSSs were conducted on silica sand with various f ($f = 0.03, 0.05, 0.1, 0.2$, and 0.5 Hz). Based on the test results, the cyclic liquefaction resistance of sand was unchanged at low f (f = 0.03 and 0.05 Hz). In this study, the optimum loading frequency was about 0.1 Hz. When the loading frequency was greater than the optimum loading frequency, the increasing of loading frequency increased the liquefaction resistance of sand.

Acknowledgements This work was supported by the National Research Foundation of Korea (NRF) grant funded by the Korea Government (MSIT) (Nos. NRF-2021R1I1A3059731 and NRF-2018R1A5A1025137).

References

1. Peacock, W.H., Seed, H.B.: Sand liquefaction under cyclic loading simple shear conditions. J. Soil Mech. Found. Div. **94**, 689–708 (1968). https://doi.org/10.1061/JSFEAQ.0001135
2. Yoshimi, Y., Oh-Oka, H.: Influence of degree of shear stress reversal on the liquefaction potential of saturated sand. Soils Found. **15**, 27–40 (1975). https://doi.org/10.3208/sandf1972.15.3_27
3. Polito, C.: The effects of non-plastic and plastic fines on the liquefaction of sandy soils. Ph.D. Thesis, Virginia Tech. **274** (1999)
4. Zhu, Z., Zhang, F., Peng, Q., Dupla, J.C., Canou, J., Cumunel, G., Foerster, E.: Effect of the loading frequency on the sand liquefaction behaviour in cyclic triaxial tests. Soil Dyn. Earthq. Eng. **147**, 106779 (2021). https://doi.org/10.1016/j.soildyn.2021.106779
5. Lee, K., Fitton, J.: Factors affecting the cyclic loading strength of soil. In: Vibration effects of earthquakes on soils and foundations, pp. 71–95. ASTM International, 100 Barr Harbor Drive, PO Box C700, West Conshohocken, PA 19428–2959 (1969). https://doi.org/10.1520/STP33637S.

6. Nong, Z., Park, S.S., Jeong, S.W., Lee, D.E.: Effect of cyclic loading frequency on liquefaction prediction of sand. Appl. Sci. (Switzerland). **10** (2020). https://doi.org/10.3390/app10134502.
7. Mulilis, J.P.: The effects of method of sample preparation on the cyclic stress-strain behaviour of sands. Tech. Rep. Univ. of California at Berkeley **75** (1975)
8. Dash, H.K., Sitharam, T.G.: Effect of frequency of cyclic loading on liquefaction and dynamic properties of saturated sand. Int. J. Geotech. Eng. **10**, 487–492 (2016). https://doi.org/10.1080/19386362.2016.1171951

An Automatic Method for Clay Minerals Extraction from Landsat 8 OLI Data. A Case Study in Chi Linh City, Hai Duong Province

Trinh Le Hung, Nguyen Sach Thanh, and Vuong Trong Kha

Abstract Landsat satellite data have been effectively used for mineral resources extraction. This paper presents an automatic method for clay minerals extraction from Landsat 8 OLI image based on band rationing and principal component analysis methods. Firstly, the Landsat 8 data is used to calculate the NIR/RED (band5/band4) and SWIR1/SWIR2 (band6/band7) band rationing images. To highlight the distribution of clay minerals, these band rationing images are further used to multiply the digital number values of NIR and SWIR2 bands, then obtaining (band5/4) × band5 and (band6/7) × band7 images. Finally, the principal component analysis (PCA) method is used to calculate the principal components, then select the 2nd principal component (PC2) to detect clay minerals by the automatic thresholding method. The results in this research can be used to provide input information for mineral exploration.

Keywords Remote sensing · Clay minerals · Band rationing · Principal component analysis · Landsat 8 · Chi Linh city

1 Introduction

Located in Southeast Asia, Vietnam is rich in mineral resources, possessing some 60 kinds, making it the seventh ranking country in the top 15 of basic resource countries. Vietnam's main mineral resources consist of coal, phosphates, rare earth elements, bauxite, chromate, copper, gold, iron, manganese, silver, zinc, offshore oil, and gas

T. Le Hung (✉) · N. S. Thanh
Le Quy Don Technical University, Hanoi, Vietnam
e-mail: trinhlehung@lqdtu.edu.vn

N. S. Thanh
e-mail: thanhns.geo@lqdtu.edu.vn

V. T. Kha
Hanoi University of Mining and Geology, Hanoi, Vietnam
e-mail: vuongtrongkha@humg.edu.vn

L. Q. Nguyen et al. (eds.), *Advances in Geospatial Technology in Mining and Earth Sciences*, Environmental Science and Engineering,
https://doi.org/10.1007/978-3-031-20463-0_7

deposits. Some minerals in Vietnam have important reserves such as bauxite (672.1 million tons), apatite (0.778 million tons), titanium (15.71 million tons), coal (3520 million tons), rare earth (1.1 million tons) and granite (15 billion m^3) [1, 2]. Mineral resources play an important role and have become one of the resources for the socio-economic development of Vietnam. Despite having rich mineral resources, mineral reserves in Vietnam are not large and are being depleted due to the mining process for economic development. Therefore, the exploration and detection of mineral-containing areas is a matter of urgent significance today in Vietnam.

Remote sensing data has been widely used in the world in monitoring and detection of mineral-containing areas [3, 4]. The main remote sensing data in mineral monitoring and detection is multispectral images such as Landsat 5 TM, Landsat 7 ETM+ , Landsat 8 OLI, Aster, and Sentinel 2 MSI [5–9]. Many studies have used these remote sensing images to detect areas containing clay minerals, iron oxide, copper…. [6, 10–12]. Based on the spectral reflectance characteristics of minerals, some techniques have been proposed to detect mineral-containing areas on multispectral satellite images such as band rationing method [13, 14], the method based on principal component analysis (PCA), the method based on spectral indices (clay mineral index, iron oxide index, Chica-Olma index, Kaufmann index, Abrams index). Crosta et al. (1989, 1993) used principal component analysis and color composite techniques to detect clay minerals and iron oxides on Landsat TM satellite images [13, 15]. The principal component analysis method and spectral indices are also used in the study of Mia and Fujimitsu (2012), in which the authors used Landsat 7 ETM+ multispectral images for mapping thermal alteration minerals in and around Kuju volcano, Kyushu, Japan [16]. Pour and Hashim (2015) have integrated optical and radar images (PALSAR and ASTER data) for mineral deposits exploration in tropical environments (Central Belt, Peninsular Malaysia) [17].

Fraser and Green (1997) [18] developed DPCA (directed principal component analysis) method for monitoring hydrothermal minerals distribution. The DPCA method is built on the combination of advantages of band rationing and PCA methods, thereby improving the accuracy in detecting exposed minerals from satellite imagery. PCA and DPCA methods are also used in the studies of Trinh et al. [19–21] in some areas of northern Vietnam (Thai Nguyen province, Vinh Phuc province), in which the authors used Landsat TM, Landsat 8 OLI, and Sentinel 2 image data. In their studies, Trinh et al. built the RS-MINERALS computer program based on MATLAB language with tools such as band rationing, principal component analysis, directed principal component analysis for minerals detection and classification from Landsat imagery. The results obtained in these studies show that multispectral image data are effective for detecting areas containing clay and iron minerals, especially areas without vegetation cover [19–21].

The vegetation cover has a great influence on the accuracy of minerals detection results on optical satellite images. Traditional techniques such as band rationing, and principal component analysis methods are highly effective in detecting minerals in areas without vegetation cover. The accuracy of mineral detection from optical satellite images is greatly reduced in areas with vegetation cover. To overcome this limitation, this study uses a combination of band rationing, principal component analysis methods, and image multiplication technique. The image multiplication technique allows highlighting mineral distribution positions on band rationing images. Meanwhile, the principal component analysis method helps to accurately identify mineral-containing areas on band rationing images. The simultaneous combination of these techniques allows limiting the influence of vegetation cover on mineral detection results from optical satellite images such as Landsat 8 OLI.

The focus of this paper is to propose an automatic method for clay minerals extracting from Landsat 8 OLI multispectral image data. In this study, the red (band 4), near-infrared (band 5), and short-wave infrared bands (band 6 and band 7) Landsat 8 OLI image were used to calculate the band rationing images, and then image multiplication and principal component analysis techniques were used to highlight the clay minerals containing areas. Finally, the automatic thresholding method is used to classify clay minerals containing areas from principal component (PC) images.

2 Study Area and Materials

2.1 Study Area

Chi Linh is a city located in the north of Hai Duong province, Vietnam. The north and northeast of the city are mountainous areas belonging to the Dong Trieu arc, the other three sides are surrounded by Kinh Thay, Thai Binh, and Dong Mai rivers. Chi Linh city has a natural area of 282.91 km, and is the district-level administrative unit with the largest area in Hai Duong province. The population of Chi Linh city in 2018 is about 220,400 people [22].

The terrain of Chi Linh city is diverse with an area of hills, alternating plains, and sloping terrain from the north to the south. North of the city is a mountainous area with natural and planted forests, the highest peak is Day Dieu with a height of 616 m. The southern part of the study area has a flat terrain with alluvial flats [22].

Chi Linh city is located in the tropical monsoon climate with 2 distinct seasons: the dry season from October to April next year and the rainy season from April to September every year. The average annual temperature is 23°C, the average annual rainfall is 1463 mm [22].

Mineral resources in Chi Linh city are not of many types, but some types of minerals have large reserves and economic value such as: Kaolin (reserve of 40000 tons), refractory clay (8 million tons), stone, construction yellow sand, brown coal mine (reserves in billions of tons). Currently, there are dozens of mining sites in Chi Linh city and the mining industry plays an important role in local socio-economic development [22].

The geographical location of Chi Linh city (Hai Duong province) is shown in Fig. 1.

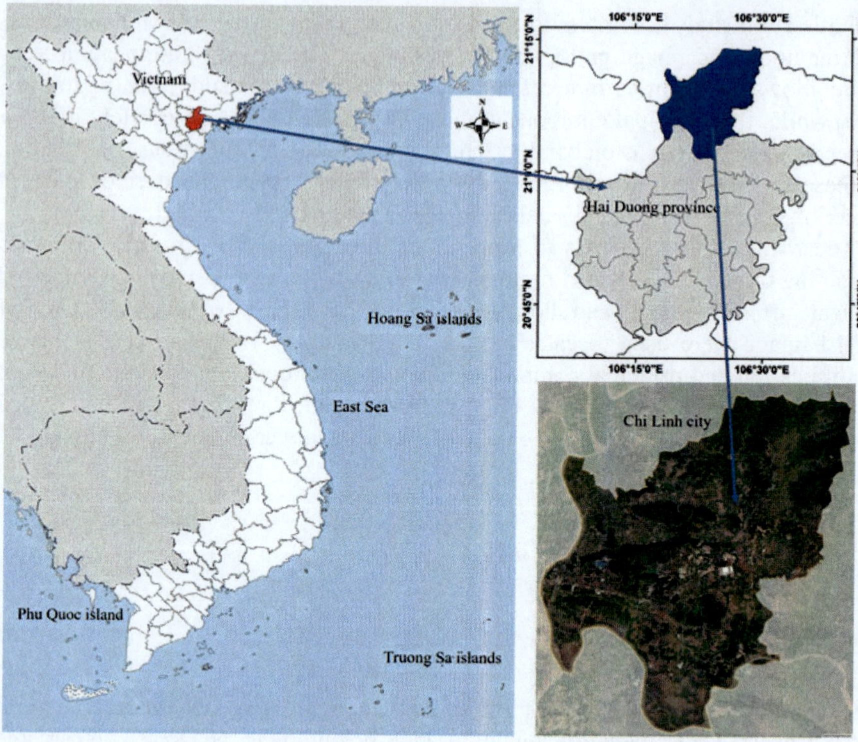

Fig. 1 Study area, Chi Linh city, Hai Duong province

2.2 *Materials*

LANDSAT 8 is the 8th generation satellite of the LANDSAT program (NASA, USA), using 2 sensors: Operational Land Imager (OLI) and Thermal InfraRed (TIRS). LANDSAT 8 was launched into orbit on February 11, 2013. LANDSAT 8 provides images in 11 spectral bands, including 9 multispectral bands with 30 m spatial resolution, 1 panchromatic band with 15 m resolution, and 2 thermal infrared bands at 100 m resolution (Table 1). The temporal resolution of the Landsat 8 image is 16 days, and with the successful launch of the Landsat 9 satellite (September 27, 2021) with completely similar characteristics to the Landsat 8, the temporal resolution of the Landsat 8/9 image is reduced to 8 days. Data from Landsat 9 was publicly available from USGS in early 2022. The shortening of image acquisition process allows improving the efficiency of the application of Landsat 8/9 satellite image data in Earth observation [23, 24].

Until September 2021, Landsat 8 had added 1.86 million images to the archive (about 20% of the total archive holdings) and each day since, Landsat 8 has added another ~ 700 new scenes. Landsat 9, like Landsat 8, is both radiometrically and geometrically better than earlier generation Landsats. Landsat data in the U.S. archive

Table 1 Landsat 8 OLI_TIRS bands characteristics

No	Landsat 8 OLI_TIRS bands	Wavelength (μm)	Resolution (m)
1	Band 1–Coastal/Aerosol	0.433–0.453	30
2	Band 2–Blue	0.450–0.515	30
3	Band 3–Green	0.525–0.600	30
4	Band 4–Red	0.630–0.680	30
5	Band 5–Near Infrared (NIR)	0.845–0.885	30
6	Band 6–Middle Infrared (MIR)	1.560–1.660	30
7	Band 7–Middle Infrared (MIR)	2.100–2.300	30
8	Band 8–Panchromatic (PAN)	0.500–0.680	15
9	Band 9–Cirrus	1.360–1.390	30
10	Band 10–Thermal Infrared (TIR)	10.30–11.30	100
11	Band 11–Thermal Infrared (TIR)	11.50–12.50	100

contributed by each Landsat satellite as of September 30, 2021, is shown in Fig. 2 [23].

In this study, the multispectral cloud-free Landsat 8 OLI image with a spatial resolution of 30 m (multispectral bands) and 100 m (thermal infrared bands), acquired on December 01, 2021, in Chi Linh city (Hai Duong province) was used for extracting clay minerals containing areas. The Landsat 8 data was the L2 level product, downloaded from the United States Geological Survey website (https://glovis.usgs.gov). Landsat 8 Level 2 provides global surface reflectance and surface temperature science

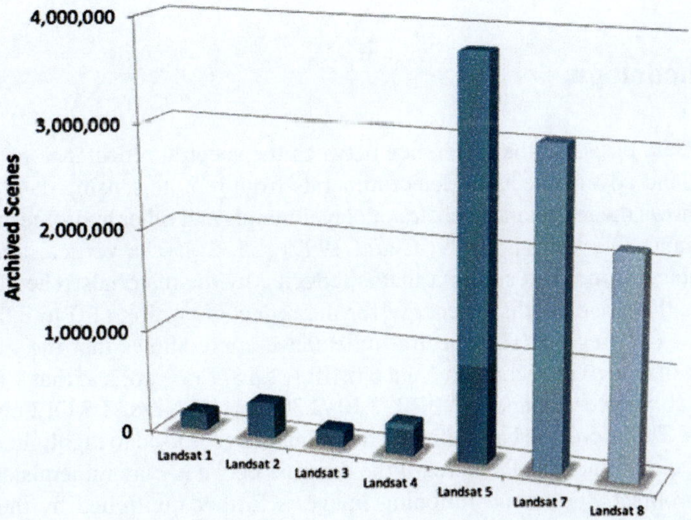

Fig. 2 Landsat data in the U.S. archive contributed by each Landsat satellite as of September 30, 2021 [23]

Fig. 3 Landsat 8 multispectral image, December 01, 2021

products. Level 2 science products are generated from Collection 2 Level-1 inputs that meet the < 76 degrees Solar Zenith Angle constraint and include the required auxiliary data inputs to generate a scientifically viable product [24].

Landsat 8 image data used in this study is presented in Fig. 3 (in the natural color band combination with red (band 4), green (band 3), and blue (band 2) bands).

3 Methodology

All studies are based on the difference between the spectral reflectance of minerals and other land cover objects to detect minerals from remote sensing data [25, 26]. Figure 4 shows the reflectance spectra of clay minerals and other hydrothermal alterations (source: Clark et al., 1989; Clark, 1999) [25, 27]. The vertical axis shows the percentage of incident sunlight that is reflected by the materials. The horizontal axis shows the wavelengths of energy for the reflected portion (1.0 to 3.0 μm) of the infrared (IR) region. The spectral reflectance curve shows that the maximum reflectance of iron oxide occurs in band 6 (MIR1, 1.56–1.66 μm) and that reflectance is considerably lower in band 7 (MIR2, 2.10–2.30 μm) LANDSAT 8 OLI multispectral images. Therefore, the band6/band7 rationing image is used to highlight the areas containing clay minerals. To increase the contrast between clay minerals and other land cover objects, this band rationing image is further multiplied by the band 7 digital number values.

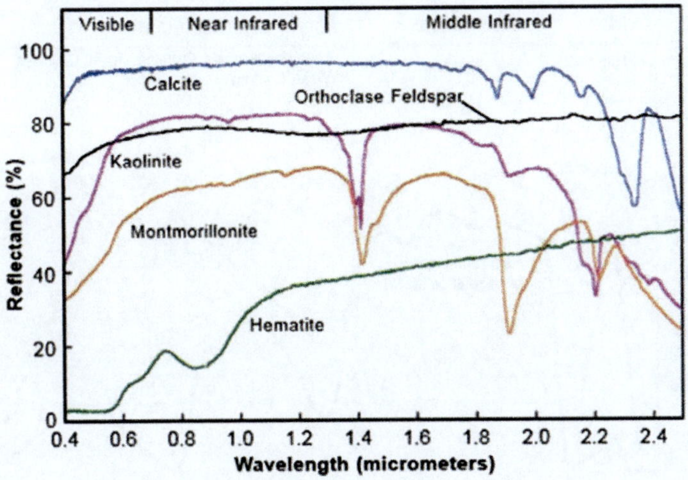

Fig. 4 Spectral reflectance of different clay minerals [25, 27]

On the other hand, vegetation cover absorbs electromagnetic radiation energy in the red spectral band and strongly reflects it in the near infrared spectral band (Fig. 5) [28]. Therefore, to eliminate the influence of vegetation cover on mineral detection results from remote sensing multispectral images, in this study, the band rationing image between near-infrared band (band 5) and red band (band 4) of Landsat 8 OLI data is used. To further increase the contrast between vegetation and other land cover objects, this band rationing image is multiplied by the digital number value of the near-infrared band (band 5), the result is an image (band5/4) × band5.

In the next step, principal components analysis method is used to calculate principal components (PC) from two band rationing images: (band5/4) × band5 and (band6/7) × band7. The PCA method uses the principal components transformation technique for reducing dimensionality of correlated multispectral data [29]. The analysis is based on multivariate statistical technique that selects uncorrelated linear combinations of variables in such a way that each successively extracted linear combination, or PC, has a smaller variance. Eigenvalues give information using magnitude and sign about which spectral properties of vegetation, rocks, and soils are responsible for the statistical variance mapped into each PC [16]. Finally, an automatic thresholding method is used to extract the clay minerals containing areas from PC image. The accuracy of detection areas containing clay minerals is evaluated based on Google high spatial resolution satellite image data and geological maps of the study area. The flowchart for the methodology used in this study to extract shoreline changes based on Landsat multi-temporal data is shown in Fig. 6.

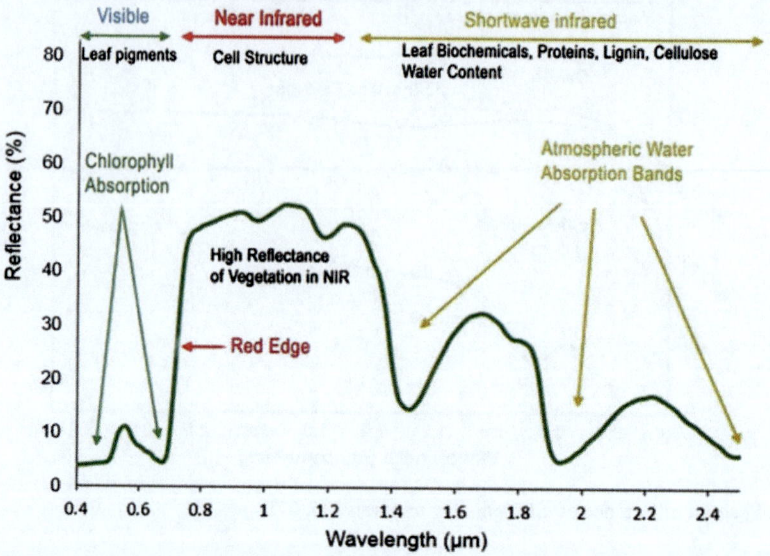

Fig. 5 Spectral characteristics of vegetation cover [28]

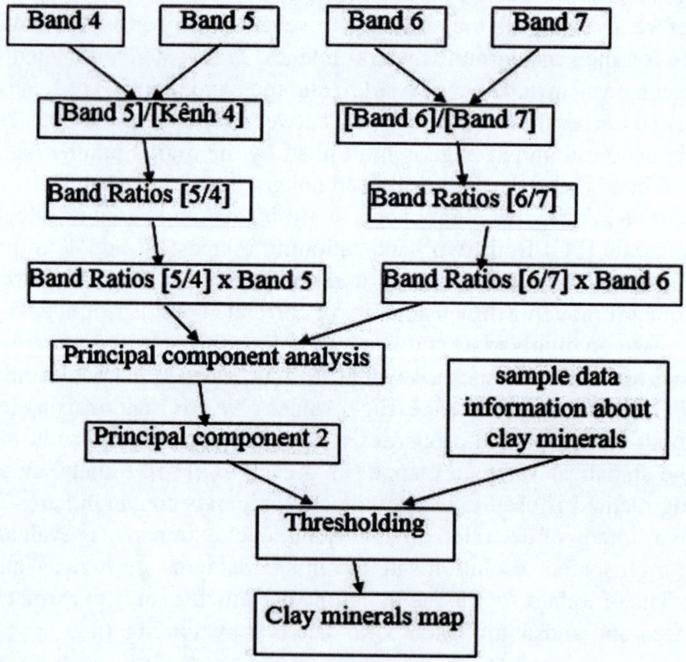

Fig. 6 Flowchart of the methodology for clay minerals extraction from Landsat 8 OLI multispectral data

4 Results and Discussion

Landsat 8 OLI data is collected at the L2A processing level (Bottom of atmosphere reflectance value), so in this study, only geometric correction was carried out to convert to the local coordinate system VN-2000. The red, near infrared and two middle infrared bands of Landsat 8 OLI image taken on December 01, 2021, are used to calculate the band rationing images. The band rationing images: Band5/Band4 and Band6/Band7 calculated from Landsat 8 OLI data in this study are shown in Fig. 7. Vegetation cover is shown in bright white pixels on band5/band4 rationing image due to strong reflection of electromagnetic radiation energy in near infrared band (band 5) and absorption in the red band (band 4) of Landsat 8 OLI image. In band6/band7 band rationing image, clay minerals are represented by bright white and are difficult to distinguish from vegetation cover.

Figure 8 shows images (Band5/4) × Band5 and (Band6/7) × Band6 after using image multiplication technique to better highlight the vegetation cover and clay minerals from band rationing images.

Table 2 shows the eigenvector matrix values and eigenvalues of the PCA for the (Band5/4) × Band5 and (Band6/7) × Band6 images. The analysis of eigenvalues and eigenvectors shows that the first principal component (PC1) contains information about the land surface cover, focusing on the vegetation cover. Meanwhile, the second principal component (PC2) contains information about the distribution of clay minerals. As can be seen, PC1—the "albedo" image, is about 97.405% of eigenvalue of the total variance for unstretched data PCA. PC2 contains 2.595% information of two band rationing images. In this study area, PC2 highlights clay minerals as bright white pixels because of the greatest loading of (Band6/7) × Band6 image (0.965478) and (Band5/4) × Band5 image (−0.260485) (Table 2).

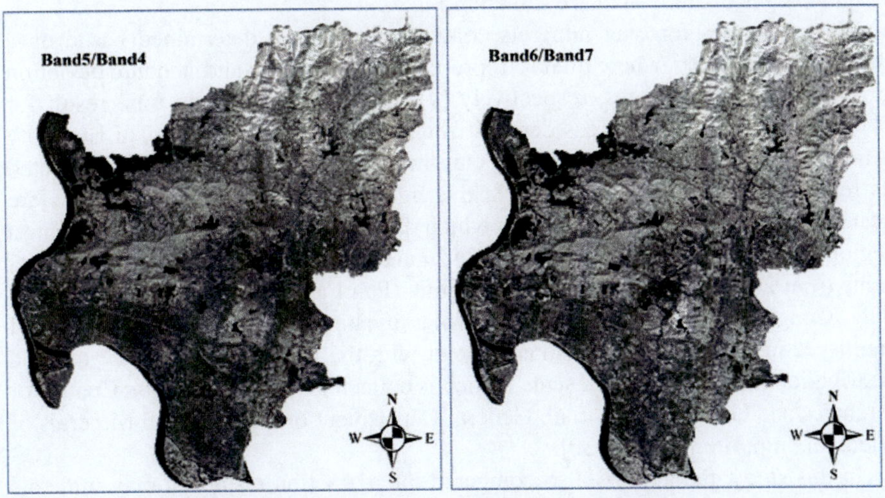

Fig. 7 The band rationing images: Band5/Band4 and Band6/Band7

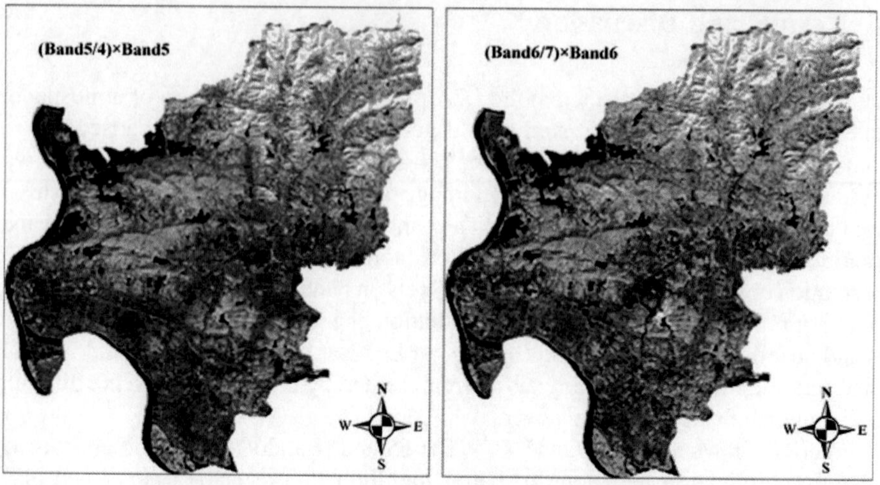

Fig. 8 The band rationing images: (Band5/4) × Band5 and (Band6/7) × Band6

Table 2 The eigenvector matrix values and eigenvalues of the PCA for the (Band5/4) × Band5 and (Band6/7) × Band6 images

Principal component	Eigen matrix		Eigenvalues (%)
	(Band5/4) × Band5	(Band6/7) × Band6	
PC1	0.965478	0.2604845	97.405
PC2	– 0.260485	0.965478	2.595

The two PC transformation on unstretched (Band5/4) × Band5 and (Band6/7) × Band6 images of study area are shown in Fig. 9.

The anomalies for clay minerals containing areas are determined based on a threshold of $\mu + 2^*\sigma$, where μ and σ represent the mean value and standard deviation of the relevant PC images, respectively [7]. Figure 10 shows the final result for clay minerals containing areas derived from Landsat 8 OLI data in Chi Linh city (Hai Duong province), in which the clay minerals containing areas are depicted as red color. The results presented in this figure show that the clay minerals are scattered throughout Chi Linh city, in which the most concentration is in the central region. Large clay mineral deposits in Chi Linh city such as Co Kenh anthracite coal mine (Van Duc ward), Phao Son kaolin mine (Pha Lai ward), Phuc Son refractory clay (Cong Hoa ward) are similar to the classification results from Landsat 8 OLI multispectral image. This is also consistent with the Hai Duong province mineral distribution map of 1:200,000 scale, which is collected from Information Center for Archives and Geological Journal, General Department of Geology and Minerals of Vietnam, http://idm.gov.vn [30].

In this study, the authors also compared the extraction results of clay minerals containing areas in Chi Linh city and ESRI high spatial resolution images (Table 3).

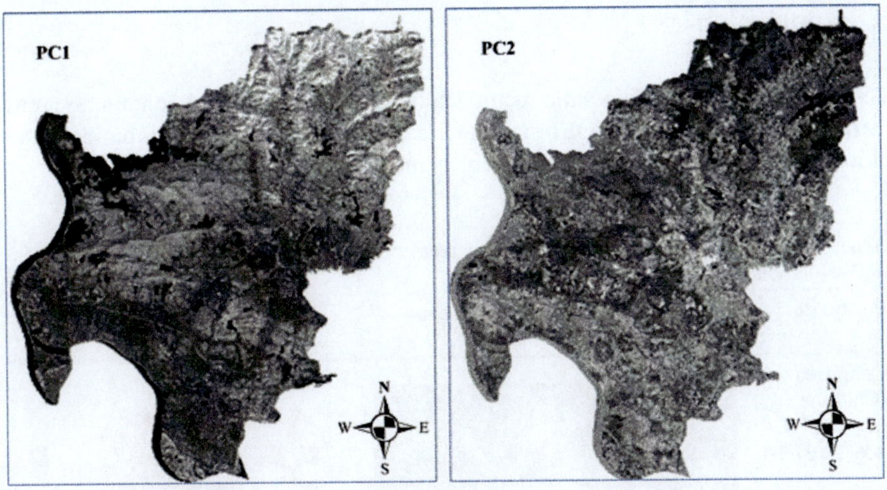

Fig. 9 First and second principal components images

Fig. 10 Result of clay minerals extraction from Landsat 8 data (Chi Linh city, Hai Duong province)

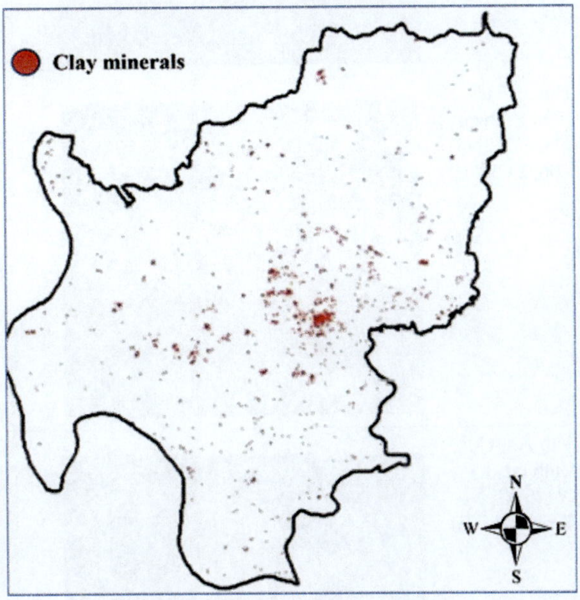

Table 3 shows that, the clay minerals containing areas such as Cong Hoa, Hoang Tien, Van An locations and refractory brick factory have been accurately classified from the Landsat 8 OLI data based on proposed method.

5 Conclusion

This study presents an automatic method solution for clay mineral containing areas extraction from Landsat 8 OLI satellite images on the basis of combining band rationing and principal component analysis method.

Table 3 Compare the classification results of areas containing clay minerals and images from high-resolution satellite images

Position	ESRI high spatial resolution images	Areas containing clay mineral classified from Landsat 8 data
Cong Hoa (Chi Linh city) (21°7′13″N, 106°22′27″E)		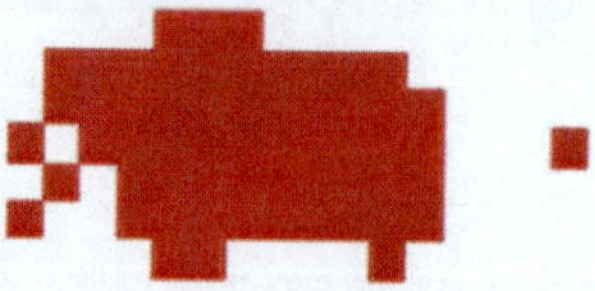
Hoang Tien (Chi Linh city) (21°8′35″N, 106°27′39″E)		
Van An (Chi Linh city) (21°6′37″N, 106°20′46″E)		

(continued)

Table 3 (continued)

Position	ESRI high spatial resolution images	Areas containing clay mineral classified from Landsat 8 data
Refractory brick factory (21°8′02″N, 106°24′24″E)		

The Landsat 8 OLI image acquired on December 1, 2021 is analyzed to map the spatial distribution of the clay minerals in Chi Linh city, Hai Duong province (northern Vietnam). Four Landsat 8 OLI bands including red (band 4), near-infrared (band 5), and two middle infrared bands (band 6 and band 7) were used to calculate band rationing images, then use principal component analysis method to select the principal component containing a lot of information about clay minerals containing areas. In this study, clay minerals containing areas were extracted based on the automatic thresholding method.

The results obtained in this study show that, Landsat 8 OLI data can be used effectively in extracting and mapping clay minerals containing areas distribution. Along with the successful launch of the Landsat 9 satellite in 2021, the Landsat 8/9 imagery is an effective data source in mineral exploration and discovery.

References

1. Luu, D.-H., Nguyen, T.H.-L.: Renewable energy policies for sustainable development in Vietnam. Vietnam National Univ. J. Sci. Earth Environ. Sci. **25**(3), 133–142 (2009)
2. Vietnam National Coal-Mineral Industries Holding Corporation Limited (Vinacomin). Retrieved from http://www.vinacomin.com.vn
3. Goetz, A.-F., Rock, F.-H., Rowa, B.-N.: Remote sensing for exploration: an overview. Econ. Geol. **78**, 573–590 (1983)
4. Rajesh, H.-M.: Application of remote sensing and GIS in mineral resource mapping–An review. J. Mineral. Petrol. Sci. **99**(3), 83–103 (2004)
5. Abrams, M.-J.: Remote sensing of porphyry copper in Southern Arizona. Econ. Geol. **78**, 591–604 (1983)
6. Abrams, M.-J., Hook, S.-J.: Simulated Aster data for geologic studies. IEEE Trans. Geosci. Remote Sens. **33**, 692–699 (1995)

7. Hu, B., Xu, Y., Wan, B., Wu, X., Yi, G.: Hydrothermally altered mineral mapping using synthetic application of Sentinel-2A MSI, ASTER and Hyperion data in the Duolong area, Tibetan Plateau, China. Ore Geol. Rev. **101**, 384–397 (2018)
8. Roman, A., Ursu, T.: Multispectral satellite imagery and airborne laser scanning techniques for the detection of archaeological vegetation marks. In: Landscape archaeology on the northern frontier of the Roman Enpire at porolissum–an interdiscliplinary research project. Mega Publishing House (2016)
9. Van der Meer, F.D., van der Werff, H.M.A., van Ruitenbeek, F.J.A.: Potential of ESA's Sentinel-2 for geological applications. Remote Sens. Environ. **148**, 124–133 (2014)
10. Alasta, A.-F.: Using remote sensing data to indentify iron deposits in central western Libya. In: International Conference on Emerging trends in Computer and Image processing, Bangkok, 56–61 (2011)
11. Bennett, S.-A., Atkinson, W.-W., Kruse, F.-A.: Use of Thematic Mapper imagery to identify mineralization in the Santa Teresa district, Sonora, Mexico. Int. Geol. Rev. **35**, 1009–1029 (1993)
12. Carranza, E.J.-M., Hale, M.: Mineral imaging with Landsat Thematic Mapper data for hydrothermal alteration mapping in heavily vegetated terrane. Int. J. Remote Sens. **23**, 4827–4852 (2002)
13. Crosta, A.-P., Moore, J.-M.: Enhancement of LANDSAT Thematic Mapper imagery for residual soil mapping in SW Minas Gerais State Brazil: a prospecting case history in greenstone belt terrain. In: Proceedings of the 9th Thematic Conference on Remote Sensing for Exploration Geology, Calgary (Ann Arbor, MI: Environmental Research Institute of Michigan), 1173–1187 (1989)
14. Kaufman, H.: Mineral exploration along the Agaba-Levant structure by use of TM-data concepts, processing and results. Int. J. Remote Sens. **9**, 1630–1658 (1988)
15. Crosta, A.-P., Rabelo, A.: Assessing Landsat TM for hydrothermal mapping in central-western Brazil. In: Proceedings of the Ninth Thematic Conference on Geologic Remote Sensing, Pasadena, California, USA, 1053–1061 (1993)
16. Mia, M.-B., Fujimitsu, Y.: Mapping hydrothermal altered mineral deposits using LANDSAT 7 ETM+ image in and around Kuju volcano, Kyushu, Japan. J. Earth Syst. Sci. **121**(4), 1049–1057 (2012)
17. Pour, A.-B., Hashim, M.: Integrating PALSAR and ASTER data for mineral deposits exploration in tropical environments: a case study from Central Belt, Peninsular Malaysia. Int. J. Image Data Fusion **6**(2), 170–188 (2015)
18. Fraster, S.-J., Green, A.-A.: A software defoliant for geological analysis of band ratio. Int. J. Remote Sens. **8**, 525–532 (1987)
19. Trinh, L.H., Zablotskii, V.R.: The method of detection of clay minerals and iron oxide based on Landsat multispectral images (as exemplified in the territory of Thai Nguyen province, Vietnam). Min. Sci. Technol. **4**(1), 65–75 (2019)
20. Trinh, L.H., Nguyen, V.N.: Mapping coal files using Normalized Difference Coal Fire Index (NDCfI): case study at Khanh Hoa coai mine, Vietnam. Min. Sci. Technol. **6**(4), 233–240 (2021)
21. Trinh, L.H.: Application of band ratio method to detect iron oxide, clay minerals and ferrous minerals. Min. Ind. J. **4**, 19–24 (2013)
22. Electronic portal of Chi Linh city. Retrieved from https://chilinh.haiduong.gov.vn/
23. National Aeronautics and Space Administration. Retrieved from https://landsat.gsfc.nasa.gov/satellites/
24. United States Geological Survey. Retrieved from https://glovis.usgs.gov
25. Clark, R.-N., Swayze, G.-A., Wise, R., Livo, K.-E., Hoefen, T.-M., Kokaly, R.-F., Sutley, S.-J.: USGS digital spectral library. USGS Open File Rep (1989)
26. Hunt, G.-R., Ashley, R.-P.: Spectra of altered rocks in the visible and near infrared. Econ. Geol. **74**, 1613–1629 (1979)
27. Clark, R.-N.: Spectroscopy of rock and minerals and principles of spectroscopy. In: Rencz, A.-N. (ed.) Remote Sensing for the Earth Sciences, Manual of Remote Sensing 3, pp. 3–58. Wiley, New York (1999)

28. Mohamed, E.-S., Saleh, A.-M., Belal, A.-B., Gad, Abd. -A.: Application of near-infrared reflectance for quantitative assessment of soil properties. Egypt. J. Remote Sens. Space Sci. **21**(1), 1–14 (2018)
29. Loughlin, W.-P.: Principal component analysis for alteration mapping. Photogram. Eng. Remote Sens. **57**(g), 1163–1169 (1991)
30. Inf. Cent Arch. Geol. J. General Department of Geology and Minerals of Vietnam. Retrieved from http://idm.gov.vn

Evaluation of the Precision of SARAL/AltiKa and Sentinel-3A Satellite Altimetry Data Over the Vietnam Sea and Its Surroundings

Do Van Mong, Nguyen Van Sang, Khuong Van Long, and Luyen K. Bui

Abstract Satellite altimetry has proven to be a useful tool to measure sea surface height, which is of diverse applications in oceanography, geodesy, among others. However, the precision of satellite altimetry data is different between missions and areas. This article evaluates the precision of observed data received by two satellite altimeters of SARAL/AltiKa and Sentinel-3A in the Exact Repeat Mission mode over the Vietnam Sea and its surroundings. The precision of the data is assessed based on height differences at intersection points between ascending and descending tracks. First, the positions of each intersection point between the ascending and descending tracks are interpolated from measure points with the second-order polynomial model. Then, the standard deviation of the height difference is estimated from all intersection points for each of 34 repeat cycles (SARAL/AltiKa) and 28 repeat cycles (Sentinel-3A) over the study area. The results show that the standard deviations of the SARAL/AltiKa data are between ± 4.5 cm and ± 7.5 cm, with an average of ± 5.9 cm, while those of Sentinel-3A range from ± 4.4 cm to ± 7.7 cm, with an average of ± 5.9 cm. Of both datasets, the height differences are greater at points located close to the coastlines and islands.

Keywords Satellite altimetry · SARAL · AltiKa · Sentinel-3A · Vietnam Sea

D. Van Mong · K. Van Long
Vietnam's People Naval Hydrographic and Oceanographic Department, Hai Phong 04000, Vietnam

N. Van Sang (✉) · L. K. Bui
Faculty of Geomatics and Land Administration, Hanoi University of Mining and Geology, Hanoi 100000, Vietnam
e-mail: nguyenvangsang@humg.edu.vn

L. K. Bui
e-mail: buikhacluyen@humg.edu.vn

L. Q. Nguyen et al. (eds.), *Advances in Geospatial Technology in Mining and Earth Sciences*, Environmental Science and Engineering,
https://doi.org/10.1007/978-3-031-20463-0_8

1 Introduction

Starting from the 1980s, satellite altimeter has been developed rapidly and become a useful tool for marine research. Since the first satellite platform Geodetic/Geophysical Satellite (Geosat) was launched in 1985, there have been many satellite missions working on their orbits; 10 inactive satellite missions are Geosat, the European Remote Sensing satellite (ERS-1/2), TOPEX/Poseidon (T/P), the US navy Geosat Follow On (GFO), Jason-1/2, Environmental Satellite (Envisat), Hai Yang 2A (HY-2A), Satellite pour l'Observation de la Terre (SPOT), and 9 active satellite missions being CryoSat, SARAL/AltiKa, Jason-3, Sentinel-3A/B, Chinese-French Oceanography Satellite (CFOSAT), HY-2B/2C, Jason-CS. The Surface Water and Ocean Topography (SWOT) satellite mission was scheduled to be launched in 2022. The satellites have created a rich database that has been applied in various fields, such as oceanography, hydrology, land and coastal studies, ice and cryosphere, climate, atmosphere, wind and waves, geodesy, and geophysics [1]. Satellite altimeters data are available in the exact repeat mission (ERM) or geodetic mission (GM) mode. In the ERM mode, the observations are made repeatedly at a regular time interval but with a low spatial resolution. This kind of data is useful for oceanography and climate sciences, but not applicable for the determination of high-resolution gravity field modeling. The data received from the geodetic mission (GM) can be used for various applications in geodesy. In this mode, the satellite flies in a non-repeated orbit where the observations are taken once only at each location, but with a much higher spatial density [2].

In the Vietnam Sea and the surrounding area, bounded by [5°N, 25°N] in latitude and [105°E, 120°E] in longitude, there have been several studies applying satellite altimeter data. For example, in 2000, Hwang et al. [3] used the Topex/Poseidon satellite altimeter data to determine surface currents in the East Sea. In 2012, marine gravity anomalies were determined from the ENVISAT data combined with *in-situ* gravity data measured at the coast and islands in the East Sea, in which an accuracy of about $\pm$ 6 mGal was obtained [4]. In 2019, Tran et al. improved the accuracy of marine gravity anomalies derived from satellite altimetry by fitting with the ship-derived gravity anomaly in the East Sea. The results showed the accuracy of the gravity anomaly from $\pm$ 1.21 to $\pm$ 9.36 mGal [5]. In 2020, the accuracies of the global gravity anomaly models DTU10GRAV, DTU13GRAV, DTU15GRAV were evaluated in the Vietnam Sea by comparing with ship-derived gravity anomaly data, which were found to be $\pm$ 5.80, $\pm$ 5.73, and $\pm$ 5.63 mGal, respectively [6]. Also in 2020, the marine gravity anomaly over the Gulf of Tonkin was determined from the CryoSat-2 and SARAL/AltiKa satellite altimeter data with the accuracies achieved before and after fitting with shipborne gravity data being $\pm$ 3.36 and $\pm$ 2.63 mGal, respectively [7]. These studies indicated that satellite altimeter data are of great interest in the study area.

The precision of satellite altimetry data is different between missions and study areas [2], depending on the characteristics of the sea, the corrections to the satellite altimeter data according to the global correction models, which may not be accurate enough for regional areas. Therefore, it is necessary to evaluate the precision of satellite altimeter data at the regional scale before applying them. As a result, several studies have been conducted to evaluate the precision of satellite altimeter data. For instance, in 2012, altimetric sea levels in coastal areas were validated using available tide gage records in the Nordic, Barents, and Kara seas [8]. The comparison results showed that, at most locations in the Nordic Sea, the satellite altimeter data are in good agreement with the tide gage data, but the agreement is low at the shallows of the Barents and Kara Seas due to the seasonal presence of sea ice [8]. In 2017, the main range and geophysical corrections of TOPEX/Poseidon, Jason-1, and Jason-2 satellite altimetry data were evaluated over the Indonesian Sea, in which the distances from the coast at altimeter crossovers were used to assess the quality of the various corrections and mean sea surface models [9]. Also in 2017, the precision of CryoSat-2 satellite altimeter data was evaluated in the waters around the Spratly Islands, Vietnam, in which the results showed that the precision of CryoSat-2 satellite altimeter data from periods 31 to 43 is $\pm$ 3.5 cm [10]. Location and height difference at crossover points in satellite altimetry data processing were determined using a direct method in 2017 [11]. In 2020, Sentinel-3A and Jason-3 satellite data in the coastal area of the European Sea were evaluated by comparison with tide gage measurements, with the results showing that the root mean square deviation (RMSD) of Sentinel-3A is 13% lower than that of the Jason-3 [12].

Sentinel-3A was launched on February 16, 2016, and is now working in the ERM mode. As a result, its measurements are suitable for studying ocean currents. The mission measurements have been collected since March 12, 2016, with a repeated cycle of 27 days [13]. The SARAL/AltiKa satellite, a product made under the cooperation between the Indian Space Research Organization (ISRO) and Centre national d'études spatiales (CNES), was launched on February 25, 2013. The SARAL/AltiKa satellite collected the measurements in the ERM mode from March 14, 2013, to July 4, 2016, with a repeated cycle of 35 days, which is suitable for studying ocean currents. It then changed into the GM mode from July 4, 2016 onward, which is suitable for gravity field research [14].

Up to now, there have no studies on the assessment of the precision of the SARAL/AltiKa or Sentinel-3A data over the Vietnam Sea and its surroundings. This paper conducts an evaluation of the precision of their data based on height differences at intersection points between the ascending and descending tracks and experimentally evaluates the precision of the SARAL/AltiKa and Sentinel-3A data in the ERM mode. The research is a preliminary step in a further planned project to apply the SARAL/AltiKa and Sentinel-3A satellite altimeter data in studying the ocean current over the Vietnam Sea and its surroundings.

2 Research Area and Data

2.1 Research Area

The study area is the Vietnam Sea and the surrounding area, bounded by [5°N, 25°N] in latitude and [105°E, 120°E] in longitude (Fig. 1). In this area, tides, waves, and currents vary significantly, which affect the precision of the corrections in the satellite altimeter data. The sea of Vietnam and the surrounding area is an area with different tidal regimes; in the North, the tidal regime is the regular diurnal tide. In the Central region, there is an irregular diurnal tide. In the South, the tidal regime is quite complicated from semi-diurnal to diurnal tide with large tidal amplitudes [15].

The surface current in the East Sea area is clearly shown by the seasonal regime and two large-scale vortices. The winter current system is dominated by the prevailing wind field in the area and partly influenced by the geomorphic current system caused by temperature fields and sea salinity. In addition, the most prominent feature of the surface current in this period is the presence of a large cyclone over the entire sea [16]. For the surface laminar current system, which is formed mainly by the southwest wind field, with the characteristics of being strongly differentiated by the impact of the tropical convergence band. In addition, with the intensification of warm waters according to the deep sea off the Southeast, anti-vortexes which counterbalance with winter vortexes have been formed [17].

The wave height properties of this area are also extremely diverse. The smallest wave height is less than 3 m and located in the western part of the Gulf of Thailand. The area with the highest wave height of more than 12 m can be found in the northeast area of the East Sea due to frequent storm and strong northeast wind activities. The central area of the Gulf of Tonkin has the highest wave height of about 9–10 m. The central area of the East Sea has the highest wave height of about 11 m [18].

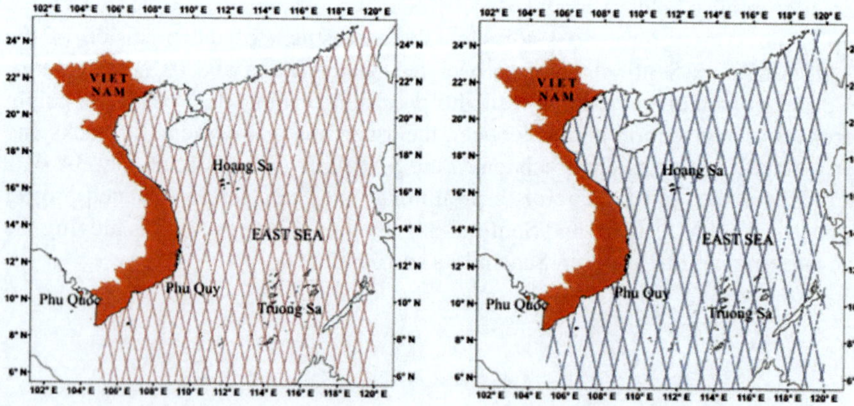

Fig. 1 Track distribution of SARAL/AltiKa (left) and Sentinel-3A (right) satellites

The composition of marine life in this area is of the tropical nature that mainly consists of tropical species distributed in the Western Indo-Pacific region. In addition, in the northern border region of Vietnam, where the tropical monsoon climate is cold, there are also a number of subtropical species distributed from the subtropical China Sea. In terms of ecology, the East Sea is also extremely diverse, with typical ecosystems such as submerged forests, coral reefs, and seagrasses [18].

2.2 Research Data

The data used in this study are obtained from two satellite altimeter systems of SARAL/AltiKa and Sentinel-3A in the ERM measurement mode, within the spatial limit from 5 to 25°N in latitude and from 105 to 120°E in longitude. These data are provided by the Radar Altimeter Database System [19].The SARAL/AltiKa data were collected and aggregated over 34 cycles (from 1st cycle to 34th cycle), corresponding to a measurement period of over 3 years (from March 14, 2013, to June 16, 2016) with a total of 306,223 measurement points. Figure 1 (left) shows the track distribution of the SARAL/AltiKa satellite.

The Sentinel-3A data were collected and aggregated over 28 cycles (from 48th cycle to 75th cycle) in the ERM mode, corresponding to a time span of over 2 years (from August 7, 2019, to September 2, 2021) with a total of 306,223 measurement points. Figure 1 (right) shows the track distribution of the Sentinel-3A satellite. The information regarding the used data is summarized in Table 1. The corrections used in calculating sea surface height (SSH) include orbit, dry and wet tropospheric, ionospheric, solid earth tide, ocean tide, load tide, pole tide, and sea state bias [19].

Table 1 Summary of collected satellite altimeter data

Data	Repeat cycle (day)	Number of cycles	Measurement time	Measurement points
SARAL/AltiKa	35	34	from March 14, 2013, to June 16, 2016	306,223
Sentinel-3A	27	28	from August 7, 2019, to September 2, 2021	210,375
Total		62		516,598

3 Precision Evaluation

3.1 Locating and Computing the Height Difference at the Intersection

According to the operating principle of satellite altimeter systems, when encountering sea water, the wave beams from the satellite reflect back to the receivers on the satellites. The tracks can be classified into ascending and descending tracks. The intersection of the ascending and descending tracks forms the intersection points. The analysis of data sources received from the SARAL/AltiKa and Sentinel altimeter satellites shows that the position of the intersection point between the arcs formed from the dataset often does not coincide with the location of the received data points. Therefore, in order to evaluate the accuracy of satellite altimeter data, the research task requires determining the exact positions of the intersection points and computing the differences in sea level in these positions [20]. In the research content, we use the second-order polynomial simulation method to solve the above two problems [21]. Suppose on a data track, the point has coordinates of (ϕ_i, λ_i). This track will be simulated by a quadratic polynomial [21]:

$$\lambda = a\varphi^2 + b\varphi + c \tag{1}$$

where a, b, c are the parameters to be determined, in which at least three known coordinate points (i.e., the data points from the altimeter satellites over the track) are required. If there are more than three points, then the parameters are determined according to the principle of least squares.

If the ascending track is simulated by a polynomial model by:

$$\lambda = a_a\varphi^2 + b_a\varphi + c_a \tag{2}$$

And the descending track is simulated by a polynomial model by:

$$\lambda = a_d\varphi^2 + b_d\varphi + c_d \tag{3}$$

Then, the position of the intersection point is the solution of the system of equations:

$$\begin{cases} \lambda = a_a\varphi^2 + b_a\varphi + c_a \\ \lambda = a_d\varphi^2 + b_d\varphi + c_d \end{cases} \tag{4}$$

The system of Eqs. (4) has two solutions, that is, there will be two intersection points located in the two halves of the parabolic graph of the quadratic polynomial. The intersection point is detected based on the beginning and the ending points of the track. Comparing these two points with the beginning and the end points of the track

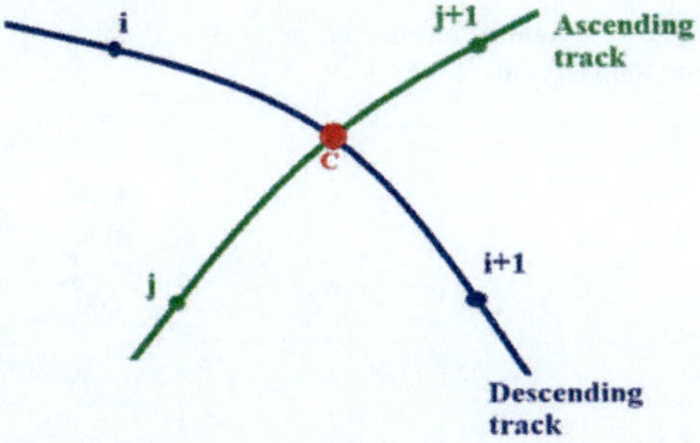

Fig. 2 Exact intersection location

helps to find a suitable intersection. The position of the found intersection point is not an exact location, but an approximate location. After determining an approximate intersection point, four neighborhood points of the intersection i, $i+1$, j, and $j+1$ from the ascending and descending tracks are determined (Fig. 2). From these 4 points, we will determine the position of the intersection point (C) more accurately.

After the position of the intersection is determined, the SSH at the intersection point according to the ascending track $\left(\mathrm{SSH}_{C,a}\right)$ and the descending track $\left(\mathrm{SSH}_{C,d}\right)$ will be interpolated from neighboring measurement points i, $i+1$, j, and $j+1$. The height difference at the intersection point is computed by:

$$\mathrm{dH} = \mathrm{SSH}_{C,a} - \mathrm{SSH}_{C,d} \tag{5}$$

3.2 *Evaluating the Precision of Satellite Altimeter Data Based on the Height Differences at the Intersection Points*

$\mathrm{SSH}_{C,a}$ and $\mathrm{SSH}_{C,d}$ are the SSHs at the intersection point C computed according to the two ascending and descending tracks. If there is no error, then the two values are identical (i.e., $\mathrm{dH} = 0$). In fact, the value of $\mathrm{dH} \neq 0$ is due to measurement errors (Fig. 3). Therefore, based on the dH value at the intersection point, we can evaluate the altimeter satellite measurement accuracy.

Let dH_i be the measured value at the i-th intersection. Over the entire study area, we have a series of measurement values $(\mathrm{dH}_1, \mathrm{dH}_2, \ldots, \mathrm{dH}_m)$. The true values of these measurements are all zero. The mathematical expectation of this range of values is calculated by:

$$E(\mathrm{dH}) = \overline{\mathrm{dH}} = \frac{[\mathrm{dH}]}{m} \tag{6}$$

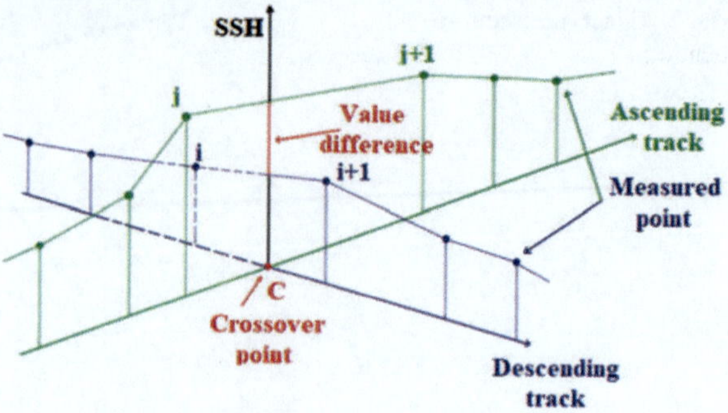

Fig. 3 Height difference at the intersection

+If the mathematical expectation E(dH) = 0, then there is no systematic error in the measurements, then the precision of the measured values is determined by:

$$\mathrm{RMS_{dH}} = \pm\sqrt{\frac{[\mathrm{dH.dH}]}{m}} \tag{7}$$

According to the principle of equal influence, from Eq. (5), we have:

$$\mathrm{RMS}^2_{\mathrm{dH}} = \mathrm{RMS}^2_a + \mathrm{RMS}^2_d = 2\mathrm{RMS}^2_{\mathrm{SSH}} \tag{8}$$

Substituting to Eq. (7), we have:

$$\mathrm{RMS_{SSH}} = \frac{\mathrm{RMS_{dH}}}{\sqrt{2}} = \pm\sqrt{\frac{[\mathrm{dHdH}]}{2m}} \tag{9}$$

+If the mathematical expectation E(dH) ≠ 0, then there is systematic error in the measurements, and the precision of the measured values is determined according to the standard deviation by the Bessel equation:

$$\mathrm{STD_{dH}} = \pm\sqrt{\frac{[vv]}{m-1}} \tag{10}$$

where given is the calculated correction:

$$v = \mathrm{dH} - \overline{\mathrm{dH}}$$

Then, the precision of the SSH is determined by:

$$\mathrm{STD_{SSH}} = \frac{\mathrm{STD_{dH}}}{\sqrt{2}} = \pm\sqrt{\frac{[vv]}{2(m-1)}} \tag{11}$$

4 Results of Precision Evaluation of SARAL/Altika and Sentinel-3A Data in the Waters of Vietnam and Surrounding Areas

4.1 Results of Precision Evaluation of the SARAL/Altika Satellite Altimeter Data

In this section, we evaluate the precision of the SARAL/AltiKa satellite altimeter data for 34 cycles with the results shown in Table 2, Figs. 4 and 5.

The results in Table 2 show that, over the study area, the standard deviations of the SARAL/AltiKa data are between ± 4.5 cm and ± 7.5 cm, with an average of ± 5.9 cm. The average deviation at the intersections is small, ranging from – 0.033 m to 0.004 m. This indicates that the systematic error in the SARAL/AltiKa satellite altimeter data over study area is small.

4.2 Results of Precision Evaluation of Sentinel-3A Satellite Altimeter Data

The results of precision evaluation of Sentinel-3A satellite altimeter data are shown in Table 3, Figs. 6 and 7.

Maximum, minimum, and average values of precision of the Sentinel-3A satellite altimeter data are ± 4.4 cm, ± 7.7 cm, and ± 5.9 cm, respectively. The average deviation at the intersection is small, ranging from – 0.022 m to 0.023 m. This proves that the systematic error in the Sentinel-3A satellite altimeter data is small over the study area.

In order to identify the areas with large height differences at the intersections (dH), we show the average height difference of all cycles of the SARAL/AltiKa satellite in Fig. 8 and the Sentinel-3A satellite in Fig. 9. The results for both data show that the height differences are larger at points located close to the coastlines and islands. Additionally, Fig. 9 shows that not only in the coastal zones, large SSH differences at the intersections can also be found in the area between the Hoang Sa and Truong Sa Islands. This is likely that, in this area, the corrections to the satellite altimeter data according to the global correction models are not very appropriate. It is necessary to study and correct local factors to improve the precision of satellite altimeter data in these areas.

Table 2 Results of evaluating of the precision of the SARAL/AltiKa satellite altimeter data

Cycle	Number of intersection	dH_{max} (m)	dH_{min} (m)	dH_{mean} (m)	RMS_{SSH} (m)	STD_{SSH} (m)
CK01	169	0.393	− 0.949	− 0.001	0.075	0.075
CK02	149	0.217	− 0.311	0.000	0.050	0.050
CK03	129	0.482	− 0.225	− 0.007	0.060	0.060
CK04	153	0.225	− 0.331	− 0.011	0.059	0.058
CK05	156	0.204	− 0.314	− 0.003	0.055	0.055
CK06	155	0.542	− 0.377	− 0.016	0.075	0.074
CK07	172	0.374	− 0.301	0.004	0.073	0.073
CK08	165	0.428	− 0.22	− 0.007	0.066	0.066
CK09	174	0.471	− 0.347	− 0.033	0.072	0.064
CK10	170	0.281	− 0.259	− 0.004	0.058	0.058
CK11	178	0.208	− 0.236	− 0.006	0.050	0.050
CK12	140	0.199	− 0.214	0.003	0.046	0.046
CK13	136	0.258	− 0.273	− 0.013	0.054	0.053
CK14	161	0.367	− 0.282	− 0.013	0.061	0.059
CK15	164	0.319	− 0.189	0.004	0.057	0.057
CK16	142	0.244	− 0.251	− 0.006	0.054	0.053
CK17	108	0.287	− 0.266	− 0.014	0.070	0.068
CK18	170	0.322	− 0.220	0.002	0.056	0.056
CK19	171	0.483	− 0.325	− 0.024	0.074	0.070
CK20	179	0.186	− 0.284	− 0.021	0.056	0.052
CK21	169	0.165	− 0.193	− 0.012	0.047	0.045
CK22	172	0.413	− 0.280	− 0.017	0.057	0.054
CK23	146	0.196	− 0.264	− 0.006	0.049	0.049
CK24	138	0.355	− 0.301	0.001	0.065	0.065
CK25	135	0.287	− 0.385	− 0.015	0.060	0.058
CK26	149	0.278	− 0.349	− 0.011	0.061	0.061
CK27	146	0.468	− 0.324	− 0.005	0.074	0.074
CK28	162	0.349	− 0.269	− 0.018	0.059	0.056
CK29	165	0.273	− 0.450	− 0.001	0.067	0.066
CK30	178	0.304	− 0.408	− 0.024	0.073	0.069
CK31	177	0.309	− 0.295	− 0.009	0.056	0.056
CK32	171	0.292	− 0.232	− 0.007	0.050	0.050
CK33	154	0.220	− 0.282	− 0.005	0.053	0.053
CK34	153	0.343	− 0.381	− 0.010	0.059	0.058
Max	179	0.542	− 0.189	0.004	0.075	0.075
Min	108	0.165	− 0.949	− 0.033	0.046	0.045

(continued)

Table 2 (continued)

Cycle	Number of intersection	dH_{max} (m)	dH_{min} (m)	dH_{mean} (m)	RMS_{SSH} (m)	STD_{SSH} (m)
Mean	157.5	0.316	– 0.311	– 0.009	0.060	0.059

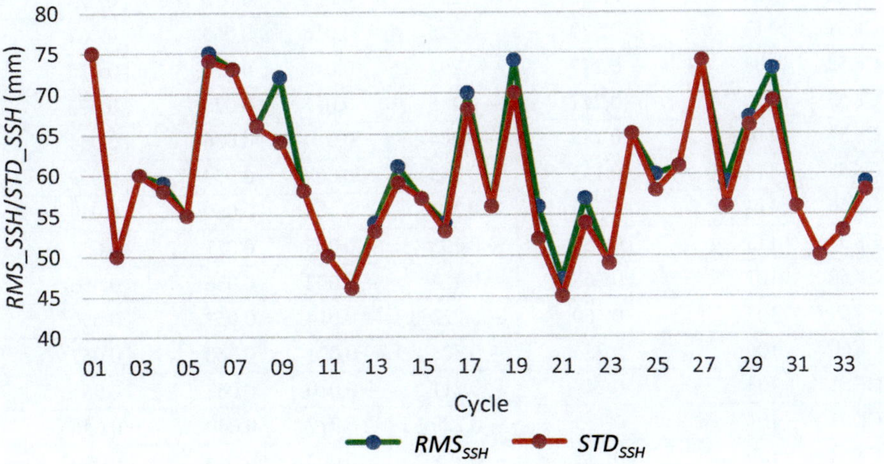

Fig. 4 RMS_{SSH} and STD_{SSH} of SARAL/AltiKa satellite data for 34 cycles

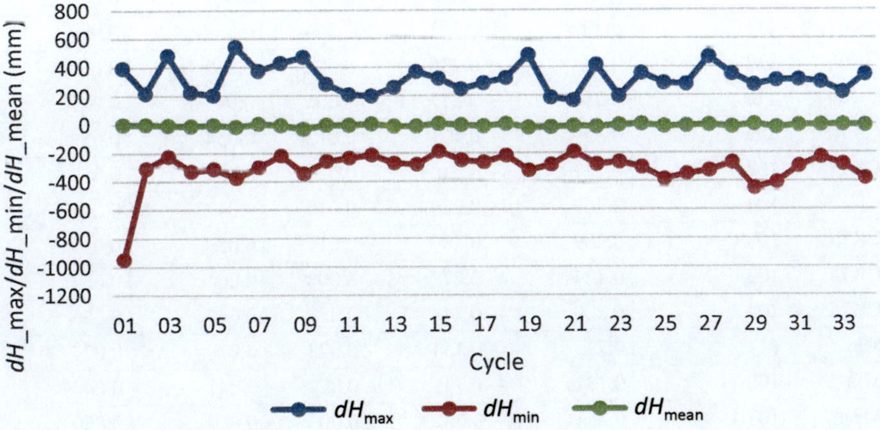

Fig. 5 dH_{max}, dH_{min}, and dH_{mean} of SARAL/AltiKa satellite data for 34 cycles

Table 3 Results of evaluating the precision of Sentinel-3A satellite altimeter data

Cycle	Number of intersection	dH_{max} (m)	dH_{min} (m)	dH_{mean} (m)	RMS_{SSH} (m)	STD_{SSH} (m)
CK48	110	0.185	− 0.315	− 0.016	0.057	0.055
CK49	111	0.266	− 0.280	0.007	0.061	0.061
CK50	108	0.725	− 0.217	0.023	0.078	0.075
CK51	113	0.379	− 0.225	0.016	0.069	0.067
CK52	108	0.247	− 0.292	0.013	0.071	0.070
CK53	110	0.247	− 0.735	0.011	0.078	0.077
CK54	114	0.185	− 0.264	0.014	0.056	0.054
CK55	114	0.193	− 0.255	0.003	0.059	0.059
CK56	114	0.372	− 0.324	0.008	0.060	0.059
CK57	113	0.259	− 0.195	0.005	0.051	0.051
CK58	107	0.268	− 0.154	− 0.002	0.056	0.056
CK59	102	0.160	− 0.223	− 0.014	0.052	0.050
CK60	106	0.256	− 0.306	− 0.021	0.060	0.057
CK61	103	0.248	− 0.211	− 0.010	0.062	0.061
CK62	101	0.135	− 0.220	− 0.017	0.049	0.046
CK63	100	0.215	− 0.390	0.002	0.059	0.059
CK64	107	0.271	− 0.302	0.004	0.070	0.070
CK65	110	0.178	− 0.387	0.012	0.062	0.061
CK66	108	0.288	− 0.477	0.022	0.067	0.064
CK67	111	0.201	− 0.193	0.004	0.059	0.059
CK68	112	0.538	− 0.236	0.005	0.062	0.062
CK69	110	0.214	− 0.237	0.006	0.049	0.049
CK70	109	0.337	− 0.171	0.012	0.054	0.052
CK71	108	0.131	− 0.316	− 0.021	0.057	0.053
CK72	104	0.203	− 0.236	− 0.016	0.047	0.044
CK73	108	0.269	− 0.209	− 0.017	0.058	0.055
CK74	109	0.424	− 0.272	− 0.008	0.064	0.063
CK75	104	0.200	− 0.244	− 0.022	0.060	0.056
Max	114	0.725	− 0.154	0.023	0.078	0.077
Min	100	0.131	− 0.735	− 0.022	0.047	0.044
Mean	108.4	0.271	− 0.282	0.000	0.060	0.059

5 Conclusion

This paper has presented the evaluation of the precision of SARAL/AltiKa and Sentinel-3A satellite altimeter data over the Vietnam Sea and surroundings based on height differences at the intersection points. The evaluation was conducted with

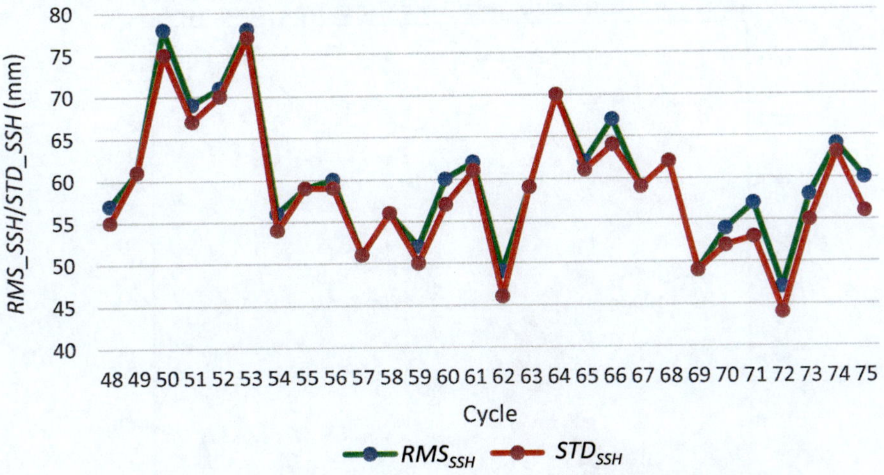

Fig. 6 RMS_{SSH} and STD_{SSH} values of Sentinel-3A satellite data for 28 cycles

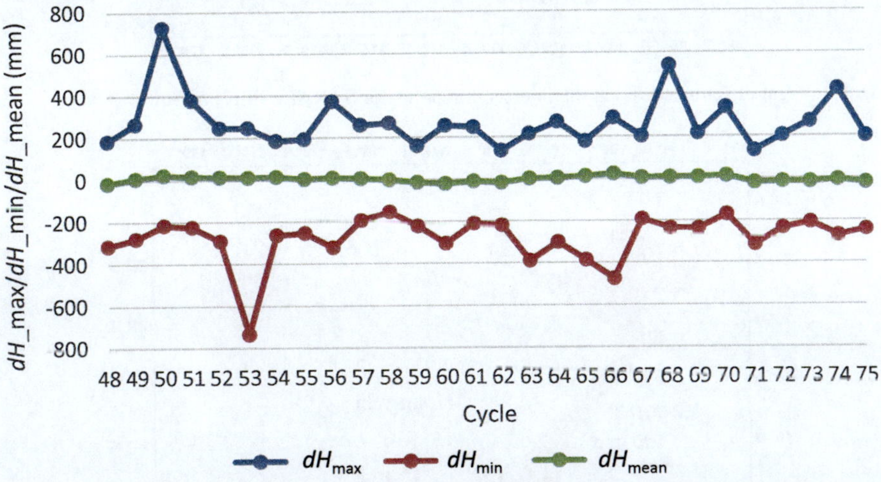

Fig. 7 dH_{max}, dH_{min}, and dH_{mean} values of Sentinel-3A satellite data for 28 cycles

34 cycles of the SARAL/AltiKa satellite data (from 1st to 34th cycle), corresponding to an over 3-year period, and the Sentinel-3A satellite data at 28 cycles (from 48 to 74th cycle), corresponding to a more than 2-year time span. Both evaluations were performed with data in the ERM mode.

The standard deviations of the SARAL/AltiKa data are between ± 4.5 cm and ± 7.5 cm, with an average of ± 5.9 cm, while that of the Sentinel-3A satellite altimeter data was found between ± 4.4 and ± 7.7 cm, with a mean value of ± 5.9 cm. The mean height difference at the intersection points is small indicating a

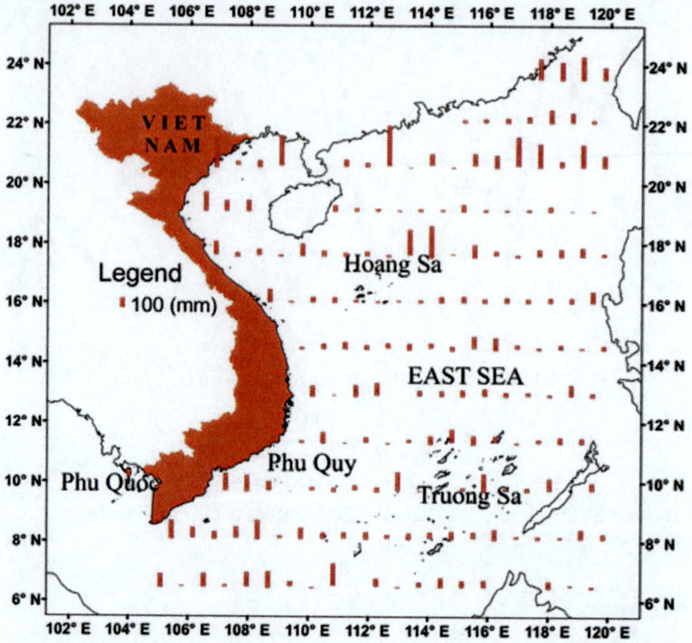

Fig. 8 Height difference (dH) at the intersection points of SARAL/AltiKa satellite

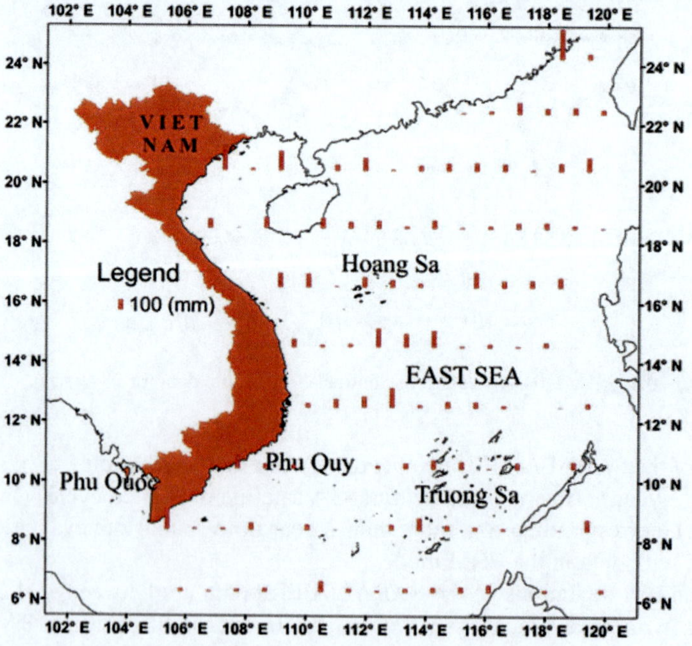

Fig. 9 Height difference (dH) at the intersection points of Sentinel-3A satellite

small systematic error in satellite altimeter data over the study area. Additionally, the height differences at intersection points are found higher in coastal areas and the areas near the archipelagos than those in offshore areas. This difference is also large with Sentinel-3A satellite data in the area between the Hoang Sa and Truong Sa Islands.

Acknowledgements This research has been supported by the project “Research on determining seafloor depth for the East Sea using gravity anomaly data”, code B2021-MDA-06 of the Vietnam Ministry of Education and Training and the project "Research to determine surface currents in the East Sea using satellite altimeter data", code 07/2021/Đ6-DATS of Vietnam’s People Naval Hydrographic and Oceanographic Department. We would also like to thank Radar Altimeter Database System for providing satellite altimeter data used in this study.

References

1. Shum, C.K., Ries, J.C., Tapley, B.D.: The accuracy and applications of satellite altimetry. Geophys. J. Int. **121**, 321–336 (1995)
2. Andersen, O.B.: Marine Gravity and Geoid From Satellite Altimetry. Springer, Geoid Determination (2013)
3. Hwang, C., Chen, S.-A.: Circulations and eddies over the South China Sea derived from TOPEX/POSEIDON altimeter data. J. Geophys. Res. Atmos. **105**, 23 (2000)
4. Nguyen, V.S.: Determination of gravity anomalies for Vietnamese waters by satellite altimeter results, vol. Ph.D.,. Moscow State University of Geodesy and Mapping, Russian Federation (2012)
5. Tran, T.D., Kulinich, R.G., Nguyen, V.S., Bui, C.Q., Nguyen, B.D., Nguyen, K.D., Tran, T.D., Tran, T.L.: Improving accuracy of altimeter-derived marine gravity anomalies for geological structure research in the Vietnam south-central continental shelf and adjacent areas. Russ. J. Pac. Geol. **13**, 11 (2019)
6. Nguyen, V.S.: Evaluation of the accuracy of the global gravity anomaly model determined from satellite altimeter over the East Sea. Min. Ind. J. **4** (2020)
7. Nguyen, V.S., Pham, V.T., Nguyen, V.L., Andersen, B.O., Forsberg, R., Bui, T.D.: Marine gravity anomaly mapping for the Gulf of Tonkin area (Vietnam) using Cryosat-2 and Saral/AltiKa satellite altimetry data. Adv. Space Res. (2020)
8. Volkov, D.L., Pujol, M.I.: Quality assessment of a satellite altimetry data product in the Nordic, Barents, and Kara seas. J. Geophys. Res. **117** (2012)
9. Eko, Y.H., Maria, J.F., Clara, L.: Assessment of altimetric range and geophysical corrections and mean sea surface models—impacts on sea level variability around the Indonesian seas. Remote Sens. **9** (2017)
10. Nguyen, V.S., Vu, V.T.: Evaluation of accuracy of the altimetry data in the sea area around the Truong Sa Archipelago. J. Min. Earth Sci. **58**, 5 (2017)
11. Nguyen, V.L., Nguyen, V.S., Tran, T.T.T., Le, T.T.T.: Determination of location and height difference at crossover points in satellite altimetry data processing using direct method. In: International Conference on Geo-spatial Technologies and Earth Resources, pp. 279–282. Publishing House for Science and Technology (2017)
12. Sánchez-Román, A., Pascual, A., Pujol, M.-I., Taburet, G., Marcos, M., Faugère, Y.: Assessment of DUACS Sentinel-3A altimetry data in the coastal band of the European seas: comparison with tide gauge measurements. Remote Sens. **3970** (2020)
13. ESA. Retrieved from https://sentinels.copernicus.eu/web/sentinel/missions/sentinel-3

14. Verron, J., Bonnefond, P., Aouf, L.: The benefits of the Ka-Band as evidenced from the SARAL/AltiKa altimetric mission: scientific applications. Remote Sens. **163** (2018)
15. Ha, M.H.: Research and assessment of sea level standards according to modern geodetic, hydrographic and tectonic methods for the construction of works and planning of Vietnam's coastal zone in the changing trend climate change. In: General report on results of scientific research and technological development of the project KC.09.19/11–15 (2015)
16. Rong, Z.M.: Analysis on the surface current features in the South China Sea in winter. Mar. Forecast. **B11**, 5 (1994)
17. Xu, X.Z., Qiu, Z., Chen, H.C.: The general descriptions of the horizontal circulation in the South China Sea. In: In Proceedings of the 1980 Symposium on Hydrometeorology of the Chinese Society of Oceanology and Limnology, pp. 137–145. Science Press (1982)
18. Le, D.T.: The Marine management. Hanoi National University of Vietnam (2005)
19. Radar Altimeter Database System. Retrieved from http://rads.tudelft.nl/rads/rads.shtml
20. Nguyen, V.S.: Adjustment of ENVISAT altimetry data crossover for water area adjacent to Vietnam. Geodesy Aerophotogr. **5** (2012)
21. Nguyen, V.S.: Determination crossover point location by simulating quadratic-equation in processing satellite altimetry data. J. Min. Earth Sci. **41**, 5 (2013)

Detection of GNSS-TEC Noise Related to the Tonga Volcanic Eruption Using Optimization Machine Learning Techniques and Integrated Data

Nhung Le, Benjamin Männel, Luyen K. Bui, Mihaela Jarema, Thai Chinh Nguyen, and Harald Schuh

Abstract Total Electron Content (TEC) is the integral of the electron density along the path between receivers and satellites. TEC measured from Global Navigation Satellite Systems (GNSS) data is valuable to monitor space weather and correct ionospheric models. TEC noise detection is also an essential channel to forecast space weather and research the relationship between the atmosphere and natural phenomena like geomagnetic storms, earthquakes, volcanos, and tsunamis. In this study, we apply optimization machine learning techniques and integrated GNSS and solar activity data to determine GNSS-TEC noise at the International GNSS Service (IGS) stations in the Tonga volcanic region. We investigate 38 indices related to the geomagnetic field and solar wind plasma to select the essential parameters for forecast models. The findings show the best-suited parameters to predict vertical TEC time series: plasma temperature (or Plasma speed), proton density, Lyman alpha, R sunspot, Ap index (or Kp, Dst), and F10.7 index. Applying the Ensemble algorithm to build the TEC forecast models at the investigated IGS stations gets the accuracy from 1.01 to 3.17 TECU. The study also shows that machine learning combined

N. Le (✉) · B. Männel · T. C. Nguyen · H. Schuh
GFZ German Research Centre for Geosciences, Potsdam, Germany
e-mail: nhung@gfz-potsdam.de

T. C. Nguyen
e-mail: nguyenthaichinh@humg.edu.vn

N. Le · H. Schuh
Technische Universität Berlin, Berlin, Germany

N. Le
Hanoi University of Natural Resources and Environment, Hanoi, Vietnam

L. K. Bui
National Center for Airborne Laser Mapping, University of Houston, Houston, USA
e-mail: buikhacluyen@humg.edu.vn

L. K. Bui · T. C. Nguyen
Hanoi University of Mining and Geology, Hanoi, Vietnam

M. Jarema
MathWorks, Munchen, Germany

L. Q. Nguyen et al. (eds.), *Advances in Geospatial Technology in Mining and Earth Sciences*, Environmental Science and Engineering,
https://doi.org/10.1007/978-3-031-20463-0_9

with integrated data can provide a robust approach to detecting TEC noise caused by seismic activities.

Keywords Machine learning · GNSS-TEC forecast · GNSS · Solar activity · Tonga volcanic eruption

1 Introduction

Continuous Global Navigation Satellite Systems (GNSS) data can be used for various applications. Satellite signal propagation in space depends uniquely on electron density in the ionosphere [1–3]. Thus, the estimation of Total Electron Contents (TEC) in the ionosphere can provide valuable information to correct errors in GNSS positioning. Furthermore, solar activity is the main factor causing fluctuations in the Earth's electrical and magnetic fields [4]. The hot plasma makes energetic charged particles in space escape from the Sun's gravity and interacts with the Earth's magnetic field. The interaction between the solar wind plasma and Earth's magnetic field leads to some natural phenomena in the atmosphere like auroras, geomagnetic storms, and ionospheric anomalies [5–8]. Monitoring TEC disturbances thereby reflects the solar activity and is an important channel to forecast space weather.

Some other factors also result in ionospheric anomalies in the short term, for example, nuclear explosions [9–12] and rocket launching [13, 14]. Thanks to an increasing number of continuous GNSS stations, the research stream that has been attractive to scientists for almost two decades is TEC anomalies associated with seismic activities [15–19]. Determination of the TEC disturbances related to seismic events applies different methods and monitoring instruments. The French low orbit satellite DEMETER[1] was launched in 2004 to investigate ionospheric disturbances related to earthquakes and volcanos [20]. The multi-purpose GNSS networks like the GEONET in Japan and the SEALION in Southeast Asia have been attached to ionosondes, scintillation monitors, and magnetometers to observe the effect of seismic events on the atmosphere [21]. Some studies revealed the signs of earthquake precursors linked to ionospheric perturbation [22–25]. The Global Ionospheric Maps (GIM) are also a valuable data source for detecting TEC anomalies caused by these seismic activities [26].

So far, there have been two approaches to studying ionospheric fluctuations related to seismic events. The first one is based on the physical mechanism of the seismic wave generation into the atmosphere [27]. The other relies on analyses of statistics and the probability of TEC anomalies in the epicenter regions and time of earthquake occurrences (mainshocks) [28]. However, there is no absolutely certain guarantee about the coincidence between observed ionospheric anomalies in the location and time of the earthquakes with other non-seismic activities. Monitoring TEC noise during periods of low solar activity to study the effect of seismic activities is a proper solution [19]; thus, many remarkable earthquakes resulting in TEC anomalies are

[1] https://directory.eoportal.org/web/eoportal.

skipped in investigations. Besides, there is also no common standard to measure ionospheric noise levels. Hence, applying Machine Learning (ML) and integrated data to distinguish TEC noise sources and extract TEC noise caused by seismic events will be carried out in this study.

ML has been a current trend applied to multidisciplinary research fields, especially for space weather forecasts and hazard warnings. The solar wind plasma and geomagnetic data have been used to predict TEC models in a few studies in the literature [29–31]. For example, Claudio Cesaroni et al. [32] used neural networks to predict global TEC at a daily sampling rate with the forecast accuracy of approximately 3 to 5 TECU. Xu Lin et al. [33] implemented the networks of convLSTM (convolutional Long Short-Term Memory) and PredRNN (Predictive Recurrent Neural Network) to correct errors of the delays. However, these criteria have not yet met the requirements of TEC anomaly detection related to seismic activity. Since TEC noise caused by seismic events often remains within a few minutes to a few hours and forecast accuracy of under 3 TECU can overcome TEC disturbances on the equator area or TEC variations at a low active time [34]. Other literary studies used the solar indices in their forecast models, such as Ap and F10.7, to correct ionospheric delays [35] or A.E. and SYM/H indices for TEC nowcasting [36]. Nevertheless, there has been no consistency in the selected indices among the studies.

Therefore, this study combines the optimization ML techniques with statistical hypothesis tests to determine suitable parameters related to the solar and geomagnetic activities for the TEC forecast models. These ML models will be the basis for separate TEC noise sources. Hence, we use the trained ML models to extract vertical TEC (VTEC) noise related to the Tonga volcanic eruption on 15 January 2022. Finally, we apply statistical and spectral techniques to analyze GNSS-VTEC noise at the International GNSS Service (IGS) stations nearby Tonga.

2 Study Area, Data, and Methodology

2.1 Study Area

The Tonga-Hunga Ha'apai includes small islands along the caldera rim in the Western-South Pacific Ocean. The Tonga volcano has experienced seven Holocene eruptive periods, with the first recorded eruption ~900 years ago [37]. For the latest period, it woke up by 20 December 2021 and ended after a massive explosion with a height of ~30 km at 04:10 UTC on 15 January 2022 [38]. The Tonga volcanic eruption triggered a tsunami with the waves observed thousands of miles away from the Caribbean and Alaska. After four minutes, a shallow earthquake of 5.7 M_w occurred near the epicenter at 20.536°S and 175.382°W [39]. It is considered one of the few volcanoes tracked in detail and with different methods and technologies.

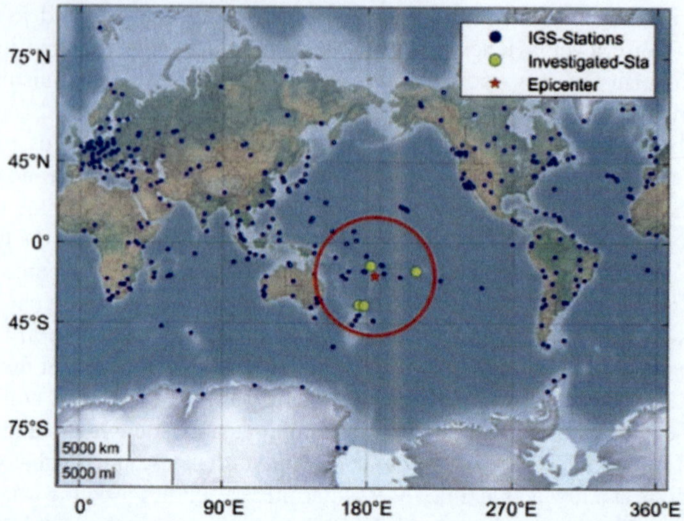

Fig. 1 Investigated TEC anomalies related to the 2022 January Tonga volcano at the IGS stations

2.2 Data

We use the GNSS data from four IGS stations surrounding the epicenter of the 2022 January Tonga volcano to study the effects of these seismic events on the ionosphere (Fig. 1). These selected IGS stations must ensure conditions like being located within the radius of perception, the equivalent accuracy, and continuously monitored data. The stations FTNA and THTI are located in French Polynesia, and AUCK and WARK are in New Zealand. The GNSS data are available at the data center of the Crustal Dynamics Data Information System (CDDIS).[2] GNSS observations are the major initial data to compute the TEC time series for building forecast models.

Thirty-eight solar wind plasma and geomagnetic field parameters are analyzed to determine the best-suited predictors for the ML models (Table 4). These data are taken from the world data bank: the space weather prediction center NOAA,[3] USA; the world data center for Geomagnetism,[4] Kyoto, Japan; and the space weather live,[5] Belgium. In addition, the seismic data are collected from the data center GEOFON,[6] GFZ Potsdam, Germany and the U.S. geological survey center USGS.[7] Figure 2 shows eight main parameters of the geomagnetic field and solar wind. The level of solar activity is usually defined via indices such as the sunspot number and the

[2] https://cddis.nasa.gov/.

[3] https://www.swpc.noaa.gov/.

[4] http://wdc.kugi.kyoto-u.ac.jp/.

[5] https://www.spaceweatherlive.com/.

[6] https://geofon.gfz-potsdam.de/.

[7] https://www.usgs.gov/.

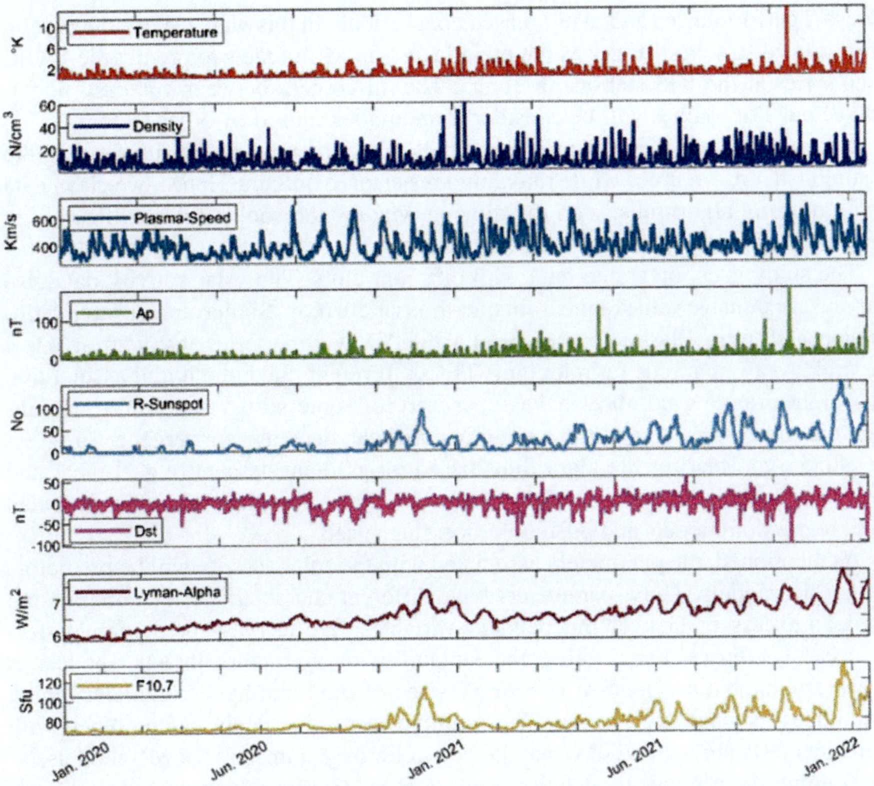

Fig. 2 Eight specific features in 38 investigated parameters of the geomagnetic field and solar wind (plasma temperature, plasma density, Plasma speed, Ap index, R sunspot, Dst index, Lyman alpha, and F10.7 index)

solar radio flux at 10.7 cm (F10.7 index). Indices of global geomagnetic activity like Kp, Ap, Cp, and C9 are provided by the German Research Center for Geosciences (GFZ). Dst values of the disturbance storm time index are obtained from the world data center for Geomagnetism in Kyoto, Japan.

2.3 Methodology

Regression analysis is a mathematical method that describes the relationship between one or many independent variables and the dependent variable. In machine learning, the independent variables are factors to predict the dependent variable. Therefore, changes in the independent variables will result in changes in the dependent variable. Usually, only main factors should be included in the regression models to optimize forecast performances and avoid biased conclusions. In this way, we use the parameters related to solar activity as the main factors in ML models to predict the VTEC time series at the IGS stations in Tonga. The differences between forecast models and actual TEC values will be considered anomalies caused by other factors.

ML models used for ionospheric anomaly detection must be sensitive enough to distinguish noise sources while remaining resistant to outliers. Hence, we clean data using filtering algorithms, with the moving window thresholds selected flexibly in the sampling rates and data characteristics.

The study uses integrated data with different units. The solar activity data also vary in an extensive value range from one-thousandth (e.g., Sigma alpha/proton ratio) to thousands (e.g., Plasma temperature), while VTEC time series change from a few (at midday) to hundreds (at midnight) TECU. It might make the initial assumption that higher ranging numbers have superiority of some sort. Furthermore, significant differences in value range among features can decrease convergence progress or saturate too fast for the algorithms based on gradient descent (e.g., Linear and Gaussian) and distance (e.g., SVM). Therefore, these input data should be re-scaled to fit regression models and push up processing speed.

As mentioned, the parameters associated with the solar activity will be predictors in the ML models. These parameters have different characteristics. The regression model's excess or lack of independent variables can decrease prediction performances. Therefore, determining the suitability of predictors in ML models is necessary, known as *"feature selection"*—one of the main hyperparameter tuning techniques in machine learning. The multiple regression analyses combined with statistical tests are applied to select the best-relevant parameters for ML models.

Training the forecast models is based on four ML algorithms, including Linear Regression, Support Vector Machine (SVM), Tree Ensemble, and Regression Trees using the ML toolbox in MATLAB® to select the optimal models. VTEC disturbances caused by seismic activities can last from some minutes to a few hours [34]. To capture the most negligible variations in the VTEC time series, we investigate two cases: (1) using hourly time series of two-year data and (2) using one-minute time series of the 15-day data to predict one day. We extract VTEC noise based on the trained ML models and analyze its physical characteristics by the spectrum method [40]. To this end, we apply Welch's algorithm to estimate the power spectral density [41] and the continuous wavelet transform (CWT) method to compute the spectral magnitude in GNSS-VTEC noise [42, 43].

Figure 3 shows the methodology mentioned above with three main steps. The first one is pre-processing with cleaning raw data, testing characteristics, and re-scaling data. The second step is feature selection using two statistical tests, analysis of variance (ANOVA), and Fisher test. The final step includes preparing input data, splitting data, training forecast models, optimization processing to get the highest performance models, and extracting and analyzing GNSS-TEC noise at the investigated IGS stations.

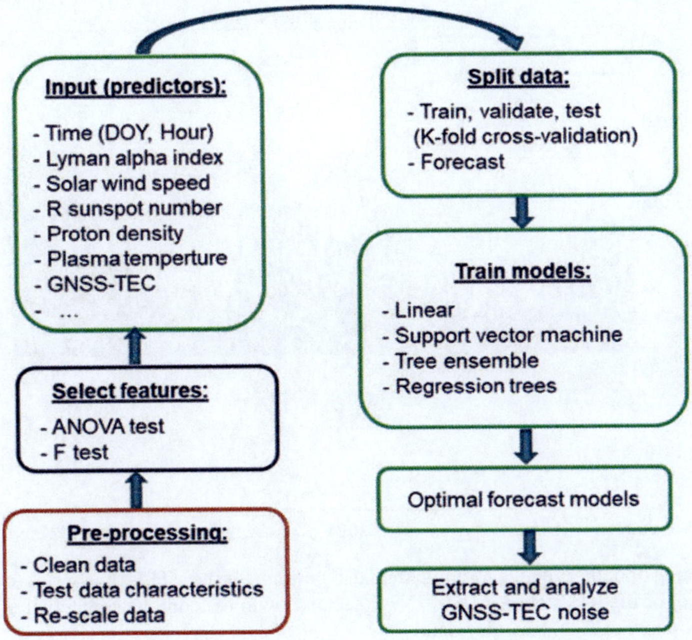

Fig. 3 Flowchart of GNSS-TEC noise detection based on ML techniques and the integrated data of GNSS and solar activity

3 Results and Discussions

3.1 Data Pre-Processing

The Moving Median filters outliers at the same thresholds and sliding window size (Fig. 4). The Augmented Dickey-Fuller (ADF) tests the stationarity of the VTEC time series at the IGS stations, and the details are shown in Table 1. The absolute values of the ADF test (in bold italics) are larger than the critical values t_{critical} for all statistical t-test levels (1, 5 and 10%) at the significant statistics (see Table 1). Hence, the VTEC time series appear to be stationary, and they can be used to train the forecast models to detect ionospheric anomalies.

3.2 Feature Selection

We apply statistical tests and analysis techniques to select the best-suited features from 42 input parameters, in which 38 parameters of the solar wind plasma and geomagnetic field, three parameters of time (Hour, Day of Year, and Year) and one lagged VTEC time series. F-test is used to measure the feature importance via the

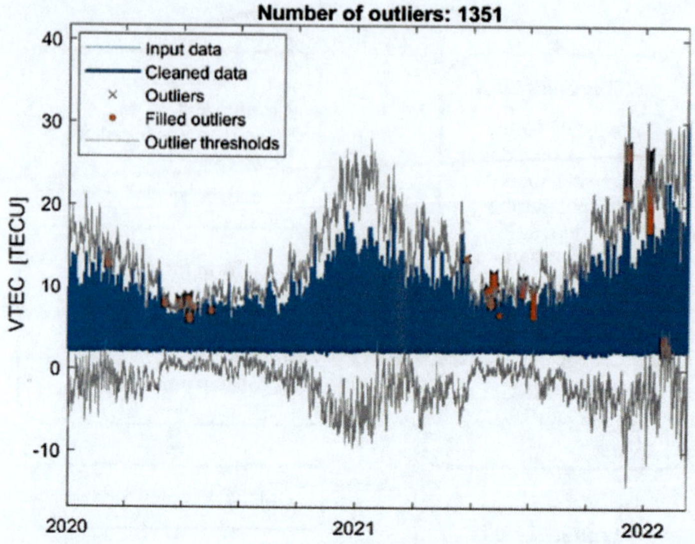

Fig. 4 Filtering of outliers by the Moving Median algorithm for the VTEC time series (one-minute sampling rate) of the IGS station AUCK (New Zealand) with the confidence interval of 99.7% at the sliding window of 2880 (two-day data)

Table 1 Test the stationarity of the VTEC time series at the IGS stations AUCK, FTNA, THTI, and WARK using the Augmented Dickey-Fuller algorithm

Test levels (%)	AUCK		FTNA		THTI		WARK	
	t	p-value	t	p-value	t	p-value	t	p-value
	– 5.03	1.80E-05	***– 5.63***	8.95E-07	***– 5.99***	1.32E-07	***– 4.73***	7.21E-05
1	– 3.43		– 3.43		– 3.43		– 3.43	
5	– 2.86		– 2.86		– 2.86		– 2.86	
10	– 2.57		– 2.57		– 2.57		– 2.57	

score ranking. Figure 5 shows the classification results of the univariate features at the IGS stations, with the nine highest-score features displayed on the horizontal axes of the graphs. The scores at the high-latitude stations (WARK and AUCK) reach up to 449.58 and 479.27, while those (THTI and FTNA) are only 217.87 and 268.40, respectively. This finding indicates that solar activity has a more significant impact on TEC observed at high-latitude stations. In addition, the effect of the variables like Kp, Ap, and Lyman alpha remains at the highest level for all the IGS stations, in which Ap and Kp are of identical scores because of their correlation. The detailed results of the feature importance score are presented in Table 4. Regression model predictors should be independent variables to improve processing speed and avoid biased conclusions. Hence, we employ analysis of variances (via ANOVA tests) to detect multicollinearity in the variables under consideration.

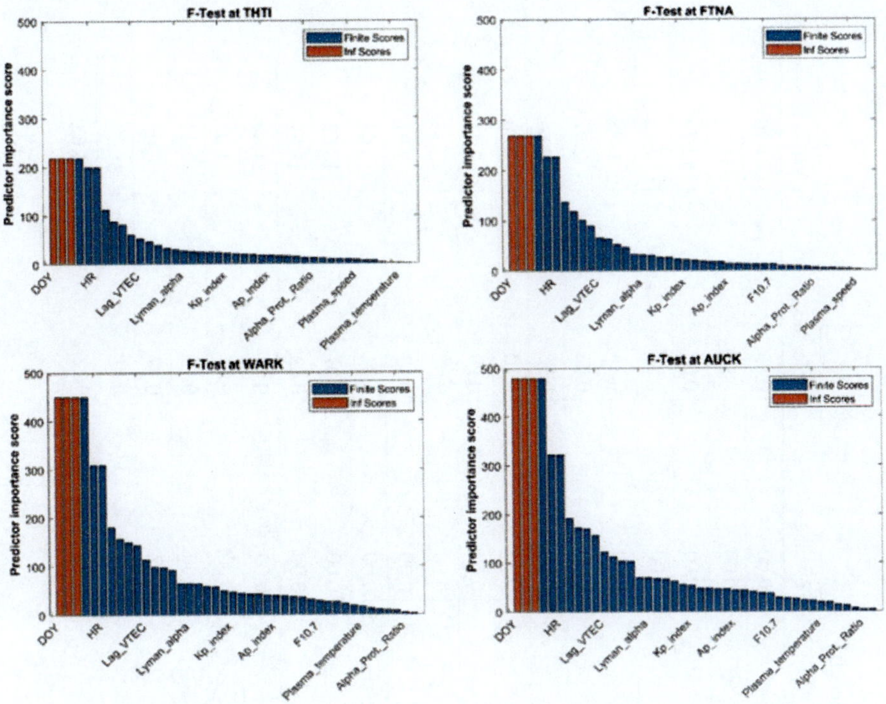

Fig. 5 Univariate feature ranking for regression models using F-test at four IGS stations AUCK, FTNA, THTI, and WARK. The horizontal axis indicates the specific score features

Table 2 lists 20 features with significant statistics (p-values < 0.05) using the ANOVA tests. The coefficients (Beta) of the t-test show the correlations among the independent variables in the regression models. Tolerance and VIF indicate information of multicollinearity.

Four features have a high potential for multicollinearity, including plasma speed, flow pressure, IMF magnitude average, and Magnitude IMF vector. Besides, the features with tolerance in the interval from 0.1 to 0.2 should also be checked cross-correlation to enhance forecast performance. The correlation matrix heatmap reveals the pairs of the linear relevant features, such as Proton quasy invariant and plasma temperature, Dst (or Kp) and Ap index (Fig. 6). The detailed information on the correlation matrix of 20 significant features is shown in Table 5. The redundant variables (e.g., IMF magnitude vector, IMF magnitude average, plasma flow pressure) should be rejected before training forecast models.

Overall, there are two outstanding advantages of feature selection based on statistical tests. The first one is a clear classification of the importance of each feature in a regression model. This helps analysts decide which parameters should be used to train forecast models. The second advantage is to detect relevant features without training test models (trial steps) as other feature selection methods (e.g., K nearest neighbor,

Table 2 ANOVA test of significant features at the IGS station WARK (New Zealand)

Features	Coefficients (Beta)	t-test	p-values	Tolerance	VIF
Lag_VTEC	0.86	212.73	0.00	0.83	1.20
HR	− 0.05	− 11.98	0.00	0.88	1.13
Plasma_speed	− 0.12	− 9.11	0.00	0.08	12.79
Kp_index	0.08	7.65	0.00	0.14	7.22
Plasma_temperature	0.07	7.58	0.00	0.14	6.99
YEAR	− 0.05	− 7.42	0.00	0.34	2.91
Proton_quazy_invariant	− 0.04	− 6.66	0.00	0.35	2.86
BZ_GSE	− 0.07	− 6.04	0.00	0.10	9.80
DOY	− 0.04	− 5.76	0.00	0.29	3.41
Flow_pressure	0.06	4.28	0.00	0.07	15.35
Lyman_alpha	0.04	3.94	0.00	0.13	7.91
Sigma_Np	− 0.03	− 3.90	0.00	0.33	3.07
IMF_magnitude_avg	0.15	3.78	0.00	0.01	118.63
Proton_density	− 0.04	− 3.28	0.00	0.08	13.27
Dst_index	− 0.02	− 3.27	0.00	0.53	1.89
Plasma_lat_angle	− 0.01	− 3.26	0.00	0.97	1.03
F10.7_index	0.03	2.83	0.00	0.13	7.43
Ap_index	0.02	2.30	0.02	0.17	6.05
Magnitude_IMF_vector	− 0.08	− 2.23	0.03	0.01	85.37
R_sunspot	0.02	2.12	0.03	0.22	4.63

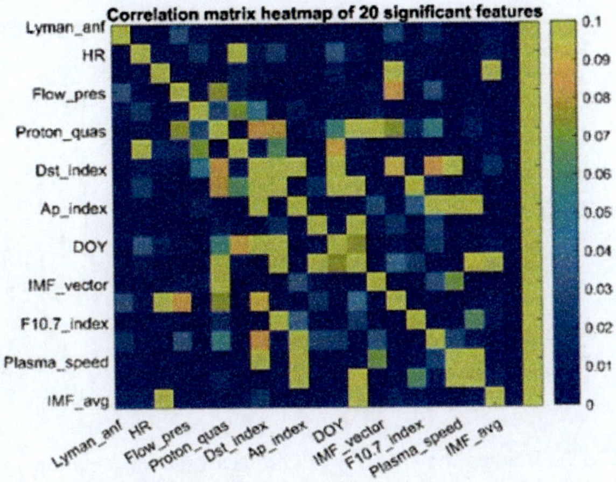

Fig. 6 Correlation matrix heatmap of 20 significant features using the ANOVA test at the IGS station WARK

Gaussian process regression, or neighborhood component analysis). Forecast performance will be significantly improved because performing trial steps via loops on all 42 features is time-consuming. In this study, statistical tests have pointed out the best-suited features to build the regression ML models: Plasma temperature (or plasma speed), proton density, Ap index (or Kp, Dst index), F10.7 index, Lyman alpha, and R sunspot. Together with the infinite variables (H.R., DOY, Lag VTEC), these six features will be used as the predictors (independent variables) in the regression ML models to forecast the GNSS-VTEC time series.

3.3 Detection and Analysis of GNSS-TEC Noise

We investigate 15 mathematical models based on four ML methods (Linear Regression, Regression Trees, SVM, and Tree Ensemble) to predict the VTEC time series for one day. Optimization processing is performed to select the best models for GNSS-VTEC noise detection. Figure 7 presents two forecast models using the one-minute and hourly time series at the THTI station. VTEC prediction using the hourly time series produces the ML models with greater generalizability and robustness to extreme values and outliers. These models will have higher reliability for anomaly detection in deformation analyses and long-term predictions. Nevertheless, based on noise characteristics caused by seismic activities, assessments on both cases (one-minute and hourly time series) should be performed.

Table 3 lists the VTEC forecast accuracy (in root mean square error, RMSE) of the ML models at four IGS stations. The accuracy of the Ensemble algorithm outperforms others. The forecast performance based on the Ensemble reaches the highest, from 1.01 TECU at the WARK station (hourly) to 3.17 TECU at the FTNA station (hourly). In contrast, the linear algorithm shows the lowest, from 1.31 TECU at the AUCK station (hourly) to 5.07 TECU at the FTNA station (one minute). The accuracy of the

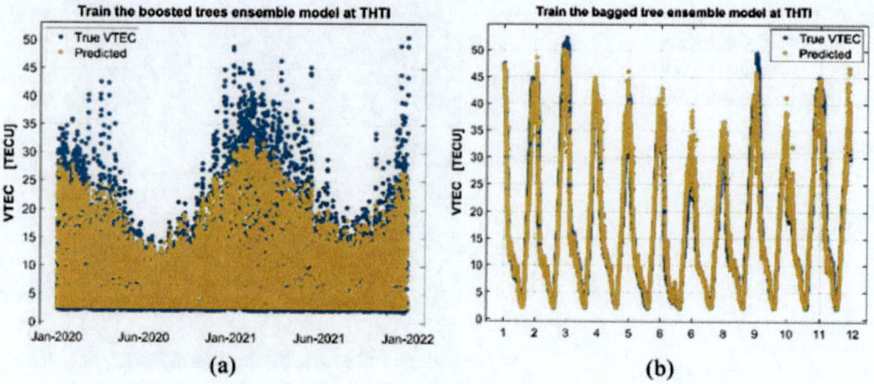

Fig. 7 Training the GNSS-VTEC forecast model at station THTI: **a** the Boosted Tree Ensemble algorithm using data from 01 January 2020 to 01 January 2021 and **b** the Bagged Tree Ensemble algorithm using data from 01st to 12th January 2022

two offshore stations (FTNA and THTI) is lower than of the inshore stations (WARK and AUCK). It is likely due to poor input data quality, which indicates RMSE of the VTEC time series of 2.86, 2.67, 1.74, and 1.71 TECU for FTNA, THTI, WARK, and AUCK, respectively.

Forecasts of VTEC at the FTNA station (French Polynesia) on 15th January 2022 are shown in Fig. 8. Given the global seismic data (GEOFON and USGS), there were no other remarkable seismic events in the 7000-km radius (from the volcano epicenter in Tonga) within three days 13th, 14th, and 15th of January. The solar activity data have been included to predict the VTEC time series. As a result, the Tonga seismic events on 15th January 2022 are believed to cause the VTEC disturbances at the investigated IGS stations.

To eliminate the effect of systematic errors, we extract the VTEC noise based on the same ML algorithm for all the IGS stations (Fig. 9). The VTEC noise at the THTI

Table 3 Accuracy of the ML models for one day forecast at four IGS stations

Methods	FTNA RMSE (TECU)	THTI RMSE (TECU)	AUCK RMSE (TECU)	WARK RMSE (TECU)	Input data	Sampling rate
Ensemble	3.17	2.86	1.02	1.01	2 years	Hourly
Coarse tree	3.81	3.49	1.22	1.21		
SVM	3.67	3.56	1.22	1.27		
Linear	3.83	3.78	1.31	1.32		
Ensemble	2.92	2.67	1.20	2.07	12 days	One minute
Coarse tree	3.14	3.55	1.53	1.33		
SVM	4.54	5.28	2.38	2.65		
Linear	5.07	5.89	2.86	2.86		

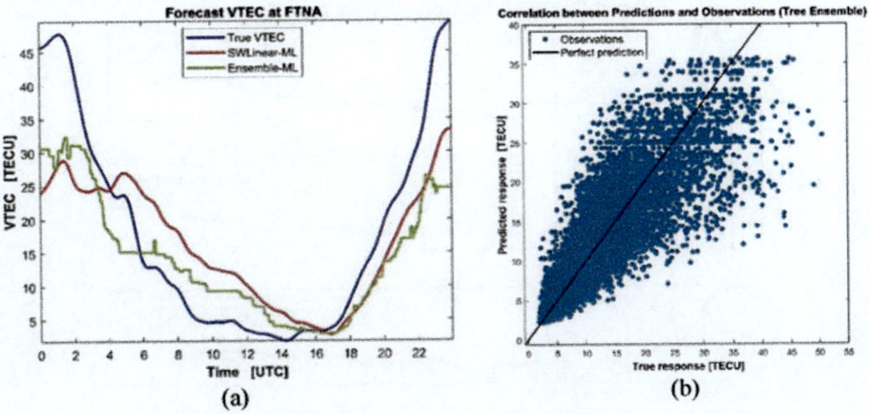

Fig. 8 Forecast of VTEC at the station FTNA on 15 January 2022 **(a)** and test the correlation between predictions and observations of the Tree Ensemble model **(b)**

station shows the highest variation, up to 34.47 TECU, followed in turn by FTNA, AUCK, and WARK with corresponding values of 25.09, 21.99, and 17.41 TECU. At the same time, the forecast accuracy (i.e., RMSE) ranges from 1.01 to 3.17 TECU. The TEC fluctuations reach up to 6.5 times the RMSE values of the ML models. This finding shows a correlation between the occurrence of the seismic and the ionosphere anomalies on 15th January. These VTEC anomalies occurred a few hours around the mainshock (at 4.10 UTC, 15 January 2022). However, no positive/negative linear relationship between the TEC fluctuation amplitudes and the distances from the IGS stations to the earthquake epicenter has been seen on the ML models in Fig. 9.

Spectral methods are used to analyze the GNSS-VTEC noise at the stations in Tonga's volcanic eruption region and assess VTEC anomalies before and after the mainshock (at 4.10 UTC, 15th January 2022).

Based on Welch's segment averaging estimation at the overlap of 50%, we determine the power spectral density (PSD), in which the power values are computed as follows:

$$y_{\mathrm{dB}} = 10 \times \log_{10}(x) \tag{1}$$

where x is the power spectral density computed by the Welch's method.

The frequencies (f_i) of the VTEC noise are converted into the normalized frequency (F_s) ranging from 0 to 1, to measure the variations of the power spectrum (Figs. 10 and 11) as follow:

$$F_s = \frac{f_i \times \pi}{1440} \tag{2}$$

where f_i is the frequency of the VTEC noise at the IGS stations, and 1440 is the sample number in the VTEC time series.

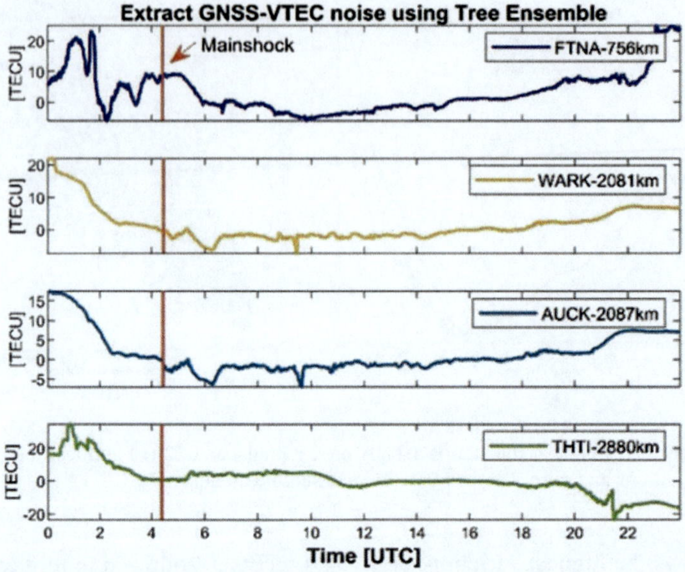

Fig. 9 Extraction of the GNSS-VTEC noise using the Tree Ensemble models at four stations FTNA, WARK, AUCK, and THTI on 15 January 2022

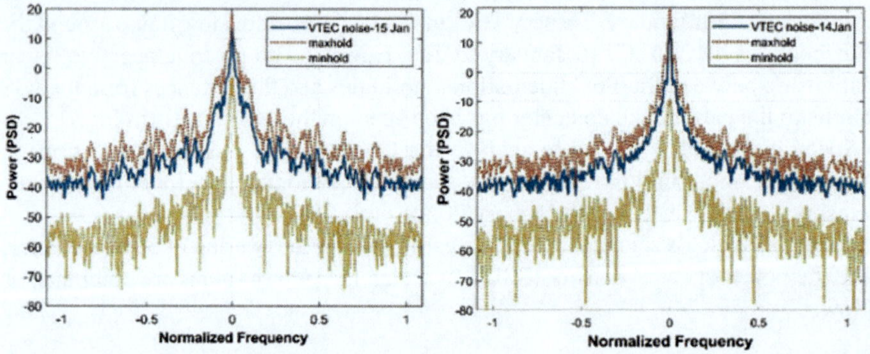

Fig. 10 Applying the Welch algorithm to estimate the power spectral density of the GNSS-VTEC noise at the station FTNA on 14 and 15 January 2022

At the same normalized frequency, the PSD pattern of VTEC noise on the 15th is rougher than on the 14th of January (Fig. 10).

The fluctuations of PSD at the FTNA station on 15 January 2022 was more significant than others, ranging from − 85.18 to 31.44 (15th January 2022); − 77.16 to 30.00 (14th January 2022); and from − 78.97 to 30.96 (13th January 2022), respectively (Fig. 11). The pattern of the GNSS-VTEC noise variations on the volcanic eruption is denser and more significant compared to other days.

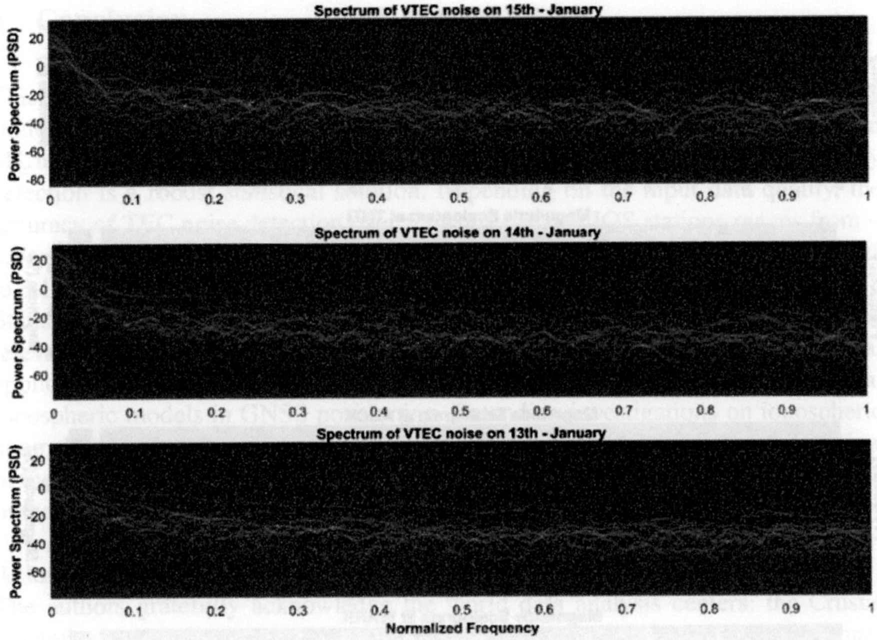

Fig. 11 Power spectrum density of VTEC at the FTNA station on 15th (top), 14th (middle), and 13 (bottom) January 2022

The continuous wavelet transformation (CWT) on the same sampling frequency band $\left(F_s = \pi / 1440\right)$ is applied to compute the spectral magnitude of the GNSS-VTEC noise at the stations over 24 h. Figure 12 describes the magnitude scalograms of the VTEC noise at four IGS stations on 15th January, in which scalogram is the CWT absolute value. The spectrum magnitude at the FTNA station reaches the highest level (5%), followed by THTI, AUCK, and WARK (spectrum magnitude ranging from 2 to 3%). Besides, the scalogram maps also present earlier and more significant fluctuations for the offshore IGS stations (FTNA and THTI in French Polynesia) compared to the inshore stations (AUCK and WARK in New Zealand). This phenomenon may be the wave consonance of the earthquake and tsunami following the volcanic eruption in the ocean. However, more research is needed to gain a complete picture of the cause-effect relationship between time, space, and noise levels in seismic areas.

Table 4 F-test of the feature importance to select the best-fitted predictors for the regression ML models corresponding to four IGS stations AUCK, FTNA, THTI, and WARK

Features (Predictors)	AUCK	FTNA	THTI	WARK
	Scores	Scores	Scores	Scores
DOY (Day of year)	*Inf*	*Inf*	*Inf*	*Inf*
H.R. (Hour)	*Inf*	*Inf*	*Inf*	*Inf*
Lag_VTEC	*Inf*	*Inf*	*Inf*	*Inf*
Lyman_alpha	479.27	268.40	217.87	449.58
Kp_index	321.47	227.27	199.24	309.34
Ap_index	321.47	227.27	199.24	309.34
F10.7_index	192.34	136.94	52.56	180.41
Plasma_temperature	173.27	63.10	80.56	156.56
Alpha/Proton Density Ratio	170.07	65.89	59.91	98.17
Plasma_speed	156.17	101.20	87.59	142.80
R_sunspot	122.22	89.63	24.98	113.19
Sigma-Alpha/Proton_ratio	112.66	65.89	59.91	98.17
IMF_magnitude_avg	103.27	26.44	26.96	96.15
Sigma_T	102.00	32.50	38.01	91.54
Sigma_IMF_vector	69.08	17.10	17.37	63.60
Magnitude_IMF_vector	68.92	18.21	17.17	63.07
Flow_pressure	67.52	22.68	14.91	63.83
Sigma_V	66.09	20.90	23.20	57.63
RMS_BZ_GSE	61.25	12.31	10.81	56.62
Plasma_beta	54.64	30.49	19.17	48.88
Magnetosonic_mach_num	53.18	31.62	22.57	46.36
RMS_BY_GSE	46.98	12.72	15.71	43.73
Sigma_flow_latitude	46.93	6.03	6.46	41.63
Proton_quasy_invariant	45.98	13.04	13.97	42.10
Dst_index	45.42	52.58	17.37	63.60
Elecrtric_field	43.30	26.85	31.70	39.02
Alfven_mach_num	42.33	12.19	11.78	38.27
RMS_BX_GSE	39.55	7.71	11.50	35.00
Sigma_flow_longitude	37.27	8.10	7.14	31.32
Plasma_flow_latitude	36.72	17.13	28.88	39.77
BY_GSE	28.07	2.61	9.29	27.89
Bx_GSE/GSM	26.79	12.99	24.15	25.42
BY_GSM	25.46	2.78	9.27	25.11
YEAR	22.07	47.07	19.02	10.06

(continued)

Table 4 (continued)

Features (Predictors)	AUCK	FTNA	THTI	WARK
	Scores	Scores	Scores	Scores
BZ_GSM	21.95	12.08	18.38	21.92
RMS_magnitude	18.38	4.97	0.89	16.45
BZ_GSE	18.06	8.62	2.40	18.14
Plasma_flow_longitude	13.03	0.43	1.24	8.58
Proton_density	12.23	20.16	9.04	12.57
Long_Avg_IMF	6.47	4.59	1.37	4.16
Sigma_Np	4.05	9.18	6.38	2.55
Lat_Avg_IMF	3.64	1.68	21.60	8.11

Table 5 The correlation matrix of 20 significant features at station WARK

Features	Feature codes	1	2	3	4	5	6	7	8	9	10	11	12	13	14	15	16	17	18	19	20
Lyman_alpha	1	1.000	-0.022	-0.004	0.031	0.010	-0.070	-0.050	-0.132	-0.680	-0.028	-0.005	-0.643	-0.097	-0.005	0.030	-0.492	0.018	0.001	-0.021	0.009
HR	2	-0.022	1.000	0.004	0.003	0.022	-0.004	0.339	0.011	0.024	-0.007	-0.003	0.033	0.008	-0.010	0.006	-0.032	0.002	-0.011	-0.006	0.010
Lat_Avg_IMF	3	-0.004	0.004	1.000	-0.023	-0.016	-0.081	0.014	0.006	-0.012	-0.126	-0.002	0.012	-0.012	-0.080	0.115	0.009	-0.587	-0.001	-0.013	0.097
Flow_pressure	4	0.031	0.003	-0.023	1.000	0.006	0.074	0.013	-0.023	-0.030	-0.203	-0.051	0.005	-0.267	0.007	0.086	-0.071	0.032	-0.115	-0.797	-0.117
Plasma_lat_angle	5	0.010	0.022	-0.016	0.006	1.000	0.028	0.071	0.048	-0.037	0.013	-0.009	-0.038	-0.027	0.010	-0.008	0.002	-0.015	-0.002	0.004	-0.018
Proton_quazy_invariant	6	-0.070	-0.004	-0.081	0.074	0.028	1.000	-0.015	0.082	0.080	-0.055	0.010	0.056	0.281	0.120	0.073	0.017	0.055	-0.318	0.013	-0.173
Lag_VTEC	7	-0.050	0.339	0.014	0.013	0.071	-0.015	1.000	0.009	0.063	-0.015	-0.001	0.087	0.009	-0.001	-0.016	-0.097	-0.003	-0.034	-0.020	0.005
Dst_index	8	-0.132	0.011	0.006	-0.023	0.048	0.082	0.009	1.000	0.103	0.127	-0.031	0.194	-0.019	-0.021	0.090	-0.002	0.084	0.107	-0.067	0.020
YEAR	9	-0.680	0.024	-0.012	-0.030	-0.037	0.080	0.063	0.103	1.000	-0.010	0.019	0.672	0.101	-0.008	0.003	0.201	0.009	-0.014	0.005	-0.010
Ap_index	10	-0.028	-0.007	-0.126	-0.203	0.013	-0.055	-0.015	0.127	-0.010	1.000	-0.066	-0.005	-0.054	0.031	-0.733	0.036	0.102	0.128	0.211	-0.032
Sigma_Np	11	-0.005	-0.003	-0.002	-0.051	-0.009	0.010	-0.001	-0.031	0.019	-0.066	1.000	0.005	0.235	0.008	0.022	-0.002	0.028	-0.022	-0.157	-0.017
DOY	12	-0.643	0.033	0.012	0.005	-0.038	0.056	0.087	0.194	0.672	-0.005	0.005	1.000	0.074	-0.009	0.003	-0.026	0.025	-0.014	-0.045	0.006
Plasma_temperature	13	-0.097	0.008	-0.012	-0.267	-0.027	0.281	0.009	-0.019	0.101	-0.034	0.235	0.074	1.000	-0.095	0.041	0.026	-0.037	-0.597	0.094	0.167
Magnitude_IMF_vector	14	-0.005	-0.010	-0.080	0.007	0.010	0.120	-0.001	-0.021	-0.008	0.031	0.008	-0.009	-0.095	1.000	-0.042	0.000	0.031	0.069	-0.008	-0.960
Kp_index	15	0.030	0.006	0.115	0.086	-0.008	0.073	-0.016	0.090	0.003	-0.733	0.022	0.003	0.041	-0.042	1.000	-0.009	-0.120	-0.206	-0.156	0.009
F10.7_index	16	-0.492	-0.032	0.009	-0.071	0.002	0.017	-0.097	-0.002	0.201	0.036	-0.002	-0.026	0.026	0.000	-0.009	1.000	0.003	0.004	0.063	0.010
BZ_GSE	17	0.018	0.002	-0.587	0.032	-0.015	0.055	-0.003	0.084	0.009	0.102	0.028	0.025	-0.037	0.031	-0.120	0.003	1.000	0.028	0.010	-0.053
Plasma_speed	18	0.001	-0.011	-0.001	-0.115	-0.002	-0.318	-0.034	0.107	-0.014	0.128	-0.022	-0.014	-0.597	0.069	-0.206	0.004	0.028	1.000	0.325	-0.161
Proton_density	19	-0.021	-0.006	-0.013	-0.797	0.004	0.013	-0.020	-0.007	0.005	0.211	0.157	-0.045	0.094	-0.008	-0.156	0.063	0.010	0.325	1.000	0.009
IMF_magnitude_avg	20	0.009	0.010	0.097	-0.117	-0.018	-0.173	0.005	0.020	-0.010	-0.032	-0.017	0.006	0.167	-0.960	0.009	0.010	-0.053	-0.161	0.009	1.000

References

1. Nathan, B., Christos, C.: Radio propagation and adaptive antennas for wireless communication links: terrestrial, atmospheric and ionospheric. Wiley, Hoboken, New Jersey (2007)
2. Robert, D.H., Jonh, K.H.: The high-latitude ionosphere and its effects on radio propagation. Cambridge University Press, New York (2003)
3. Hornbostel, A.: Propagation problems in satellite navigation. Proc. WFMN07 Chemnitz Ger., pp. 42–49, (2007). Retrieved from http://archiv.tu-chemnitz.de/pub/2007/0210/
4. Medžida, M., Natraš, R., Džana, H., Dževad, K.: Investigation of ionospheric variations and sudden disturbances as a source of GNSS errors and earthquake precursor. Sci. J. Civ. Eng. (2017). Retrieved from https://publik.tuwien.ac.at/files/publik_270748.pdf
5. Meyer-Vernet, N.: Basics of the Solar Wind. The United States of America by Cambridge University Press, New York (2007)

6. Gurnett, D.A., Amitava, B.: Introduction to Plasma Physics with Space and Laboratory Applications, vol. 17997, no. 383 (2017)
7. Hoffert, M.I.: The effects of solar variability on climate, vol. 19, no. C. The National Academies Press, Washington, D.C. (2012)
8. Ninla Elmawati Falabiba: The sun solar wind heliosphere. Springer, Dordrecht Heidelberg London New York (2019)
9. Jin, S., Jin, R., Liu, X.: GNSS Atmospheric Seismology. Springer Nature Singapore Pte Ltd. (2019)
10. Huang, C.Y., Helmboldt, J.F., Park, J., Pedersen, T.R., Willemann, R.: Ionospheric detection of explosive events. Rev. Geophys. **57**(1), 78–105 (2019). https://doi.org/10.1029/2017RG000594
11. Obayashi, T.: Upper atmospheric disturbances due to high altitude nuclear explosions. Planet. Space Sci. **10**, 47–63 (1963). https://doi.org/10.1016/0032-0633(63)90006-0
12. Park, J., Grejner-Brzezinska, D.A., Von Frese, R.R.B., Morton, Y., Gaya-Pique, L.R.: On using traveling ionospheric disturbances to detect underground nuclear tests. Inst. Navig. Int. Tech. Meet. (ITM) **2**, 1581–1589 (2012)
13. Mabie, J., Bullett, T., Moore, P., Vieira, G.: Identification of rocket-induced acoustic waves in the ionosphere. Geophys. Res. Lett. **43**(20), 11024–11029 (2016). https://doi.org/10.1002/2016GL070820
14. Lin, C.H., et al.: Ionospheric shock waves triggered by rockets. Ann. Geophys. **32**(9), 1145–1152 (2014). https://doi.org/10.5194/angeo-32-1145-2014
15. Heki, K., Fujimoto, T.: Atmospheric modes excited by the 2021 August eruption of the Fukutoku-Okanoba volcano, Izu–Bonin Arc, observed as harmonic TEC oscillations by QZSS. Earth Planets Sp., **74**(1) (2022). https://doi.org/10.1186/s40623-022-01587-5
16. Hasbi, A.M., et al.: Ionospheric and geomagnetic disturbances during the 2005 Sumatran earthquakes. J. Atmos. Solar Terr. Phys. **71**(17–18), 1992–2005 (2009). https://doi.org/10.1016/j.jastp.2009.09.004
17. Pulinets, S.A., Legen'ka, A.D., Hegai, V.V., Kim, V.P., Korsunova, L.P.: Ionosphere disturbances preceding earthquakes according to the data of ground based station of the vertical ionospheric sounding wakkanai. Geomagn. Aeron. **58**(5), 686–692 (2018). https://doi.org/10.1134/S0016793218050110
18. Korsunova, L.P., Khegai, V.V.: Possible short-term precursors of strong crustal earthquakes in japan based on data from the ground stations of vertical ionospheric sounding. Geomagn. Aeron. (2018). https://doi.org/10.1134/S0016793218010085
19. Shi, K., Liu, X., Guo, J., Liu, L., You, X., Wang, F.: Pre-earthquake and coseismic ionosphere disturbances of the Mw 6.6 Lushan earthquake on 20 April 2013 monitored by CMONOC. Atmos. (Basel) **10**(4), 1–21 (2019). https://doi.org/10.3390/ATMOS10040216
20. Zlotnicki, J., Li, F., Parrot, M.: Ionospheric disturbances recorded by DEMETER satellite over active volcanoes: from august 2004 to december 2010. Int. J. Geophys. **2013** (2013). https://doi.org/10.1155/2013/530865
21. Ishii, M.: Extreme Space Weather Research in Japan, vol. 1957. Elsevier Inc. (2018)
22. Akyol, A.A., Arikan, O., Arikan, F.: A machine learning-based detection of earthquake precursors using ionospheric data. Radio Sci. **55**(11), 1–21 (2020). https://doi.org/10.1029/2019RS006931
23. Sharma, G., Champati ray, P.K., Mohanty, S., Kannaujiya, S.: Ionospheric TEC modelling for earthquakes precursors from GNSS data. Quat. Int. **462**, 65–74 (2017). https://doi.org/10.1016/j.quaint.2017.05.007
24. Ulukavak, M., Yalcinkaya, M.: Precursor analysis of ionospheric GPS-TEC variations before the 2010 M7.2 Baja California earthquake. Geomatics Nat. Hazards Risk, **8**(2), 295–308 (2017). https://doi.org/10.1080/19475705.2016.1208684
25. Goto, S.I., Uchida, R., Igarashi, K., Chen, C.H., Kao, M., Umeno, K.: Preseismic ionospheric anomalies detected before the 2016 Taiwan earthquake. J. Geophys. Res. Sp. Phys. **124**(11), 9239–9252 (2019). https://doi.org/10.1029/2019JA026640
26. Tariq, M.A., Shah, M., Hernández-Pajares, M., Iqbal, T.: Pre-earthquake ionospheric anomalies before three major earthquakes by GPS-TEC and GIM-TEC data during 2015–2017. Adv. Sp. Res. **63**(7), 2088–2099 (2019). https://doi.org/10.1016/j.asr.2018.12.028

27. Nina, A. et al.: Variation in natural short-period ionospheric noise, and acoustic and gravity waves revealed by the amplitude analysis of a VLF radio signal on the occasion of the Kraljevo earthquake (Mw = 5.4). Sci. Total Environ. **710**, 136406 (2020). https://doi.org/10.1016/j.scitotenv.2019.136406
28. Zhao, S., Shen, X.H., Zhima, Z., Zhou, C.: The very low-frequency transmitter radio wave anomalies related to the 2010 Ms 7.1 Yushu earthquake observed by the DEMETER satellite and the possible mechanism. Ann. Geophys. **38**(5), 969–981 (2020). https://doi.org/10.5194/angeo-38-969-2020
29. Sun, W. et al.: Forecasting of ionospheric vertical total electron content (TEC) using LSTM networks. Proc. 2017 Int. Conf. Mach. Learn. Cybern. (ICMLC) **2**, 340–344 (2017). https://doi.org/10.1109/ICMLC.2017.8108945
30. Liu, L., Zou, S., Yao, Y., Wang, Z.: Forecasting global ionospheric TEC using deep learning approach. Sp. Weather **18**(11), 1–12 (2020). https://doi.org/10.1029/2020SW002501
31. Ruwali, A., Kumar, A.J.S., Prakash, K.B., Sivavaraprasad, G., Ratnam, D.V.: Implementation of hybrid deep learning model (LSTM-CNN) for ionospheric TEC forecasting using GPS data. IEEE Geosci. Remote Sens. Lett. **18**(6), 1004–1008 (2021). https://doi.org/10.1109/LGRS.2020.2992633
32. Cesaroni, C. et al.: Neural network based model for global total electron content forecasting. J. Sp. Weather Sp. Clim. **10** (2020). https://doi.org/10.1051/swsc/2020013
33. Lin, X. et al.: A Spatiotemporal Network Model for Global Ionospheric TEC Forecasting (2022)
34. Heki, K.: Advances in Ionospheric Research: Current Understanding and Challenges - Ionospheric Disturbances Related to Earthquakes. Wiley/AGU Online Library (2021)
35. Mallika, L.I., Ratnam, D.V., Raman, S., Sivavaraprasad, G.: Machine learning algorithm to forecast ionospheric time delays using Global Navigation satellite system observations. Acta Astronaut. **173**, 221–231 (2020). https://doi.org/10.1016/j.actaastro.2020.04.048
36. Zhukov, A., Sidorov, D., Mylnikova, A., Yasyukevich, Y.: Machine learning methodology for ionosphere total electron content nowcasting. Int. J. Artif. Intell. **16**(1), 144–157 (2018). https://doi.org/10.13140/rg.2.2.19349.83685
37. Global Volcanism Program | Raikoke. Retrieved from https://volcano.si.edu/volcano.cfm?vn=290250
38. Ripple effect_ What the Tonga eruption could mean for tsunami research _ National Oceanic and Atmospheric Administration. Retrieved from https://volcano.si.edu/volcano.cfm?vn=243040
39. GEOFON Program GFZ Potsdam: 9C Seismic Network. Retrieved from http://geofon.gfz-potsdam.de/eqinfo/list.php?datemin=2022-01-15&datemax=2022-01-15&latmax=&lonmin=&lonmax=&latmin=&magmin=&fmt=html&nmax=
40. Stoica, P., Moses, R.L.: Spectral Analysis of Signals, vol. 4, no. 1. Pearson Prentice Hall (2005)
41. Welch, P.: The use of fast Fourier transform for the estimation of power spectra: a method based on time averaging over short, modified periodograms. IEEE Trans. Audio Electroacoust. **15**(2), 70–73 (1967). https://doi.org/10.1109/TAU.1967.1161901
42. Lilly, J.M.: Element analysis: a wavelet-based method for analyzing time-localized events in noisy time series. Proc. R. Soc. A Math. Phys. Eng. Sci. (2017). https://doi.org/10.1098/rspa.2016.0776
43. Lilly, J.M., Olhede, S.C.: Generalized morse wavelets as a superfamily of analytic wavelets. IEEE Trans. Signal Process. **60**(11), 6036–6041 (2012). https://doi.org/10.1109/TSP.2012.2210890
44. Hudson, T.S., Horseman, A., Sugier, J.: Diurnal, seasonal, and 11-yr solar cycle variation effects on the virtual ionosphere reflection height and implications for the Met Office's lightning detection system, ATDnet. J. Atmos. Ocean. Technol. **33**(7), 1429–1441 (2016). https://doi.org/10.1175/JTECH-D-15-0133.1
45. Zheng, W., et al.: Diurnal, seasonal, annual, and semi-annual variations of ionospheric parameters at different latitudes in East Asian sector during ascending phase of solar activity. Solar Terr. Phys. **3**(2), 45–53 (2017). https://doi.org/10.12737/22594

Fig. 1 Republic road III-708 near Elhovo, Bulgaria

and is unstable in areas of provoked landslides. The overall safety, including the risk of destabilization of the soils and the risk of road accidents, is quite unsatisfactory.

The second factor is the quality of execution of bulk earthworks. In some sections, the embankments are not sufficiently compacted. In this case, even small degrees of oversaturation result in a loss of stability [2]. The third factor is related to the destabilization of the embankments which stems from soils of black and Pliocene clays. When oversaturated, these clays show volumetric instability [3] and become a sliding surface that provokes landslides in the road embankments.

So far, no radical measures have been taken in the problem areas to stabilize and restore the functioning of the drainage facilities. Only local patches of no use and short duration are made on the surface layer of the asphalt concrete. The partial or complete scraping of the embankment with possible reduction of its height and the replacement of the compromised material with a rock fraction in combination with geogrids is considered to be a more serious measure in practice [4]. These measures give results only in the case of poor quality bulk earthworks in combination with relatively strong soils, i.e., in the case of weak soils, and they do not have a long-term effect.

In the case of clays at the base of the embankment, which are unstable because of oversaturation and cannot be scraped up, there is no possibility for the road to bypass them, and the use of a weighted compacted embankment of rock materials may result in new landslides with sudden loss of stability. The use of applied nanotechnologies is a modern approach, but has not yet applied in the mass practice [5]. The present study aims to provide practical recommendations for solving the case with the stability of embankments on weak soils that cannot be removed. These recommendations are based on the possible technical means applied in practice and are aimed at preventing future accidents, given the real and pessimistic perspective that roads and drainage systems will not be repaired and maintained.

2 Study of the Stability of Oversaturated Embankments on Weak Soils

Oversaturated embankments, in which the destabilization starts not from the embankment body but from the soil foundation, are considered. Usually the weak foundation is made of black clays, which overlay on Pliocene brown clays. During the construction or previous repairs of the embankments, unsuccessful attempts were made to partially remove the black clays. The ineffectiveness of removal is also due to the fact that brown Pliocene clays in depth exhibit the same properties as black clays when being saturated.

When performing stability calculations, all sliding surfaces pass under the heel of the embankment. The plastic zones are found in the base of the embankment, and due to their presence, the safety factor of the slopes at the time of destabilization is less than 1. These facts are the evidence that the cause of deformation of the embankment is in the base beneath it. This requires the repair activities to be concentrated on its reinforcement, combined with drainage measures by building drainages and restoring dewatering ditches and gutters. During the construction works, it is not necessary to remove and replace the material of the whole embankment. Only the uncompacted material in the compromised part must be removed.

Provided that the drainage and dewatering systems will be fully repaired and maintained, the effective solution is, in addition to restoring the damaged embankment, to envisage drainage ribs in the slope with the lost stability. These drainage ribs should be combined in a collecting drainage ditch designed to drain the soil and be anchored in a functioning gutter (Fig. 2).

When the condition for maintenance is fulfilled, the above-mentioned possibility is the most feasible and economically viable solution of the issue with the stability and

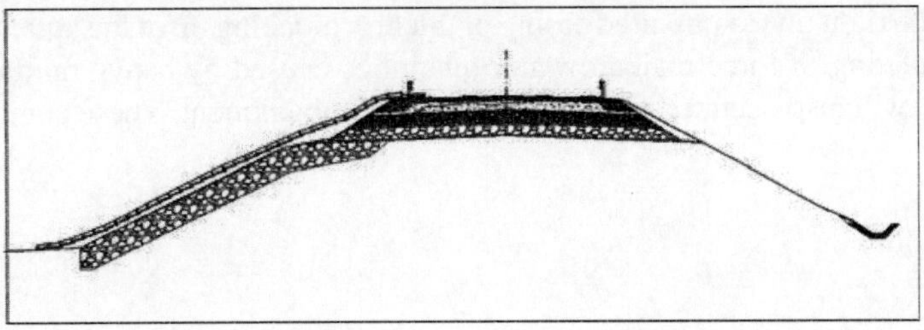

Fig. 2 Option with one-sided drainage and ribs

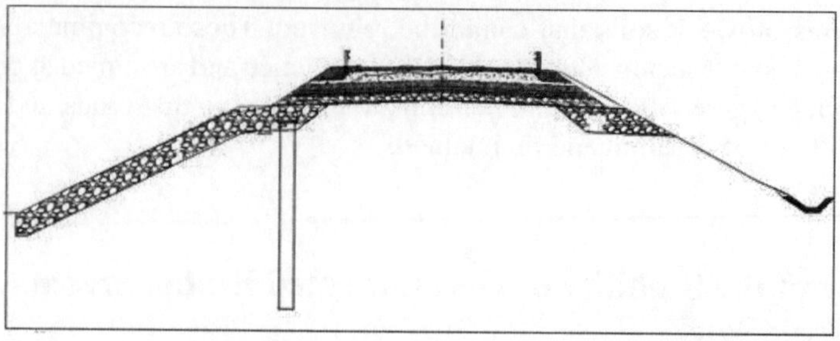

Fig. 3 Option with one-sided drainage and ribs combined with unilateral piles

safety of the road traffic. Unfortunately, this type of decision is not accepted by the road administration due to the fact that there will be no complete road maintenance. The intervention will be reduced to patching parts of the roadway, which does not guarantee drainage of the weak geological foundation. Due to the prospect of lack of maintenance in the future, the authorities usually add a requirement to reinforce the slopes damaged by oversaturation of the base with piles (Fig. 3).

This study is devoted to the problems which result from the uncertainty of providing some piles. There are two border cases for the study of this type of bulk slopes on a weak foundation. In the first case, when there is road maintenance and the drainage facilities are operational, there is no need for piles. In the other case, the drainage ceases to act and a pile structure should be provided to permanently resist the aggression of the oversaturation of the soils.

The authorities' decision for the drainage measures in the damaged slope on the left to be supplemented by a single line pile system implies the long-term recurrent oversaturation at the base of the embankment. This oversaturation causes destabilization of the embankment and failure to the left of the profile initiated by a watering source located on the right. The collapse is on the left and it is supposed that the piles will be triggered and there will be no further accident on the left side of the road. Such a solution is in the direction of wishful thinking and, in solving such cases, three geotechnical circumstances should be taken into account.

Firstly, the source of saturation may not be of the same type such as a discontinuous gutter on the left zone designed to conduct the waters coming to the right of the profile. As practice shows that, in the next 10 years, it is likely that saturation will occur on the right due to impaired casing of ditches, blocking up of the gutter's inflow and impounding of force majeure water quantities caused by heavy rainfall. Thus, destability will also occur on the right side of the embankment. These combinations are fully possible in emerging climate change and virtually missing maintenance of road facilities especially in third- and fourth-grade roads.

Secondly, placing one row of piles only to the left will cause partial backwatering of the waters entering the foundation, which will further contribute to the complete saturation of the soils and will provoke destabilization on the right.

Thirdly, in modeling the stability, it should be kept in mind that the deformations of the embankment are caused by its foundation. As mentioned above, this is evidenced by the presence of plastic zones under the embankment and sliding surfaces, which are beneath the embankment's heel despite the attempts for the black clays at the base to be removed.

The physico-mechanical properties of soils are strongly influenced by the period in which the studies are made. In this regard, the study should not be based only on the soil parameters measured during the last geological study. During saturation, the dimensioning parameters of the soil worsen. Reverse calculations of the stabilized assessment should be made. The purpose of these reverse calculations is to model the state of the soils which is closest to the time when the large destruction of the embankment happens (Figs. 4 and 5). The sliding in reverse calculations is reproduced in a model for which the soil characteristics at the base—angle of inner friction and cohesion—are reduced by 30%. This finding corresponds to the approach applied in underground construction for decreasing the rock mass indicators in the disturbed areas.

On the basis of the above-mentioned circumstances, it can be concluded that the piles should not be provided only on the left where the destabilization occurs, but to be bilaterally placed on the roadway (Fig. 6). In this way, the embankment's stability can be secured until removing the causes for the failure of the drainage facilities. The significant number of registered failures of geotechnical constructions on the country's road infrastructure due to non-maintenance of the drainage facilities should result in more radical measures to avoid the repeated increase in capital expenditures caused by roadway damage.

The study of the stability of the left and right parts of the problematic profile is shown in the following schemes (Figs. 7 and 8).

Formally, for the society and the authorities, the proposed technical solution is more radical and expensive than the recommendations of the road administration. And the contradiction between reality and conjuncture in this case is easily explained—the recommendations of the administration stem from the reality that there are no planned costs for maintenance and the aspiration during the approval is for the construction to remain longer in service. This approach is noble in its essence. The contradiction is that it is not about the normal duration of exploitation of at least 100 years, but of imaginary time after which the measures and participants

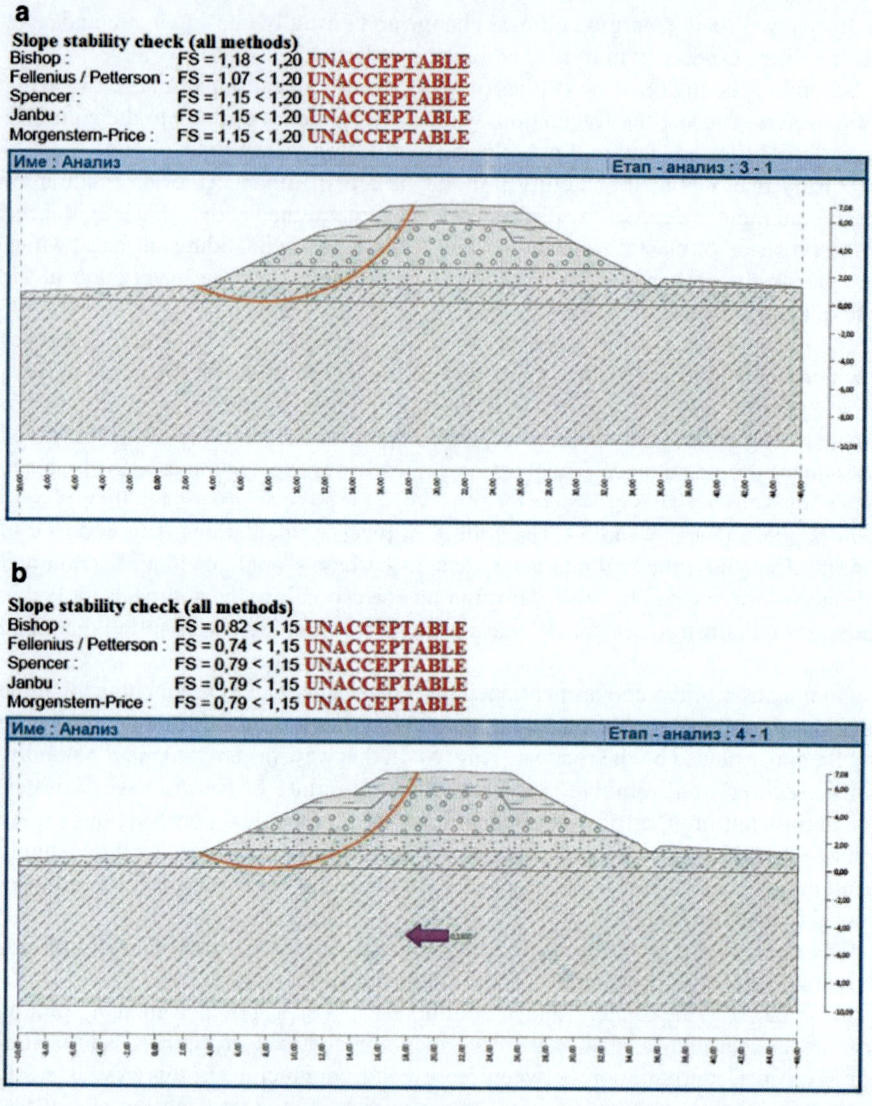

Fig. 4 a, b: Computing schemes of recovered conditions close to the time of destabilization in basic and earthquake combinations for the left part

in the investment process become void by prescription and the meaningless provision of funds without any concept about the future of road infrastructure will continue. Such an approach contradicts all global practice where historical transport facilities

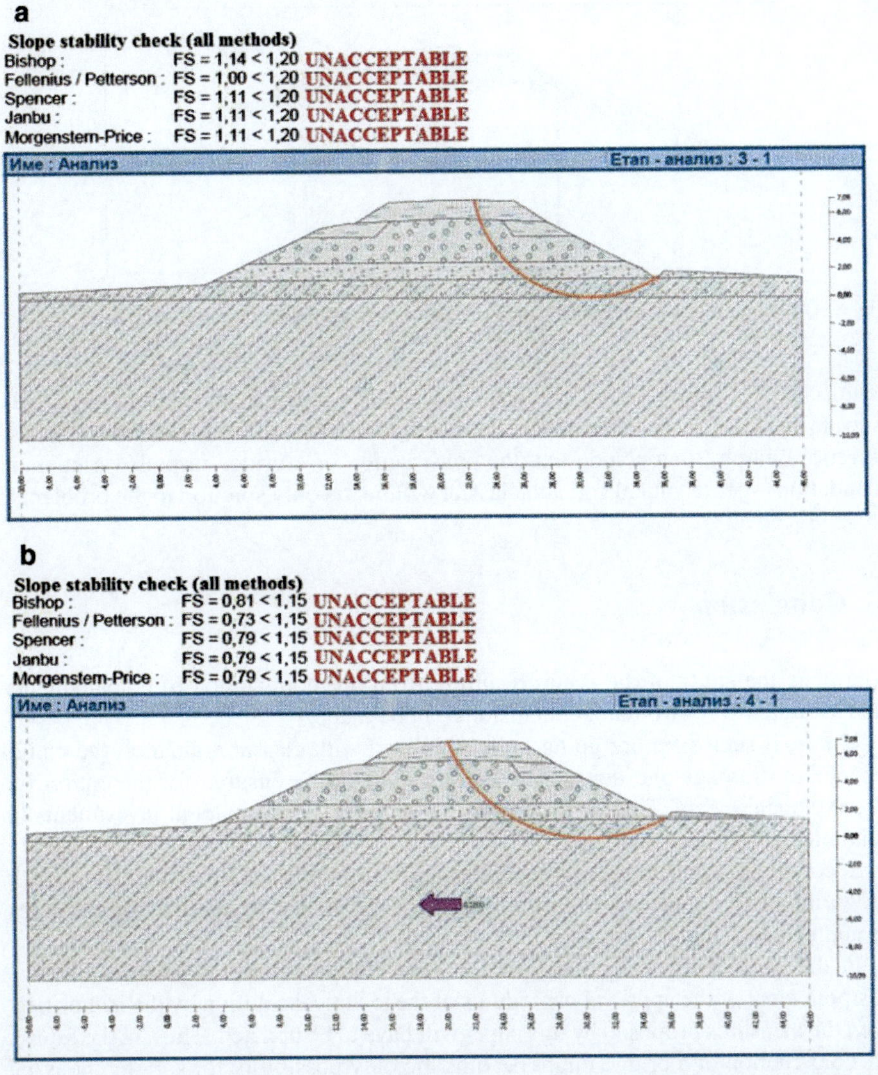

Fig. 5 a, b: Computing schemes of recovered conditions close to the time of destabilization in basic and earthquake combinations for the right part

exist since the beginning of the industrial age and millennial road infrastructures are upgraded.

In cases where such problems need to be solved, Eurocode requirements should be followed. These requirements oblige the designer to foresee the most unfavorable

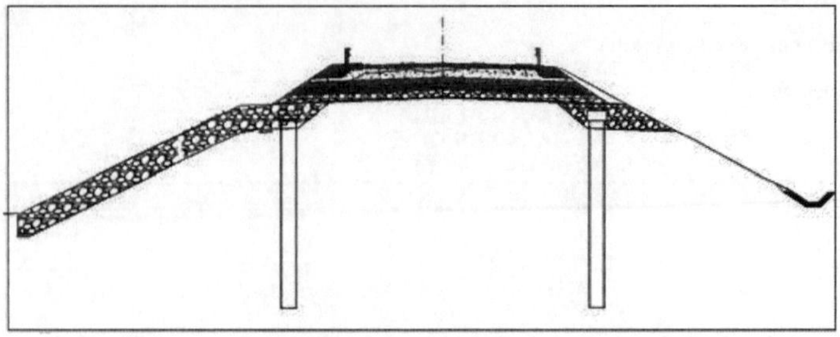

Fig. 6 Option with one-sided drainage and ribs combined with bilateral piles

conditions for the operation of the facility. Following this logic, it is fully realistic, in the case of a delay of the stabilizing activities similar to the ones in the current research, to conclude that the construction of a bridge in both varieties of foundation—geotechnical soil and classic, will be the only solution to the problems.

3 Conclusion

Following the study of the issues resulting from the stabilization of embankments on a saturated soil, two main conclusions can be made.

Firstly, if such sites are going to be stabilized with classic drainages, the maintenance of drainage and dewatering facilities should be ensured for the entire life of the embankments. Then there will be no need for serious capital investments to guarantee the slopes' stability.

Secondly, in order to make a reliable stabilization of the embankments on weak soils without a guarantee for the current drainage maintenance, stabilization measures should be envisaged as proposed in this study.

In the case of recurring delay of the stabilization activities due to subjective postponement of the approval procedures of the technical solution by the authorities, even the measures proposed by this paper will have to be updated. Thus, in the future, the construction of a bridge might become the only possibility for solving the issue with weak soils under the bulk body.

Hence, the making of piles cannot be used as a panacea by every specialist who is not familiar with the specificity of geotechnics everywhere and without reason only because such compromise and technically unjustified solutions will be in compliance with the conjunctural criteria of the road administration.

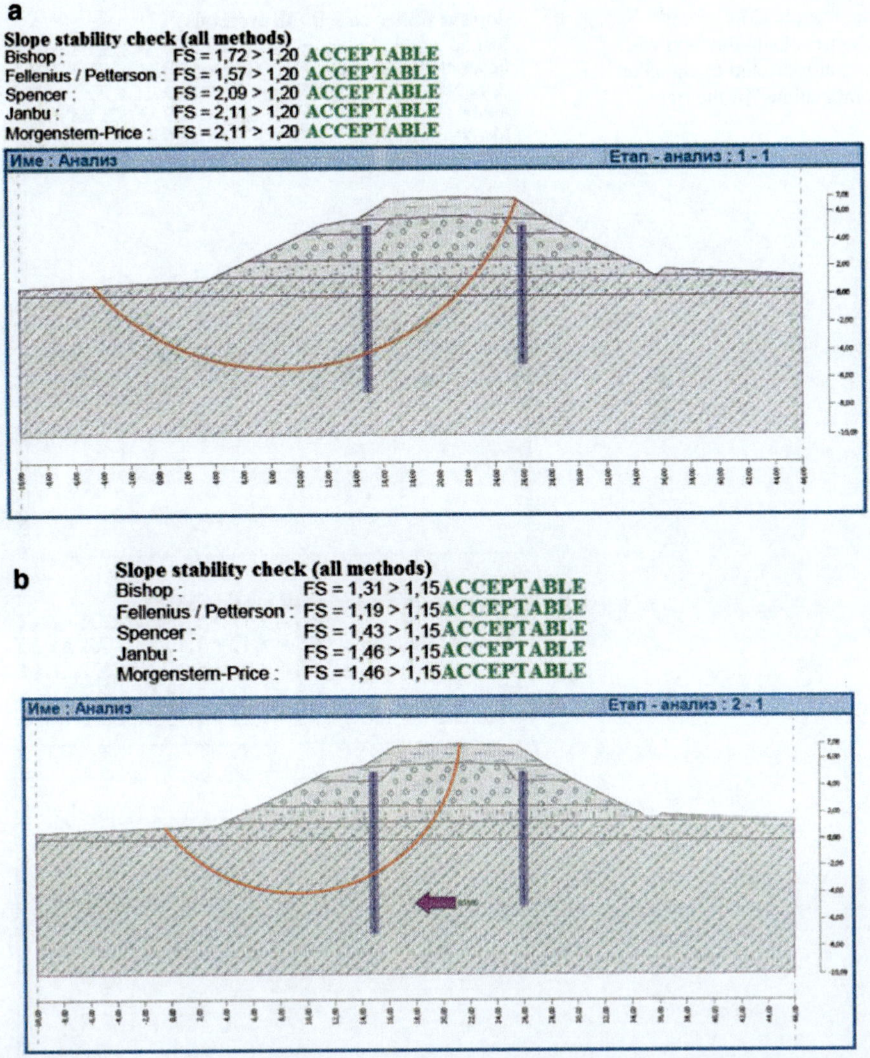

Fig. 7 a, b: Computing schemes of stabilization with piles at basic and earthquake combinations for the left part

Fig. 8 a, b: Computing schemes of stabilization with piles at basic and earthquake combinations for the right part

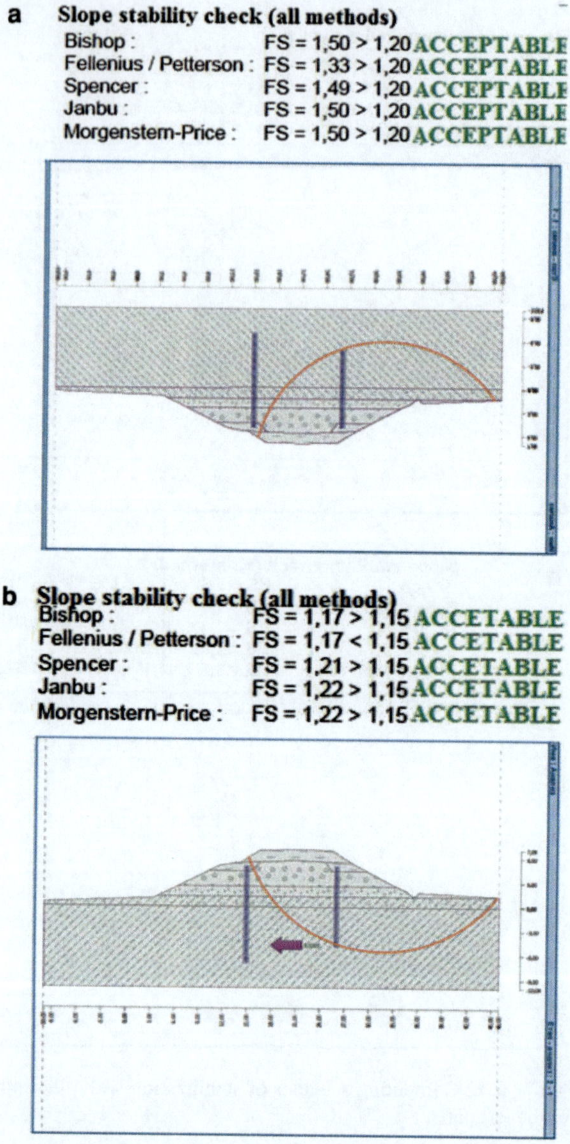

References

1. Jechev, N.: Socio-ecological issues caused by landslides and rock falls. Geology and Mineral Resources, ISSN:1310-2265, issue 4–5 (2021)
2. Mitev, I., Pavlov, P., Totev, L.: Elasticity and deformation module elaluation using a pressure slab in construction works. Safety Environment – Working and Ambient, RIFREN, Sofia, ISBN:978-954-353-083-0, pp. 241–246 (2008)
3. Ivanova, V., Karacgolov, P., Malinova, D., Pavlov, P., Balev, V.: Geotechnical investigation of

the clays from lower level of the internal waste dump of "Troyanovo-1" mine – Branch of "Mini maritsa iztok" e EAD". In: VII International Geomechanics Conference, Varna, ISSN:1314-6467, pp. 34-44 (2016)
4. Tamaskovics, N., Tondera, D., Pavlov, P.: Interaction behaviour of geosynthetics in cohesive soils. In: Annual of the University of Mining and Geology "St. Ivan Rilski", vol. 56, part II, Mining and Mineral processing. ISSN:1312-1820, pp.178–183 (2013)
5. Mitev, I.: Applied nano - technology for strengthening of soils in the ground base of engineering facilities. In: XIV International Conference of the Open and Underwater Mining of Minerals. ISSN:2535-0854, pp. 405–411 (2017)

Indirect Georeferencing in Terrestrial Laser Scanning: One-Step and Two-Step Approaches

Dung Trung Pham, Long Quoc Nguyen, Tinh Duc Le, and Ha Thanh Tran

Abstract The georeferencing procedure is to transform geospatial data from a local coordinate system to a global coordinate system, notably geodetic coordinate system on a geocentric datum. In this paper, both the one-step and two-step approaches of indirect georeferencing of 3D point cloud from terrestrial laser scanning are investigated. The georeferencing procedure is applied to a real dataset acquired by a Faro Focus^{3M} X130 laser scanner and the control points and targets are measured by total station TS06 plus. Five scenarios are used for the comparison between the one-step and two-step approaches in terms of both accuracies of the 3D model and time consumption. Besides, the influence of the target's configuration on the 3D model is evaluated by changing either the number or the position of the targets. The results suggest that the 3D model's accuracies when using both the one-step and two-step approaches of indirect georeferencing are comparable. Additionally, the target's configuration greatly affects the 3D model with the one-step approach of indirect georeferencing. From the rigorous analyses of the benefit and drawbacks of both approaches evaluated on the real dataset, the paper significantly contributes to the indirect georeferencing procedure when transforming the 3D point cloud acquired from terrestrial laser scanning into the geodetic coordinate system.

D. T. Pham (✉) · L. Q. Nguyen · T. D. Le · H. T. Tran
Faculty of Geomatics and Land Administration, Hanoi University of Mining and Geology, Hanoi 100000, Vietnam
e-mail: phamtrungdung@humg.edu.vn

L. Q. Nguyen
e-mail: nguyenquoclong@humg.edu.vn

T. D. Le
e-mail: leductinh@humg.edu.vn

H. T. Tran
e-mail: tranthanhha@humg.edu.vn

L. Q. Nguyen
Innovations for Sustainable and Responsible Mining (ISRM) Research Group, Hanoi University of Mining and Geology, Hanoi 100000, Vietnam

L. Q. Nguyen et al. (eds.), *Advances in Geospatial Technology in Mining and Earth Sciences*, Environmental Science and Engineering,
https://doi.org/10.1007/978-3-031-20463-0_11

Keywords Terrestrial laser scanner · Indirect georeferencing · One-step approach · Two-step approach

1 Introduction

Terrestrial laser scanner (TLS) is an instrument for fast collection of 3D point cloud (PC) data with high accuracy. The 3D PC of a scan object in many cases is needed to be determined in a reference coordinate system. The procedure to transform a 3D PC into a reference coordinate system (RCS) is known as a georeferencing procedure. Generally, georeferencing can be divided into two basic methods that are direct and indirect georeferencing. Conventionally, the ground RCS can be established by total station or global navigation satellite system (GNSS) before scanning the field work.

The direct georeferencing method is based on additional sensors and equipment to provide the coordinates and orientation at the moment of data acquisition. The additional sensors and equipment normally are low-cost sensors, e.g., an inertial measurement unit (IMU), continuously referencing station, or a telescope for azimuth direction orientation. The direct georeferencing of the TLS has been investigated in many studies [1–5] and can be simply done by using an optical plummet to centering over a control point (CP) and a telescope that is mounted into the TLS to directly determining the orientation [1]. A continuously operating referencing station is also used for the georeferencing procedure in which GNSS receivers are mounted at the center of both the scanner and targets [2]. Furthermore, Schuhmacher and Böhm [3] used a digital compass with low-cost GNSS and tilt sensor for georeferencing. Together with the development of direct georeferencing approaches by using additional sensors, deep analyses about the error sources are investigated by Lichti et al., Reshetyuk, Pandžić et al. [6–8]. Although the direct georeferencing method is available for reducing time of data post-processing in the office, the further equipment leads to an increase in the budget [7]. The difficult and time-consuming calibration procedure in direct georeferencing is the main disadvantage [9]. Moreover, the low accuracy achieved by the direct georeferencing method is also a great consideration when applying this method.

By contrast, the indirect georeferencing method is suitable for obtaining a higher accurate 3D PC data [7]. Several researchers contributed to this method that can be found in [3, 10–14]. The indirect georeferencing method can be classified into two approaches: one-step approach and two-step approach [15]. The two-step approach is based on the registration procedure that PCs from multiple scans are transformed into a common coordinate and then the registered PCs are transformed into the external coordinate system based on ground control points (GCPs). The main advantage of the two-step approach is that it uses a moderate number of CPs. But at least 30% overlap between two adjacent scans is still required that lead to more time for scanning, especially for a large scan object. The one-step approach involves independent georeferencing of individual scans using at least three CPs. No overlap between difference scans is needed for this method that reduces the time for scanning.

However, the georeferencing independently carried out for each scan leads to more extract survey work [7].

Currently, the indirect georeferencing is known as an efficient method in TLS. However, the main problem of this method is inconsistency in the PC accuracy and just few studies have addressed the effect of target's configuration on the accuracy of 3D model. This study addresses two important questions on (1) which approaches in the indirect georeferencing method among the one-step and two-step approaches are better for the accuracy of the 3D model and (2) how the targets' configuration affects the accuracy of the 3D model and what the crucial factors are.

The paper is organized as follows: The first section gives a brief overview of georeferencing method in TLS. The second section presents a mathematical principle of indirect georeferencing for both the one-step and two-step approaches. The experiment is introduced in the third section. The fourth section is on the analysis and discussion of experimental results. Some conclusions and future works can be found in the last section.

2 Indirect Georeferencing

Georeferencing is a transformation of PCs from the scanner coordinate system to the external coordinate system (normally a national or local coordinate system) based on known points. To transform between two 3D coordinate systems, seven parameters need to be determined. If two PCs are obtained by the same scanner, the scale transformation parameter equals one. Therefore, in TLS two scanner coordinate systems can be transformed by six parameters. As a result, in indirect georeferencing, six transformation parameters, including three angles (φ, ω, κ) and three coordinates (X, Y, Z), need to be determined. To obtain six parameters, at least six coordinates in both systems (i.e., the scanner and external coordinate systems in Fig. 1) or three points must be known. These points are normally GCPs of which the coordinates are determined by a total station or a GNSS receiver. It is noted that these GCPs should be well distributed vertically and horizontally (not on the same line or plane) [3].

Indirect georeferencing can be realized by two basic ways that are one-step and two-step approaches. The one-step approach means transforming the scan data directly from the scanner coordinate system to the external coordinate system of geodetic network. In contrast, in the two-step approach, the coordinates of points are transformed through two independent steps. The following section will discuss these two approaches in detail.

2.1 Two-Step Approach

The two-step approach (Fig. 2) is performed in the following sequence. In the 1st step, the PCs obtained from multiple scanners are registered in a common (or global)

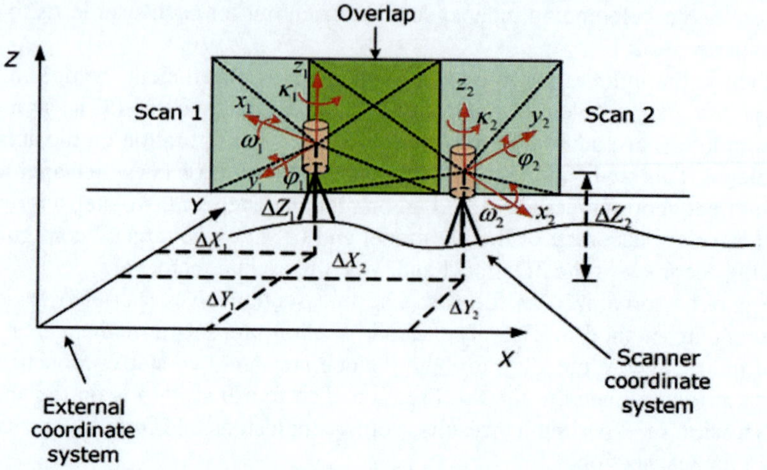

Fig. 1 Relationship between the scanner and external coordinate systems [7] shown by six transformation parameters

coordinate system. In this registration procedure, several methods such as target-based, natural point features, surface matching, and common geometrical objects can be applied separately or based on their combination. In the 2nd step, the PCs in the common coordinate system are georeferenced in a geodetic control network. At least three CPs must be used in this case.

The mathematical model of the two-step approach can be described as follows: In the first step, the coordinates of the PC in the scanner $\mathbf{X}_i$ are transformed into the common coordinate system $\mathbf{X}_g$ by:

$$\boldsymbol{X}_g = \Delta \boldsymbol{X}_{ig} + \boldsymbol{R}_{ig} \boldsymbol{X}_i \tag{1}$$

where $\boldsymbol{R}_{ig}$ and $\Delta \boldsymbol{X}_{ig}$ are the rotation matrix and the translation vector from the scanner to the common system, respectively. In the second step, the coordinates of the PC from the common system are transformed into the geodetic control network as:

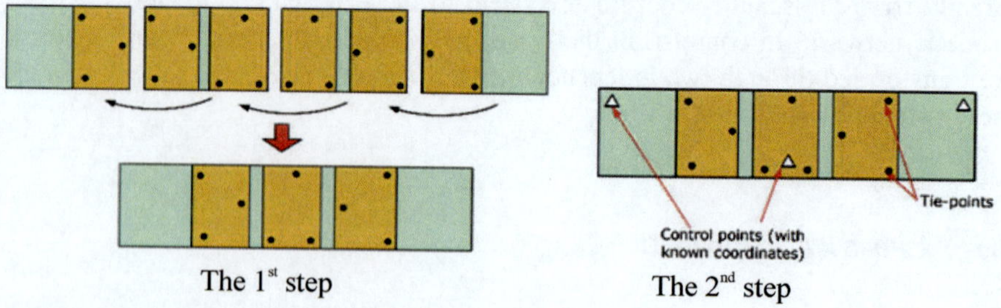

Fig. 2 Two-step approach of indirect georeferencing [7]

$$\boldsymbol{X}_e = \Delta\boldsymbol{X}_{ge} + \boldsymbol{R}_{ge}\boldsymbol{X}_g \tag{2}$$

where $\boldsymbol{R}_{ge}$ and $\Delta\boldsymbol{X}_{ge}$ are the rotation matrix and the translation vector from the common system to the geodetic system, respectively.

Replacement of Eq. (1) in (2) yields:

$$\boldsymbol{X}_e = \Delta\boldsymbol{X}_{ge} + \boldsymbol{R}_{ge}\left(\Delta\boldsymbol{X}_{ig} + \boldsymbol{R}_{ig}\boldsymbol{X}_i\right) \tag{3}$$

The rotation matrix is the function of the rotation angles ω, ϕ, and κ about the x, y, and z coordinates, respectively. The rotation matrix is calculated by:

$$\boldsymbol{R} = \boldsymbol{R}_3(\boldsymbol{\kappa})\boldsymbol{R}_2(\boldsymbol{\phi})\boldsymbol{R}_1(\boldsymbol{\omega}) \tag{4}$$

where

$$\boldsymbol{R}_1(\omega) = \begin{pmatrix} 1 & 0 & 0 \\ 0 & \cos\omega & \sin\omega \\ 0 & -\sin\omega & \cos\omega \end{pmatrix} \tag{5}$$

$$\boldsymbol{R}_2(\phi) = \begin{pmatrix} \cos\phi & 0 & -\sin\phi \\ 0 & 1 & 0 \\ \sin\phi & 0 & \cos\phi \end{pmatrix} \tag{6}$$

and

$$\boldsymbol{R}_3(\kappa) = \begin{pmatrix} \cos\kappa & \sin\kappa & 0 \\ -\sin\kappa & \cos\kappa & 0 \\ 0 & 0 & 1 \end{pmatrix} \tag{7}$$

with $\boldsymbol{R}_1(\omega)$, $\boldsymbol{R}_2(\phi)$, and $\boldsymbol{R}_3(\kappa)$ being the rotation matrices around the x, y, and z coordinates, respectively.

2.2 *One-Step Approach*

The one-step approach (Fig. 3) is to directly transform the coordinates of the PC of individual scanner into the geodetic coordinate system by using GCPs. Since each PC is registered into the geodetic coordinate system separately, the number of GCPs increases and is inconsistent by control network configurations. The mathematical model of the one-step approach can be described as:

$$\boldsymbol{X}_e = \Delta\boldsymbol{X}_{ie} + \boldsymbol{R}_{ie}\boldsymbol{X}_i \tag{8}$$

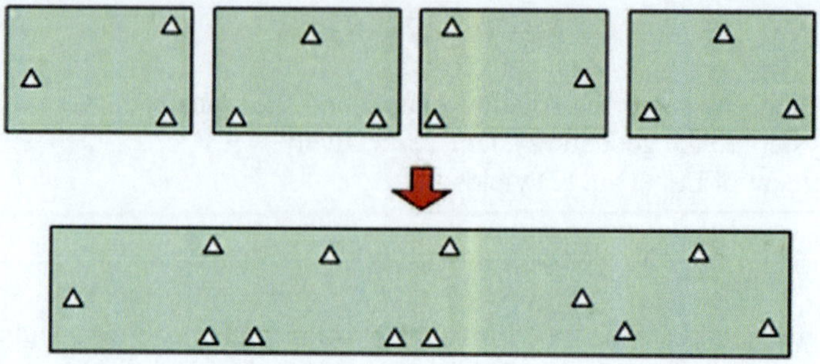

Fig. 3 One-step approach of indirect georeferencing [7]

where $\boldsymbol{R}_{ie}$ and $\Delta\boldsymbol{X}_{ie}$ are the rotation matrix and the translation vector from the scanner to the geodetic systems, respectively.

2.3 *Error Model of the Indirect Georeferencing*

The main purpose of this paper is to investigate the influence of the indirect georeferencing method on the accuracy of 3D PC. As mentioned above, indirect georeferencing can be dealt with either the one-step approach or two-step approach. According to the one-step approach, the coordinate of PCs is directly transformed into the geodetic coordinate system by Eq. (8). The error in the indirect georeferencing in this case is affected by the scanner random errors and the errors of transformation parameters. The covariance matrix of PC coordinates in geodetic coordinate system can be computed as [8]:

$$\boldsymbol{C}_{X_e} = \boldsymbol{J}_{\text{trans}}\boldsymbol{C}_{\text{trans}}\boldsymbol{J}_{\text{trans}}^T + \boldsymbol{R}_{ie}\boldsymbol{J}\boldsymbol{C}_{\text{int}}\boldsymbol{J}^T\boldsymbol{R}_{ie}^T \tag{9}$$

where $\mathbf{J}_{\text{trans}}$ is the Jacobian matrix of point coordinates in the geodetic coordinate system with respect to the transformation parameters and $\boldsymbol{J}$ is the Jacobian matrix of point coordinates in the scanner coordinate system with respect to the scan measurements (horizontal and vertical angles and distance) and is computed as [10]:

$$\boldsymbol{J} = \begin{bmatrix} \frac{\partial x_j}{\partial r_j} & \frac{\partial x_j}{\partial \varphi_j} & \frac{\partial x_j}{\partial \theta_j} \\ \frac{\partial y_j}{\partial r_j} & \frac{\partial y_j}{\partial \varphi_j} & \frac{\partial y_j}{\partial \theta_j} \\ \frac{\partial z_j}{\partial r_j} & \frac{\partial z_j}{\partial \varphi_j} & \frac{\partial z_j}{\partial \theta_j} \end{bmatrix} \tag{10}$$

$\boldsymbol{C}_{\text{trans}}$ is the covariance matrices of the transformation parameters between the scanner and geodetic coordinate systems. $\boldsymbol{C}_{\text{int}}$ is the covariance matrix that is a combination of noise measurement and laser beam width as [10]:

$$C_{\text{int}} = \textbf{diag}\Big(\sigma_r^2\ \sigma_\varphi^2 + \sigma_{\text{beam}}^2\ \sigma_\theta^2 + \sigma_{\text{beam}}^2\Big) \tag{11}$$

with σ_r^2, σ_φ^2, σ_θ^2, and σ_{beam}^2 are the standard deviations of the distance and horizontal and vertical angles, and the beam width, respectively.

3 Experiments

The experiment in this study is carried out by a real dataset collected by a TLS. The following subsection will present in detail the experiment.

Scan object. The scan object is a façade of five-story and two-story buildings located on the main campus of Hanoi University of Mining and Geology (HUMG) on 10th November 2021.

Instruments. The data are collected with a Faro Focus^{3M} X130 scanner in two different scan stations for the whole study area in 40 min. The resolution and quality parameter settings for scanner are 6 mm point spacing at a 10 m distance (or 28,000 points per one square meter), which corresponds to a 4X level of resolution of this scanner. A total station Leica TS06 plus is used to establish a control network and measured the coordinates of the targets.

Geodetic control network. To carry out indirect georeferencing of the PC, a geodetic control network is established by a traverse network including six GCPs (see Fig. 4) in the VN-2000 coordinate system.

Targets. The targets are both checkerboard and clearly natural objects, which are measured by a Leica total station TS06 plus. In these experiments, 11 checkerboards

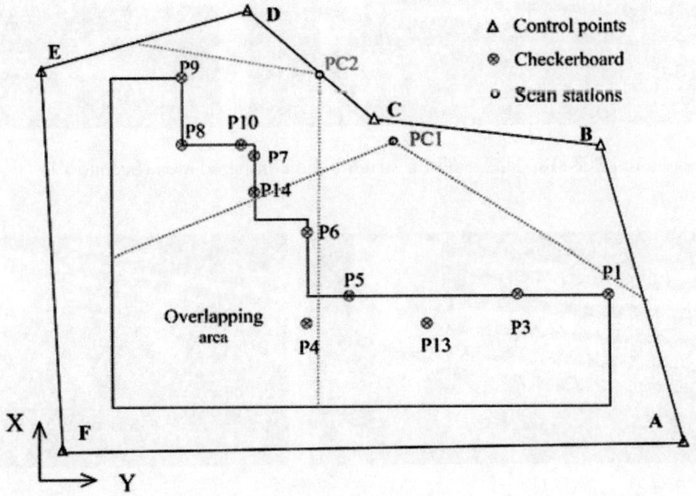

Fig. 4 Geodetic control network and checkerboard

(see Fig. 4) and 10 natural objects (see Table 2) are used. The coordinates of these points are determined in the geodetic coordinate system based on the control network shown in Fig. 4.

Scenarios. In this experiment, five scenarios are carried out for both the one-step and two-step approaches of the indirect georeferencing method. The following subsections will describe these scenarios in detail.

In Scenario 1, the two-step approach is used in which the first step is a registration procedure and the second step is a georeferencing procedure. Figure 5 shows two PCs before processing by both registration and georeferencing procedures. The align function in the Cloud Compare (CC) software is applied for both registration and georeferencing procedures. In Fig. 6 (left), the registration is carried out based on five tie points, which are distributed over the overlapping area between PC1 and PC2. These tie points are normally chosen at clear natural objects like the corners and windows. The georeferencing procedure is to transform the registered PC into the geodetic coordinate system based on five GCPs (see Fig. 6, right). Five GCPs are used since these points are well distributed around the PC and adequate for evaluating the accuracy of georeferencing procedure. The accuracy of this procedure is estimated by root mean square (RMS) by the iterative closest point (ICP) algorithm [5] in the CC software. The RMSs of the five points in the registration and georeferencing procedures are 2.5 and 1.5 cm, respectively.

In Scenario 2 to Scenario 5, the one-step approach is used that allows us to transform directly the PC into the geodetic system based on CPs. It is noted that no

Fig. 5 Unregistered PCs (PC 1-lelf and PC2-right) and compared area (Scenario 1)

Fig. 6 Registered PC based on five tie points (left) and georeferenced PC based on five CPs (right) (Scenario 1)

registration procedure is needed in these scenarios. These scenarios are to evaluate the effect of target's configuration on the 3D model, so the number and position of the targets (or CPs) are varied as follows. In Scenario 2, four CPs are distributed at the bottom of PC1 and PC2 (see Fig. 7). Scenario 3 uses five and four CPs, which are located in one side of PC1 and PC2 as illustrated in Fig. 8. In Scenario 4, four CPs spread throughout PC1 and PC2 (see Fig. 9). Finally, Scenario 5 is extended from Scenario 4 by adding two more CPs for each PC, in which six CPs and seven CPs are also spread throughout PC1 and PC2, respectively, as shown in Fig. 10.

Table 1 summarizes the number of CPs and the error of alignment of these above scenarios in which the error of alignment is computed by the RMS between targets in these scenarios and the RMSs are from 1.5 to 2.2 cm.

Fig. 7 Scenario 2—CPs are distributed at the bottom PC1 (left) and PC2 (right)

Fig. 8 Scenario 3—CPs are distributed on one side of PC1 (left) and PC2 (right)

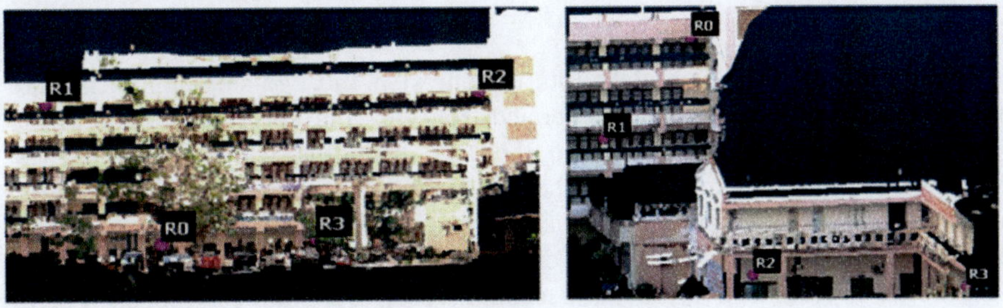

Fig. 9 Scenario 4—CPs are evenly distributed on both PC1 (left) and PC2 (right)

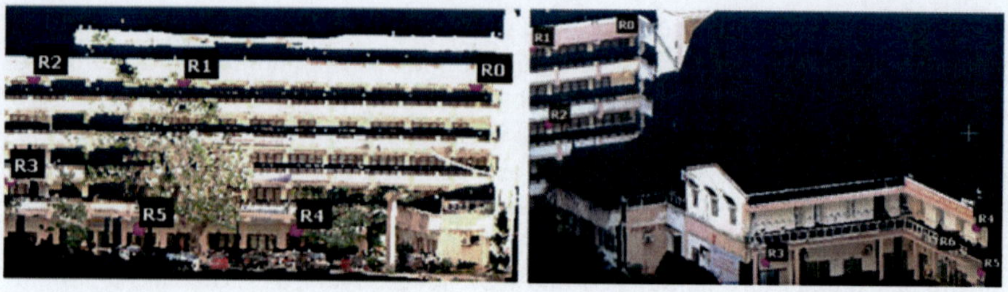

Fig. 10 Scenario 5—CPs are evenly distributed on both PC1 (left) and PC2 (right)

Table 1 Number of targets and error of alignment (RMS) in the one-step approach

Scenarios	Numbers of targets		Errors of alignment (RMS)	
	CP1	CP2	CP1 (cm)	CP2 (cm)
Scenario 2	4	4	1.8	2.2
Scenario 3	4	5	2.2	2.2
Scenario 4	4	4	1.8	1.5
Scenario 5	6	7	1.9	1.9

Table 2 Difference in 3D coordinates of natural points using the one-step approach

No.	Name	Δ_x	Δ_y	Δ_z	No.	Name	Δ_x	Δ_y	Δ_z
1	T51	– 4.0	– 1.5	2.7	11	LCR1	1.2	1.0	0.3
2	T52	– 4.6	0.5	5.4	12	LCR2	0.3	0.5	– 0.1
3	T53	– 0.3	2.9	0.8	13	PTN	– 2.0	– 0.6	– 1.2
4	T54	– 0.5	2.9	3.9	14	XD	0.5	– 1.5	– 0.1
5	T24	– 2.2	2.2	– 3.4	15	T41	– 1.1	1.5	– 5.7
6	QLDA	– 3.1	0.3	– 1.7	16	T42	0.2	4.3	– 5.7
7	A108	– 3.0	– 0.9	1.3	17	T31	– 1.2	– 0.1	2.4
8	VPD	1.3	0.9	– 0.6	18	T32	– 1.4	0.5	1.0
9	QHCC1	0.5	0.0	– 0.8	19	A105	– 2.9	0.8	– 0.2
10	QHCC2	0.3	– 1.4	0.9	20	A106	– 3.0	– 0.3	– 0.2

Evaluation of the 3D model. Both the one-step and two-step approaches are used to transform the 3D model generated from PCs into the geodetic coordinate system. To evaluate the georeferenced 3D model using the two above approaches, two investigations are carried out in the case study. First, the accuracy between the one-step and two-step approaches is compared using 20 natural points in the 3D model (see Fig. 11). The two-step and one-step approaches are described in Scenario 1 and Scenario 2, respectively. In these scenarios, each of the four CPs (checkerboard) measured by the Leica total station TS 06 plus is used for georeferencing. These above natural points are also measured by this total station in the geodetic coordinate system

Fig. 11 Twenty natural points used for comparison between the one-step and two-step approaches

and then are compared to their coordinates in the 3D model. The difference in 3D coordinate can be computed as:

$$\begin{aligned} \Delta_x &= x_{TS} - x_{\text{Model}} \\ \Delta_y &= y_{TS} - y_{\text{Model}} \\ \Delta_z &= z_{TS} - z_{\text{Model}} \end{aligned} \tag{12}$$

where x_{TS}, y_{TS}, and z_{TS} are the coordinates measured by the total station; x_{Model}, y_{Model}, and z_{Model} are the coordinates measured in the 3D model. The accuracy of the 3D model is assessed by RMSs as:

$$\begin{aligned} RMS_x &= \pm\sqrt{\Delta_x \Delta_x / n} \\ RMS_y &= \pm\sqrt{\Delta_y \Delta_y / n} \\ RMS_z &= \pm\sqrt{\Delta_z \Delta_z / n} \end{aligned} \tag{13}$$

where Δ_x, Δ_y, and Δ_z are the differences in the x, y, and z coordinates in Eq. (12), and n is the number of compared points.

Second, the influence of target's configuration on the 3D model's accuracy is investigated from Scenario 2 to Scenario 5. In these scenarios, the target's configuration is changed by either the number of targets or the location of targets. The compared area in the overlapping area between PC1 and PC2 is shown in Fig. 5. Theoretically, no overlapping PC is needed for georeferencing when using the one-step approach. However, to evaluate the influence of target's configuration on the georeferencing error, an overlapping area between PC1 and PC2 is used. Two PCs (PC1 and PC2) are not perfectly coincided in the overlapping area because of errors in scan observations, registration and georeferencing procedures, and the control network, etc. The distance between the two point clouds is used as a parameter for

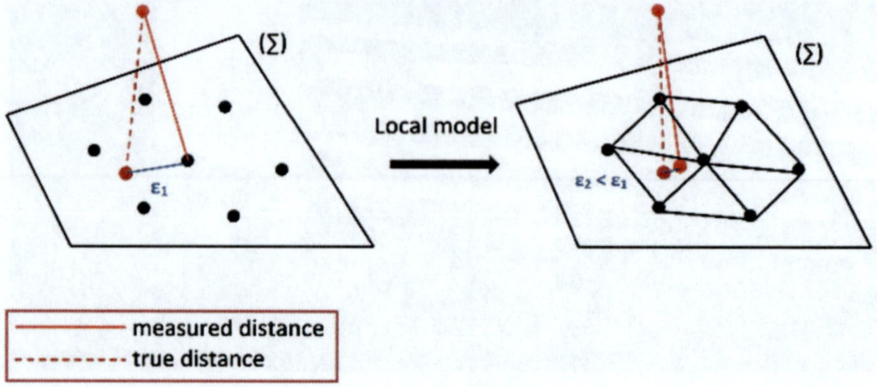

Fig. 12 Concept of local surface model of the C2C method. The local surface allows us to better approximate measured distance ($\varepsilon_2 < \varepsilon_1$) [16]

this evaluation. The distance is computed by the Cloud-to-Cloud (C2C) method in the CC software, as shown in Fig. 12.

The C2C approach applies the Hausdorff distance, which is defined by a max–min distance. The Hausdorff distance between any two finite point sets $A = \{a_1 \ldots a_p\}$ and $B = \{b_1 \ldots b_p\}$ is defined as [17]:

$H(A, B) = \max(h(A, B), h(B, A))$, (14).

where

$$h(A, B) = \max_{a \in A} \min_{b \in B} a - b, \tag{15}$$

and $\|\cdot\|$ is a norm on the points of A and B (e.g., the Euclidean norm).

Function $h(A, B)$ is called the directed Hausdorff distance from A to B. This approach is also used in cloud matching techniques such as ICP [18].

4 Results and Discussion

In this section, the results of the two abovementioned evaluations are shown. First, the one-step and two-step approaches are compared based on the accuracy of the 3D model. The accuracy of the 3D model is evaluated through 20 natural points that are measured in both the georeferenced 3D model and using the total station. The difference in coordinates of these points is summarized in Tables 2 and 3. In the one-step approach, RMSs with respect to the x, y, and z coordinates computed by Eq. (13) are 2.1, 1.6, and 2.7 cm, respectively. Similarly, RMSs are 1.8, 2.2, and 3.2 cm by using the two-step approach.

The results suggest that the accuracy of the 3D model is not significantly different when using two different approaches of the indirect georeferencing procedure. The

Table 3 Difference in 3D coordinates of natural points using the two-step approach

No.	Name	Δ_x	Δ_y	Δ_z	No.	Name	Δ_x	Δ_y	Δ_z
1	T51	1.7	1.3	– 2.0	11	LCR1	0.4	– 1.1	2.5
2	T52	2.0	4.7	– 2.7	12	LCR2	0.0	– 1.8	3.4
3	T53	0.6	0.3	– 5.1	13	PTN	– 1.3	– 2.3	– 4.2
4	T54	– 1.0	1.8	– 2.5	14	XD	0.2	– 2.2	– 2.7
5	T24	– 3.0	3.3	– 2.8	15	T41	2.5	1.1	– 3.2
6	QLDA	– 2.9	2.8	– 4.0	16	T42	2.2	3.4	– 3.4
7	A108	– 2.5	– 0.9	– 4.1	17	T31	– 0.4	0.2	– 0.2
8	VPD	– 1.5	– 0.8	– 3.7	18	T32	– 0.3	– 0.7	– 3.6
9	QHCC1	– 0.2	– 3.1	– 2.8	19	A105	– 2.7	0.8	– 2.4
10	QHCC2	– 1.1	– 3.5	– 3.8	20	A106	– 2.6	– 0.9	– 2.6

RMS values are approximately 2 cm with respect to the x and y coordinates, while these values are about 3 cm with respect to z coordinate. As a previous mention, the important advantage of the two-step approach is that the number of CPs used for georeferencing can be considerably reduced. For georeferencing two PCs, the two-step approach only uses five CPs compared to eight in the one-step approach. It is referred that when using the two-step approach in the case of many PCs, the number of CPs will be significantly reduced. However, the potential drawback of the two-step approach is that it needs more time for the registration procedure. By contrast, the one-step approach is able to directly georeferenced PC without the registration procedure. It allows reducing the time for registration processing. But more CPs need for georeferencing that is the main disadvantage of the one-step approach. Because each PC needs at least three CPs (normally four CPs in practice), the extra survey should be done by a total station or GNSS.

Second, this section also presents the effect of the target's configuration on the 3D model in the case of using the one-step approach for georeferencing procedure. Figure 13 shows that the target's configuration greatly influences the 3D model. When the CPs are in one side of the PC (scenarios 2 and 3), distances between two PCs as previously mentioned in Fig. 12 with respect to the x, y, and z coordinates are approximately from 1.5 to 2 times larger than that in the case that CPs spread around the PC (Scenario 4 and Scenario 5). The maximum values of the mean distance in the x and y coordinates are approximately 8 and 4 cm, respectively, while this value in the z coordinate is about 15 cm, which are two and three folds larger than those values in x and y coordinates.

By contrast, when the CPs are evenly distributed around the PC, the distance between two PCs becomes smaller (Scenario 4 and Scenario 5). The maximum values of the mean distance between two PCs with respect to the x and y coordinates are 3 and 1.5 cm, respectively, while the value in the z coordinate is approximately 10 cm. In these scenarios, the distance between two PCs in the z coordinate is still 3 time larger than those values in the x and y coordinates.

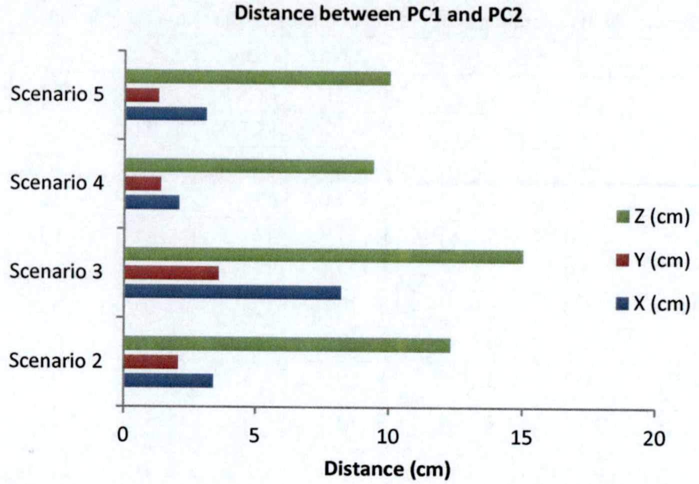

Fig. 13 Influence of target's configuration on the accuracy of the 3D model evaluated by the distance between two PCs in four different scenarios

These results can be explained by the quality of the target's configuration. In Scenario 2 and Scenario 3, when the CPs are in one side of PCs, the distance between PCs is large because of the bad target's configuration. Inversely, when the CPs are distributed throughout the PC, the distance between PCs is small because of a good target's configuration.

In Scenario 3, the distance between two PCs is largest when the CPs are selected in unfavorable positions and far from the compared area (see Fig. 5). A possible reason for the results of this scenario is the significant distortion in the georeferenced model due to unsuitable placement of CPs. These results are in agreement with the theory about the relationship between accuracy and configuration and that is consistent with the results in [19]. In addition, the distance between two PCs in the z coordinate is about from two-fold to three-fold compared to that in x and y coordinates. These results are also consistent with the result in Tables 2 and 3.

Another important result is that the accuracy of the 3D model is insignificantly influenced by adding more numbers of CPs. In Scenario 5 (see Fig. 10), some more CPs are added from Scenario 4 (Fig. 9). Although the number of CPs increases, the distance between two PCs in the compared area remains unchanged. This result can account for the unchanged quality of target's configuration and is in agreement with the result in [19]. In that study, the amount of ground control point considerably changes but RMS is very little different.

Our results show that surveyors need to consider not only the position of targets used for the indirect georeferencing by the one-step approach, but also a moderate amount of targets used to avoid the extra survey for control points. Besides, some limitation of the experiment should be presented here. First, the CPs in PC1 are distributed over the scan object, but the depth of them is quite small at 1.5 m compared to about 50 m of its length. Second, some control point targets are located far from the scanner so that the divergence of laser beam width increases. They are the main reasons why the accuracy of 3D model is centimeter level with respect to the x, y, and z coordinates in Tables 2 and 3.

5 Conclusions

The georeferencing procedure is one of the important steps to transform 3D PC acquired from TLS into the geodetic coordinate system. This paper has given a comparison between the one-step and two-step approaches of indirect georeferencing procedure. The main results of this paper can be summarized as follows.

The accuracies of 3D model when using the one-step and two-step approaches were equivalent. The benefit of the one-step approach was that it can reduce time for data post-processing in the office and the overlap between multiple scans is not need. But this approach needs more control points for georeferencing that needs extra survey of control points since each PC used at least three control points. Inversely, in the two-step approach, a certain overlap PC (30%) was necessary for registration procedure. However, the amount of control points used for georeferencing could considerably reduce especially for complex objects.

The target's configuration greatly influenced the accuracy of the 3D model. The control points should be spread over the scan object. It was recommended that the control points should not be distributed over one part of scan object. The georeferencing procedure could distort the 3D model due to poor configuration.

The combination between the one-step and two-step approaches should be considered to enhance the benefits and diminish the drawback of these approaches when applying for the complex construction like tunnels, bridges, and complex buildings in the subsequent works.

References

1. Scaioni, M.: Direct georeferencing of TLS in surveying of complex sites. Proceedings of the ISPRS Working Group **4**, 22–24 (2005)
2. Altuntas, C., Karabork, H., Tusat, E.: Georeferencing of ground-based LIDAR data using continuously operating reference stations. Opt. Eng. **53**, 114110 (2014)
3. Schuhmacher, S., Böhm, J.: Georeferencing of terrestrial laserscanner data for applications in architectural modeling. In: International Architectural of Photogrammetry, Remote Sensing and Spatial Information Sciences, vol. 36 (2005)
4. Paffenholz, J.-A.: Direct geo-referencing of 3D point clouds with 3D positioning sensors. Gottfried Wilhelm Leibniz Universität Hannover, Hannover (2012)
5. Paffenholz, J.-A., Alkhatib, H., Kutterer, H.: Direct geo-referencing of a static terrestrial laser scanner. J. Appl. Geodesy **4**(3), 115–126 (2010)
6. Lichti, D.D., Gordon, S.J., Tipdecho, T.: Error models and propagation in directly georeferenced terrestrial laser scanner networks. J. Surv. Eng. **131**, 135–142 (2005)

7. Reshetyuk, Y.: Self-calibration and direct georeferencing in terrestrial laser scanning. PhD Thesis, KTH University, Sweden (2009)
8. Pandžić, J., Pejić, M., Božić, B., Erić, V.: TLS data georeferencing-error sources and effects. In: NGEO 2017–7th International Conference on Engineering Surveying, Portugal, Lisbon, October 18–20, 2017, pp. 293–300. Laboratório nacional de engenharia civil, Lisboa (2017)
9. Khoshelham, K., Gorte, B.: Registering pointclouds of polyhedral buildings to 2D maps. In: Proceedings of the 3rd ISPRS International Workshop 3D-ARCH 2009: 3D Virtual Reconstruction and Visualization of Complex Architectures, Trento, Italy, 25–28 February 2009. International Society for Photogrammetry and Remote Sensing (2009)
10. Lichti, D.D., Gordon, S.J.: Error propagation in directly georeferenced terrestrial laser scanner point clouds for cultural heritage recording, pp. 22–27. Proc. of FIG Working Week, Athens, Greece, May (2004)
11. http://www.leicageosystems.com/hds/en/lgs_29445.htm. Last Accessed 20 May 2022
12. Tait, M., Fox, R., Teskey, W.: A comparison and full error budget analysis for close range photogrammetry and 3D terrestrial laser scanning with rigorous ground control in an industrial setting. In: Proceedings of INGEO 2004 and FIG Regional Central and Eastern European Conference on Engineering Surveying, Bratislava, Slovakia, November, pp. 11–13 (2004)
13. Bornaz, L., Lingua, A., Rinaudo, F.: Multiple scanner registration in LIDAR close-range applications. INT. ARCH. PHOTOGRAMMETRY REMOTE SENSING SPATIAL INF. SCI. **34**, 72–77 (2003)
14. Mikhail, E.M., Bethel, J.S., McGlone, J.C.: Introduction to modern photogrammetry. New York 19 (2001)
15. Reshetyuk, Y.: Investigation and calibration of pulsed time-of-flight terrestrial laser scanners. Licentiate Thesis, KTH University, Sweden (2006)
16. https://www.cloudcompare.org/doc/wiki/index.php?title=Distances_Computation#Local_modeling. Last accessed 23 March 2022
17. Huttenlocher, D.P., Rucklidge, W.J.: A multi-resolution technique for comparing images using the Hausdorff distance. https://ieeexplore.ieee.org/stamp/stamp.jsp?tp=&arnumber=341019 (1992)
18. Besl, P.J., McKay, N.D.: Method for registration of 3-D shapes. In: Sensor Fusion IV: Control Paradigms and Data Structures, pp. 586–606. Spie (1992)
19. Page, C., Sirguey, P., Hemi, R., Ferrè, G., Simonetto, E., Charlet, C., Houvet, D.: Terrestrial laser scanning for the documentation of heritage tunnels: an error analysis. 2017. In: FIG Working Week 2017, Helsinki, Finland (2017)

Technological Solutions for Fly Ash and Red Mud Upcycling Approach the Vietnam's Government Target of Net-Zero Carbon by 2050

Van-Duc Nguyen, Chang-Woo Lee, Xuan-Nam Bui, Pham Van Chung, Quang-Tuan Lai, Hoang Nguyen, Tran Thi Huong Hue, Van-Trieu Do, and Ji-Whan Ahn

Abstract At the Conference of the Parties (COP26) in 2021, over 190 world leaders came together to address climate change. In the remarks at COP26, Vietnam's Government had committed to the net-zero carbon emissions target by 2050. This can be considered as a critical goal for Vietnam Government to respond to climate change. To achieve the Vietnamese Government's goal of net-zero emissions by 2050,

V.-D. Nguyen (✉) · Q.-T. Lai · J.-W. Ahn (✉)
Center for Carbon Mineralization, Mineral Resources Research Division, Korea Institute of Geosciences and Mineral Resources, 124 Gwahak-Ro, Yeseong-Gu, Daejeon 34132, South Korea
e-mail: ahnjw@kigam.re.kr

C.-W. Lee
Department of Future Energy Engineering, College of Engineering, Dong—A University, Busan 49315, South Korea
e-mail: cwlee@dau.ac.kr

X.-N. Bui · H. Nguyen
Mining Faculty, Department of Surface Mining, Hanoi University of Mining and Geology, 18 Vien Street, Hanoi 100000, Vietnam
e-mail: buixuannam@humg.edu.vn

H. Nguyen
e-mail: nguyenhoang@humg.edu.vn

V.-D. Nguyen · C.-W. Lee · X.-N. Bui · H. Nguyen
Innovations for Sustainable and Responsible Mining (ISRM) Research Group, Hanoi University of Mining and Geology, 18 Vien Street, Hanoi 100000, Vietnam

P. Van Chung
Department of Mine Surveying, Hanoi University of Mining and Geology, 18 Vien Street, Hanoi 100000, Vietnam
e-mail: phamvanchung@humg.edu.vn

Q.-T. Lai
Department of Resources Recycling, University of Science and Technology, 217 Gajeong-Ro, Yuseong-Gu, Daejeon 34113, Republic of Korea

T. T. H. Hue
National Academy of Public Administration, 77 Nguyen Chi Thanh, Hanoi 100000, Vietnam
e-mail: huetth@napa.vn

L. Q. Nguyen et al. (eds.), *Advances in Geospatial Technology in Mining and Earth Sciences*, Environmental Science and Engineering,
https://doi.org/10.1007/978-3-031-20463-0_12

the Vietnamese Government needs to focus on low-carbon sustainable development technologies. Carbon Capture Utilization and Storage (CCUS) technology solutions can be an appropriate option based on the experience of developed countries. Before COP26, CCUS technologies were not interested in Vietnam because of the expensive investment cost, and no emission targets had been set. However, after COP26, the Vietnam Government has begun to pay attention to advanced technologies which reduce greenhouse gas emission. Accordingly, the Decree 06/2022/ND-CP regulating the mitigation of greenhouse gas emissions and the protection of the ozone layer was issued on January 7, 2022. Advanced technologies that contribute to the reduction of greenhouse gas emissions are encouraged to apply to meet the demand for net-zero carbon by 2050. This study aims to evaluate the possibilities of applying CCUS technology to recycle fly ash and red mud as raw material for low CO_2 emission cement production. Thus, CCUS technical solutions in recycling waste of fly ash from the thermal power plants and red mud from bauxite processing plants are proposed. These technical solutions can provide significant support in achieving the overall goal of the Vietnam Government to reduce greenhouse gas emissions to the net-zero by 2050.

Keywords Red mud · Upcycling · Net-zero carbon · CCUS technology · CSA cement · Vietnam

1 Introduction

Over the decades, Vietnam has been known as the fastest-growing economy globally. Between 2002 and 2020, GDP per capita increased 2.7 times, reaching almost US$2,800. Over the same period, poverty rates (US$1.90/day) declined sharply from over 32 percent in 2011 to below 2% [1]. Fossil fuels such as coal, oil, and gas are the dominant energy sources for long-term economic growth. Specifically, the contribution of fossil fuels to Vietnam's total energy consumption increased from 27.6% in 1990 to 70% in 2016 [2]. However, fossil fuels are the leading cause of global greenhouse gas (GHG) emission, accounting for 58% [3]. Table 1 illustrates the Vietnam annual electricity generation by source from 2019 to 2020 [4]. Even though in 2021, Vietnam revised the national power development plan for the period of 2021–2030 and will prioritize the development of renewable energy power projects such as wind power and solar power; the dominant source of fuel for electricity production in Vietnam is coal. Currently, there are about 25 thermal power plants in operation in Vietnam. Every year, more than 13 million tons of coal ash are discharged into the environment [5]. The most common disposal option of coal ash is landfilling. Therefore, finding solutions to recycle industrial coal ash will significantly contribute to environmental protection and bring many economic values.

V.-T. Do
Department of Surface Mining, Institute of Mining Science and Technology, 03 Phan Dinh Giot Street, Phuong Liet, Hanoi 100000, Vietnam

Table 1 Vietnam power generation by sources of fuel in 2019–2020 [4]

Power source	2019 (%)	2020 (%)
Hydropower	36.81	29.98
Coal fired	35.83	31.10
Gas-fired and oil-fired	16.07	12.78
Wind	0.67	0.75
Solar	8.47	12.80
Rooftop solar	0.58	11.23
Biomass	0.53	0.53
Imported	1.04	0.83
Total	1.00	1.00

Figure 1 shows the CO_2 emission from fossil fuel combustion by year in Vietnam. In 2020, CO_2 emissions was 254 million tons [6], and it can be seen that Vietnam's CO_2 emissions increased dramatically from 1980 to 2020 [6]. Before COP26, the Vietnam Government paid no attention and neither took practical actions to reduce greenhouse gas emissions. However, after COP26, the Government has committed to reducing emissions to net-zero carbon by 2050. This is not just a challenge and but also an opportunity to apply technologies to reduce greenhouse gas emissions. Based on the experience of developed countries, several strategies can be applied to achieve net-zero emissions by 2050: (a) applying geoengineering approaches, (b) improving energy efficiency, (c) developing alternative fuels with lower CO_2 emissions, and (d) developing low-carbon emission technologies [7, 8]. Although CCUS has inspired both environmental and economic benefits, making it possible for industrial deployment [7, 8] and CCUS alone can contribute to at least 20% of CO_2 emission reductions worldwide by 2050 [7]. It is estimated that fossil fuels (e.g., coal, natural gas, and petroleum) are still expected to dominate the world energy supply until at least the middle of this century. CCUS plays a vital role in the transition to a low-carbon energy society and can allow continued use of fossil fuels until alternative low-carbon energy sources are ready for adoption. The general conceptual structure of the CCUS supply chain is shown in Fig. 2. It has three main components: (1) CO_2 capture from different stationary emission sources (e.g., power plants, cement plants, iron and steel manufacturing, and chemical industries), (2) transport of captured CO_2 through pipelines, ships, and road/rail tankers in dense phase, and (3) pump CO_2 into different geological reservoirs (e.g., saline aquifers, depleted oil reservoirs or gas fields, and unidentifiable coal seams) over a long period, or use CO_2 in different industrial processes.

This study evaluates and proposes the possibility of applying CCUS technology in recycling industrial waste such as coal ash and red mud. The current legal framework as well as opportunities and challenges will be discussed. The ultimate aim of this study is to suggest practical solutions that contribute to the overall goal of the Vietnam Government to reduce CO_2 emissions to net-zero by 2050. This study can open up

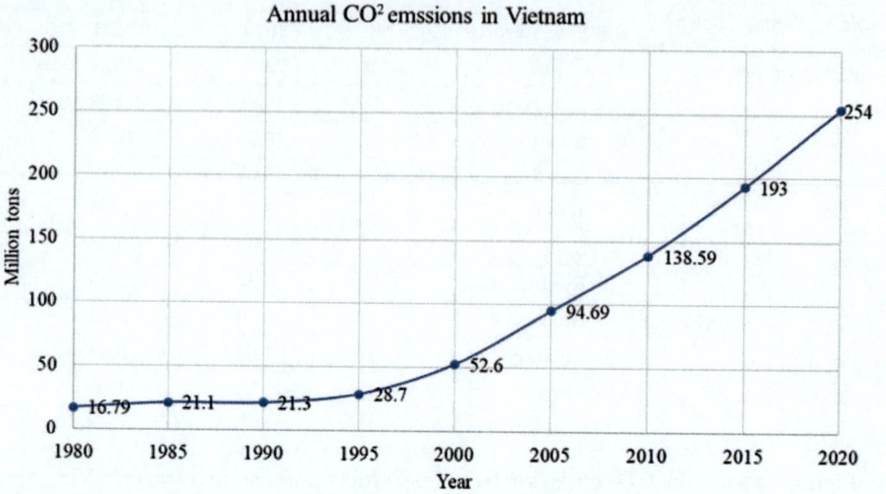

Fig. 1 CO_2 emissions from the burning of fossil fuels for energy in Vietnam [2, 3, 6]

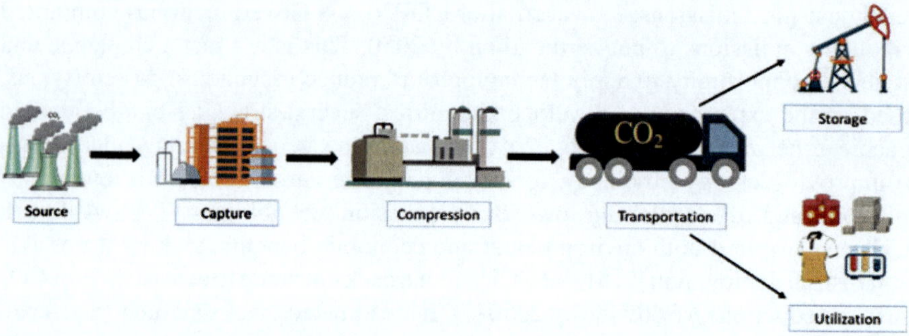

Fig. 2 CCUS supply chain scheme [9]

opportunities for cooperation on transferring CCUS technologies to Vietnam in the field of waste recycling in the general and industrial waste in particular.

2 Potential Application of CCUS Technical Solutions for Coal Ash and Red Mud Upcycling in Vietnam

2.1 *Legal Framework of Greenhouse Gas Emission*

The Vietnam Government's decree on mitigation of greenhouse gas (GHG) emissions and protection of the ozone layer is described through Articles 92, 139 of the Law on Environmental Protection 2020. The regulation regarding to climate change in the Law on Environmental Protection in 2020 is an essential legal basis for implementing measures to achieve the target of reducing greenhouse gas emissions set out in the Nationally Natural Contributions (NDC) of Vietnam; is a prerequisite for the development of the domestic carbon market; manages and eliminates ozone-depleting substances and greenhouse-effect substances controlled under the International Agreement on ozone layer protection to which Vietnam is a member. Vietnam National GHG Inventory System can be seen in Fig. 3 [10].

On January 7, 2022, the Government-issued Decree 06/2022/ND-CP is regulating the mitigation of greenhouse gas emissions and the protection of the ozone layer. Accordingly, the methods of mitigating greenhouse gas emissions include [11]:

- Policy measures, management of activities to reduce greenhouse gas emissions.
- Plan to reduce greenhouse gas emissions at the field and grassroots level.
- Technologies, production processes, and services with low greenhouse gas emissions.
- Mechanisms and methods of cooperation on mitigation of greenhouse gas emissions are consistent with the provisions of laws and international treaties to which the Socialist Republic of Vietnam is a signatory.

In addition, the subjects of mitigation of greenhouse gas emissions are specified as follows:

- Establishments that are on the list of sectors and facilities that emit greenhouse gases must carry out an inventory of greenhouse gases promulgated by the Prime Minister.
- Organizations and individuals other than those mentioned above are encouraged to reduce greenhouse gas emissions following their conditions and activities.

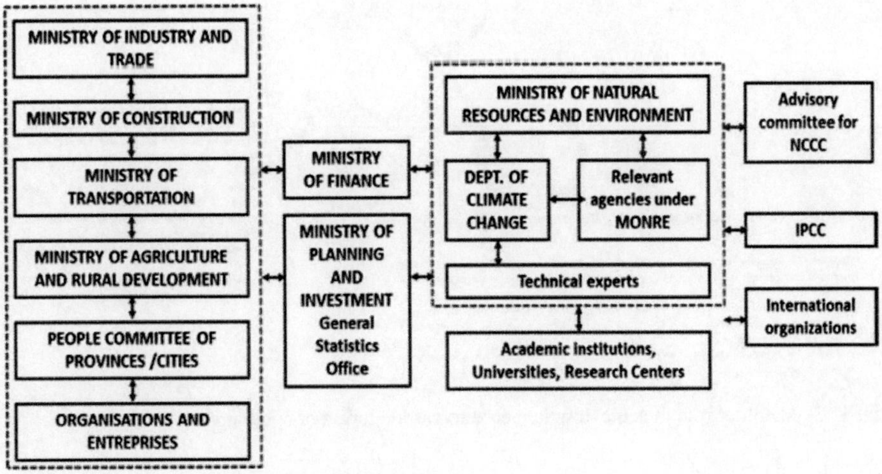

Fig. 3 Vietnam National GHG Inventory System [10]

- The ministries that manage energy, agriculture, land use and forestry, waste management, industrial processes are the ministries: Ministry of Industry and Trade, Ministry of Transport, Ministry of Agriculture and Rural Development, Ministry of Natural Resources and Environment, Ministry of Construction.

Thus, before the COP26 conference, the Vietnam Government did not have specific strategies to reduce greenhouse gas emissions. There are only a few general provisions in the 2014 environmental protection law No regarding emission reduction 55/2014/QH13 [12]. However, these regulations were still not clear with unspecific strategies and actions. The fact that many ministries and agencies are involved leads to unclear responsibilities for relevant ministries and agencies in Fig. 4. However, after COP26, the Vietnam Government issued regulations on reducing greenhouse gas emissions to reduce CO_2 emissions to net-zero by 2050. Accordingly, in the coming years, the Government will prioritize the application of advanced technologies such as CCUS. Figure 4 shows the stakeholder roles to achieve the net-zero carbon target of Vietnam. This opens up opportunities for pioneer companies in applying CCUS technology to reduce greenhouse gas emissions. This study evaluates and proposes the possibility of applying CCUS technology in recycling industrial waste such as coal ash and red mud. The current legal framework as well as opportunities and challenges will be discussed. The ultimate aim of this study is to suggest practical solutions that contribute to the overall goal of the Vietnam Government to reduce CO_2 emissions to net-zero by 2050. This study can open up opportunities for cooperation on transferring CCUS technologies to Vietnam in the field of waste recycling in the general and industrial waste in particular.

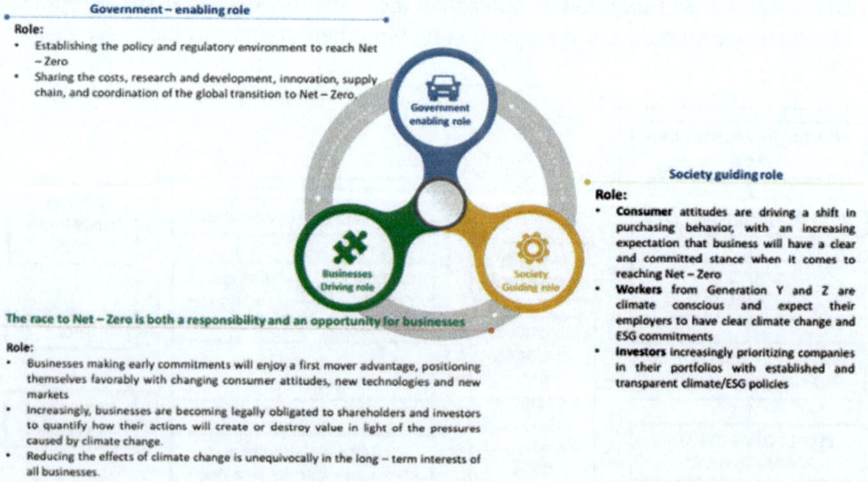

Fig. 4 Stakeholder role in achieving the net-zero carbon target of Vietnam

2.2 Current Coal Ash and Red Mud Management in Vietnam

As mentioned above, this study will evaluate the applicability of CCUS technology in recycling the industrial waste coal ash and red mud in Vietnam. Accordingly, the current status and environmental issues related to industrial waste such as fly ash and red mud are discussed for proposing suitable CCUS technical solutions.

2.2.1 Management of Coal Ash

Based on the Decision No. 452/QD-TTg of the Vietnam Prime Minister in 2017, it has approved the scheme to promote the treatment and use of ash, slag, and gypsum of thermal power plants, chemical plants, and fertilizers as raw materials for production building materials and in construction works. However, after more than five years of implementing the Prime Minister's Decision No. 452/QD-TTg although there have been many efforts, the actual results have not yet reached the set goals. Currently, Vietnam has 25 coal-fired thermal power plants in operation. The total amount of ash and slag disposal is about 13 million tons/year, of which fly ash accounts for 80 to 85% [5]. By the end of 2020, the total amount of ash and slag consumed by thermal power plants nationwide is about 44.5 million tons, equivalent to 42% of the entire disposals over the years [4]. Ash and slag are used the most in mineral additives for cement, estimated at 24 million tons, accounting for 70%. Ash and slag production of baked clay bricks and unburnt bricks is estimated at 4 million tons, accounting for 12%. Civil constructions are estimated at 3 million tons, accounting for 8%. Making materials for leveling and filling roads of all kinds is about 3.5 million tons, accounting for 9% [13]. The chemical composition of fly ash from three thermal power plants in Vietnam was described in Table 2. Based on TCVN 10,302:2014, the fly ash at the most thermal power plants in Vietnam is classified as F class, with total oxide content of SiO_2, Al_2O_3, and Fe_2O_3 over 70 weight % (wt.%) [13].

However, the total amount of ash and slag stored at the storage yards of coal-fired power plants is still about 47.65 million tons [14]. However, the Government has issued regulations on the reuse of ash slag for other industries under regulation No.

Table 2 Chemical composition of fly ash from 3 plants in Vietnam [13]

Chemical composition	Pha Lai (wt. %)	Mong Duong (wt. %)	Duyen Hai 1 (wt. %)
SiO_2	57	54.27	55.7
Fe_2O_3	4.7	4.71	6.1
Al_2O_3	23.8	25.02	23.1
K_2O	6.56	6.76	3.76
MgO	1.2	1.22	1.7
CaO	0.81	0.91	0.7
SO_3	0.53	0.73	0.15

452/QD-TTg. However, due to technological limitations, the implementation of ash slag recycling is still minimal. As a result, the remaining fly ash seriously affects the ecosystem balance and increases the natural disaster. Thus, the application of CCUS technology in recycling waste ash can be a new idea that helps the Government reduce CO_2 emissions and achieve economic efficiency.

2.2.2 Management of Red Mud

Vietnam has considerable bauxite resources, with reserves ranked third in the world (USGS, 2018). Based on the geological surveys [15, 16], bauxite reserves are distributed in many places of Vietnam, especially in Tay Nguyen [17, 18]. According to the zoning of exploration, exploitation, processing by 2025, and use of bauxite ore from 2007 to 2015, the bauxite ore reserves in Vietnam have been identified and forecasted to be about 5.5 billion tons, about 91 million tons in the Northern region and about 5.4 billion tons in the Southern and Central Highlands regions [19]. Currently, in Vietnam, there are three active bauxite mining projects including Ma Cat—Lang Son mine, Nhan Co—Dak Nong mine, and Tan Rai—Lam Dong mine. It is mainly concentrated in 2 bauxite mines, Tan Rai and Dak Nong managed by VINACOMIN, with a total mining capacity of about 3.07 million tons of concentrate per year (accounting for 98.59% of the total output of bauxite concentrate) [19].

There are many environmental problems arising in the process of bauxite ore mining and aluminum production at the two Tan Rai and Nhan Co aluminum complexes. The environmental issues with negative impacts related to production technology include arable land loss due to bauxite mining, destruction of natural water balance, tailing sludge and tailing sludge reservoir, red mud and red mud reservoir, fly ash and waste slag of thermal power plants, air pollution, water pollution, etc. The oxide composition of red mud in Vietnam compared with other countries is given in Table 3.

Thus, compared with the red mud of factories abroad, the red mud of two Vietnamese aluminum plants has higher Fe content but lower Al and Si, as well as different trace element compositions. The red mud of Tan Rai and Nhan Co aluminum plants also shows safe level of background radiation. This is the main reason for the possibility that we can utilize red mud for other practical applications. The red mud of Tan Rai aluminum plant is divided into three categories: Grain grade 1–0.05 mm accounts for 57%, grain grade 0.001 to <0.05 mm accounts for 33, 8%, grain grade <0.001 mm accounted for 9.13% [20].

Since red mud also contains a large amount of residual alkali (NaOH) from the Bayer process, the pH is usually high, in the range of 10–13 [31, 32]. The red mud of Tan Rai aluminum factory collected at red mud disposal site No. 1 has a pH of 11.12–11.23 [20], while the red mud taken from the red mud lake No. 1 of Nhan Co

Table 3 Comparison of oxide composition of red mud in Vietnam with other countries

Country	%, Composition (by dry weight)						References
	Fe_2O_3	Al_2O_3	SiO_2	TiO_2	Na_2O	CaO	
Vietnam	46.32–46.32	16.91–17.56	6.62–6.72	5.48–7.20	5.06–5.43	4.48–5.20	[16, 19, 20]
China	10.63–29.79	7.34–22.96	18.92–27.99	1.83–6.64	0.58–8.93	2.03–44.69	[21–23]
Iran	22.17	13.98	13.00	7.17	4.20	24.25	[24]
Turkey	35.04	20.20	17.29	4.00	9.40	5.30	[25]
Brazil	19.85	19.87	14.34	2.66	7.35	4.61	[26]
Spain	31.80–37.22	12.40–20.10	3.81–6.10	20.10–22.60	4.64–4.70	4.78–6.30	[27, 28]
Hungary	31.63	10.50	12.20	5.90	6.09	11.60	[29]
Jamaica	49.50	16.50	3.00	7.00	2.30	5.50	[30]

aluminum plant has a lower pH value of 10.5 [20]. Excess amount of alkali has to be removed before recirculating water from the red mud reservoir.

Currently, in Vietnam, red mud is regarded as one of the hazardous wastes that urgently need to be addressed. Based on the current plan on bauxite mining in the Central Highlands, by 2025, the amount of red mud discharged will reach 23 million tons; in ten years, it will be 230 million tons; and in 50 years, it will be 1.15 billion tons [33]. The huge volume of red mud will require a huge area of land to landfill. Therefore, finding a solution to reuse red mud in Vietnam is essential.

3 Technological Solutions for Fly Ash and Red Mud Upcycling in Vietnam

3.1 *Current Fly Ash and Red Mud Recycling*

Over the years, there have been so many studies around the world to utilize red mud and fly ash. The stabilization material for the preparation of liner can be found by [34, 35]; the removal of heavy metal was studied by [36–38]. The recovery of iron and other heavy metals has been carried out by [39–41]. The production of radiopaque materials was researched by [42]. The production of construction bricks by red mud can be found by [43–48]. The application of red mud for cement production was studied by [49–52]. Table 4 summarizes the previously studied results on the utilization of red mud combined with fly ash. It can be observed that the combination of fly ash and red mud for producing construction and building material has the considerable potential to consume large quantities of waste.

Figure 5 shows the various technological approach that can be applied for a comprehensive utilization technology of red mud in the world [58]. The widespread use of red mud as a resource opens up in various fields, such as red mud geopolymers in the construction and metal mining industries. Much research has been done to extract metals from red mud to utilize the resource [58–61].

Red mud disposal risks the release of radioactive materials. Researchers have proposed several options for treating and applying red mud from aluminum production plants. Red mud can be a significant resource in the construction industry, such as bricks and other suitable materials for building and road pavement. Red mud can be used for iron recovery or and in the metal mining and smelting sectors to produce cast iron by-products and calcium-titanium-rich compounds for titanium recovery [58].

Table 4 Previous studies on utilizing red mud and fly ash

Country	Red mud combined with	Studied parameters	Utilization sectors	References
USA	Fly ash (Class C*)	• Three different weight ratios of red mud and fly ash were 80:20, 50:50, and 20:80 to form the geopolymer • The alkaline NaOH solution was not applied	Geopolymer making	[53]
Australia	Fly ash (Class F**)	• Red mud was varied from 0 to 40 wt% in the red mud—fly ash mix • NaOH solution as an activator	Paving block—construction material	[54]
USA	Fly ash (Class F)	• The mass ratio of 1:4 (fly ash/red mud) • 50 wt% of NaOH, 2 M sodium trisilicate, and deionized water with a proportion of 3:7:3 as alkaline activator	Geopolymer making	[55]
South Korea	Fly ash (Class F)	• Red mud combined with CNF powder ($Ca(OH)_2$, Na_2CO_3, fly ash), where red mud of 0 to 30 wt% and composition of C, N, and F were changed • Red mud content was varied: 10%, 20%, 30% • NaOH used as alkaline activator with the concentration (1, 3, and 5 wt%)	Construction material	[56]

(continued)

Table 4 (continued)

Country	Red mud combined with	Studied parameters	Utilization sectors	References
India	Fly ash (Class F)	• Red mud and fly ash for manufacturing bricks with pyrophyllite • Fly ash and red mud can be used up to 50% in making brick	Construction material	[43]
China	Fly ash (Class C)	• The ratio of red mud and fly ash was 50:50 • Alkaline thermal pretreated red mud with condition (calcined at 800 oC) with different dosages of Na2O • Na2O of 5% was fixed, and the fly ash content of 0, 25, 25, 50, 75, and 100 wt% was observed	Geopolymer making	[57]

* Fly ash (Class C)—$SiO_2 + Al_2O_3 + Fe_2O_3 \geq 50\%$ (cementitious) more $CaSO_4$, free lime, C3A, calcium-rich glass, MgO

** Fly ash (Class F) – $SiO_2 + Al_2O_3 + Fe_2O_3 \geq 70\%$ (pozzolanic) more glass, alumino silicate glass, quartz

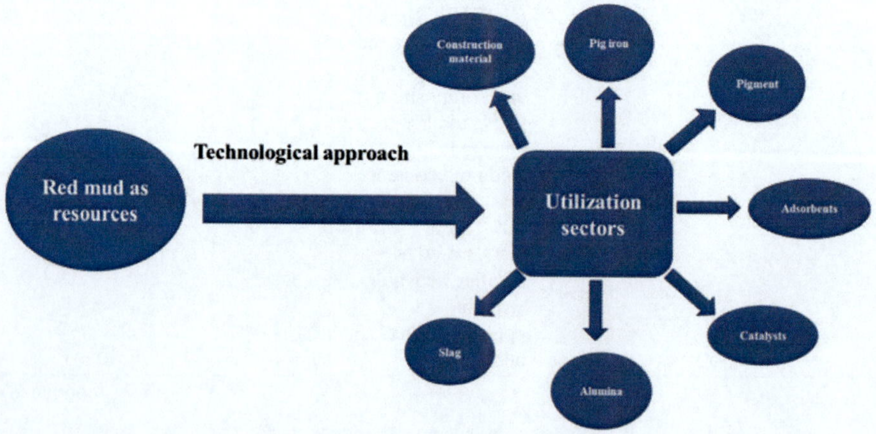

Fig. 5 Red mud upcycling technologies in various sectors [58]

3.2 *Carbon Mineralization as a Technical Solution for Coal Ash and Red Mud Upcycling in Vietnam*

3.2.1 Utilizing Fly Ash and Red Mud by Carbon Mineralization Technology as Raw Material for Green Cement

Currently, the recycling of red mud and fly ash for industrial applications is not a new approach to utilizing solid wastes in the world. However, because of technological and policy limitations, landfilling is currently the only technique of treating red mud and fly ash in Vietnam. According to the above studies, the qualities of Vietnam's red mud and fly ash are ideal for recycling and industrial application and contributing to a circular economy.

Furthermore, the alumina production complex in Vietnam today includes thermal power plants, alumina plants, and mining activities, all of which release significant amounts of CO_2. Vietnam's Government issued a new policy on reducing greenhouse gas emissions under decree No. 06/2022/ND-CP regulating the mitigation of greenhouse gas emissions and the protection of the ozone layer. It is undeniable that, in order to meet the Government's goals, measures to reduce greenhouse gas emissions will be required in the near future. This study proposes a carbon mineralization model for recycling red mud and flying ash in Vietnam, as shown in Fig. 6. Carbon from power plants and other industries will be captured and stored, then mineralized with red mud and fly ash to stabilize the heavy metals in red mud. Then, red mud is used as an additive to other industries such as construction materials and cement.

The concept of green cement has been mentioned since the 1950s. However, up to the present time, the application of green cement technology in industries is still limited. Researchers from the Korea Institute of Geosciences and Mineral

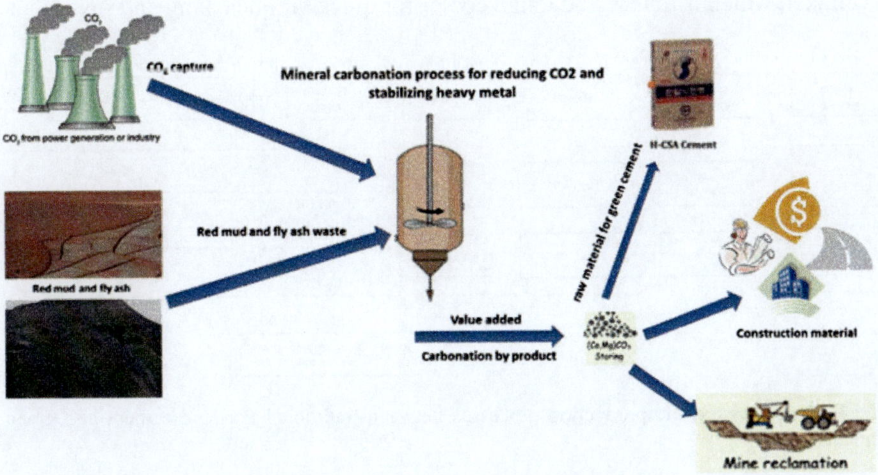

Fig. 6 Utilizing fly ash and red mud by carbon mineralization technology model in Vietnam

Science (KIGAM) collaborated with Korea Hanil Company conducted a pilot plant scale study on synthesizing calcium sulfoaluminate (CSA) with the goal of climate change mitigation. The findings from the study show that the process of synthesizing CSA has reduced the release of CO_2. Moreover, CSA cement outperforms Ordinary Portland Cement (OPC) in terms of properties.

Differences in the production processes between Portland cement and green cement can be shown in Fig. 7. One of the most distinctive features of CSA cement is the alternative materials, including fly ash from the coal power plant of Circulating Flow Bed Combustion (CFBC) and red mud. These alternative raw materials are added at the time of feeding materials for clinker production, which reduces greenhouse gas emissions (fixed combustion) by reducing substances releasing CO_2 (process emission) and coal fuel generated through decarboxylation in the clinker calcination. This technique increases the ratio of cement blended materials to reduce the emission of greenhouse gas through reduced production of clinker. The CSA production process is as follows:

- Process 1 (raw material feeding): The raw materials, including limestone, fly ash, and red mud, are blended precisely through a material mill, and the fine powder thus generated goes through the drying and grinding process inside the material mill and saved in the silo. Air is used inside the material silo to mix saved materials again for proper calcination of clinker in the process
- Process 2 (calcination): A rotary kiln is used for the calcination of clinker. Blended material manufactured in the process of material feeding is supplied from the top end of the preheat tower. This blended material is delivered to the entrance of the kiln through sufficient thermal exchange in the preheat tower. From the exit of the kiln, diesel fuel or pulverized coal is blown into the kiln burner to burn the material. Blended materials are moved gradually by the rotation of the kiln to be sintered at the temperature of 1250 °C inside the calcination kiln to form a CSA clinker, which is transferred to the cooler for quick air quenching and saved.

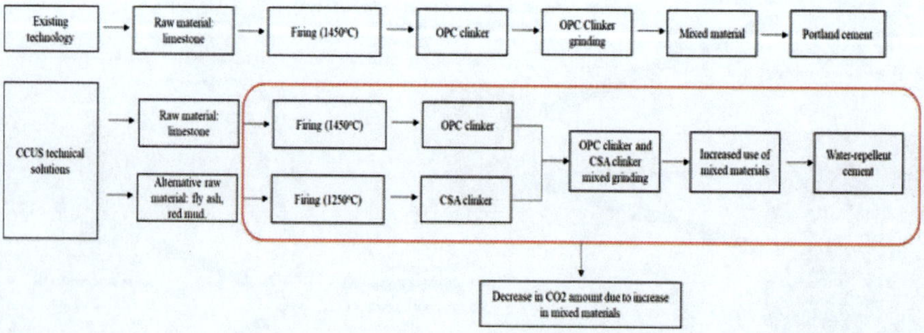

Fig. 7 Differences in the production processes between traditional Portland cement and green cement

Table 5 Required chemical composition for OPC and CSA cement and properties of fly ash and red mud in Vietnam [13, 20, 62]

Type	CaO (%)	Si_2O (%)	Al_2O_3 (%)	Fe_2O_3 (%)	SO_3 (%)
Portland cement	60–67	17–25	3–8	0.5–6	1–3
CSA cement	52.8	1.5	16.1	0.5	27.5
Fly ash in Vietnam	0.81–7.1	46.4–57	22–25	4.7–15.6	0.15–0.9
Red mud in Vietnam	4.48–5.2	6.62–6.72	16.91–17.56	46.41–46.32	0.35

- Process 3 (mixing—grinding): The CSA clinker cooled down after calcination in the kiln is grinded finely in the cement mill with gypsum, OPC clinker, and other blended materials to manufacture CSA cement.

Compared to the traditional OPC cement production process, there are several dominant features of CSA cement: (i) reduction of process emission in the calcination process based on the reduction of limestone through the feeding of alternative raw materials in the production process of clinkers; (ii) reduction of fixed emission of fossil fuel due to the lowered calcination temperature by the feeding of alternative raw material; (iii) reduction of greenhouse gas emissions as a result of reduction of clinker production by the increased ratio of blended materials. The efficiency of CO_2 emission reduction will be discussed as following section. Some advantages of CSA cement are:

- The compressive strength of concrete built with CSA is over 5000 psi within 24 h. Furthermore, CSA cement has the same hardness as Portland cement after 24 h, whereas Portland cement takes up to 28 days to fully solidify. CSA is preferred over OPC for this reason in some applications where rapid setting and high durability are required. For instance, when constructing an airport runway or repairing a bridge or highway that requires quick construction, construction speed is critical. Therefore, CSA is preferred over OPC in this kind of urgent situation.
- Portland cement requires a sintering temperature of 1450 °C in a kiln, whereas CSA cement needs only 1250 °C [62]. As a result, OPC clinker is harder than CSA clinker and consumes higher energy and power for grinding. Therefore, this can be one of the reasons CSA cement is cheaper than OPC in terms of cost.
- The required chemical composition for OPC and CSA cement is compared with fly ash and red mud properties in Vietnam in Table 5. It indicates that Vietnam's fly ash and red mud properties are suitable as alternative raw materials for green cement production. [62].

3.2.2 CO_2 Emission Reduction

During the OPC manufacturing process, 48% of CO_2 is emitted to the environment through the conversion of $CaCO_3$ to CaO [62]. Besides, an additional 42% of CO_2 is emitted to the atmosphere through fossil fuels combustion to obtain elevated temperatures within kilns. Hence, in general, CO_2 emission rate reaches 90% in OPC production. The manufacture of 1ton CSA cement, contrarily, generates just 216 kg CO_2, which is around 62% less than the OPC. Previously, the common practice of utilizing CSA cement was replacing OPC with pozzolan-based industrial wastes like fly ash and steel-furnace slag at a typical ratio of roughly 10–30% [62]. However, 100% replacement of OPC with CSA, known as green cement, will help protect the environment since reducing CO_2 emission is significant. When compared to OPC, which already contains a large quantity of green "pozzolan" substitution by fly ash and slag, CSA cement emits 2 to 6 times less CO_2 [62]. Among nine various forms of cement, including magnesia (Sorel cement), methacrylic sodium (liquid glass), and calcium aluminate cement, the CSA cement emits the least amount of CO_2.

The CSA cement has been used in versatile construction works. For instance, repairing roads, bridges, and airport runways are some of the construction works that require the application of CSA cement for rapid setting and high durability, as mentioned above. Moreover, CSA cement is mixed with Portland cement and used to shrinkage compensated concrete for subterranean pits.

3.2.3 Possibility to Commercialize the Green Cement in the Vietnam Market

The cement industry in Vietnam currently has 90 production lines, with a total capacity of 106.6 million tons/year [63]. Domestic consumption and construction of infrastructure in the country are still the targets of most domestic cement producers. According to the January 2021 statistics of the Vietnam Cement Association, cement production in the first month of the new year reached 7.16 million tons of cement. In January, consumption of cement and clinker was about 8.22 million tons, an increase of 52% over the same period in 2020 and a decrease of 15% compared to December 2020 [61]. Total consumption of cement and clinker in the first quarter of 2021 reached 21.6 million tons, up 2.6% over the same period, becoming a bright spot in the picture of construction material consumption. Up to now, there are 89 cement plants in Vietnam in which the integrated plants are 66; clinker plant one and grinding plants 22 [61]. The demand for cement production is extremely high. Thus, applying green cement technology to reduce greenhouse gas emissions is necessary. Besides, the annual volume of ash and slag discharged into the environment is very large. Recycling coal ash and red mud for green cement production brings economic and environmental benefits. The SWOT analysis results utilizing fly ash and red mud as alternative raw materials for green cement production in Vietnam are summarized in Fig. 8. Hence, the green cement with low CO_2 emissions can be commercialized in the Vietnam market to reduce the national emission target.

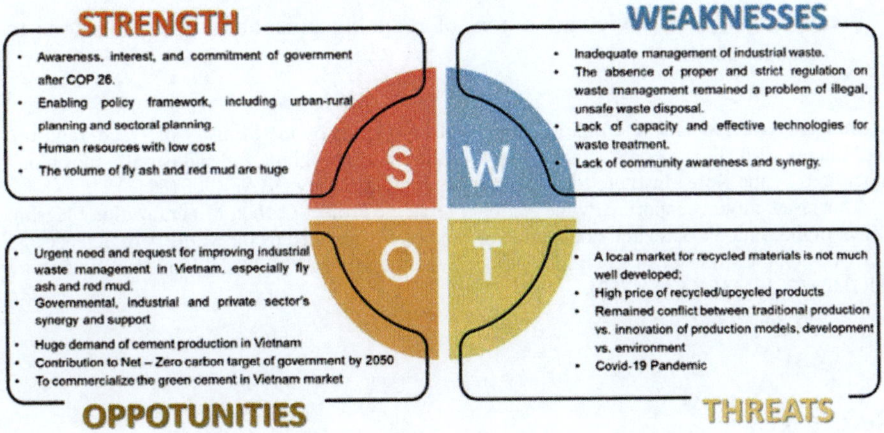

Fig. 8 SWOT analysis of utilizing fly ash and red mud as raw material for green cement

4 Conclusions

Recently, the greenhouse gas emission in Vietnam has increased considerably. The primary source of energy used is fossil fuels, and this is considered to be the main reason for the high annual increase in emissions. This study evaluates the current status of the policy and management system of application of coal fly ash and red mud in Vietnam. Several important conclusions can be summarized as follows:

- Vietnam is a country with a large amount of greenhouse gas emissions not only in the region but also globally. After COP26, Vietnam pledged to reduce emissions to zero by 2050. This opens up opportunities for cooperation on CCUS technology transfer to reduce greenhouse gas emissions in Vietnam.
- Government-issued Decree 06/2022/ND-CP regulates the mitigation of greenhouse gas emissions and protecting the ozone layer. This is an important step to force national economic groups' production units such as thermal power, cement, and bauxite to apply advanced technologies in reducing greenhouse gas emissions.
- The annual volume of coal ash and red mud discharged into the environment is very large. However, the current management and recycling of coal ash and red mud in Vietnam is limited. Up to now, there is no suitable solution for recycling ash slag and red mud. This study proposes an alternative to recycle coal ash and red mud as an additive material for green cement production after the carbonation process. This technology has a potential to make a crucial contribution to reducing CO_2 emission, which helps Vietnam Government achieve the national target of net-zero carbon in 2050 and bring economic efficiency.
- Finally, this study's results are technical solution-oriented based on a successful pilot study in Korea. Therefore, applying this technology in Vietnam requires feasibility research and cooperation between countries. However, based on practical study results, carbon mineralization solutions can certainly be contributed

greatly to achieving Vietnam's goal of reducing greenhouse gas emissions to net-zero by 2050.

Acknowledgements This research was financially supported by the Ministry of Trade, Industry, and Energy (MOTIE), Korea, under the "Infrastructure program for industrial innovation" supervised by the Korea Institute for Advancement of Technology (KIAT)(P0018011).

This research was supported by the National Strategic Project-Carbon Mineralization Flagship Center of the National Research Foundation of Korea (NRF) funded by the Ministry of Science and ICT (MSIT), the Ministry of Environment (ME) and the Ministry of Trade, Industry, and Energy (MOTIE) (2017M3D8A2084752).

References

1. The World Bank. "The World Bank in Vietnam". https://www.worldbank.org/en/country/vietnam/overview#1 Accessed 01 Jan 2022
2. Tran, Q.M.: Projection of fossil fuel demands in Vietnam to 2050 and climate change implications. Asia Pacific Policy Stud. **6**(2), 208–221 (2019)
3. EPA. Global greenhouse gas emissions data. United States Environment Proection Agency. https://www.epa.gov/ghgemissions/global-greenhouse-gas-emissions-data (2022) (Accessed 01 Jan 2022).
4. EVN.: EVN annual report 2021, Vietnam Electricity. https://en.evn.com.vn/d6/news/Annual-Report-2021-6-13-2537.aspx (2021) (Accessed 01 Dec 2021).
5. Hong, H.T.V., Ahn, J.W.: Overview of coal-fired power plant ash situation and cement industry in Vietnam. J. Energ. Eng. **27**(3), 57–62 (2018)
6. Ritchie, H., Roser, M.: "Vietnam: CO_2 country profile," Our World in Data. https://ourworldindata.org/co2/country/vietnam (2021) (Accessed Dec 01 2021).
7. Leung, D.Y., Caramanna, G., Maroto-Valer, M.M.: An overview of current status of carbon dioxide capture and storage technologies. Renew. Sustain. Energy Rev. **39**, 426–443 (2014)
8. Yang, H., Xu, Z., Fan, M., Gupta, R., Slimane, R.B., Bland, A.E., Wright, I.: Progress in carbon dioxide separation and capture: a review. J. Environ. Sci. **20**(1), 14–27 (2008)
9. Zhang, S., Liu, L., Zhang, L., Zhuang, Y., Du, J.: An optimization model for carbon capture utilization and storage supply chain: a case study in Northeastern China. Appl. Energy **231**, 194–206 (2018)
10. Decision No. 2359/QD-TTg dated December 22, 2015 of the Prime Minister: Approving the National System of GHG inventory (2015).
11. Decree No. 06/2022/ND-CP dated January 07, 2022 of the Government providing regulations on reduction of greenhouse gas emissions and protection of the ozone layer (2022).
12. Law on environmental protection No. 55/2014/QH13 dated June 23, 2014 of the Vietnam National Assembly.
13. Thuy H.: Current status of treatment of thermal power waste ash and slag. Environmental industry. https://congnghiepmoitruong.vn/thuc-trang-cong-tac-xu-ly-tro-xi-thai-nhiet-dien-7179.html (2021) Accessed 01 Dec 2021 (in Vietnamese)
14. Pham, N.Q., Le, K.A.: Coal fly ash in Vietnam and its application as a lightweight material. Chem. Eng. Trans. **83**, 31–36 (2021)
15. Liem, N.V.: Upper Paleozoic stratigraphy and the age of bauxite in Dong Dang—Lang Son. J. Geol. **57**, 25–31 (in Vietnamese) (1966)
16. Tinh, D.N., Hai, N.M., Phuc, N.H., Thong, P.V., Thuan, N.V.: Bauxite and iron laterite potential in the basaltic weathering crust of Southern Vietnam. J. Geol. **320**, 147–158 (2010). (in Vietnamese)

17. Xinh, L.T.: Geology and mineral resources of Vietnam. Gen. Dept. Geol. Min. (GDGM) Vietnam, Hanoi, 217 (1988)
18. Lazarus, K. M.: In search of aluminum: China's role in the Mekong Region, Phnom Penh: Heinrich Böll Stiftung, WWF and International Institute for Sustainable Development, Cambodia (2009).
19. Lai, Q.T., Ahn, J.W.: Bauxite developments in Vietnam: opportunities and challenges. J. Energ. Eng. **27**(3), 41–47 (2018)
20. Luu, D.H.: Alumin production technology and pressing environmental issues, Environmental economy. https://tinhtemoitruong.vn/ky-2-cong-nghe-san-xuat-alumin-va-cac-van-de-cap-bach-ve-moi-truong-14234.html (Accessed 01 Dec 2021) (in Vietnamese).
21. Zhu, X., Li, W., Guan, X.: An active dealkalization of red mud with roasting and water leaching. J. Hazard. Mater **286**, 85–91 (2015)
22. Li, H., et al.: The influence of additives (Ca2+, Al3+, and Fe3+) on the interaction energy and loosely bound extracellular polymeric substances (EPS) of activated sludge and their flocculation mechanisms. Bioresour. Technol. **114**, 188–194 (2012)
23. Chen, S., et al.: Bromate removal by Fe (II)–akaganeite (β-FeOOH) modified red mud granule material. RSC Adv. **6**(34), 28257–28262 (2016)
24. Deihimi, N., Irannajad, M., Rezai, B.: Characterization studies of red mud modification processes as adsorbent for enhancing ferricyanide removal. J. Environ. Manage. **206**, 266–275 (2018)
25. Nadaroglu, H., Kalkan, E., Demir, N.: Removal of copper from aqueous solution using red mud. Desalination **251**(1–3), 90–95 (2010)
26. Ribeiro, D.V., Labrincha, J.A., Morelli, M.R.: Potential use of natural red mud as pozzolan for Portland cement. Mater. Res. **14**(1), 60–66 (2011)
27. Rubinos, D., Spagnoli, G., Barral, M.T.: Assessment of bauxite refining residue (red mud) as a liner for waste disposal facilities. Int. J. Min. Reclam. Environ. **29**(6), 433–452 (2015)
28. Lopez, E., et al.: Adsorbent properties of red mud and its use for wastewater treatment. Water Res. **32**(4), 1314–1322 (1998)
29. Gonda1–Zsombor, A. K., Fekete1–Márton Tóth1–István Székely1–Balázs, Zákányi1–Ferenc Móricz2–Balázs Kovács.: Physical and chemical characterization of red mud in terms of its environmental effects. Geosci Eng 3, 129–137 (2014)
30. Pinnock, W.R., Gordon, J.N.: Assessment of strength development in Bayer-process residues. J. Mater. Sci. **27**(3), 692–696 (1992)
31. Gräfe, M., Power, G., Klauber, C.: Bauxite resi-due issues: III. Alkalinity Assoc Chem Hydrometallurgy **108**, 60–79 (2011)
32. Abhilash, S. S., Meshram, P., Pandey, B. D., Behera, P.K., Satpathy, B.K.: Red Mud: a secondary resource for rare earth elements. In: International bauxite, alumina and aluminium symposium, The IBAAS Binder **3**, 148-162 (2014)
33. My, H.: Turning polluting red mud into useful materials, Intellectual Property Office of Vietnam. https://www.ipvietnam.gov.vn/sang-che-viet/-/asset_publisher/XzSH8lY4WRq7/content/bien-bun-o-gay-o-nhiem-thanh-nguyen-lieu-huu-ich (Accessed 01 Dec 2021) (2020) (in Vietnamese)
34. Coruh, S., Ergun, O.N.: Use of fly ash, phosphogypsum and red mud as a liner material for the disposal of hazardous zinc leach res-idue waste. J. Hazard. Mater. **173**, 468–473 (2010)
35. Kalkan, E.: Utilization of red mud as a stabilization material for the preparation of clay liners. Eng. Geol. **87**, 220–229 (2006)
36. Apak, R., Tutem, E., Hugul, M., Julide, H.: Heavy metal cation retention by unconventional sorbents (red muds and fly ashes). Water Res. **32**, 430–440 (1988)
37. Li, Y., Liu, C., Luan, Z., Peng, X., Zhu, C., Chen, Z., et al.: Phosphate removal from aqueous solutions using raw and activated red mud and fly ash. J. Hazard. Mater. **B137**, 374–383 (2006)
38. Huang, W., Wang, S., Zhu, Z., Li, L., Yao, X., Rudolph, V., et al.: Phosphate removal from wastewater using red mud. J. Hazard. Mater. **158**, 35–42 (2008)
39. Cengeloglu, Y., Kir, E., Ersoz, M., Buyukerkek, T., Gezgin, S.: Recovery and concentration of metals from red mud by Donnan dialysis. Colloids Surf., A **223**, 95–101 (2003)

40. Liu, W., Yang, J., Xiao, B.: Application of Bayer red mud for iron recovery and building material production from alumosilicate residues. J. Hazard. Mater. **161**, 474–478 (2009)
41. Zhong, L., Zhang, Y.F., Zhang, Y.: Extraction of alumina and sodium oxide from red mud by a mild hydro-chemical process. J. Hazard. Mater. **172**, 1629–1634 (2009)
42. Amritphale, S.S., Anshul, A., Chandra, N., Ramakrishnan, N.: A novel process for making radiopaque materials using bauxite-Red mud. J. Eur. Ceram. Soc. **27**, 1945–1951 (2007)
43. Kumar, V., Nautiyal, B.D., Jha, A.K.: Use of neutralized red mud in concrete. Indian Concr. J. **63**, 505–507 (1989)
44. Sglavo, V.M., Campostrini, R., Maurina, S., Carturan, G., Monagheddu, M., Budroni, G., Cocco G.: Bauxite 'red mud' in the ceramic industry. Part 1: thermal behavior. J. Eur. Ceram. Soc. **20**, 235–244 (2000).
45. Sglavo, V.M., Maurina, S., Conci, A., Salviati, A., Carturan, G., Cocco, G.: Bauxite 'red mud' in the ceramic industry. Part 2: production of clay-based ceramics. J. Eur. Ceram. Soc. **20**, 245–252 (2000).
46. Yang, J., Zhang, D., Hou, J., He, B., Xiao, B.: Preparation of glass-ceramics from red mud in the aluminium industries. Ceram. Int. **34**, 125–130 (2008)
47. Amritphale, S.S., Patel, M.: Utilization of red mud, fly ash for manufacturing bricks with pyrophyllite. Silic. Indus. **52**, 31 (1987)
48. Yang, J., Xiao, B.: Development of unsintered construction materials from red mud wastes produced in the sintering alumina process. Constr. Build. Mater. **22**, 2299–2307 (2008)
49. Singh, M., Upadhayay, S.N., Prasad, P.M.: Preparation of special cements from red mud. Waste Manage. **8**, 665–670 (1996)
50. Singh, M., Upadhayay, S.N., Prasad, P.M.: Preparation of iron rich cements using red mud. Cem. Concr. Res. **27**, 1037–1046 (1997)
51. Tsakiridis, P.E., Agatzini-Leonardou, S., Oustadakis, P.: Red mud addition in the raw meal for the production of Portland cement clinker. J. Hazard. Mater. **B116**, 103–110 (2004)
52. Vangelatos, I., Angelopoulos, G.N., Boufounos, D.: Utilization of ferroalumina as raw material in the production of ordinary Portland cement. J. Hazard. Mater. **168**, 473–478 (2009)
53. He, J., Zhang, J., Yu, Y., Zhang, G.: The strength and microstructure of two geopolymers derived from metakaolin and red mud-fly ash admixture: a comparative study. Constr. Build. Mater. **30**, 80–91 (2012) https://doi.org/10.1016/j.conbuildmat.2011.12.011
54. Kumar, A., Kumar, S.: Development of paving blocks from synergistic use of red mud and fly ash using geopolymerization, Constr. Build. Mater. **38**, 865–871 (2013) https://doi.org/10.1016/j.conbuildmat.2012.09.013
55. Zhang, M., Zhao, M., Zhang, G., Mann, D., Lumsden, K., Tao, M.: Durability of red mud- fly ash based geopolymer and leaching behavior of heavy metals in sulfuric acid solutions and deionized water. Constr. Build. Mater **24**, 373–382 (2016)
56. Kim, S.Y., Jun, Y., Jeon, D., Oh, J.E.: Synthesis of structural binder for red brick production based on red mud and fly ash activated using $Ca(OH)_2$ and Na_2CO_3. Constr. Build. Mater. **147**, 101–116 (2017)
57. Hu, Y., Liang, S., Yang, J., Chen, Y., Ye, N., Ke, Y., Tao, S., Xiao, K., Hu, J., Hou, H., Fan, W.: Role of Fe species in geopolymer synthesized from alkali-thermal pretreated Fe-rich Bayer red mud. Constr. Build. Mater **200**, 398–407 (2019). https://doi.org/10.1016/j.conbuildmat.2018.12.122
58. Samal, S.: Utilization of red mud as a source for metal ions-a review. Materials **14**(9), 2211 (2021)
59. Jayasankar, K., Ray, P.K., Chaubey, A.K., Padhi, A., Satapathy, B.K., Mukherjee, P.S.: Production of pig iron from red mud waste fines using thermal plasma technology. Int. J. Miner. Metall. Mater **19**, 679–684 (2012)
60. Lim, K., Shon, B.: Metal Components (Fe, Al and Ti) recovery from red mud by sulfuric acid leaching assisted with ultrasonic waves. Int. J. Emerg. Technol. Adv. Eng **5**, 25–32 (2015)
61. Voßenkaul, D., Birich, A., Müller, N., Stoltz, N., Friedrich, B.: Hydrometallurgical processing of eudialyte bearing concentrates to recover rare earth elements via low-temperature dry digestion to prevent the silica gel formation. J. Sustain. Metall **3**, 79–89 (2017)

62. Thenepalli, T., Ngoc, N.T.M., Tuan, L.Q., Son, T.H., Hieu, H.H., Thuy, D.T.N., Thao, N.T.T., Tam, D.T.T., Huyen, D.T.N., Van, T.T., Chilakala, R.: Technological solutions for recycling ash slag from the Cao Ngan coal power plant in Vietnam. Energies **11**(8) (2018).
63. FiinResearch, "Vietnam cement market report 2021," https://fiingroup.vn/NewsInsights/Detail/6064924 (Accessed Dec 01 2021).

Pile–Soil Interaction Mechanism and Optimization Measures Based on Finite Element Method

Qi Xu, Remin Zhu, Jianlin Cao, Xuedong Li, Yi Zhang, and Qiang Tang

Abstract With the wide application of foundation pit engineering, the design scheme of supporting structure needs to be adjusted accordingly. The optimization of retaining structure and deformation control of foundation pit are focused on as the important problems during the construction. The finite element model of a foundation pit in Suzhou was established; the simulated results are in good agreement with the measured ones. The maximum foundation heave of foundation pit is 3.63 cm, the maximum surface settlement is 3.12 cm, and the maximum displacement of pile is 2.18 cm. The sensitivity of the parameters of the retaining pile was discussed. The maximum displacement of the pile increases significantly with the decrease of the pile diameter. Pile spacing increases from 3 to 6 m, and the maximum bending moment of pile increases by 458.971 kN·m. The insertion ratio and elastic modulus have little effect on pile position movement. The surface subsidence changes little under the influence of various parameters.

Keywords Finite element simulation · Pile · Soil interaction · Foundation pit excavation · Parameter sensitivity

1 Introduction

The structure and function of urban center buildings are upgraded rapidly, and the utilization rate of urban land is getting higher and higher. Therefore, the utilization of underground space becomes a solution to road congestion and land shortage [1–3]. There are many kinds of supporting types in foundation pit engineering, including

Q. Xu (✉) · Q. Tang
School of Rail Transportation, Soochow University, Suzhou 215131, China

Q. Tang
e-mail: tangqiang@suda.edu.cn

R. Zhu · J. Cao · X. Li · Y. Zhang
Jiangsu Suzhou Geological Engineering Survey Institute, Suzhou 215011, China

The Fourth Geological Brigade of Jiangsu Geological and Mineral Bureau, Suzhou 215004, China

L. Q. Nguyen et al. (eds.), *Advances in Geospatial Technology in Mining and Earth Sciences*, Environmental Science and Engineering,
https://doi.org/10.1007/978-3-031-20463-0_13

underground continuous wall, SMW pile, soil nailing wall, inverse arch wall and so on. Pile foundation has been widely used because of its characteristics of high bearing capacity, uniform foundation settlement, the ability to apply to complex geological environments in the process of excavation [4–6]. Due to the large difference in stiffness between pile and soil, and under the action of external forces, shear stress and relative displacement are often formed at the contact interface of pile and soil [7]. Underground foundation pit construction is prone to safety accidents; therefore, it is very important to correctly analyze the mechanism of stress and deformation on the contact surface of this kind of structure [8].

The analysis method of interaction mechanism between supporting structure and soil is the theoretical basis of classical earth pressure theory. Pile–soil interaction problem belongs to the contact problem of different media in solid mechanics. It is a typical nonlinear problem manifested by material nonlinearity (concrete, soil) and contact nonlinearity (the pile–soil contact surface has bonding, sliding, opening and closing under complex load conditions). With scientific inquiry, the current pile and soil interaction analysis mainly includes the load transfer method, elastic theory method, shear displacement method and numerical calculation method. The essence of the load transfer method is the foundation reaction method. According to the different means of obtaining the transfer function, the transfer function method contains divided into displacement coordination method and analytical method. The load transfer method can simulate the stress of a single pile, because the continuity of soil is not considered, and the load transfer method cannot be directly used in pile group. The elastic theory method assumes that the pile and soil are elastic materials, and the Young's modulus of soil is either constant or varies with the depth according to a certain law. The displacement of the pile is obtained by the compression of the pile under axial load. Shear displacement method is to introduce a change matrix in the vertical, which can easily consider the behavior of layered foundation, homogeneous soil does not need to discretize the pile model, and the analysis of pile group does not depend on many interaction coefficients, so it is easy to calculate [9–11]. The numerical calculation method becomes mainstream. The main practical applications include the finite element method, boundary element method and discrete element method [12, 13]. The finite element method is the most widely used numerical analysis method. The finite element method (FEM) is mature in numerical calculation because of its reliability and effectiveness in solving problems. The finite element method cannot only solve the linear elastic problem but also be easily used in the analysis of nonlinear elastic problems [14–16].

This paper takes a tunnel foundation pit project in Suzhou as the research object whose deformation is simulated by finite element model and verifies the accuracy of the numerical simulation by comparing it with the monitoring data. The influences of pile diameter, pile spacing, insertion ratio and elastic modulus on the deformation of foundation pit and pile were analyzed.

2 Project Profile

2.1 *Project Overview*

The proposed site of the project is shown in Fig. 1, which is located in Xiangcheng District, Suzhou, southeast of the Yangtze River Delta and in the middle of Taihu Lake water network plain. It is a delta alluvial plain landform. The quaternary strata of the site are widely distributed and thick, and the soil is soft and easy to cause construction accidents.

The soil composition of the site is many and complex. To facilitate calculation, the soil with similar mechanical properties is simplified in this paper. The simplified soil layer is divided into three layers; parameters of each soil layer which were obtained from laboratory experiments by direct shear and uniaxial compression experiments, etc. are shown in Table 1.

Fig. 1 Site of project

Table 1 Simplified soil layer parameters

Stratum	Stratum thickness (m)	Density (g/cm^3)	Cohesion (kPa)	Internal friction angle (°)	Elastic modulus (MPa)	Poisson's ratio	Hydraulic conductivity (cm/s)
Silty clay	5	1.92	34.8	16.3	25	0.30	6.52×10^{-6}
Silty sand	8	1.87	2.8	28.9	50	0.30	3.89×10^{-5}
Silty clay	30	1.96	45.4	16.2	30	0.30	4.35×10^{-6}

2.2 Pit Supporting Structure

The foundation pit in this project is strip foundation pit, with a total depth of 17.3 m and a width of 24.2 m. The first support adopts reinforced concrete support; the specification is 800 × 800 mm. The spacing of concrete supports is 5 m. Steel bracing is adopted for the second to fifth bracing of foundation pit, and the specifications of steel bracing are uniformly adopted for D609 steel pipe with 16 mm wall thickness. The support spacing is usually 3–4 m. Figure 2 shows the horizontal plane of the foundation pit, and Fig. 3 shows the section of the foundation pit. The standard pile length is 25.3 m, and the insertion ratio is 1.1. Table 2 shows the parameters of supporting structure.

The area has high annual rainfall and loose fill structures above the water table. In addition, groundwater in Suzhou area is generally shallower, so to prevent the adverse impact of groundwater on construction, the foundation pit adopts tube well

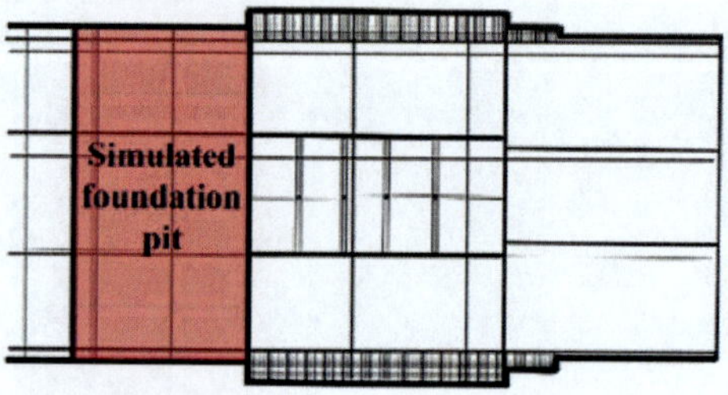

Fig. 2 Cross section of foundation pit

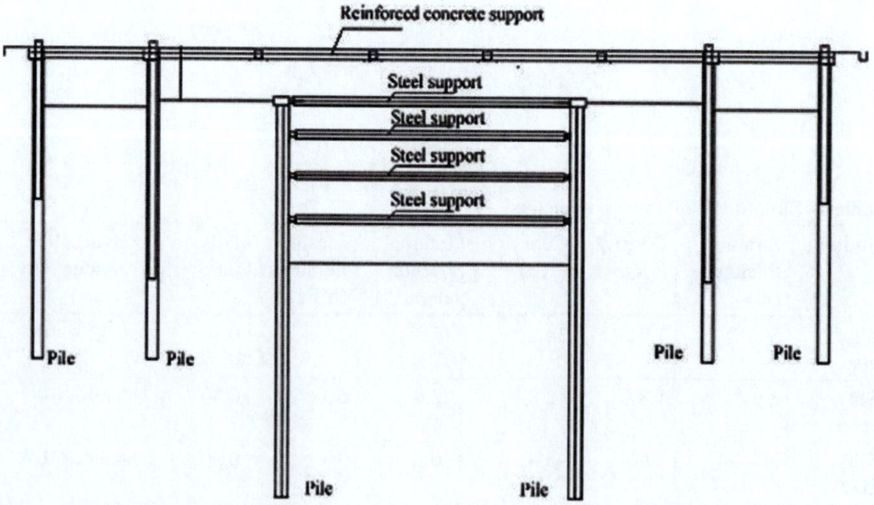

Fig. 3 Plan view of foundation pit

Table 2 Mechanical parameters of supporting structure

Supporting system	Density (kg/m^3)	Modulus of elasticity (GPa)	Poisson's ratio
Retaining piles	2400	1	0.2
Steel braces	7900	200	0.2
Concrete braces	2400	30	0.2

precipitation, and the tube well is laid horizontally along with the foundation pit. The underground water level decreases with time, and the underground water level during construction is kept below 1 m of excavation depth.

3 Numerical Analysis

3.1 Model Establishment

MIDAS/GTS effectively integrates the professional requirements of engineering with the numerical calculation and simulation technology of finite element to solve complex engineering problems that cannot be solved by traditional methods [17]. Rock and soil is a kind of multiphase medium material with complex and diverse engineering properties. It is very difficult and impractical to simulate all the characteristics of rock and soil materials completely by numerical simulation. It is desirable to simplify the model for definite analytical purposes [18]. MIDAS/GTS provides automatic grid generation command. The automatic grid generation command is invoked during the modeling process to quickly complete the establishment of complex model, which greatly improves the modeling efficiency.

For the analysis of stress, strain and internal force, the grid corresponding to the part of the model should be encrypted, while for the part not involved in the analysis, the grid division can be relatively rough. Figure 4 shows the established foundation pit profile and grid division diagram.

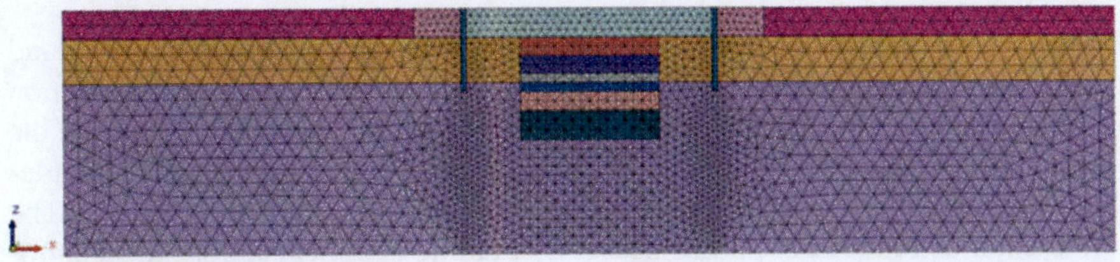

Fig. 4 Finite element meshing

Table 3 Construction stage group design

Construction steps	Contents
Step 0	Calculate site initial ground stress, construction of envelope structure
Step 1	The first excavation and activate the concrete support of the main foundation pit
Step 2	The second excavation activates the first steel support
Step 3	The third excavation activates the second steel support
Step 4	The fourth excavation activates the third steel support
Step 5	The fifth excavation activates the fourth steel support

3.2 Numerical Model Construction

The accuracy and rationality of engineering numerical simulation results are affected by different soil constitutive relations. In this paper, the soil constitutive model selected the modified Mohr–Coulomb model. Compared with the traditional Mohr–Coulomb model, this model is more consistent with the actual deformation of soil and can better reflect the surface settlement. The use of the modified Mohr–Coulomb model has more guiding value for the prediction of foundation pit deformation [19–21]. By simulating the boundary conditions of tunnel foundation pit construction, the displacement in the X, Y and Z directions of the bottom, the displacement in the X and Z directions of the left and right surfaces, as well as deformation in the Y direction of the front and rear surfaces were constrained.

In order to study separately the stress and strain law of foundation pit from bored pile construction to foundation pit excavation to bottom, the displacement should be cleared in the initial stress stage to eliminate the interference in this stage. For the sake of simulating the actual construction and excavation steps, the model is excavated step by step, and the specific construction steps are simulated following in Table 3.

3.3 Simulation Verification

The excavation of foundation pit changes the distribution of stress field, and the change in groundwater environment affects the effective stress. Foundation pit retaining structure deformation under water load will inevitably lead to the surrounding surface, roads, buildings, usually characterized by the ground settlement, basement uplift and the deformation of retaining structure, etc. [22, 23]. The deformation of foundation pit has necessary guiding significance for the safety of project construction [24].

As the construction process progresses, the gravity stress of the excavated soil is released, which leads to vertical unloading of the soil at the bottom. The original stress state of the soil at the bottom of the pit is destroyed where the soil is uplifted [25]. The rule of heave deformation of pit bottom is consistent under each procedure. From the two edges to the middle of pit bottom, the heave deformation increases continuously. At the horizontal center point, the heave deformation reaches its maximum value of about 3.63 cm. The comparison between the simulated value of basement uplift and the actual monitored value is displayed in Fig. 5. There is little difference between the simulation results and the measured ones, indicating that the simulation is good.

The variation law of surface settlement under each working procedure is shown in Fig. 6. The influence range of construction on nearby surface settlement is approximately 64 m, and the surface settlement increases gradually with the excavation, and the variation law is consistent with the distance from the longitudinal edge: At first, the ground settlement increases rapidly. After a certain distance from the vertical position of the foundation pit edge, the surface settlement reaches the maximum value and then decreases. The settlement rate decreases with the increase of the distance of foundation pit, and finally tends to be flat. The maximum settlement of the foundation reached about 3.12 cm after the excavation was completed.

As the process goes on, the horizontal displacement of retaining structure will occur which will change with the increase of excavation depth which will directly affect the displacement of soil outside the pit and also indirectly affect the safety of

Fig. 5 Foundation pit uplift

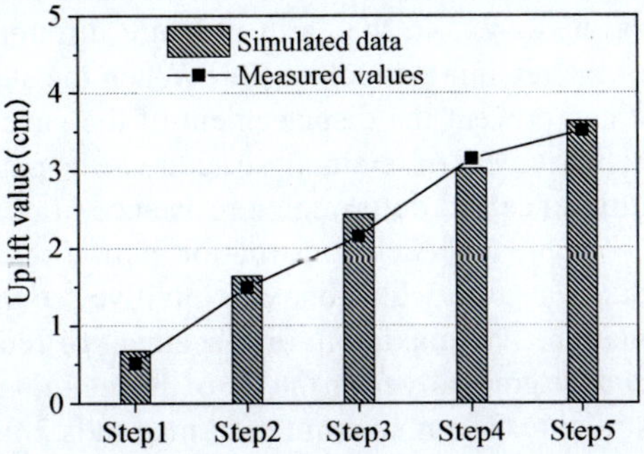

Fig. 6 Ground settlement

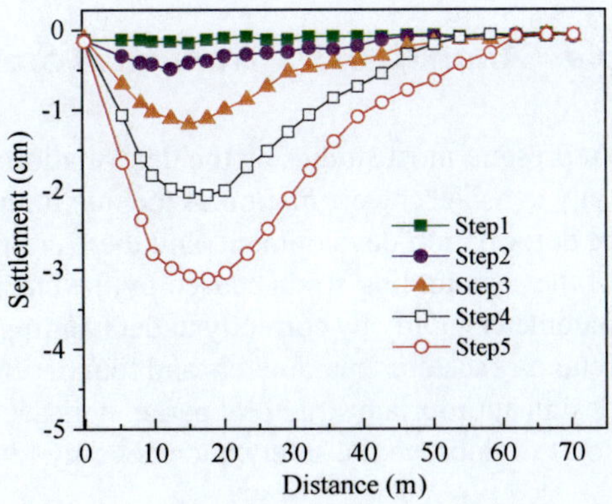

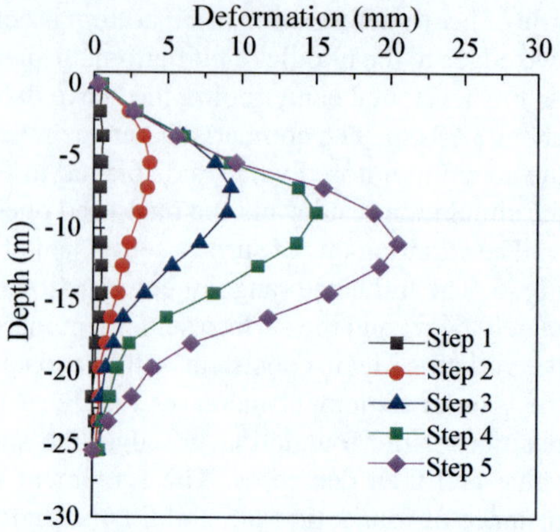

Fig. 7 Horizontal displacement of piles at different depths

buildings outside the pit and underground pipelines and facilities [26, 27]. Therefore, the horizontal displacement of the pile is the key monitoring link in foundation pit excavation.

There is an earth pressure difference between the excavated area and the unexcavated area, so the earth pressure difference causes the horizontal displacement of the retaining structure [28]. When the supporting structure acts on the envelope, it can prevent the displacement of the envelope to a certain extent. The horizontal displacement of retaining structure is the result of interaction between earth pressure difference and deformation resistance of retaining structure [29].

Figure 7 is each construction step of retaining structure horizontal displacement diagram from which one can intuitive see the palisade structure to the inside deformation. The maximum displacement of retaining structure is constantly changing, change gradually from the top of the retaining structure, near to the middle of retaining structure. The maximum deformation is 2.05 cm at a depth of about 12 m.

3.4 Analysis of the Numerical Simulation Results

At present, most studies on the deformation characteristics of foundation pit usually only consider the estimation of the maximum deformation and the dynamic process of deformation development, and there are few studies on the deformation response of the surrounding strata caused by the change of foundation pit retaining structure parameters. Only by correctly understanding the relationship between foundation pit retaining structure parameters and foundation pit deformation characteristics, can we design appropriate structural parameters according to the deformation control value, so as to improve the safety and economy [30, 31].

Pile diameter

The diameter of the retaining pile takes an important part in the stabilization and security of the whole foundation ditch. In the simulation calculation process, we assume that the material properties of the retaining pile and the specific construction process are not changed and adjust the diameter of the retaining pile. Based on the original pile diameter of 1 m, the change of pile body in different pile diameters and its influence on the surrounding environment are analyzed. Three diverse diameters of pile 0.6, 0.8 and 1.2 m were set respectively to simulate comparison.

The horizontal displacement of pile under different pile diameters is shown in Fig. 8a. If the pile diameter is larger, the ability of retaining pile to resist elastic deformation affected by construction will be improved. The comparison shows that the peak value horizontal displacement of the retaining pile is about 31.69 mm when the measured consequence of the retaining pile is 0.6 m. As the diameter of retaining pile increases, the horizontal displacement of retaining pile decreases gradually. When the diameter of the retaining pile is adjusted to 1.2 m, the maximum horizontal displacement of the whole retaining pile decreases to 1.49 cm. Compared with the pile diameter of 0.6 m, the peak value of horizontal displacement of the retaining pile under 1.2 m is reduced by 52.7%. In the case of 1 m pile diameter, the maximum horizontal displacement is about 2.05 cm, reduced by 35.3% compared with the maximum horizontal displacement. By comparing the variation trend of different pile diameters, it can be seen that with the continuous growth of pile diameters, the entire deformation range of the pile also increases. Figure 8b reflects the impact of diverse pile diameters on surface settlement. The maximum changing value of land subsidence is 2.07 mm, and the change rate is 6.62%. Pile diameter has little effect on surface settlement.

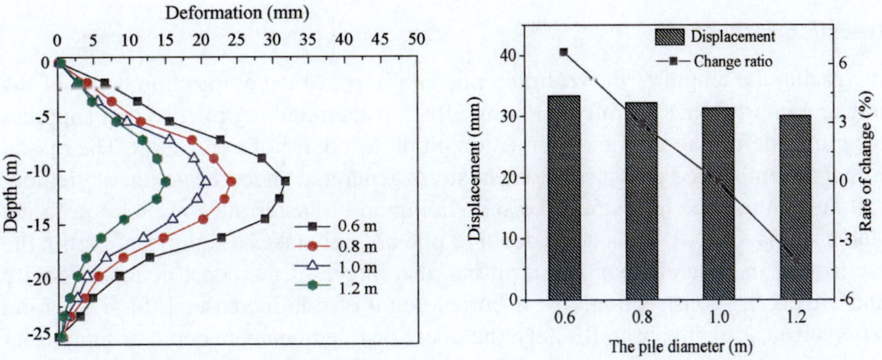

(a) Horizontal displacement of piles (b) Ground settlement

Fig. 8 Max displacement of pile and ground settlement under different pile diameters

Pile spacing

There is a certain relationship between the stress state of soil and the value of pile spacing, and the value of pile spacing also affects significantly the stability of foundation pit, and the value of pile spacing will also affect the construction technology and thus affect the project cost [32]. Therefore, pile spacing is set as a variable parameter as the research object in this section. Different pile spacing values are simulated to study the impact of pile spacing changing the environment around and the overall horizontal displacement of pile during the course of foundation pit support. The pile distances of 3, 4, 5 and 6 m were set, respectively, and the different conditions of different pile distances were numerically simulated.

Pile position shifts under different pile spacing are shown in Fig. 9a. The increase of pile spacing significantly increases the maximum displacement of pile. When the pile spacing is 6 m, the maximum horizontal displacement reaches 3.62 cm, which is 1.58 cm higher than when the pile spacing is 3 m. Figure 9b shows the comparison of pile bending moments under the two conditions. The maximum bending moment of pile spacing 6 m is 458.971 kN·m larger than that of pile spacing 3 m, both occurring near the middle position of pile. The distribution forms of bending moments are similar. If the pile spacing is too large, the bored pile cannot exert the soil retaining function fully, resulting in the soil extruding from the dowry and affecting the stability of deep foundation pit. If the pile spacing is too small, the stability of foundation pit can be increased, but the cost and construction period will increase [33]. Therefore, it is necessary to determine the appropriate pile spacing in the design of drilling and perfusion makeup, so as to reduce the project cost on the premise of ensuring the stability of deep foundation pit. Figure 9c reflects the change of surface settlement with pile diameter. The change value of settlement increases with the increase of pile spacing, the maximum is 32.53 mm and the growth rate is 4.08%. The increase in pile spacing reduces the support points of soil and increases settlement.

Insertion ratio of pile

Increasing the length of the retaining pile can improve the embedding depth of the pile and improve the overall anchoring effect of the retaining pile, so that suppress the global deformation of the foundation pit in the course of excavation. The reason is that the embedded pile uses the shear stress generated under the action of yielding soil and static zone to restrain the soil sliding and transfer the soil pressure to the stable soil layer [34]. The buried depth of pile not only takes a significant part in the security of the overall foundation pit but also increases the construction difficulty and affects the construction cost if buried depth of pile increases [35]. Therefore, based on ensuring the overall safety, the selection of embedment depth of supporting structure is particularly important.

In reality, the insertion ratio of foundation pit is about 1.1. For studying the effect of insertion ratio (i.e., embedment depth) on foundation pit, three insertion ratios of 0.7, 1 and 1.3 are elected. The relationship between horizontal displacement of pile and insertion ratio is shown in Fig. 10a. As can be seen from the comparison of horizontal displacements of different pile lengths, with the growth of pile lengths, the

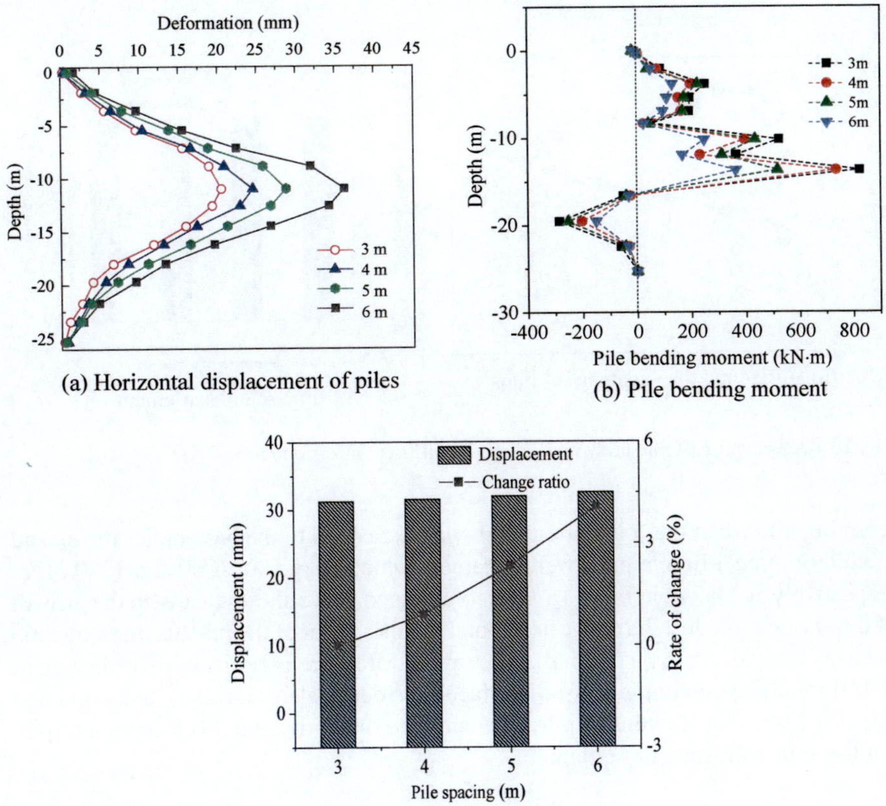

(a) Horizontal displacement of piles

(b) Pile bending moment

(c) Ground settlement

Fig. 9 Deformation of pile body and soil under different pile spacing

horizontal displacement of the pile is decreasing gradually. When the pile length of 23.3 m reaches the maximum horizontal displacement, the peak value of horizontal displacement decreases gradually with the decrease of insertion ratio. In the process of the pile length increasing from 23.3 to 29.7 m, the overall horizontal displacement of the pile decreases obviously, but the maximum displacement of the pile only decreases by 2.56 mm. The variation trends of different pile lengths are roughly coincident, indicating that with the gradual lengthening of the pile, the overall impact on the pile decreases gradually. The change rate of surface subsidence ranges from 0.01 to 0.46%, as shown in Fig. 10b. The insertion ratio is insensitive to the impact of surface subsidence.

Modulus of elasticity

The different stiffness of the envelope will affect the deformation of the envelope. If the elastic modulus is low, the deformation will easily occur [35, 36]. The calculation results of the model were extracted, and the horizontal displacement curves of the

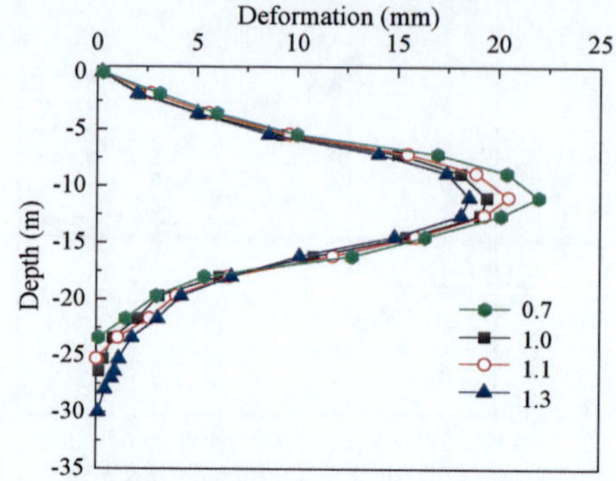

(a) Horizontal displacement of piles

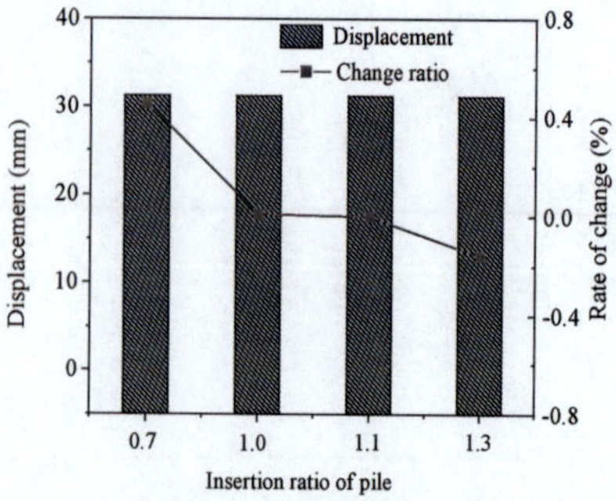

(c) Ground settlement

Fig. 10 Deformation of pile body and soil under different insertion ratio of pile

retaining pile when the foundation ditch was excavated to the base under the elastic moduli of three different piles were obtained, which were 800, 1000 and 1200 MPa, respectively, as shown in Fig. 11a. The comparison curve shows that with the growth of the elastic modulus, the entire horizontal displacement of the pile declines, the pile behaves as rigid. As in Fig. 11b, the maximum difference in horizontal displacement is 4.01 mm. The maximum value of surface subsidence also decreases, but its change rate is 0.29%. The modulus of pile is sensitive to the horizontal displacement of pile but has little influence on settlement.

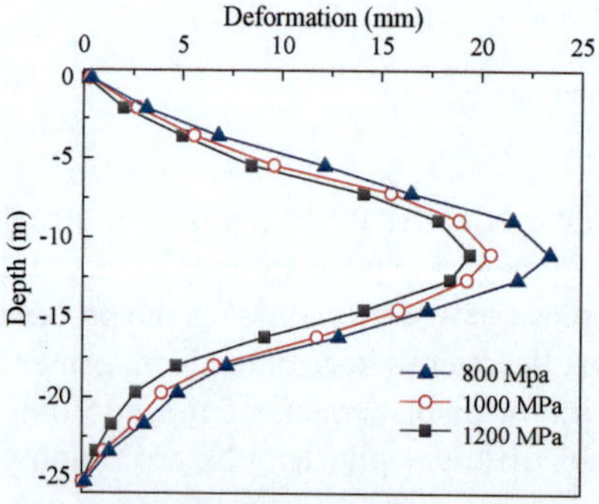

(a) Horizontal displacement of piles

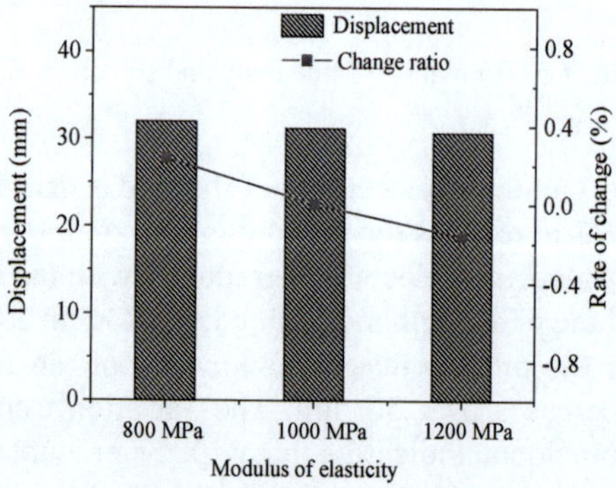

(b) Ground settlement

Fig. 11 Horizontal displacement under different pile lengths

4 Conclusion

This paper takes a foundation pit in Suzhou city as the research target, establishes a finite element model and studies the deformation characteristics of soil layer and its retaining structure under different parameters of the retaining structure according to the simulation results, so as to provide support for the foundation pit engineering. The following conclusions were obtained.

1. As the process goes on, the center deformation of the bottom of the pit is the largest in each procedure. The surface settlement increases at the early stage and then decreases with the distance from the foundation pit. The pile position shift increases with the process, and the maximum deformation depth increases.
2. Pile diameter and pile spacing are sensitive to soil and pile deformation. The increase of pile diameter restrains the overall horizontal displacement.
3. The influence of insertion ratio and elastic modulus on soil and pile deformation is relatively small. The maximum change of pile displacement is only 2.56 mm, and the influence of elastic modulus on surface settlement is small.

Acknowledgements The research presented here is supported by the National Natural Science Foundation of China (52078317), Natural Science Foundation of Jiangsu Province for Excellent Young Scholars (BK20211597), project from Bureau of Housing and Urban–Rural Development of Suzhou (2021-25; 2021ZD02; 2021ZD30), Bureau of Geology and Mineral Exploration of Jiangsu (2021KY06), China Tiesiju Civil Engineering Group (2021-19), CCCC First Highway Engineering Group Company Limited (KJYF-2021-B-19) and CCCC Tunnel Engineering Company Limited (8gs-2021-04).

References

1. Kishii, T.: Utilization of underground space in Japan. Tunn. Undergr. Space Technol. **55**, 320–323 (2016)
2. Qihu, Q.: Present state, problems and development trends of urban underground space in China. Tunn. Undergr. Space Technol. **55**, 280–289 (2016)
3. Bobylev, N.: Underground space as an urban indicator: measuring use of subsurface. Tunn. Undergr. Space Technol. **55**, 40–51 (2016)
4. Kumar, M., Bardhan, A., Samui, P.: Reliability analysis of pile foundation using soft computing techniques: a comparative study. Processes **9**(3), 486 (2021)
5. Xue, Z.: Study on new test technology of pile foundation bearing capacity. In: IOP Conference Series: Materials Science and Engineering, p. 012005. IOP Publishing (2021)
6. Lau, A.Y.A., Terry, J.P., Switzer, A.D.: Advantages of beachrock slabs for interpreting high-energy wave transport: evidence from Ludao Island in south-eastern Taiwan. Geomorphology **228**, 263–274 (2015)
7. Korff, M., Mair, R.J., Van Tol, F.A.F.: Pile-soil interaction and settlement effects induced by deep excavations. J. Geotech. Geoenviron. Eng. **142**(8), 04016034 (2016)
8. Ye, S., Zhao, Z., Wang, D.: Deformation analysis and safety assessment of existing metro tunnels affected by excavation of a foundation pit. Undergr. Space **6**(4), 421–431 (2021)
9. Chore, H.S., Ingle, R.K., Sawant, V.A.: Building frame-pile foundation-soil interaction analysis: a parametric study. Interact. Multiscale Mech **3**(1), 55–79 (2010)

10. Editorial Department of China Journal of Highway and Transport. Review of Academic Research on Subgrade Engineering in China 2021. China J. Highw. Transp. 34(3), 49. (2021) (in Chinese)
11. Dias, T.G.S., Bezuijen, A.: Load-transfer method for piles under axial loading and unloading. J. Geotech. Geoenviron. Eng. **144**(1), 4017096 (2018)
12. Józefiak, K., Zbiciak, A., Maślakowski, M.: Numerical modelling and bearing capacity analysis of pile foundation. Procedia Eng. **111**, 356–363 (2015)
13. Hoffman, J.D., Frankel, S.: Numerical methods for engineers and scientists. CRC press (2018)
14. Brezeanu, L.C.: Contact stresses: analysis by finite element method (FEM). Procedia Technol. **12**, 401–410 (2014)
15. Han, S.Y., Xiong, J., Zhang, Y.: Supporting measure and the effect evaluation of the unstable foundation pit. Appl. Mech. Mater. 744, 647–651, Trans Tech Publications Ltd (2015)
16. Yu, C.: Numerical analysis of deep foundation pit excavation supported by large diameter ring beam. In: IOP Conference Series: Earth and Environmental Science, 062033. IOP Publishing (2019)
17. Sekisov, A.N., Varvarkina, A.V., Gura, D.A.: Pit design taking into account water level reduction using the software package MIDAS GTS NX. IIOAB J. **10**(S2), 93–100 (2019)
18. Fairhurst, C., Long, L.: Improved design in rock and soil engineering with numerical modelling. In: Distinct Element Modelling in Geomechanics. Routledge (2018)
19. Robert, D.J.: A modified Mohr-Coulomb model to simulate the behavior of pipelines in unsaturated soils. Comput. Geotech. **91**, 146–160 (2017)
20. Xiang, X., Zi, D.: Numerical implementation of a modified Mohr-Coulomb model and its application in slope stability analysis. J. Modern Trans. **25**(1), 40–51 (2017)
21. Zhang, D., Liu, E.L., Liu, X.Y.: A damage constitutive model for frozen sandy soils based on modified Mohr-Coulomb yield criterion. Chin. J. Rock Mechan. Eng. **37**(4), 978–986 (2018)
22. Liu, J., Shi, C., Cao, C.: Improved analytical method for pile response due to foundation pit excavation. Comput. Geotech. **123**, 103609 (2020)
23. Yu, X., Jia, B.: Analysis of excavating foundation pit to nearby bridge foundation. Procedia Earth Planet. Sci. **5**, 102–106 (2012)
24. Zhao, Z., Y. Zhu., Ye, S.: Study on settlement deformation of high fill foundation in large thickness loess area. Arabian J. Geosci. 14, 1–12 (2021)
25. Feng, S., Wu, Y., Li, J.: The analysis of spatial effect of deep foundation pit in soft soil areas. Procedia Earth Planetary Sci. **5**, 309–313 (2012)
26. Muszyński, Z., Rybakm J.: Horizontal displacement control in course of lateral loading of a pile in a slope. In: IOP Conference Series: Materials Science and Engineering, 032002. IOP Publishing (2017)
27. Loria, A.F.R., Vadrot, A., Laloui, L.: Analysis of the vertical displacement of energy pile groups. Geomech Energ Environ **16**, 1–14 (2018)
28. Clayton, C.R.I., Woods, R.I.: Milititsky J. Earth pressure and earth-retaining structures. CRC press (2014)
29. Pantelidis, L.: The generalized coefficients of earth pressure: a unified approach[J]. Appl. Sci. 9 (24), 5291 (2019)
30. Zhou, Y., Su, W., Ding, L.: Predicting safety risks in deep foundation pits in subway infrastructure projects: support vector machine approach. J. Comput. Civ. Eng. **31**(5), 04017052 (2017)
31. Yu, L., Yan, S., Hai, L.: Research on deformation control and prevention measures of pile anchor support system in foundation pit. In: IOP Conference Series: Earth and Environmental Science 510 (5), 052054. IOP Publishing (2020)
32. Garala, T.K., Madabhushi, G.S.P.: Role of pile spacing on dynamic behavior of pile groups in layered soils. J. Geotech. Geoenviron. Eng. **147**(3), 04021005 (2021)
33. Magade, S.B., Ingle, R.K.: Influence of clear edge distance and spacing of piles on failure of pile cap. Iranian J. Sci. Technol. Transac. Civil Eng. **44**(4), 1265–1281 (2020)
34. Huang, H., Quan, Z., Liu, B.: Mechanical behavior and influencing factors of different pile-anchor supporting structures. J. Highw Transp. Res. Dev (English Edition) **10**(3), 41–50 (2016)

35. Petrie, C. M., Niyanth, S., Frederick, C.: Embedded Fiber-Optic Sensors for In-Pile Applications. Oak Ridge National Lab. (ORNL), Oak Ridge, TN (United States) (2019)
36. Prendergast, L.J., Gavin, K.: A comparison of initial stiffness formulations for small-strain soil–pile dynamic Winkler modelling. Soil Dyn. Earthq. Eng. **81**, 27–41 (2016)

Determination of Illegal Signs of Coal Mining Expansion in Thai Nguyen Province, Vietnam from a Combination of Radar and Optical Imagery

Tran Van Anh, Tran Hong Hanh, Nguyen Quynh Nga, Le Thanh Nghi, Truong Xuan Quang, Khuc Thanh Dong, and Tran Trung Anh

Abstract Vietnam has diverse and rich natural resources with over 5000 mines and 60 different types of minerals. Mineral activities in Vietnam are licensed by the Government and must ensure to protect the environment and daily activities. However, due to the high profits generated from mining, there have been many illegal mining activities taking place all over Vietnam. One of the causes leading to this situation is the insufficiency in monitoring and managing resources and minerals. Additionally, many mines are located in the forest which is difficult to detect. Therefore, in this study, we adopt Sentinel-1 radar satellite image and Sentinel-2 multi-temporal satellite image to determine illegal coal mining activities at the Minh Tien

T. Van Anh (✉) · T. H. Hanh · L. T. Nghi · T. T. Anh
Hanoi University of Mining and Geology, 18 Vien Street, Hanoi 100000, Vietnam
e-mail: tranvananh@humg.edu.vn

T. H. Hanh
e-mail: tranhonghanh@humg.edu.vn

L. T. Nghi
e-mail: lethanhnghi@humg.edu.vn

T. T. Anh
e-mail: trantrunganh@humg.edu.vn

N. Q. Nga
National Academy of Public Administration, Hanoi 100000, Vietnam
e-mail: nganq@napa.vn

T. X. Quang
Hanoi University of Natural Resources and Environment, 41A Phu Dien Road, Hanoi 100000, Vietnam
e-mail: txquang@hunre.edu.vn

K. T. Dong
Hanoi University of Civil Engineering, 55 Giai Phong Street, Hanoi 100000, Vietnam
e-mail: dongkt@huce.edu.vn

T. Van Anh
Innovations for Sustainable and Responsible Mining (ISRM) Research Group, Hanoi University of Mining and Geology, Hanoi 100000, Vietnam

L. Q. Nguyen et al. (eds.), *Advances in Geospatial Technology in Mining and Earth Sciences*, Environmental Science and Engineering,
https://doi.org/10.1007/978-3-031-20463-0_14

coal mine, Thai Nguyen province, Vietnam. The research method includes applying Google Earth Engine (GEE) for processing the Sentinel-1 images in both ascending and descending directions and Sentinel-2 optical images to calculate the Normalized Difference Vegetation Index (NDVI) series to control the different trends from 2016 to 2021. NDVI is then employed to mask the areas with vegetation that witness no abrupt changes in the land cover. This NDVI also acts as the basis for collecting samples for Random Forest classification of the Sentinel-2 images. The results from Sentinel-1 and Sentinel-2 combined with MASK from NDVI have determined the expansion area of the Minh Tien coal mine. The results are compared with those published on the website of Thai Nguyen Portal and have a significant similarity.

Keywords Illegal mining · Thai Nguyen · Sentinel-1 · Sentinel-2 · NDVI · Google earth engine

1 Introduction

In Vietnam, most minerals are distributed scatteredly over a large area, mainly in the midland and mountainous areas where the terrain is complex and travel is arduous, which is difficult to manage and protect. Furthermore, the inspection, examination, and supervision of mining activities are costly in regard to money, time, and human effort. In 2019, the General Department of Geology and Minerals made a record of sanctioning administrative violations against 65 enterprises for their illegal mining, hence the application of modern technology in supporting mineral extraction management is considered indispensable [1].

The application of remote sensing technology in recent years has become the major method in monitoring natural resources and environment in general and monitoring mineral activities in particular. The advantage of large spatial coverage and the multi-temporal properties of satellite images allows researchers to record changes in land cover. Utilizing different spectral bands to accurately record objects is extremely crucial in the study of mining activities, especially through their signatures on the earth's surface [2]. In addition, the size and location of the mine areas as well as the land-cover changes due to mineral extraction can be identified from satellite images [3]. Remote sensing can make mining planning easier, increase safety during and after mining, and monitor environmental impacts in surrounding mine areas [4].

There have been some studies using remote sensing to monitor issues related to mining operations, such as capability assessment of remote sensing and GIS techniques for monitoring reclamation success in coal mine degraded lands [5]. In other papers such as [6–8], the authors have applied multi-temporal radar satellite images to monitor land subsidence at the quarries in Binh Duong, Vietnam. LaJeunesse [9] has used free satellite imagery and open-source software to detect and asses mining expansion in Myanmar. Likewise, Li et al. [10] applied multi-temporal Landsat data to track the expansion of coal mining in northern China. Demirel et al. [11] also used high-resolution satellite imagery to detect land-use change in coal mine surfaces at the Goynuk and Bolu surface mines in Turkey. Another study by Guan et al. [12] investigated land-use change in a surface coal mine located in northeastern China and assessed the environmental impact using Landsat satellite imagery. In addition, Latifovic et al. [13] used Landsat images to present a method for determining land-cover change in the Athabasca Oil Sands, northeastern Alberta, Canada. The research team combined the use of Landsat images with National Oceanic and Atmospheric Administration (NOOA) images to further evaluate the land cover changes by building the Normalized Difference Vegetation Index (NDVI) time series. Maxwell et al. [14] used very high-resolution imaging and LiDAR data to map the land cover of a coal mine site in the southern coal mines of West Virginia, USA. Demirel et al. [15] applied multi-temporal satellite imagery and machine-learning classifier (Support Vector Machine-SVM) to classify the surface at an open-pit mine area. Townsend et al. [16] introduced the application of Landsat images to determine land-use changes and land-cover changes due to mining in the Central Appalachian Mountains, Eastern United States, during the period 1976–2006.

In accordance with the above analysis, the changes in the land cover around the mine area have been a great interest to many scientists, and there have been many applications using multi-temporal satellite images to monitor such changes. Specifically, most of them used either free or commercial software to process multi-temporal images. In this study, we exploit cloud base processing engine and Google Earth Engine (GEE) to determine the mining expansion for image processing purposes. GEE utilizes JavaScript or Python programming language, a cloud computing platform to store and develop geographic data, thus it is easier to create maps with multi-temporal analysis with satellite image data source available in the cloud. GEE capabilities have been used for a variety of applications [17], including vegetation mapping and monitoring, land-use/land-cover change mapping [18]. In [19], the GEE platform was applied to process Sentinel-2, Sentinel-1, and Landsat 8 multi-temporal images in determining land cover. The study demonstrates temporal aggregation as a promising tool in integrating large amounts of data, which is efficient and that can compensate for the lower quality of automatic image selection and cloud masking. In [20], multi-temporal Sentinel-2 image series have been used in studying the changes in Amazon mines to assess the impact of mining on the environment.

With regard to the fore-mentioned research, the convenience of remote sensing, as well as the GEE platform, has proved its ability in determining the change of land cover and mine expansion. The application of such platform for the purpose of identifying illegal mining in Vietnam is time-efficient and assist managers to have a better overview of the area they manage. Within this article, we focus on processing Sentinel-1 and Sentinel-2 images simultaneously to detect coal mining extent areas that affect the environment and landscape around the Minh Tien coal mine, Thai Nguyen province, Vietnam.

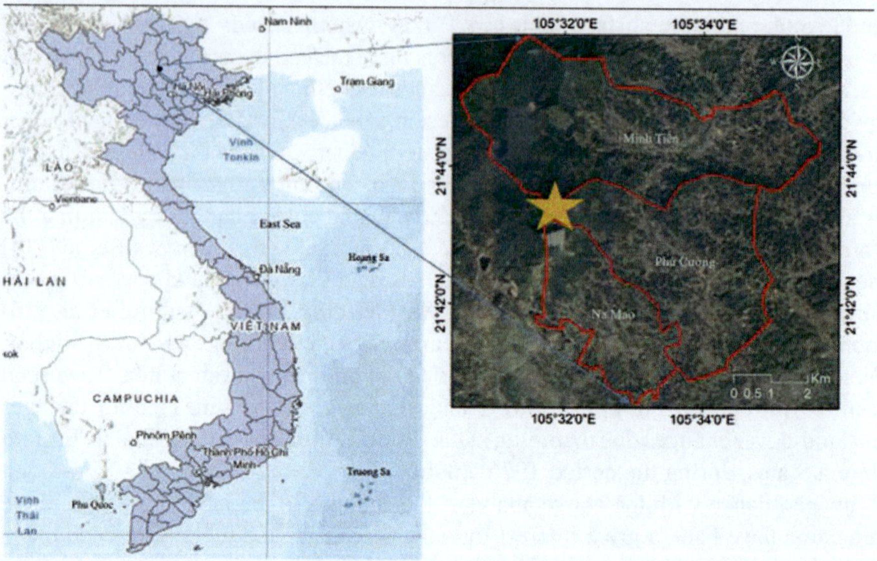

Fig. 1 Location of the Minh Tien coal mine (yellow star) in Dai Tu District, Thai Nguyen Province, Vietnam

2 Materials and Methods

2.1 *Study Area*

The Minh Tien coal mine is located in Minh Tien and Na Mao communes, Thai Nguyen Province, Vietnam. This coal mine was licensed to operate and exploit in 2014, with the allowed reserve of 8500 tons/year; however, the mining company did not comply properly. Since 2018, the company started mining coal in the open-pit form and has exploited its annual capacity numerous times, exceeding the licensed production [30]. Existing mines have been greatly expanded, of which the mine waste has been dumped near the farming areas, affecting the environment and threatening people's lives. Figure 1 is the mining location, where the yellow star represents the location of the Minh Tien coal mine.

2.2 *Materials*

To detect changes in the coal mine, we select Sentinel-1 and Sentinel-2 time-series images. Sentinel-1 images are processed at level 1 with ground range detected (GRD) and dual polarization (VV and VH). The Sentinel-1 image has a spatial resolution of

Table 1 Information of images

Sensor	Date range	Spatial resolution (m)	Number of images
Sentinel-2 (previous)	May 2016–May 2019	10, 20	33
Sentinel-2 (post)	May 2019–December 2021	10, 20	30
Pre-sentinel-1 GRD (VV) (ascending)	May 2016–May 2019	10	70
Post-sentinel-1 GRD (VV) (ascending)	May 2019–December 2021	10	62
Pre-sentinel-1 GRD (VV) (descending)	May 2016–May 2019	10	78
Post-sentinel-1 GRD (VV) (descending)	May 2019–December 2021	10	59

10 m and a repeat period of 12 days. The image acquiring period is from May 2016 to December 2021.

Sentinel-2 images are supplied at L2A product which has been corrected for surface reflectance using the Dense Dark Vegetation (DDV) and the Atmospheric Pre-corrected Differential Absorption (APDA) algorithms [21]. Table 1 shows the images selected for processing.

3 Methodology

Mining activities always leave an impact that changes the landscape and environment, such as vegetation cover, surface water, and soil cover surrounding the mining site. By monitoring and analyzing surface cover changes, signs of mining activities can be detected.

In this study, we combine two types of images as mentioned above, which are Sentinel-1 and Sentinel-2 images. The data processing is shown in Fig. 2. There are two blocks of work that are done with those datasets. The first block is used to process Sentinel-1 and Sentinel-2 images on GEE. After image processing, the second block is done in the ArcGIS Desktop software.

The study focuses on processing two types of images to determine the land-cover changes caused by mineral extraction and evaluate such changes in pixel values between two time periods. Those change values can be the abrupt type due to deforestation or urbanization. In addition, mining signs will cause these abrupt changes [22, 23]. However, just examining the difference between two image periods can be confusing because the vegetation is changed by the seasons. That is the reason why we have added more procedures to analyze the Sentinel-2 time-series images so as to determine the NDVI, which is an indicator of the most pronounced seasonal changes in vegetation. In addition, this index also tells us about the trend change

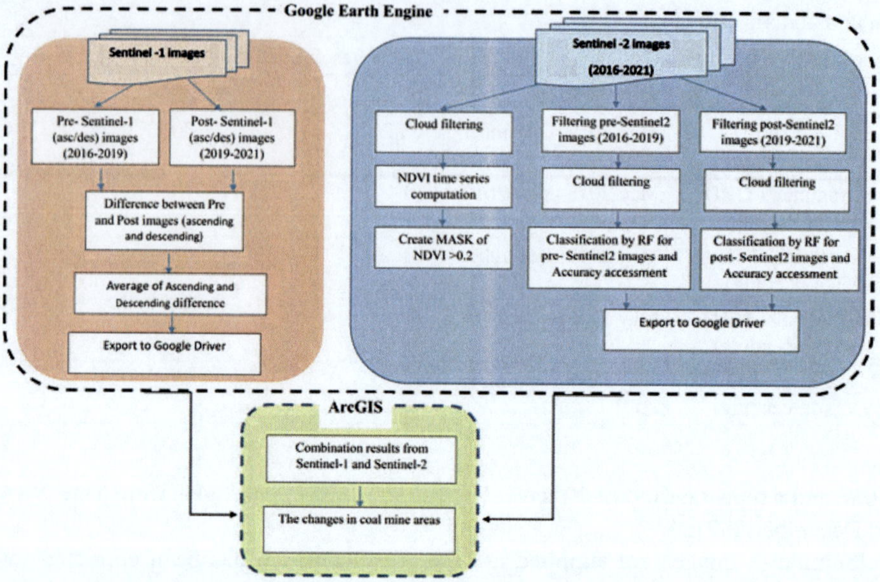

Fig. 2 Flowchart of data processing

of vegetation cover over the growth periods of the crop, thereby extracting abrupt changes of vegetation. The image processing is done on the GEE platform.

GEE works through a JavaScript Application Online Interface (API). By using this interface, the scripts can be run or shared and repeat geospatial data processing and analysis processes. Code Editor helps users to perform all the functions available in Earth Engine. Furthermore, the input data can be imported directly from the data WEB pages without downloading to the computer, which is a strength of GEE, allowing it to quickly analyze data without depending on the data and memory capacity of the computer. Figure 3 is the interface of GEE.

3.1 *Data Processing on Google Earth Engine*

See (Fig. 4)

3.1.1 Building NDVI Series to Detect Areas with Vegetation Cover

NDVI is calculated for the Sentinel-2 time-series images in the period of 2016–2021. The total number of images during this period is 63 images. However, the images for processing have been selected to reduce cloud cover. Selected images have less than 10% cloud coverage; thus, there are only 25 images that satisfy the set conditions.

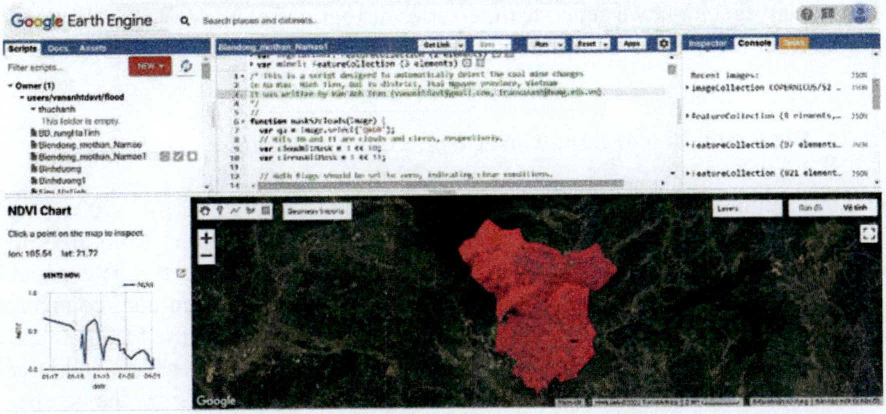

Fig. 3 GEE interface and coal mine location on sentinel-2 image

```
// create NDVI through all images
// Colect the time series images for checking the changes on images
var S2_SR = ee.ImageCollection("COPERNICUS/S2")
.filterBounds(shp)
.filter(ee.Filter.lt('CLOUDY_PIXEL_PERCENTAGE', 10))
.filter(ee.Filter.date('2016-10-01', '2021-05-01'));

//print (S2_SR.size());
print (S2_SR, 'S2_SR');
// Create a function that adds an NDVI band to a Sentinel-2 image
var addNDVI = function(image) {
  var ndvi = image.normalizedDifference(['B8', 'B4']).rename('NDVI');
  return image.addBands(ndvi);
};

// Apply this function across the image collection ////
var S2_NDVI = S2_SR.map(addNDVI);

// Sort the image collection by date and select the most recent image
var recent_S2 = ee.Image(S2_NDVI.sort('system:time_start', false).median());

// Print the image's metadata to the console to view its metadata
print('Recent image: ', recent_S2);

// Select just the NDVI band and add the NDVI image to the map using the NDVI color scheme
var recent_S2_NDVI = recent_S2.select('NDVI').clip(shp);
Map.addLayer(recent_S2_NDVI, {}, 'Recent Sentinel NDVI');
print('recent_S2_NDVI: ', recent_S2_NDVI);
```

Fig. 4 Code of NDVI calculation in the time series

The reason for determining the NDVI time series is that where cropland changes with growth cycles and seasons will be clearly defined and thereby confirm where there are sudden changes of vegetation such as deforestation to switch to mining. The relevant spectral bands for NDVI are the Red and the Nir InfraRed (NIR) bands. The code for the NDVI calculation is shown in Fig. 4. NDVI is a simple yet effective and commonly used indicator for greening determination. The NDVI value ranges from -1 to $+1$. Values above 0.2 usually indicate the existence of green vegetation [2]. As a basis for image classification sampling and to separate out areas of unchanged or seasonal vegetation, we generate time-series histograms of NDVI values. Figure 5 shows NDVI histograms for different land-cover types.

3.1.2 Determination of Land Cover by Sentinel-2 Images

Sentinel-2 time-series images had the acquisition in the period of 2016—2021. The series images are divided into two periods from 2016 to 2019 and from 2019 to 2021 (Table 1). These two periods are the two times before and after the changes of mine surface. This threshold is chosen because the Minh Tien coal company has started to increase exploitation many times more than the licensed amount and switched to open-pit mining in 2019 [30]. The Sentinel-2 images before and after the event are median filtered to enhance the image quality and minimize the number of images, because this function in GEE calculates the median of all pixel values across the stack of all matching bands to create one output image. We use a machine-learning algorithm Random Forest (RF) to classify the land cover. This method uses group learning, which is a technique that combines multiple 'decision trees' to solve complex problems. Each tree is established from randomly selected training pixels (bootstrap). There are two parameters that need to be decided in this classification algorithm, including the number of trees (ntrees) and the number of variables to be divided at each node [24]. The selected ntrees depend on the shortest processing time to get the lowest error, which runs from 1 to 500 trees as default, and the number of variables varies from the minimum number of independent variables (equal to 1) to the maximum number of independent variables that utilized in the classifier [25].

Sentinel-2 images, after median filtering and NDVI calculation, are used for RF classification. Selected bands include B2, B3, B4, B8, and NDVI. Figure 6 shows the code in GEE used to classify the two image periods before and after the land-cover changes.

Land cover in the study area has been divided into four types: water, vegetation, urban, and mining areas. The selection of sample regions is based on the composite image, google earth visualization, and the time series of NDVI charts. The classification is carried out on the GEE platform, so the sampling method is fulfilled directly on the screen.

Figure 7 illustrates the results of RF image classification of the two periods before and after the change in the coal mine area, and Table 2 is the result of classification accuracy when choosing the number of trees (ntrees) is 500 for RF classification.

3.2 Determine the Difference Between Two Periods Using Sentinel-1 Images

The main goal of using Sentinel-1 images is to determine the change in the surface roughly between two times when the change occurs. In the flow chart, the principle

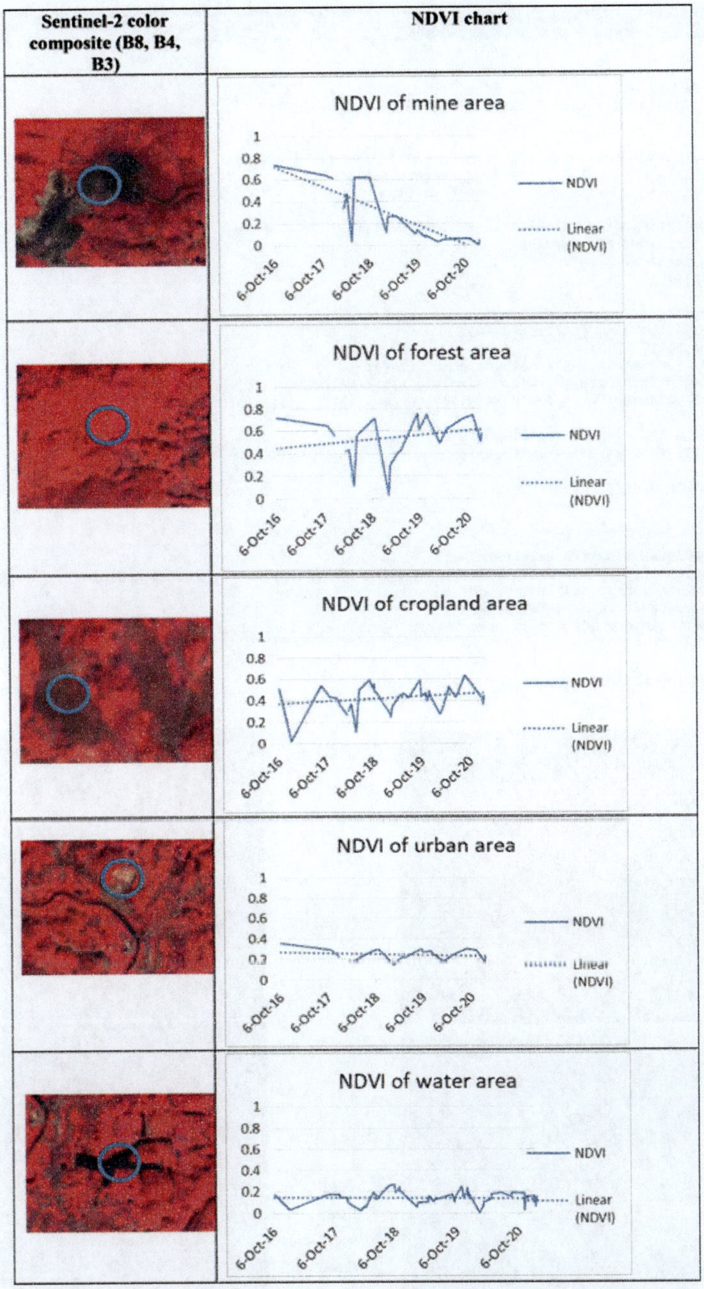

Fig. 5 Samples of different land-cover types and histogram of NDVI values over time of these types

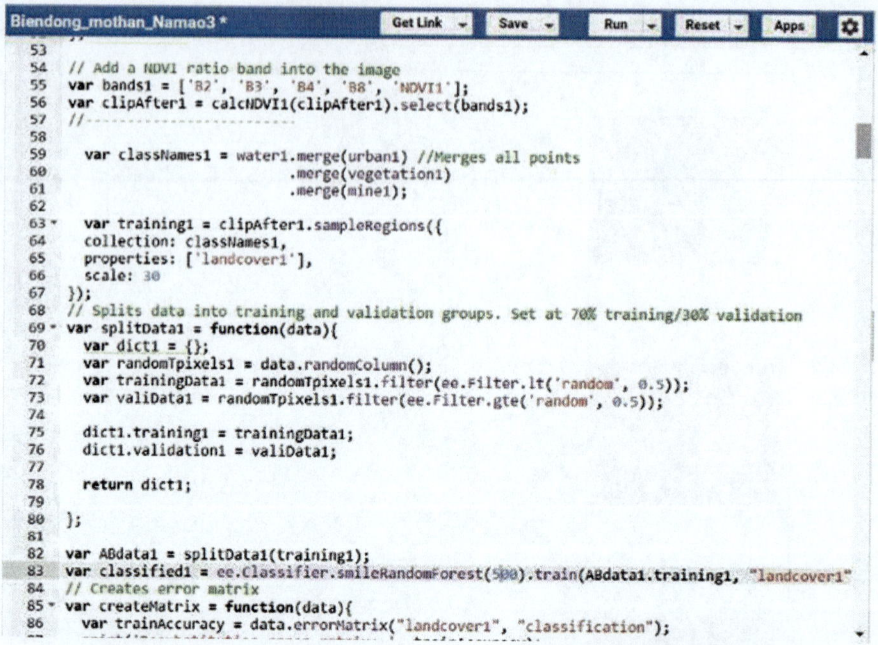

Fig. 6 Code in GEE used to classify the images

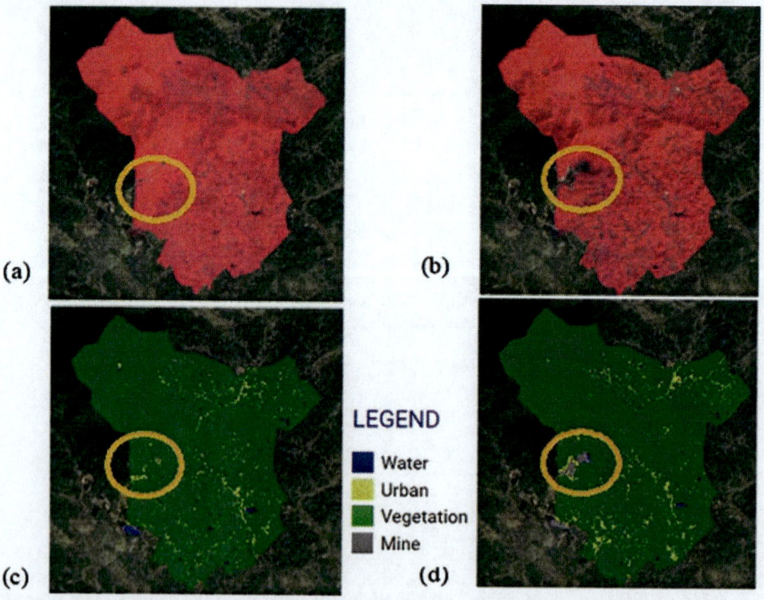

Fig. 7 **a** Sentinel-2 image (2016–2019) median filter, **b** Sentinel-2 images (2019–2021) median filter, **c** RF classification image (2016–2019), and **d** RF classification image (2019–2021)

Table 2 Accuracy of two data sets before and after 2019

Sentinel-2 images	OA (%)	Kapa
2016–2019	97	0.95
2019–2021	94.2	0.92

of determining the change between two periods with Sentinel-1 GRD images is detecting the amplitude changes, A_change, defined as the stack before the change event minus the stack after the change event (Eq. 1), which is a change detection method based on amplitude is most commonly used [26–28]. The changes can be either positive or negative, corresponding to changes in land-cover patterns. Positive values correspond to the areas of land-cover loss [27, 29]. Due to the oblique view of the radar satellite, it will have a difference in the amount of backscattering when the image is in the ascending or descending direction; therefore in this study, both directions of radar images are used and then we take the average for detecting the changes.

$$\text{A\_change} = \text{Apre} - \text{Apost} \tag{1}$$

Apre Backscatter values of the images before the event occurs.
Apost Backscatter values of the images after the event occurs.

Radar images are not affected by weather conditions, atmosphere, and clouds; thus, the change in land cover determined between two image periods is reliable. However, these changes also represent many different factors, not only the vegetation cover changes but also changes in urban areas or other objects. So, for further confirmation about the vegetation changes related to mine, the result will be combined with analysis from multi-temporal Sentinel-2 images. The code for image processing on GEE is available at: https://code.earthengine.google.com/?scriptPath=users%2Fvananhtdavt%2Fflood%3ABiendong_mothan_Namao3.

As mentioned above, the difference between the two periods of the Sentinel-1 images represents the changes of the land cover, where blue is a negative value and red is a positive value (Fig. 8). The positive value depicts the loss of the land cover, so it is clear that, in the mining area (inside the blue ellipse), there is a large change in vegetation cover related to mining.

3.3 Determining Mining Extension at the Minh Tien Coal Mine by Combining Results from Sentinel-1 Images, Image Classification from Sentinel-2, and NDVI Images

All results performed on GEE have been exported to Google Driver for further processing on the ArcGIS desktop. The two resulting layers of Sentinel-2 image classification at the previous time (2016–2019) and the latter time (2019–2021) are

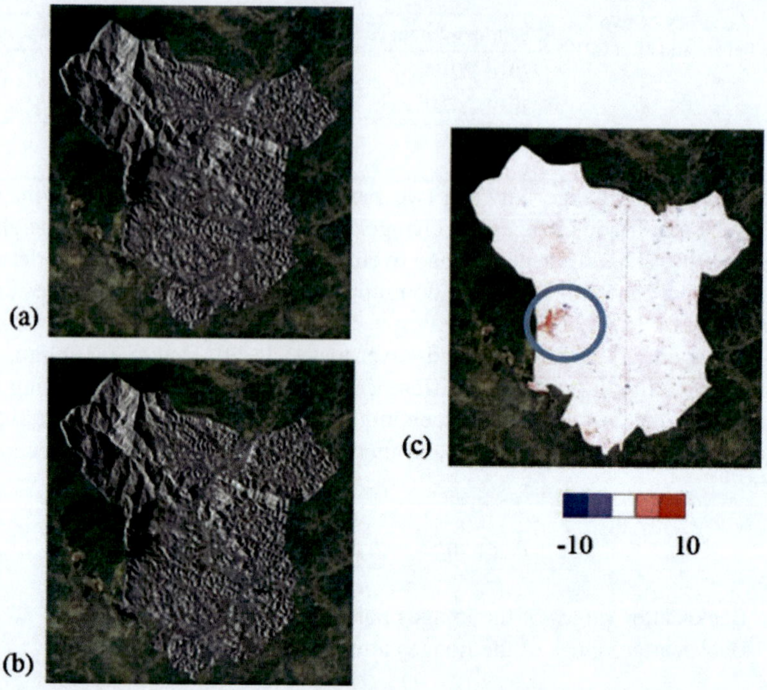

Fig. 8 **a** Sentinel-1 image before 2019, **b** Sentinel-1 image after 2019, and **c** subtraction between the two times of sentinel-1 images

overlaid to detect the changes. Among the land-cover changes, the change from vegetation to mining will be of interest. Figure 9 shows the change from vegetation to mining activity. Vegetative change to mining is detected in the relatively large Minh Tien coal mine area (Fig. 9a); however, some non-mining outside locations are also mistakenly detected. The NDVI series for the period 2016–2021 is also used to overcome this issue.

Based on the spectral reflectance chart of different locations on the image (Fig. 5), we found that at locations with suspected mining, the NDVI value changes from 2016 to 2021 with a downward trend. The average NDVI value at this location is less than 0.2. The time-series NDVI values in forested areas also vary between seasons, but the mean value is still greater than 0.56. In arable areas, the NDVI value curve most clearly shows the seasonal change of crops because NDVI values increase and decrease cyclically. The mean NDVI in these areas is still greater than 0.5. In the lake area, the NDVI varies between 0 and 0.25. At the location of water, the signal with much noise also causes an increase in the value of the NDVI. The average value of the NDVI here is less than 0.14. By analyzing the values of the NDVI time series, it is reasonable to use MASK for forest cover and agricultural lands with NDVI values greater than 0.2. In the sites where vegetation loss is related to coal mining, the value

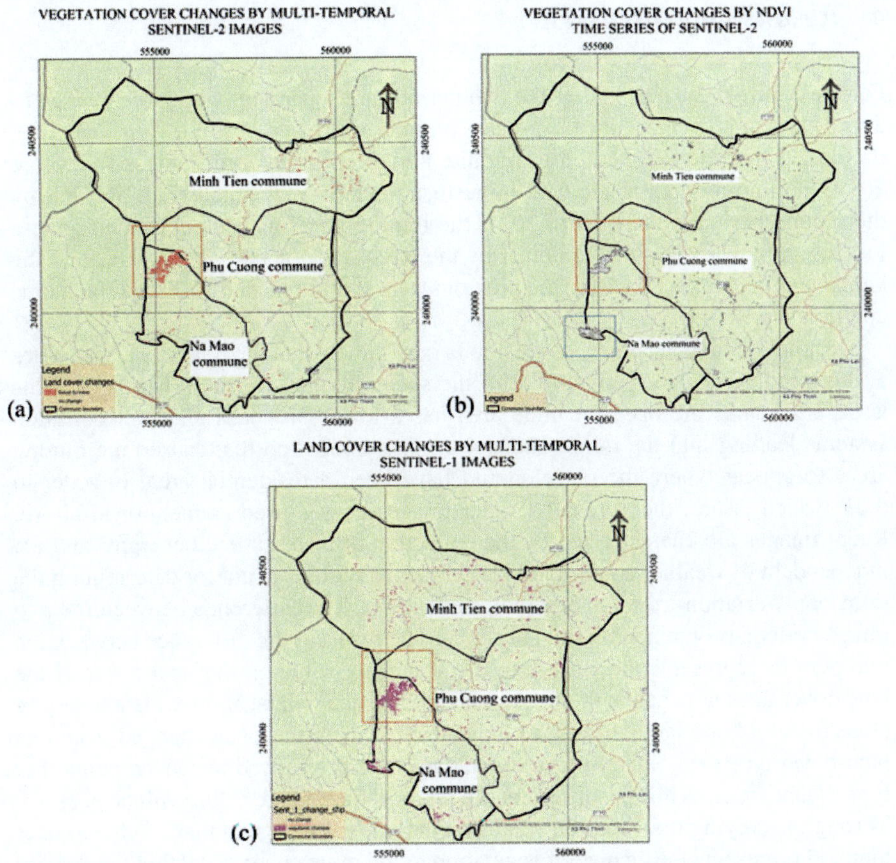

Fig. 9 **a** Land-cover changes by Sentinel-2, **b** NDVI MASK, and **c** land-cover changes by Sentinel-1 images

will be less than 0.2. Figure 9b shows a MASK map made from the mean NDVI of the time series.

The results from the temporal difference of the Sentinel-1 images are also used to further precisely define the extent of the coal mine. The results from the Sentinel-1 images shown in Fig. 8 are used to further define the extent of the coal mine expansion. This resulting image has been converted to vector. In the regions with a lot of vegetation changes, the value is greater than 1.7. Thus, the binary image is generated when change values are selected greater than 1.7. Figure 9c shows the final change from the Sentinel-1 images.

4 Results and Discussion

For the Sentinel-2 dataset, after the two periods are separated, taking the year 2019 as the landmark to determine the process of mine expansion of Minh Tien coal mine, it is noted that the RF method gave quite a high accuracy with images set before 2019, the mining area at the coal mine location was very small (Fig. 5c), but by the second period from 2019 to 2021, the mining area has been expanded quite a lot (Fig. 5d). The classification error by the RF method achieved was 94% and the kappa coefficient was 0.92 for the previous period and reached 97% and the kappa coefficient was 0.95.

Despite the high accuracy, confusion is inevitable when the image spectra values at the mining sites are very similar to the spectral value for the urban layer. This issue is because the open-pit mine area has a lot of pit roads and transportation systems leading into the mine. There was some confusion that caused the mining areas to appear where the place should have been a residential area. In order to reduce such issues, the Sentinel-1 radar image set was used as mentioned above. Radar images are characterized by the collection of backscattered energies and are independent of weather conditions. These images will be an aid for determining the mine area variation from the Sentinel-2 image set. The difference between the two image periods is obtained by Eq. (1) above. The result of the difference between the two periods represents the change in the scattering value on the land cover. If the land cover does not change much, then this difference will be approximately zero or close to zero, but a large difference indicates that there has been changed from one land-cover type to another. In case vegetative cover is removed to convert to another type of land, such as mining or residential regions, the change values will be positive. Through surveying the variation values on the image, we found that a value greater than 2.0 represents the change of vegetation to the mine cover. Selecting the density slice method to separate areas with such changes and exclude other changed areas will greatly assist the process of determining variation from Sentinel-2 images.

Also, another support is the NDVI series of the period 2016–2021, which is averaged. In places with high NDVI values, vegetation exists, and in locations with NDVI lower than 0.2 vegetation is not possible. Figure 9 shows the land-cover change of Sentinel-2, Sentinel-1, and NDVI MASK images. In all three of these maps, it is noticed that, at the location of Minh Tien coal mine marked with a red square, there are many change points. At this position, the differences between the two Sentinel-2 and Sentinel-1 image periods are quite similar. In addition, some surrounding locations have also detected small changes but are not located near the coal mine area, so these points need to check for the field.

The NDVI mask map has removed values greater than 0.2, so locations around the mine will be the most suspicious. In Minh Tien (where is in the orange box), the suspected changes are similar to the position of the two maps above. The location at the Na Mao mine, which is in blue square, also found the existence of mining (Fig. 9b). The discovery of this location is slightly different from the two change maps made from Sentinel-1 and Sentinel-2 images. The map created from the NDVI over time from 2016 to 2021 is an average representation of all NDVIs, so determining the change using those values will be more difficult than determining the changes by different period.

The final result was received by using the overlaying method from the results of Sentinel-1, Sentinel-2, and NDVI MASK together. Figure 10 illustrates the expanded mining area at Minh Tien and also the field check at this region. The total area of mine expansion is 105225 m^2 (inside the red frame on the map). After overlaying the three layers, it is easy to see the sites that are not coal mines outside the red frame no longer appear as change points, but the extension only focuses around the location of Minh Tien coal mine. This demonstrates that with the combination of both results from Sentinel-1, Sentinel-2 with the NDVI time series has proven effective in excluding urban or farmland variations.

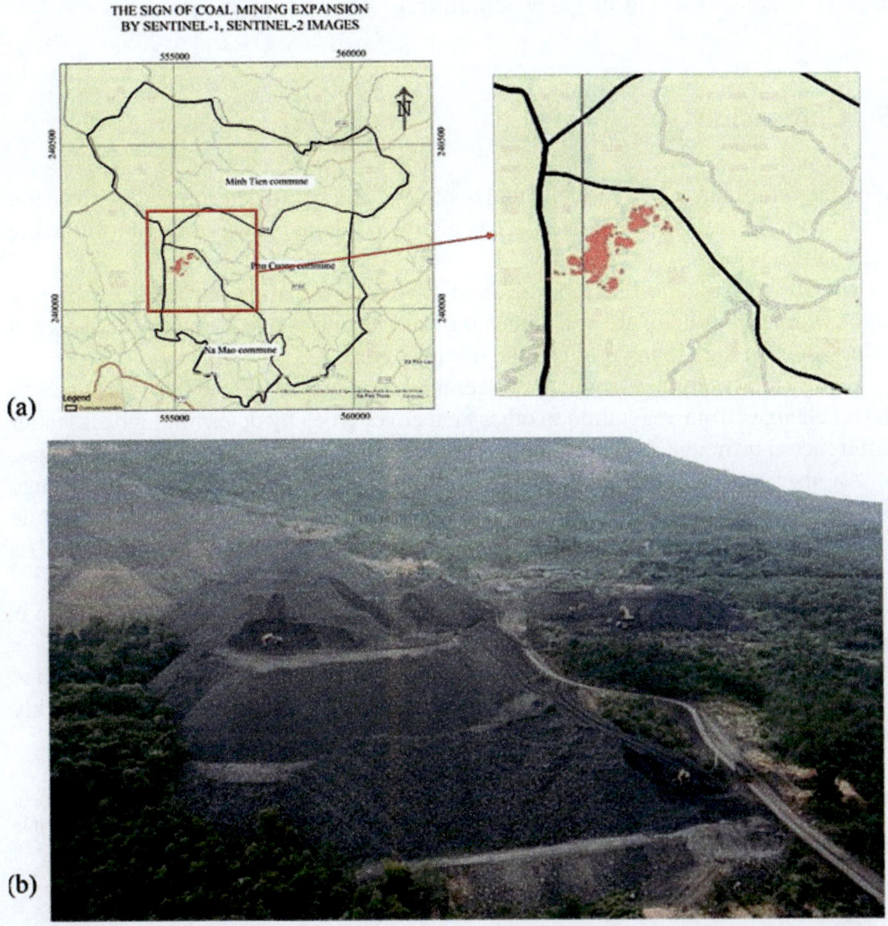

Fig. 10 **a** Expansion of coal mine area from 2016 to 2021, **b** a photo at the location of the Minh Tien coal mine taken in September 2021 [30]

This mining expansion position is very suitable for the survey position, which is reflected on the Web Portal of Thai Nguyen province. The photo in Fig. 10b was taken at the coal storage yard of the Minh Tien coal mine [30] by the Thai Nguyen journal reporter. According to the information on this page, from March 2019 to August 2021, the Dong Bac Hai Duong Company exploited about 2 million tons of raw coal. On average, each year, the amount of coal illegally exploited by this company is more than 120 times of the license. Even this company did not fully comply with the regulations on the observance of the provisions of the law on environmental and mineral protection for the Minh Tien coal mine. Specifically, since this company went into operation, it has caused a series of systemic consequences for the locality, such as mining outside the licensed mine boundaries, rainwater causing coal to overflow on the ground without treatment and into rivers and streams, and overflowing into people's fields with an area of about 7.5 hectares. Therefore, this image can also be called a demonstration of illegal coal mining.

5 Conclusion

With the combination of Sentinel-2 and Sentinel-1 images to determine the expansion of the mining area in Minh Tien, Thai Nguyen, Vietnam, we have the following comments:

The RF classification method for land-cover classification has achieved quite a high accuracy, but there are still some confused locations due to the similar spectral value between the urban layer and the mining layer.

Using a multi-temporal sequence of Sentinel-1 images separated into two phases to detect changes from vegetation to other land cover types by determining backscatter differences is efficient, however, there is still confusion with non-vegetation changes.

Another important factor that helps to improve the ability to determine the change of vegetation cover to the mining expansion is the NDVI time series index. As the value of this index decreases over time it represents areas that are certain to be converted from vegetation to mining.

The combination of image types and support algorithms on the GEE platform is a revolution in multi-source, multi-resolution data analysis, helping users not have to struggle to download and process data on personal computers while saving time for users. The application of GEE helps managers have an overview and quickly monitor changes on the earth's surface in near real-time.

Acknowledgements We would like to thank Google provided Google Earth Engine development platform for processing data. Thank to ISRM Reasearch Group for providing a lot of valuable information about the mining situation in Thai Nguyen area.

References

1. Mossa, J., James, L.: 13.6 Impacts of mining on geomorphic systems. (2013)
2. Kotaridis, I., Lazaridou, M.: Delineation of open-pit mining boundaries on multispectral imagery. Kotaridis, I. and Lazaridou, M.(2020).'Delineation of Open-Pit Mining Boundaries on Multispectral Imagery'. in Remote Sensing. IntechOpen. doi **10** (2020)
3. Werner, T.T., Mudd, G.M., Schipper, A.M., Huijbregts, M.A., Taneja, L., Northey, S.A.: Global-scale remote sensing of mine areas and analysis of factors explaining their extent. Glob. Environ. Chang. **60**, 102007 (2020)
4. Molaei, F., Rahimi, E., Siavoshi, H., Afrouz, S.G., Tenorio, V.: A comprehensive review on internet of things (IoT) and its implications in the mining industry. Am. J. Eng. Appl. Sci. **13**, 499–515 (2020)
5. Karan, S.K., Samadder, S.R., Maiti, S.K.: Assessment of the capability of remote sensing and GIS techniques for monitoring reclamation success in coal mine degraded lands. J. Environ. Manage. **182**, 272–283 (2016)
6. Tran, V.A., Bui, X.N., Nguyen, Q.L., Tran, T.-A.: Land subsidence detection in Tan My-Thuong Tan open pit mine and surrounding areas by time series of sentinel-1 images. Inżynieria Mineralna (2020)
7. Nam, B.X., Anh, T.V., Bui, L.K., Long, N.Q., Thu Ha, T.L., Goyal, R.: Mining-induced land subsidence detection by persistent scatterer InSAR and sentinel-1: application to Phugiao Quarries, Vietnam. In: Proceedings of the International Conference on Innovations for Sustainable and Responsible Mining, pp. 18–38. Springer (2020)
8. Nguyen, Q.L., Tran, V.A., Bui, K.L.: Determination of ground subsidence by sentinel-1 SAR data (2018–2020) over Binh Duong Quarries, Vietnam. VNU J. Sci.: Earth Environ. Sci. **37** (2021)
9. LaJeunesse Connette, K.J., Connette, G., Bernd, A., Phyo, P., Aung, K.H., Tun, Y.L., Thein, Z.M., Horning, N., Leimgruber, P., Songer, M.: Assessment of mining extent and expansion in Myanmar based on freely-available satellite imagery. Remote Sens. **8**, 912 (2016)
10. Li, N., Yan, C., Xie, J.: Remote sensing monitoring recent rapid increase of coal mining activity of an important energy base in northern China, a case study of Mu Us Sandy Land. Resour. Conserv. Recycl. **94**, 129–135 (2015)
11. Demirel, N., Düzgün, Ş, Emil, M.K.: Landuse change detection in a surface coal mine area using multi-temporal high-resolution satellite images. Int. J. Min. Reclam. Environ. **25**, 342–349 (2011)
12. Guan, C., Zhang, B., Li, J., Zhao, J.: Temporal and spatial changes of land use and landscape in a coal mining area in Xilingol grassland. In: IOP Conference Series: Earth and Environmental Science, pp. 012052. IOP Publishing, (2017)
13. Latifovic, R., Fytas, K., Chen, J., Paraszczak, J.: Assessing land cover change resulting from large surface mining development. Int. J. Appl. Earth Obs. Geoinf. **7**, 29–48 (2005)
14. Maxwell, A.E., Warner, T.A., Strager, M.P., Pal, M.: Combining RapidEye satellite imagery and Lidar for mapping of mining and mine reclamation. Photogramm. Eng. Remote. Sens. **80**, 179–189 (2014)
15. Demirel, N., Emil, M.K., Duzgun, H.S.: Surface coal mine area monitoring using multi-temporal high-resolution satellite imagery. Int. J. Coal Geol. **86**, 3–11 (2011)
16. Townsend, P.A., Helmers, D.P., Kingdon, C.C., McNeil, B.E., de Beurs, K.M., Eshleman, K.N.: Changes in the extent of surface mining and reclamation in the Central Appalachians detected using a 1976–2006 Landsat time series. Remote Sens. Environ. **113**, 62–72 (2009)
17. Mutanga, O., Kumar, L.: Google earth engine applications. vol. 11, pp. 591. Multidisciplinary Digital Publishing Institute (2019)
18. Midekisa, A., Holl, F., Savory, D.J., Andrade-Pacheco, R., Gething, P.W., Bennett, A., Sturrock, H.J.: Mapping land cover change over continental Africa using Landsat and google earth engine cloud computing. PLoS ONE **12**, e0184926 (2017)

19. Carrasco, L., O'Neil, A.W., Morton, R.D., Rowland, C.S.: Evaluating combinations of temporally aggregated Sentinel-1, Sentinel-2 and Landsat 8 for land cover mapping with google earth engine. Remote Sens. **11**, 288 (2019)
20. Lobo, F.D.L., Souza-Filho, P.W.M., Novo, E.M.L.D.M., Carlos, F.M., Barbosa, C.C.F.: Mapping mining areas in the Brazilian amazon using msi/sentinel-2 imagery (2017). Remote Sensing **10**, 1178 (2018)
21. Sentinel, E.: User Handbook. ESA Standard Document 64
22. Arévalo, P., Bullock, E.L., Woodcock, C.E., Olofsson, P.: A suite of tools for continuous land change monitoring in google earth engine. Frontiers Clim. **26** (2020)
23. Zhu, Z., Woodcock, C.E.: Continuous change detection and classification of land cover using all available Landsat data. Remote Sens. Environ. **144**, 152–171 (2014)
24. Van Anh, T., Le, L.T., Hung, N.N., Nghi, L.T., Hanh, T.H.: Monitoring vegetation cover changes by sentinel-1 radar images using random forest classification method. Inżynieria Mineralna 1 (2021)
25. Breiman, L.: Random forests. Mach. Learn. **45**, 5–32 (2001)
26. Akcin, H., Kutoglu, H., Deguchi, T., Koksal, E.: Monitoring subsidence effects in the urban area of Zonguldak Hardcoal Basin of Turkey by InSAR-GIS integration. Nat. Hazard. **10**, 1807 (2010)
27. Adriano, B., Yokoya, N., Miura, H., Matsuoka, M., Koshimura, S.: A semiautomatic pixel-object method for detecting landslides using multitemporal ALOS-2 intensity images. Remote Sens. **12**, 561 (2020)
28. Mondini, A.C., Santangelo, M., Rocchetti, M., Rossetto, E., Manconi, A., Monserrat, O.: Sentinel-1 SAR amplitude imagery for rapid landslide detection. Remote Sens. **11**, 760 (2019)
29. Rignot, E.J., Van Zyl, J.J.: Change detection techniques for ERS-1 SAR data. IEEE Trans. Geosci. Remote Sens. **31**, 896–906 (1993)
30. Thai Nguyen Portal, 12/6/2022, https://thainguyen.gov.vn/bai-viet-ve-thai-nguyen/-/asset_publisher/L0n17VJXU23O/content/thai-nguyennhieu-sai-pham-tai-mo-than-minh-tien-a-uoc-huyen-ai-tu-canh-bao-truoc/20181

Evaluation of Coal Reserve Reliability in the Nui Beo Mine, Quang Ninh Province Based on the Statistical and Stable Random Function Methods

Khuong The Hung, Vu Thai Linh, Pham Thanh Tinh, and Nguyen Khac Duc

Abstract The Nui Beo is a famous coal mine in the Hon Gai coal zone, Quang Ninh province, where contains many coal resources in Vietnam. In this study, synthesizing and processing data of the mining geological parameters by using the stable random function and statistical methods are used to evaluate the coal reserve reliability in the Nui Beo mine. The results show that the coal reserve reliability depends on geological conditions, especially in the distances between exploration works and parameters of calculating coal reserves (i.e., thickness, density, and declination angle on coal seams). Most exploration efforts in the Nui Beo mine satisfy the suitable pattern exploration grid for reserve level 122, according to the stable random function used to calculate the affected range. Statistical models based on a confidence probability of 0.95 are used to evaluate coal reserve dependability in the research region, resulting in $\pm$ 10.12 $\div$ $\pm$ 16.46% of coal reserve errors. Estimating coal resources during the exploratory process is sufficient for the exploitation design. The findings also give an overview of the parts of coal resource dependability that have been impacted and a foundation for assessing the coal resource error in the Nui Beo mine. These can be used on other coal resources with general comparable geological and mining conditions.

Keywords Coal reserves · Statistical method · Stable random function · Nui Beo mine

K. T. Hung (✉) · V. T. Linh
Hanoi University of Mining and Geology, 100000 Hanoi, Vietnam
e-mail: khuongthehung@humg.edu.vn

P. T. Tinh
Vinacomin—Informatics, Technology, Environment Joint Stock Company, 100000 Hanoi, Vietnam

N. K. Duc
Vietbac Geological Joint Stock Company, 100000 Hanoi, Vietnam

L. Q. Nguyen et al. (eds.), *Advances in Geospatial Technology in Mining and Earth Sciences*, Environmental Science and Engineering,
https://doi.org/10.1007/978-3-031-20463-0_15

1 Introduction

The orientation exploration grid is frequently used at the start of any mineral exploration based on geological similarities [1–4]. However, ore bodies are generally seldom the same size, thickness, internal orebody structure, etc., and implementations of this approach have considerable limitations. As a result, the use of stable random function and statistical methods in research-exploration data processing is the most effective approach for determining the most appropriate exploration grid for each mineral item and mining group of deposits [4–6]. Using statistical approaches to set up an exploration grid, numerous projects have had great success in geological research [2, 4, 7], meanings that this step is essential.

In Vietnam, mineral coal distributed on the mainland occurs in six major coal basins: Quang Ninh, Thai Nguyen-An Chau, Da River, Red River, Nghe Tinh, Nong Son, and scattered in small coal-bearing bands and areas. The Quang Ninh coal basin has an arc shape extending a length of 250 km from the Linh Duc mine in the west of Tam Dao mountain range (Tuyen Quang province), through Dong Trieu to Hon Gai, Cam Pha, and Ke Bao areas (Quang Ninh province). It is filled with Norian-Rhaetian coal-bearing sediments of the Hon Gai formation. The Nui Beo mine belongs to the Hon Gai coal zone of the Quang Ninh coal basin, where many coal resources have been estimated.

According to the report on summarizing and recalculating the coal resources of Ha Lam, the Nui Beo coal mine has been additionally explored to − 350 m with the networks in level 122, and level 333 are 75 ÷ 100 m × 120 ÷ 150 m and 250 m × 250 m, respectively [8, 9]. Exploration results have provided necessary documents on the quality, coal reserves, and mining conditions for mining for recent years. Currently, the open-pit method exploits the mine to the level of − 350 m. Analysis of exploratory documents shows that detailed research mainly focuses on the outcrop to the −300 m level. The study is still preliminary from −300 m to − 500 m levels. Only a few works have drilled controlling the coal seams. Therefore, it is necessary to evaluate the reliability of exploration documents in general and coal reserves in the exploration areas as a basis for deep exploration orientation. The stratigraphy of the Nui Beo area contains coal seams from V4 to V14 seams and some associated lenticular sub-seams. Furthermore, the V11, V10, V9, and V7 coal seams are considered promising for exploitation, and in fact, most the mining pits are currently focusing on these seams. Wherefore, the paper will focus on the coal seams, as mentioned earlier.

In real exploration, the reliability of mineral reserves can be assessed by various methods; in which statistics, geostatistics, or stable stochastic function are methods of the geological mathematical model applied indirectly to determine the reliability of the reserve through evaluating the reliability of the exploration grid and the errors of the parameters affect the reliability of the reserve calculation. All reserve calculation errors are divided into three primary groups: geological, technical errors, and errors related to selecting reserve calculation methods [4, 10–14]. The geological error is mainly related to the extrapolation of actual documents collected in the exploration work to neighboring areas (method of zoning mineral bodies, relating sections according to the linear exploration system), determining the spatial distribution of beneficial and harmful components, etc.). The geological error is often the largest in the reserve calculation, but it is of little interest. According to many researchers [4–6, 11–16], the geological error in the reserve calculation can be up to 10 ÷ 15%. Technical errors are mainly related to measuring techniques and determining initial parameters (thickness, sample analysis results, etc.) to calculate reserves, such as thickness measurement, chemical analysis, determining contents of valuable ingredients, weight density, measuring area, etc. The technical errors related to the reserve calculation include: The error in determining body weight density (Δ_d) ranges from 5% [16] to 10% [11], and the plot's error of area determination (Δ_s) is from 2 to 3%. Moreover, the thickness measurement error (Δ_m) varies depending on the thickness of the ore body (coal seam), type of exploration work, and measurement method. According to Vasiliev [16], the thickness error is usually from 2 to 3%. However, Kreige et al. [11] argued that the error of measuring the thickness of the ore veins in the excavation work and the borehole is usually from 2 to 10% and 20 to 30%, respectively. The largest error is mainly related to the ore veins with small thickness. The average allowable random error in the chemical analysis (Δ_c) of the coal ash content usually ranges from 1 to 20%, sometimes up to 30%. In the general case, this error is generally in the range of ± (2 ÷ 5%) and up to ± (25 ÷ 30%) for coal seams with low ash content and high ash coal seams, respectively.

This study aims to evaluate the coal reserve reliability of the Nui Beo mine based on the statistical and stable random function methods.

2 Overview of the Nui Beo Coal Mine

The Nui Beo mine is located in Ha Long city, Quang Ninh province. The mine is about 4 km far from the southwest of Ha Long city center, on the left side of National Highway 18A from Ha Long to Mong Duong (Fig. 1a).

In the mine site, there are Triassic sediments of Hon Gai formation, middle sub-formation, and loose deposits of the Quaternary system [17] (Fig. 1b). The petrographic composition of the middle Hon Gai sub-formation includes layers of siltstone, sandstone, claystone, coal clay, and coal seams lying alternately, with a stratigraphic thickness of about 1800 m. The middle Hon Gai formation contains industrial coal seams (Fig. 2).

Quaternary sediments directly cover the middle Hon Gai formation, distributed in the lower areas and valleys around the Nui Beo mine. Sedimentary compositions include pebbles, gravels, sands, loose clay, sometimes rolling boulders, and weathering products from pre-existing rocks.

Folds and fault systems are developed in the Nui Beo mine area, making it complicated to identify the seam and mining coal.

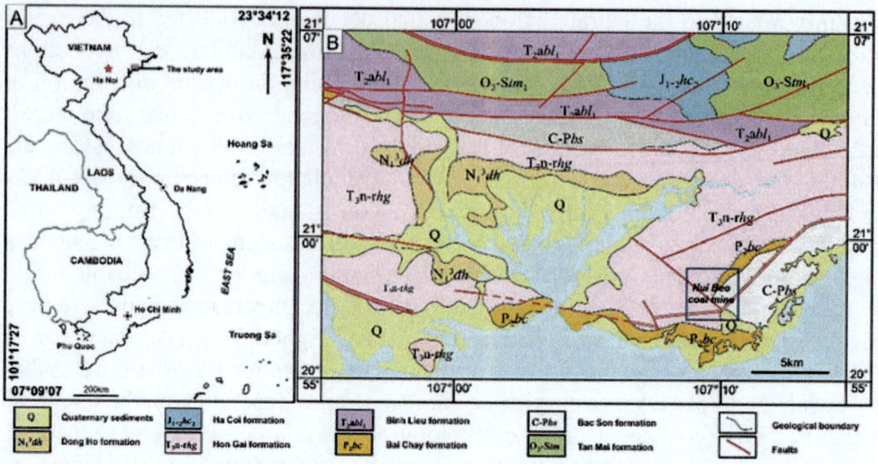

Fig. 1 **A** Sketch map of Vietnam and location of the study area, **B** geological map of Ha Long area showing the location of the Nui Beo mine [17]

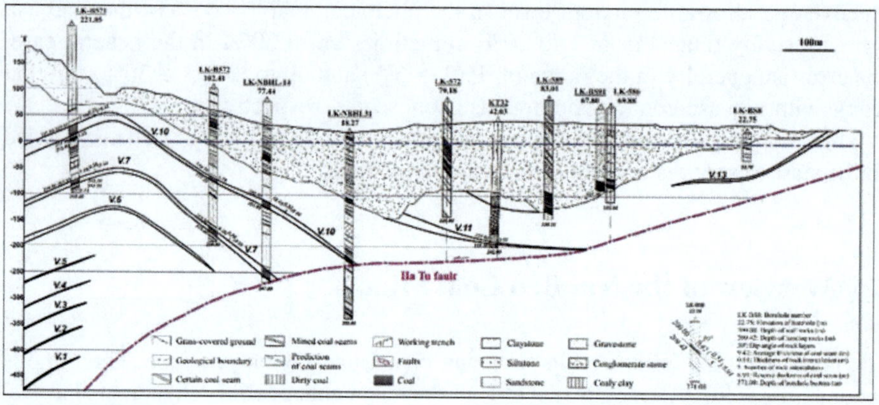

Fig. 2 Geological cross-section of line T.VA showing the coal seams and structure of the Nui Beo coal mine [8, 9]

3 Materials and Methods

3.1 Database

Documents used in this study are mainly from the report on the exploration results of the Ha Lam coal mine edited by Hung [8], Anh [9], combined with new exploratory drilling documents in the greater depth area of the Nui Beo mine recently. In particular, other data collected from the report on exploration results and recalculation of reserves of the Ha Lam coal mine are also used in this paper.

Statistical calculation data and random function construction are based on the data collected from 544 drilling and trenching works that have been constructed through periods and exploration stages within the Nui Beo mine area.

3.2 Reserve Calculation Conventional Method

Based on geological characteristics, distribution, and positions of coal seams, the coal reserve of the Nui Beo coal mine is calculated according to the Secang method [8, 9].

Each block's coal reserve/resource is calculated according to the following formula.

$$Q_i = S_i \times m_i \times D = S_b \times \sec\alpha_i \times m_i \times D \tag{1}$$

$$\text{Reserve of seam} : Q = \sum Q_i \tag{1a}$$

where Q_i—coal reserves of the i-th reserve calculation block (ton), Q—total coal reserves of the seam (ton), S_i—actual area in i-th reserve calculation block (m^2), S_b—area of the plan view of seam's cylinder (m^2), α_i—the average angle of the seam in the i-th reserve calculation block (degrees), m_i—the actual average thickness of the i-th reserve calculation block (m), and D—coal weight density ($D = 1.39$ ton/m^3).

However, the method is often limited to calculating steeply inclined seams' coal seam reserver (the angle > 45°). In addition, the process also encounters specific errors due to factors encountered when determining the area, thickness, etc., and not showing the shape of the coal seam in space.

3.3 Research Methods

Stable stochastic function model

Using the stochastic function model allows solving many problems related to exploration and mining. There are two main types of stochastic function models: stable and unstable stochastic functions. The characteristic of a stable stochastic function is that the mathematical expectation M(x) and the variance D(x) is a quantity that is unchanged with any spatial coordinates, i.e., independent of the position of the observation point, the correlation function K(h) depends on the observation step (h) and is a vector depending on the observation direction h. In the scope of this study, a stable stochastic function model is used to assess the reliability of coal reserves by evaluating the grid of exploration works conducted at the Nui Beo mine.

The stable stochastic function is characterized by the correlation function K(h) depending on the observation step h, the observation direction r(h) [4]. The formula determines the correlation function.

$$K(h) = \frac{1}{n-h}\sum_{i=1}^{n-h}[f(x_i) - X][f(x_i + h) - X] \tag{2}$$

where *n*—number of observation points, *h*—distance between observation points is expressed as the number of distances between adjacent points, and *X*—mean value (mathematical expectation) of research parameters.

We often use the norm correlation function r(h) when solving practical tasks.

$$r(h) = \frac{K(h)}{\sigma^2} \tag{3}$$

The norm correlation function allows separating the original variance σ^2 into two parts: spatial correlation variance: $\sigma^2{}_K = \sigma^2 \times r^2(h)$; and the random variance components: $\sigma^2{}_H = \sigma^2 \times [1 - r^2(h)]$.

Formula (3) shows that, as the observation step (h) increases, the spatially correlated variance part decreases, while the random (non-spatial correlation) component increases gradually to the limit σ^2. The observation step where the value of parameter r(h) is suppressed and the random variance coincides with σ^2 ($\sigma^2{}_H = \sigma^2$) is called the auto-correlation limit R. This limit characterizes the size of the influence zone of the observation points.

When the observation step h < R, the parameter's value is a dependent random quantity, and then the coefficient of variation of the research parameter (V_H) is determined by the formula.

$$V_H = \frac{\sigma_H}{X} = \frac{\sigma}{X}\sqrt{1 - r^2(h)} = V\sqrt{1 - r^2(h)} \tag{4}$$

When the step h > R, the correlation coefficient r(h) = 0 and $V_H = V$, the variability is characterized by the normal coefficient of variation, and statistical models can be used to study the characteristics of metallization. In the case, where R(h) > 0 for all h, determine the size of the influence zone H according to the following method.

Make a graph of the norm correlation coefficient.

$$R^*(h) = e^{-\alpha.h};\ \alpha = \frac{\sum_{i=1}^{m}\ln|R(h_i)|}{\sum_{i=1}^{m}h_i} \tag{5}$$

where *m*—number of observation steps; *h*—the value of observation steps.

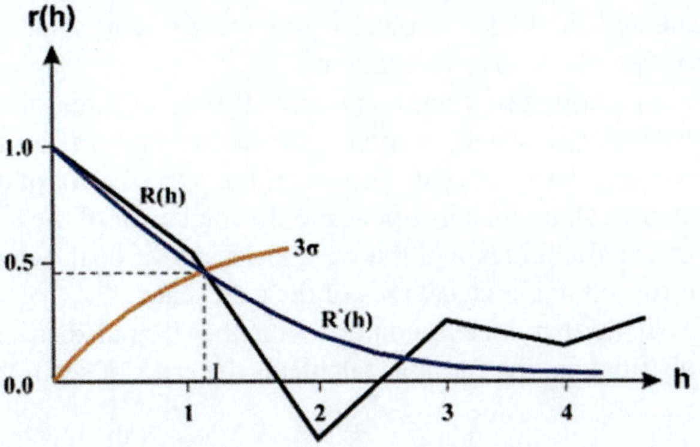

Fig. 3 Spatial correlation function model R(h) [4]

Based on the calculation results $R(h_i)$ và $R^*(h_i)$ build a spatial correlation graph; the horizontal axis represents the observation step distance (h_i), and the vertical axis represents R(h) và $R^*(h_i)$.

At the position $R^*(h)$ intersects the graph $2\sigma_r$ or $3\sigma_r$ ($\sigma_r = \frac{1}{\sqrt{N}} \times [1 - R^*(h)]$) make a perpendicular line to the horizontal axis, assuming the horizontal axis at point i. The distance OI is the size of the influence zone to be determined (H = OI) (Fig. 3).

Based on the size of the influence zone (H), it is possible to determine the anisotropy index and choose the distance to arrange the exploration work.

To characterize the degree of anisotropy, it is possible to use the anisotropy coefficient (A) determined by the formula.

$$A = \frac{I_d}{I_{df}} \tag{6}$$

where I_d—effect size (H) to the dip direction, I_{df}—effect size (H) to the strike line.

Based on the anisotropy index A, select the layout shape of the exploration work. When A $\cong$ 1) it is best to use a square network, and when A $\neq$ 1, a rectangular or linear network should be used.

Stable stochastic optimization techniques are fragile, sensitive to stepsize selection and other algorithmic factors, and unstable outside well-behaved goal families. Stable stochastic methods may be rendered stable, provably convergent, and asymptotically optimum with properly precise models; merely modeling that the target is nonnegative is sufficient for this stability. We emphasize the necessity of resilience and precise modeling with experimental evaluations of convergence time and algorithm sensitivity.

Statistical Methods

Statistical approaches have done more than only make it possible to assess the consistency of results. They have provided techniques for testing and made it feasible to create more efficient and thorough trials. Accordingly, statistical and mathematical methods are widely applied in geological research, including evaluating the reliability of mineral reserve calculation.

According to formulas (1) and (1a), the coal reserve of the Nui Beo mine is calculated by the Secang method, so the error of reserve is mainly the error of volume and weight density of coal. Therefore, the overall error of coal reserve assessment in the statistical model corresponds to the total error of the volume of the reserve/resource calculation block and the weight density of coal, where volume error includes the error of thickness and area of the coal seam.

If the parameters comply with the normal distribution or refer to the normal distribution, the formula calculates the error of seam thickness.

$$\Delta_m = \frac{tV}{\sqrt{N}} \tag{7}$$

where Δ_m—thickness error, t—the coefficient of probability $t = 1.96$, corresponding confidence probability is 0.95; V—coefficient of variation of seam thickness, and N—number of exploration works.

According to Kazdan [4], the total technical error in reserve calculation in each block is determined by the following formula:

$$\sum \Delta = \sqrt{\Delta_m^2 + \Delta_s^2 + \Delta_d^2} \tag{8}$$

The total error calculated by the formula (8) can account for 12–15% or more.

4 Results and Discussions

4.1 Factors Affecting Reserve Accuracy

The accuracy of calculating mineral reserves in the ground depends on many factors: geological factors, system, the density of the exploration grid, method of interpolation of geological documents, and determining geological-industrial parameters (parameters for calculating reserves).

The geological factors that affect the accuracy of the reserve calculation are quite diverse, including the geological structure of the mine, the type of mineral, the shape, the bedding conditions and the internal structure of the mineral body, the number of ore bodies, and the degree of variation of geological-industrial parameters. If the coal mine (coal seam) is more complicated, then the reserve calculation error is greater, and on the contrary, it is less reserve calculation error.

The system and density of the exploration grid are important factors affecting the accuracy of the calculation of underground mineral reserves. When the exploration system, weight density, and orientation of the exploration grid are selected inappropriately, it can lead to a misperception about the geological characteristics of the mine, shape, depth of position, and minerals' distribution law; as a result, this is the cause of increasing errors in reserve calculation. In general, if the structural coal mine and quantity of reserves are determined more precisely, then the density of the exploration grid is higher, and coal exploration cost is also greater. Therefore, in order to improve the efficiency of exploration work and the accuracy of reserve calculation, it is necessary to rationally select the system and shape of the exploration grid based on a complete analysis of the group of exploration mines, the space homogeneity, and anisotropy of minerals (coal seams).

The methods of interpolating geological data include zoning the mineral bodies, relating the cross-sections according to the exploration line system, and determining the spatial distribution of useful and harmful components. The interpolation results of geological data are the basis for determining all the initial documents for the reserve calculation: area, thickness, weight density, and content of useful components. Therefore, interpolating geological data is considered one factor that generates errors and decreases or increases the determination of underground mineral reserves. Geologists need to analyze and select suitable interpolating geological data for the exploration object to solve this problem.

Determining geological-industrial parameters (reserve calculation parameters) directly affects the contouring of the coal seam shape and the number of coal reserves in the ground. Their influence is especially large when systematic errors occur during the sampling, processing, and analysis of samples or the determination of other parameters. These errors can only be overcome by completing the method and technique of sampling, measuring thickness, area, and anomalous processing samples.

According to Hung (2019) [8], the assessment of the remaining reserves for the Nui Beo coal seams by the end of 2019 is summarized in Table 1.

Table 1 Remaining coal reserves are in the Nui Beo mine [8]

Name of seams	Reserves (ton)		Resources (ton)		
	122	Total	333	334a	Total
V13	150,825	854,219	1,005,044	532,442	532,442
V11	2,284,517	8,235,274	10,519,791		
V10	2,056,707	11,440,714	13,497,421	5,847,015	5,847,015
V9	1,497,738	5,565,200	7,062,938	4,968,159	4,968,159
V7	2,504,758	4,759,936	7,264,694	8,033,175	8,033,175
V6		409,727	409,727	542,478	542,478
V5				3,635	3,635
Total	8,494,545	31,265,070	39,759,615	19,926,904	19,926,904

4.2 *Assessment of the Reliability of the Nui Beo Mine Exploration Grid*

The Nui Beo mine has a synclinal structure; exploration lines are arranged on the map in azimuth 10–190°, 130 m apart, and drilling works on the defined route are 80 m apart on average [8, 9]. There are longitudinal and transverse lines concepts because the coal seams are located in the Nui Beo synclinal structure. Therefore, to determine the distance between the exploration works along the line in the azimuth 80–260°, linking the works were carried out on the exploration lines in the azimuth 10–190°. Thus, the dimension of actual exploration work on the network map is 80 m × 130 m.

To evaluate the reliability of the exploration grid, a stable stochastic function model is applied to the exploration lines with the number of works crossing the seam V11, V10, V9, and V7.

Calculating the stable stochastic function along the exploration lines for some selected seams is presented in Figs. 4 and 5.

Summary of survey results of stable stochastic function along exploration lines with azimuths of 80 ÷ 260° and 10 ÷ 190° corresponding to some coal seams are presented in Table 2.

From Table 2, it can be seen that the size of the influence zone to the strike direction varies from 65 to 208 m, the size of the dip direction varies from 32 to 56 m, and has obvious anisotropy (A = 0.41). Thus, the exploration grid applied (80 m ×

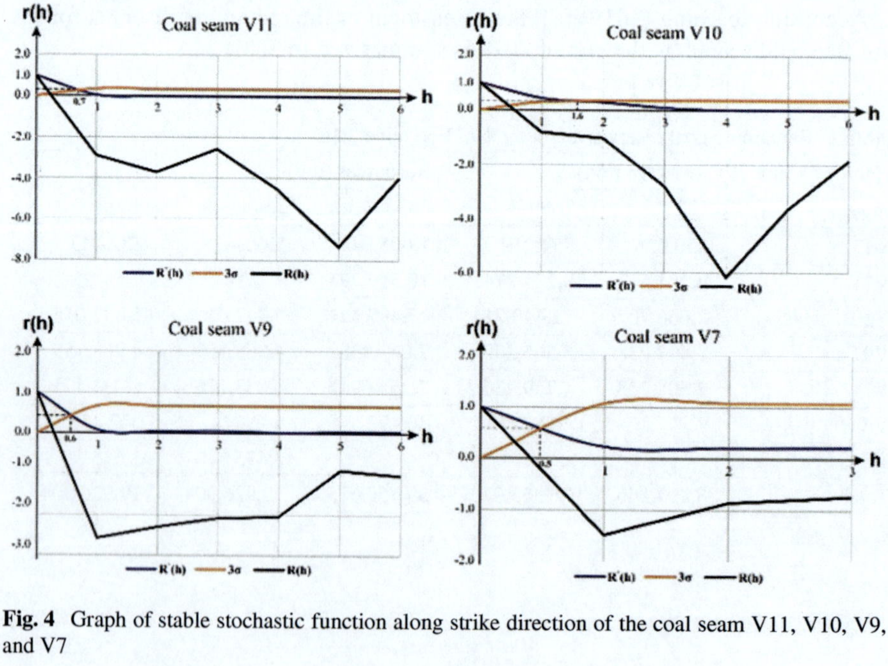

Fig. 4 Graph of stable stochastic function along strike direction of the coal seam V11, V10, V9, and V7

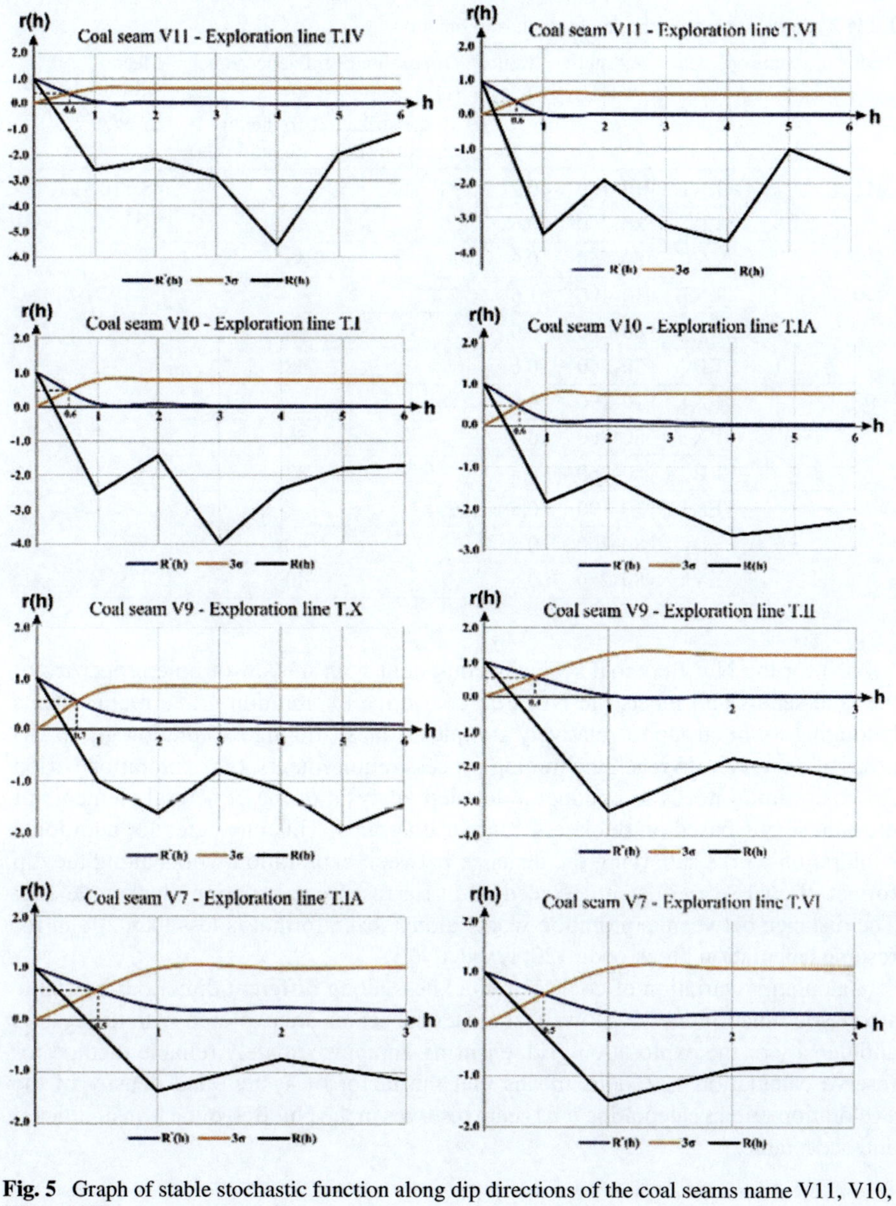

Fig. 5 Graph of stable stochastic function along dip directions of the coal seams name V11, V10, V9, and V7

Table 2 Survey results of the stable stochastic function

Name of seams	Route	Azimuth (deg.)	Value h	Dimension of influence zone (m)		Mean anisotropic coefficient (A)
				To the strike direction (I_{df})	To the dip direction (I_d)	
V11	T.NT	10–190	0.7	91		45:110.5 = 0.41
	T.IV	80–260	0.6		48	
	T.VI	80–260	0.6		48	
V10	T.NT	10–190	1.6	208		
	T.I	80–260	0.6		48	
	T.IA	80–260	0.6		48	
V9	T.NT	10–190	0.6	78		
	T.X	80–260	0.7		56	
	T.II	80–260	0.4		32	
V7	T.NT	10–190	0.5	65		
	T.IA	80–260	0.5		40	
	T.VI	80–260	0.5		40	

130 m) for the Nui Beo coal seams is consistent with the anisotropic properties of the coal seams and meets the Nui Beo coal mine exploration grid's requirements belonging to the group of relatively complex mines (mining exploration group III) [18, 19]. However, to reach the mining process requirements, the exploration period for coal mining needs to conduct a detailed analysis of the structural elements of the coal seams based on thickness contour diagrams. This one helps for additional exploration works satisfying the distance between exploration works along the dip format of coal seam does not exceed 32 m for reserve calculation block code 122. The distance between exploration works along strike format is less than 208 m for reserve calculation block code 122, respectively.

Calculating variation of coal seam thickness along different directions by stable stochastic function model shows that most coal seams are explored with the system and density of the exploration grid, ensuring an approximately reliable method for reserve calculation 122. This means that the factor of system and density of the exploration grid in calculating coal seam reserves in the Nui Beo mine is insignificant and acceptable.

4.3 *Assessment of the Reliability of Reserves by Statistical Methods*

Statistical methods are widely applied in geological research, including reliable and accurate assessment of mineral reserves. As mentioned above, the coal reserves of the

Table 3 Coal seam thickness error

Number	Name of coal seam	Work numbers	V_m (%)	Thickness error (%)
1	11	91	46.79	9.61
2	10	88	50.84	10.62
3	9	50	58.02	16.08
4	7	48	45.82	12.96

Nui Beo mine are calculated by the Secang method, implying the error of the reserve is mainly the error of the seam volume and the weight of coal. According to formula (7), the error of seam thickness is evaluated based on the calculated parameters, as shown in Table 3.

The error of seam area (Δ_s): Because coal seams rarely show wavy folds, when determining the area on the planar projection by isometric lines, there are usually small thanks to the application of specialized software (such as AutoCAD or Mapinfor programs) to measure the area of the coal seam. Therefore, in the case of the Nui Beo mine, the research team chose an error of 3% on the map.

The weight error of coal calculated for each seam is applied according to the formula (7), similar to the thickness error. The calculation results are summarized in Table 4.

To determine the error of coal reserves in the detailed areas, the formula (8) was applied, and reserve error ΔQ (%) was calculated and summarized in Table 5.

The calculated data show that with a reliability probability of 0.95, the reserves' error varies from ± 10.12 to ± 16.46%, which is reasonable and acceptable for reserves class 122 for the mining exploration group III, such as the Nui Beo mine [7, 15].

Table 4 Error on coal weight density

Number	Name of coal seam	Work numbers	V_d (%)	Weight density error (%)
1	11	91	4.67	0.96
2	10	88	5.03	1.05
3	9	50	6.57	1.82
4	7	48	6.30	1.78

Table 5 Error on coal reserves

Number	Coal seam number	*t*	Δ_m	Δ_s	Δ_d	ΔQ (%)
1	11	1.96	9.61	3.0	0.96	± 10.12
2	10		10.62	3.0	1.05	± 11.09
3	9		16.08	3.0	1.82	± 16.46
4	7		12.96	3.0	1.78	± 13.42

The assessment results of geological-industrial parameters of coal seams, such as thickness, area, and weight density of the Nui Beo coal seams, are within the allowable error threshold of reserves of grade 122, mining exploration group III. Thus, the statistical method's application shows that the coal seams' geological-industrial parameters have little influence on the reliability of the Nui Beo mine reserve calculation. Therefore, the Nui Beo mine estimate of reserves meets the feasible design and exploitation reliability shortly.

Evaluating the reserves reliability of the Nui Beo coal mine based on statistical and stable stochastic function methods shows that the factors affecting the reliability of coal reserves are within the allowable limit. Therefore, the assessment of coal reserves of the remaining mine in the range from − 350 to − 500 m is sufficient for the next underground mining design. However, to ensure objectivity and comparability with computer-aided modeling and simulation methods in the near future, the modeling of coal seams in a three-dimensional model should be implemented. Modeling the ore body in a 3D model is also a method widely applied in mining worldwide because of its intuitiveness and ability to update the reservoir parameters quickly and conveniently.

5 Conclusion

Some conclusions are drawn from the data collecting, analysis, and reliability assessment of coal reserves in the Nui Beo mine, Quang Ninh province.

According to exploration lines with azimuth 80–260° and 10–190° for coal seams as the V11, V10, V9, and V7 seams, survey results show that the coal seams have anisotropic properties with properties an anisotropy coefficient of 0.41. Therefore, in the exploration process for mining, it is necessary to adjust the exploration grid of the Nui Beo coal mine to ensure that the distance between the exploration works in the dip direction is 0.41 times the distance between the exploration lines.

The error assessment of coal reserves in the exploration areas of the Nui Beo mine shows that with a reliability probability of 0.95, the error in the reserve varies from ± 10.12 to ± 16.46% is acceptable. Within the allowable threshold for reserves of 122, the calculated reserves in the basic exploration phase meet the reliability for exploitation.

Acknowledgements We are deeply grateful to the leaders and staff of HUMG for providing us with excellent facilities in research and unconditional help in carrying out the study. The paper was presented during the International Conference on Geospatial Technologies and Earth Resources (GTER 2022), 13-14.10.2022, HUMG, Hanoi, Vietnam.

References

1. Albarède, F.: Introduction to Geochemical Modeling. Cambridge University Press, Cambridge (1995)
2. Battalgazy, N., Madani, N.: Categorization of mineral resources based on different geostatistical simulation algorithms: a case study from an iron ore deposit. Nat. Resour. Res. **28**, 1329–1351 (2019)
3. Hung, K.T., Khang, L.Q., Sang, P.N., Van Vuong, H.: Establishing a tungsten deposit group and a pattern grid exploration in the Nui Phao Area, Northeastern Vietnam. In: Proceedings of the International Conference on Innovations for Sustainable and Responsible Mining, pp. 58–78. Springer International Publishing (2021)
4. Kazdan, A.B.: Prospecting and Exploration of Mineral Deposits. Nedra Publishers, Moscow (1997)
5. Pogrebiski, E.O.: Prospecting and Exploration of Mineral Deposits. Nedra Publishers, Moscow (1973)
6. Wellmer, F.W.: Statistical evaluations in exploration for mineral deposits. In: Wellmer, F.W. (ed.) Statistical evaluations in exploration for mineral deposits. Springer—Verlag Berlin Heidelberg, Germany (1998)
7. Saikia, K., Sarkar, B.C.: Exploration drilling optimisation using geostatistics: a case in jharia coalfield. India. Applied Earth Science **115**, 13–22 (2006)
8. Hung, N.M.: Report on the results of the exploration of Ha Lam coal mine, Ha Long, Quang Ninh (2019)
9. Anh, P.T.: Report on data collecting and recalculating mineral reserves in Ha Lam, Ha Long, Quang Ninh (2010)
10. Gandhi, S.M., Sarkar, B.C.: Essentials of Mineral Exploration and Evaluation, pp. 406. Elsevier (2016)
11. Kreige, A.A., Rozin, A.A., Pojariski, K.L.: Basis of Mineral Exploration Method. Nedra Publishers, Moscow (1940)
12. Flores, R.M.: Chapter 6—Resource Evaluation Methodologies. In: Flores, R.M. (ed.) Coal and Coalbed Gas, pp. 301–368. Elsevier, Boston (2014)
13. Agafonov, V.V.: Integrated study of permanent conditions of coal reserves. Russian Coal J. 82–85 (2019)
14. Anfyorov, B.A., Kuznetsova, L.V.: Development of hard-to-remove coal reserves of steep and steeply inclined seams. Geotechnology (underground, open, and construction), 91–101 (2021)
15. Sarkar, B., Saikia, K., Sarma, M., Pandey, S., Paul, P.R.: A geostatistical approach to the estimation of coal bed methane potentiality in a selected part of Jharia coalfield, Jharkhand. J. Mines Metals Fuels **55**, 586–594 (2007)
16. Vaxiliev, I.X., Kazakovski, D.A.: Processing Geochemical Data In Mineral Exploration. Nedra Publishers, Moscow (1929)
17. Hung, L.: Report on Geological and mineral resource mapping at 1:50.000 scale of the Hon Gai - Cam Pha sheet group. (1996)
18. MNRE: Promulgating the Regulation on solid-mineral deposits and resources classification. Ministry of Natural resources and Environment Vietnam (2006)
19. MNRE: Regulation on decentralization of coal reserves and resources. Ministry of Natural Resources and Environment, Vietnam (2007)

Chromite Ore Modeling Based on Detailed Gravity Method in Pursat, Cambodia

Trong Cao Dinh, Hung Pham Nam, Thanh Duong Van, Luc Nguyen Manh, Bach Mai Xuan, Trieu Cao Dinh, and Hung Luu Viet

Abstract Researching mineral deposits is an essential task for Cambodia when its resources are gradually being exhausted. Among them, Chromite ore mine is one of the top priorities. This recent study presents an analysis process based on 25 highly detailed gravimetric profiles for solving the 2.5D/3D inverse gravity problems to image chromite ore structure at the Pursat area based on the Bouguer gravity anomaly map. As a result, a relatively realistic model of chromite ore distribution in the northeast region of Phnum Kri Mountain has been built. This area has extensive reserves of chromite ore and a fungal structure with a 200–300 m depth to the upper boundary of the core layer from the outcrop on the ground surface.

Keywords Highly detailed gravity method · 2.5D/3D inverse gravity problem · Chromite ore · Phnum Kri Mountain · Pursat province · Cambodia

T. C. Dinh (✉) · H. P. Nam · T. D. Van · L. N. Manh · B. M. Xuan
Institute of Geophysics, VAST, A8/18 Hoang Quoc Viet Road, Hanoi, Vietnam
e-mail: trongcd3284@gmail.com

T. C. Dinh · H. P. Nam · T. C. Dinh
University of Science and Technology, VAST, A28/18 Hoang Quoc Viet Road, Hanoi, Vietnam

T. C. Dinh
Institute for Applied Geophysics, VUSTA, 210 Doi Can Road, Hanoi, Vietnam

Vietnam National University Ho Chi Minh City, 227 Nguyen Van Cu Road, Ho Chi Minh City, Vietnam

H. L. Viet
Ho Chi Minh City University of Technology and Education, 1 Vo Van Ngan Street, Ho Chi Minh City, Vietnam
e-mail: hungluu@hcmute.edu.vn

T. C. Dinh
Innovations for Sustainable and Responsible Mining (ISRM) Research Group, Hanoi University of Mining and Geology, Hanoi 100000, Vietnam

L. Q. Nguyen et al. (eds.), *Advances in Geospatial Technology in Mining and Earth Sciences*, Environmental Science and Engineering,
https://doi.org/10.1007/978-3-031-20463-0_16

1 Introduction

Gravity exploration is a geophysical method that applies gravitational field measurement to study geological structures and explore minerals in the Earth's crust. Researching subjects have a density difference compared to the surrounding environment. The difference in density values of objects relative to their surroundings causes gravity anomalies. This anomaly value will include the sum of all things located at different depths and sizes. For each density boundary surface, there can be other gravity effects, so the application of processing and analysis methods for these gravity anomalies could help us better understand the structure and average depths to sources in the Earth's crust.

In the world, Sabah and Al-Rahim [1] and Kanthiya et al. [2] have used the 3D inversion gravity method to study the geological structure and surface depth of the rocks. The corresponding study area is Mesopotamia Basin, Mae Suai Basin. In addition to applying 3D processing and analysis methods, 2D processing methods are also widely used to study geological structures, fault systems, and source depths in the works of Ghazala et al. [3] for Sohag Governorate, Upper Egypt [3] and Kebede [4] for Ziway–Shala Basin [4]. Gravity data is also used to determine the crystalline foundation's structure, architecture, and composition as works by Lenhart et al. [5] for the western Norwegian Sea area [5].

The gravity method with analytical techniques to study geological structures and investigate minerals in Vietnam has also been studied and applied to meet the needs of human life, avoid risks and disasters such as earthquakes, tectonic activities. Cao et al. [6] have processed gravity data to image geological boundary and fault systems for determining the characteristics of earthquake activity for the South Central region [6]. Cao et al. [7] also applied their gravity processing workflow to assess the earthquake hazard in Thuong Tan–Tan My quarry cluster. The interest area Thuong Tan–Tan My belongs to Di An commune, Binh Duong province, Vietnam [7]. Mineral deposits discovered by gravity data in Vietnam, such as Thach Khe iron mine in Ha Tinh. It has been determined that the elongated part of its iron ore body located at the depth of about 700 m or the Ban Bo—Cao lead–zinc mine area. Bang, the copper-nickel mine area, Phan Thanh–Cao Bang, is categorized in copper-nickel minerals with sulfide ones diffused in the lherzolite bodies [8].

The detailed gravity method is used to increase resolution and accuracy of the gravimetric data and to effectively identify local small-scale structures and sources. Typically, the work of Gozlan [9] uses the detailed gravity method to identify small fault zones and sulfide deposits in the northeast region of Bendigo city, Central Victoria. Ercan et al. [10] used the detailed gravity method to search for minerals, specifically gold and silver deposits in Kısladag province, western Turkey. The distance between the above measurement lines is 50 m. Research results have identified locations with high-density values corresponding to those with gold and silver [10].

The mining sector offers enormous potential for the Cambodian economy [11]. Since the second half of the nineteenth century, French and Chinese geologists have conducted geological studies and mineral investigations in Cambodia. Their results show that Cambodia has significant mineral potential such as gold, iron, bauxite, manganese, silica sand, kaolin, limestone, phosphate, jade, rubies, coal, building materials, and other minerals. However, due to the decades-long civil war, the mining sector in Cambodia is still primitive, undeveloped, and underinvested. Since the Law on Mineral Resources Exploitation and Management was promulgated in 2001, the government has licensed many domestic and foreign companies to explore and exploit bauxite, gold, iron, antimony, chromium, etc.... Pursat province has extensive reserves of minerals chromium (Cr) and antimony (Sb).

Vietnamese Mineral Company (Geosimco, https://geosimco.com.vn) has indicated the intrusive ultramafic formations of Phnum Kri Mountain, Pursat province, Cambodia [12]. One of its peak locations sharing its name has the highest elevation of 540 m above the sea level. Being along the southeast zone, scattered over about 16 km^2 and at the 100–300 m height, 07 chromite ore outcrops were discovered. The Institute of Geophysics in the Vietnam Academy of Science and Technology was assigned to measure highly detailed gravity in a preliminary study on the chromite ore distribution in this area.

The paper's objective is to use modern approaches based on detailed gravity data to construct chromite ore. The combination of wavelet problems in determining the center of the object, specifying the object boundary, and the inverse gravity data in 2.5D analysis was first used to limit the multi-solution of the 2.5D linear decomposition problem. In addition, correcting the mine surface depth by the structure inversion problem also helps to be more accurate in building the depth map of the mine surface, improving the ability to exploit in the future.

2 Study Area and Data

2.1 Geological Features

In the area of Phnum Kri Mountain, Phnum Kra district, Pursat province, Cambodia, a series of geological formations is listed below [12]:

(1) *Strata*: The study area strata age can be arranged from old to young order as follows:

(i) Devonian–Carboniferous (D–C): The Devonian–Carboniferous terrigenous-carbonate sedimentary formations are widely distributed in the area. Quaternary (aQ) sedimentary formations cover these formations in many places. The main petrographic composition is sandstone, shale, marl, and limestone. In the design of these formations, in addition to the common rocks mentioned above, lenses of siliceous schist and siliceous rock are accompanied by breccia and conglomerate with low content. There is rhyolitic tuff and tuffaceous sandstone in the upper part of the section. In general, the rock is strongly folded, with the steep attitude of the layers, mainly extending in the latitudinal–sub-latitudinal direction to the northeast–southwest.

(ii) Undivided quaternary (aQ): This formation is distributed in the lower part of hillsides and along the stream valleys in the study area that commonly cover the older formations. The main petrographic composition is pebble, grit, gravel, sand, sandy clay, and laterite. These formations possibly contain chrome spinel placer.

(2) *Intrusive magma (σPR)*: Survey results by Geosimco Company in early December 2007 have recorded the presence of intrusive ultramafic formations at the peak Phnum Kri and other peaks of Phnum Kri Mountain. This complex structure includes a combination of rocks with ultramafic to a mafic composition. It is strongly metamorphosed and gradually change to green amphibolite, including pyroxenite–amphibolite, peridotite (dunite), and gabbro–amphibolite. Mineral composition mainly includes olivine, diopside, and other minor minerals. The olivine grains undergo mosaic serpentinization in the diopside. Few chromspinel grains and biotite flakes are in the rock-forming mineral composition. Small non-transparent mineral grains can be observed in the green antigorite, possibly garnet grains. In the results of petrographic analysis of rock samples around chromspinel ore bodies, they are strongly serpentinized and talc-altered to form serpentine—talc schist. The serpentine accounts for 70–80% of mineral composition, talc with 17–23%, and ore minerals with 3–8%. These formations are ancestrally related to chromium.

The study area and its vicinity are in the Pre-Cambrian lithospheric intraplate Indosinian, covered by the Early Paleozoic to Cenozoic terrigenous-carbonate and extrusive sedimentary formations, and penetrated by the Late Paleozoic–Mesozoic intrusive magmas [13, 14]. The Paleozoic-Cenozoic intrusive formations can be formed in the tectonic settings of Early Paleozoic back-arc basin, Late Paleozoic plutonic, volcanic arc-magmatic arc, Early–Middle Mesozoic thermally rejuvenated region, Early–Middle Jurassic foreland residual basin, Late Mesozoic active continental margin and block-type domal uplift along with strong denudation. The study area occupies the southwest part of the Indosinian block. Hence, its geological evolution is inseparable from the tectonic development of the southwest edge of this block. Generally, the tectonic development of the study area consists of the following main stages:

(i) Devonian–Carboniferous: In this stage, the study area has the tectonic regime of the southwest passive continental margin of the Indosinian continent approaching Meso-Tethys. Note that Meso-Tethys is located between Indosinian and Sibumasu continents.

(ii) Late Carboniferous–Permian: The Meso-Tethys oceanic crust was subducted below the southwest edge of the Indosinian continent, forming the Dak Lin–Ben Giang volcanic belt in Vietnam.

(iii) Middle Triassic: It is the stage of thermally rejuvenating tectono-magmatic activation of the crust which develops widely in the northeast and southwest of the study area. There are Mang Yang-type extrusives and Van Canh-type granite as in Vietnam.

(iv) Early–Middle Jurassic: In this stage, the area is subsided and is a part of the Early–Middle Jurassic basin extending from Da Lat through the study area, the post-plate collision, residual foreland basin.
(v) Late Mesozoic: Due to the influence of subduction from the east of the Pacific oceanic crust to the Southeast Asian continental crust, the Late Jurassic–Cretaceous calc-alkaline is extrusive and intrusive formations. However, it is the inner zone, very far from the subduction zone; therefore, these magmatic formations probably have a deeper source than the Late Mesozoic magma in the Da Lat zone in Vietnam.
(vi) Paleogene–Early Miocene: In this stage, the study area is the central part of the continuously denudated and uplifted block. Here, the sedimentary formations and magmatic activity do not develop.
(vii) Late Cenozoic: The study area is in the stage of block-type domal uplift, accompanied by alluvial sediments along the valleys.

According to the data collection in the study area (Fig. 1), seven chromite ore outcrops have been detected, taken as the standard sample for gravity data analysis. In the area of Phnum Kri Mountain, some rocks related to ores have been found (Table 1). The density values of rock and ores in the study area are shown in Table 1 [12, 15] with the significance of italics mainly used when defining the geological structure.

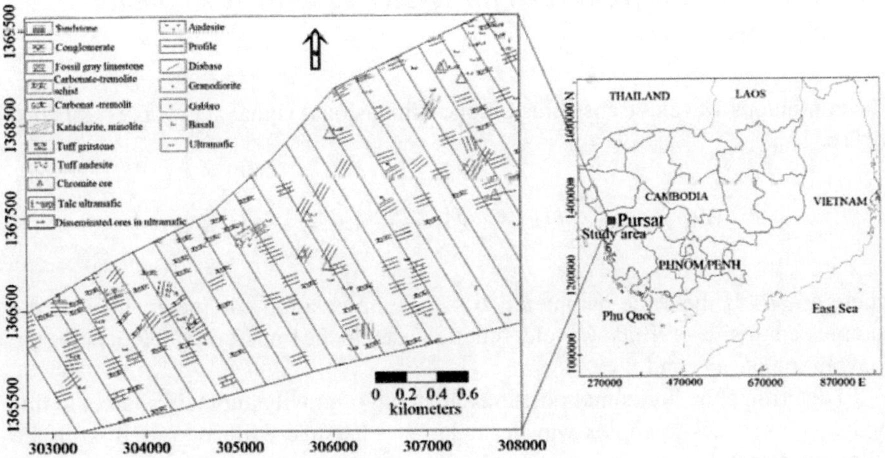

Fig. 1 Geological features of the study area

Table 1 Density values of rocks and ores in the study area

No	Rock and ore	Density value (g/cm^3)
1	Chromite ore	3.9–4.2 *(advange 4.05)*
2	Ultramafic rock	1.7–2.2 *(advange 1.9)*
3	Limestone	2.7–2.8 *(advange 2.75)*
4	Sulfide rock	2.68–2.8 *(advange 2.74)*
5	Gabbro	2.74–2.92 *(advange 2.82)*

2.2 Gravity Data

The ZLS-B37 (http://zlscorp.com) gravity machine with a resolution of 0.015 mGal and an error of ±0.02 mGal was used in measuring 1324 points along 25 profiles over the researching zone (Fig. 2). The distance between two points and two profiles is 40 m and 250 m, respectively. The Bouguer gravity anomaly was calculated with a maximum accuracy of ±0.2 mGal, meeting the technical requirements for detailed gravity measurement at the scale of 1:25,000, and the density value in terrain and Bouguer correction was 2.63 g/cm^3 [15].

3 Methodology

3.1 Wavelet Transform (2D) for Detecting Gravity Anomaly Source

The continuous wavelet transforms on one-dimensional signal $f(x)$ was established by [16, 17]:

$$W(a, b) = \int_{-\infty}^{+\infty} f(x)\frac{1}{\sqrt{a}}\overline{\Psi}\left(\frac{b-x}{a}\right)dx = \frac{1}{\sqrt{a}}\left(f^{*}\overline{\Psi}\right) \quad (1)$$

where a $\in R^{+}$ is the scale parameter; $b \in R$ is the position parameter; $\overline{\Psi}(x)$ is the complex conjugate of $\Psi(x)$, wavelet function used in the transform; $f^{*}\overline{\Psi}$ denotes the convolution of $f(x)$ and $\overline{\Psi}(x)$.

To determine the horizontal position and depth of gravity anomaly source, in this study, we use a new complex wavelet function—Farshad-Sailhac [16–18] with the following form:

$$\Psi^{(FS)}(x) = \Psi^{(F)}(x) + i\Psi^{(S)}(x) \quad (2)$$

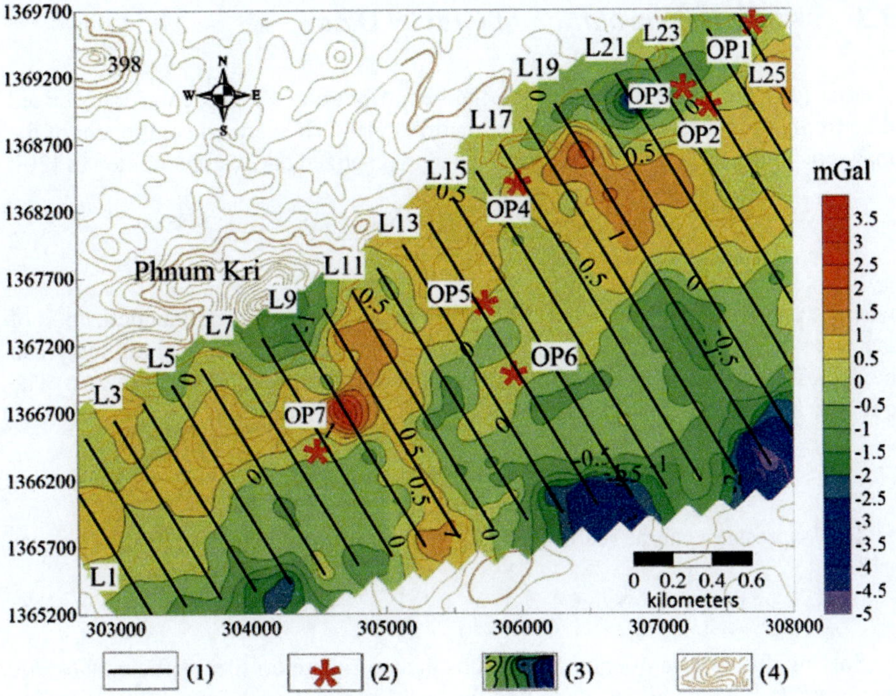

Fig. 2 Study area covered the chromite ore distribution and gravity data: 1–Gravity measurement profile (L1 to L25); 2–Chromite ore outcrop on the surface (OP1 to OP7); 3–Bouguer gravity anomaly; 4–Terrain contour line

where

$$\Psi^{(F)}(x) = \frac{4 - 2x^2}{\left(x^2 + 2^2\right)^{\frac{5}{2}}} - \frac{1 - 2x^2}{\left(x^2 + 1^2\right)^{\frac{5}{2}}} \tag{3}$$

$$\Psi^{(S)}(x) = Hilbert\left(\Psi^{(F)}(x)\right) \tag{4}$$

In the wavelet transform, the scale is related to the depth of the anomaly source. However, the scale factor is not the depth and does not provide direct information about the depth. There is an almost linear correlation between the depth of the source z and the product of the scale a and the measurement step Δ through the scale factor k [18]:

$$z = k * (a * \Delta) \tag{5}$$

The factor k depends on the structural index of the source which is determined and applied to estimate the depth of anomaly sources in the analysis of actual data.

3.2 Werner Deconvolution of Gravity Data

Werner [19] proposed separating the field contribution of a particular magnetized dike from the interference of neighboring dikes [19]. The vertical component of the magnetic fields of an arbitrarily oriented and magnetized dike can be written as [20]:

$$Y_a = \frac{A(x - x_0) + B(y - y_0)}{(x - x_0)^2 + (y - y_0)^2} \tag{6}$$

where A and B are constants that depend upon the orientation and magnetization of the dike; the interference from a neighboring anomaly or a regional anomaly can be expressed as a simple polynomial field added to the irregularity of interest. The quantity measured is shown, therefore:

$$Y_{field} = Y_a + c_0 + c_1 + c_2x^2 + \ldots \tag{7}$$

This expression can be rearranged to give the following equation [21]:

$$x^2Y_{\text{field}} = a_0 + a_1x + b_0Y_{\text{field}} + b_1Y_{\text{field}} + k_0x^2 + k_1x^3 + \ldots \tag{8}$$

Solving this system of equations results in an estimate for the coefficients b_0, b_1, a_0, a_1, k_0, $k_{1\ldots}$ which allows us to find the position and depth of the dike.

3.3 Constrained Two-Dimensional Gravity Inversion

Constrained inversion is an inversion algorithm to recover 2D contrast density distribution from gravity data which was presented. The algorithm incorporates constraints about the directions of the mass concentration to obtain a mathematically stable and physically and geologically meaningful solution. To simplify the problem, the applied constraint was given explicitly as the coordinate points of the axes. The versatility of such constraints in the algorithm was exemplified by inversion of synthetic data [22].

Constrained two-dimensional gravity inversion was applied in analyzing the gravity and magnetic data in 1984 by Guillen and Menichetti [23]. Later, Barbosa, Silva, Mendonca and Blakely [24–28] demonstrated that it somewhat limits the multi-solution of gravity and magnetic modeling problem. The structural section containing the anomaly sources is divided into rectangular prisms in the x and z axes (the y-axis object length is considered infinite, i.e., 2D case), the density contrast is defined as a constant value inside each prism and can vary between prisms. Using matrix notation, gravity anomaly vector $d = [di]$; $i = 1, 2, \ldots, N$ (N is the number of data) is given by the function:

$$d = G \cdot m \tag{9}$$

where $G = [g_{ij}]$; $i = 1, 2, \ldots, N$; $j = 1, 2, \ldots, M$ is the kernel matrix. G plays role as a modeling operator; g_{ij} is the contribution of the jth prism to the gravity value on the ith observation point. The model parameter vector represents the density contrast of the prism $m = [m_j]$; $j = 1, 2, \ldots, M$; M is the number of model parameters (the number of prisms in the x and z directions). The gravity component of an elementary prism is the gravitational force of a polygon in the 2D problem [27]. In this case, the polygon has four perpendicular sides coinciding with the model grid.

The value m can be determined by solving the formula [25, 26]:

$$m = G^T\left[GG^T + \lambda I\right]^{-1} d \tag{10}$$

where λ is damping factor, I unitary matrix, and T matrix transposition.

The iterative solution is performed in 2D conditional inversion for determining the density parameter [24, 28–30]. The maximum positive and minimum negative density contrast values will be of interest as they represent the structural blocks [29, 30]. This method is used not only for 2D gravity and magnetic analysis but also for 3D case with other geophysical data, such as resistivity.

3.4 The 2.5D Gravity Inversion

In the case of a two-dimensional model where the object structure is an n-sided polygon, the horizontal and vertical components of gravity anomaly are determined based on the formula [31]:

$$\Delta g_z = 2G\rho \sum_{i=1}^{n} Z_i \tag{11}$$

$$\Delta g_x = 2G\rho \sum_{i=1}^{n} X_i \tag{12}$$

where Z_i and X_i are line integrals along the ith side of the polygon, G is the gravitational constant, and ρ is the density of the polygons.

3.5 The 3D Gravity Inversion

The gravity structural inversion function modifies the elevation of the selected layer to minimize the gravity misfit. Inversion updates the calculated response and error to reflect the structural changes. GM–SYS (https://www.seequent.com) contains two types of inversion density inversion and structural inversion used in the present study.

Firstly, a forward calculation must run to create the calculated and error grids required to optimize the inversion method. After running a forward calculation, structural inversion without constraints run, replace the relief surface with the selected relief layer [32].

In the present study, 3D gravity inversion is made by GM–SYS, which depends on the equation described by Cordell and Henderson [33] to calculate the gravity anomaly caused by a uniform layer of material using Fourier transforms [33]. This expression, in its one-dimensional form, is defined as [32]:

$$F(\Delta g) = -2\pi G\rho^{(-k_{z0})} \sum_{n-1}^{\infty} \frac{k^{n-1}}{n!} F\left[h^n(x)\right] \tag{13}$$

where $F(\Delta g)$ is the Fourier transform of the gravity anomaly, G is the gravitational constant, ρ is the density contrast across the interface, k is the wavenumber, $h(x)$ is the depth to the interface (positive downwards), and z_0 is the mean depth of the horizontal interface. Oldenburg [34] rearranged this equation to compute the depth to the undulating interface from the gravity anomaly profile and is given as below [34]:

$$F[h(x)] = -\frac{F\left[\Delta g(x)\right] e^{(-k_{z0})}}{2\pi G\rho} - \sum_{n-2}^{\infty} \frac{k^{n-1}}{n!} F\left[h^n(x)\right] \tag{14}$$

This expression allows us to determine the topography of the interface density through an iterative inversion procedure. This procedure assumes the mean depth of the interface, z_0, and the density contrast associated with two media ρ. The gravity anomaly is first demeaned before the Fourier transforms calculation. Then, the first term of Eq. (14) is computed by assigning $h(x) = 0$. The inverse Fourier transform for Eq. (14) will first approximate the topography interface $h(x)$. This value of $h(x)$ is then used in Eq. (14) to evaluate a new estimate of $h(x)$. This process is continued until a reasonable solution is achieved.

4 Results and Discussion

The analyzing process for highly detailed gravity data studying chromite ore distribution includes two steps: (1) Building the initial model and solving the 2.5D gravity problem along with the measurement profiles; (2) solving the 3D gravity problem based on the result of step 1 as an initial model for the 3D gravity inversion. The software Geosoft Oasis Montaj 7.0.1 [35] was used to correct the terrain parameters, establish the Bouguer gravity anomaly map, and solve the 2.5D/3D inverse problems on detecting the chromite ore structure of this study area [36, 37].

4.1 Some Initial Modeling Results of Chromite Ore Structure

The 2.5D analysis results along 25 profiles to a depth of 400 m (Fig. 3) show that the density value of chromite ore is in the range of 3.9–4.1 g/cm^3 and the density value of surrounding rocks 2.65–2.75 g/cm^3. The ore is distributed from the terrain surface to the depth −400 m. The 2.5D model along these 25 profiles is used to solve the 3D gravity problem using the GM-SYS program. 2.5D analysis results are presented for two feature routes L14 and L19. L14 passes through two outcrops, which is the basis used in the initial model building process, while L19 passes through the structure with the Bourguer gravity anomaly having the greatest variation.

(1) *Building the initial model by solving the 2.5D inverse gravity problem*: To limit solutions on the inverse problems of gravity, an initial model which represents mostly close to the characteristics of the geological structure is built. The analysis methods from the third part of this work (3.1 to 3.5 underpart) are carried out to solve the following tasks: (a) The center of the extraneous body is determined based on application of wavelet transform (Fig. 3), (b) the structural boundary of the object is defined based on Werner deconvolution results (Fig. 3), and (c) the analysis program ZondGM2D [29] is used to invert the gravity data and to preliminarily analyze for identifying the initial structural model (Fig. 3) [29]. The characteristics of geological structure and positions of chromite ore outcrops are precious actual data for building an initial model of geological structure density map.

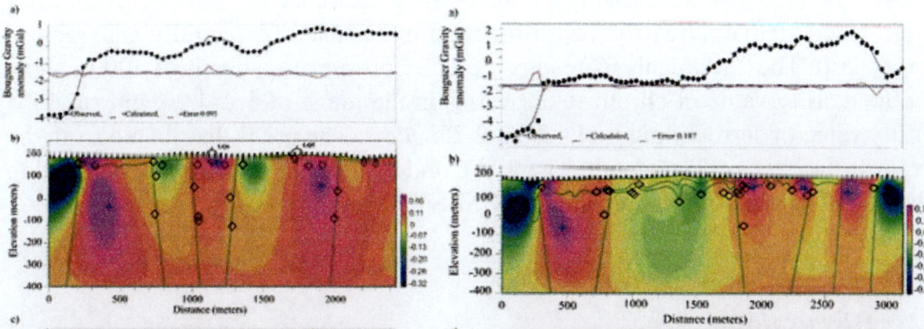

Fig. 3 Geological structure density of L14 and L19 cross sections: **a** Bouguer gravity values (observation, calculation, and error); **b** initial model and result of solving 2.5D inverse gravity problem; **c** geological structure density. (1–Object center, according to result of wavelet analysis; 2–geological structure margin according to Werner method; 3–chromite ore outcrops on the surface)

In Fig. 3a, the position of the inflection points, where the sign change of gravity anomaly coincides with the boundary of the division of structural blocks, clearly shows the color change in the constrain inversion results (see Fig. 3b).

Considering the results of the combined analysis between wavelet and constrain inversion, the wavelet values are distributed within the center position of the structure block, the value located in the area with constrain density is the largest (Fig. 3b). In addition, the Weiner calculation results also show that the boundary range coincides with the boundary of the anomaly band of the constrain density (Fig. 3b).

The 2.5D analysis results along 25 profiles to a depth of 400 m (Fig. 2) show that the density value of chromite ore varies in the range of 3.9–4.1 g/cm^3 and the density value of surrounding rocks 2.65–2.75 g/cm^3. The ore is distributed from the terrain surface to the depth −400 m. The 2.5D model along these 25 profiles is used to solve the 3D gravity problem using the GM-SYS program.

(2) *Using the GM-SYS tool to solve the 2.5D inverse problem of gravity*: The initial structural model combined with the available geological data is used to solve the 2.5D inverse problem of gravity by the GM-SYS tool. As a result, 25 measurement profiles were built, evenly distributed in the study area. For example, L14 and L19 profiles are shown in Fig. 3. The depth along the profiles is 400 m, the density value of chromite ore is 3.9 - 4.2 g/cm^3, and the density of the surrounding sediment layer is about 2.63 – 2.80 g/cm^3 (Table 1).

The used 3D inversion program is GM–SYS, a surfaced-based frequency domain gravity and magnetic forward and inverse modeling program that runs under Oasis Montaj. Note that the model is defined by some surfaces or geologic horizons. Each surface is defined by grid. Each layer is either assigned a constant density, given series of sub-layers define a density–depth function or assigned by laterally changing in density grid. The 2.5D analysis results along 25 profiles to a depth of 400 m show that the density value of chromite ore varies in the range of 3.9–4.1 g/cm^3 and the density value of surrounding rocks 2.65–2.75 g/cm^3. The ore is distributed from the terrain surface to −400 m depth. The 2.5D model along these 25 profiles is used to solve the 3D gravity problem using the GM-SYS program.

4.2 Discussion

With the support of the 3D GM-SYS tool and the input data (Fig. 4a), we have built the 3D model of chromite ore distribution (Fig. 4b) and the error in determining depth to the upper surface of the chromite ore layer (Fig. 4c). The distribution map of the upper boundary of the chromite ore layer in the study area is presented in Fig. 5. 3D gravity data analysis results show that the southeast area of Phnum Kri Mountain (Pursat province, Cambodia) has extensive reserves of chromite ore, with the average density value of 3.9–4.1 g/cm. The chromite ore body here has a fungal structure with the 200–300 m depth to the upper boundary of the ore layer from its outcrop on the ground surface and with the density of surrounding rocks of about 2.65–2.75 g/cm^3.

It could not be denied that the inverse problem of gravity data is a multi-solution problem; therefore, the establishment of an initial model for solving this problem is significant. To limit the multi-solution, it usually requires a lot of detailed research results on the stratigraphic structure of the study area. The process of building the initial model for solving the gravity inverse problem to study chromite ore distribution along 25 highly detailed gravity measurement profiles in this paper is quite rigorous. The detection of 7 ore outcrops and the specific determination of density value of chromite ore and surrounding rock has partly minimized the multi-solution of 2.5D gravity inverse problem along with the measurement profiles. Using Oasis Montaj

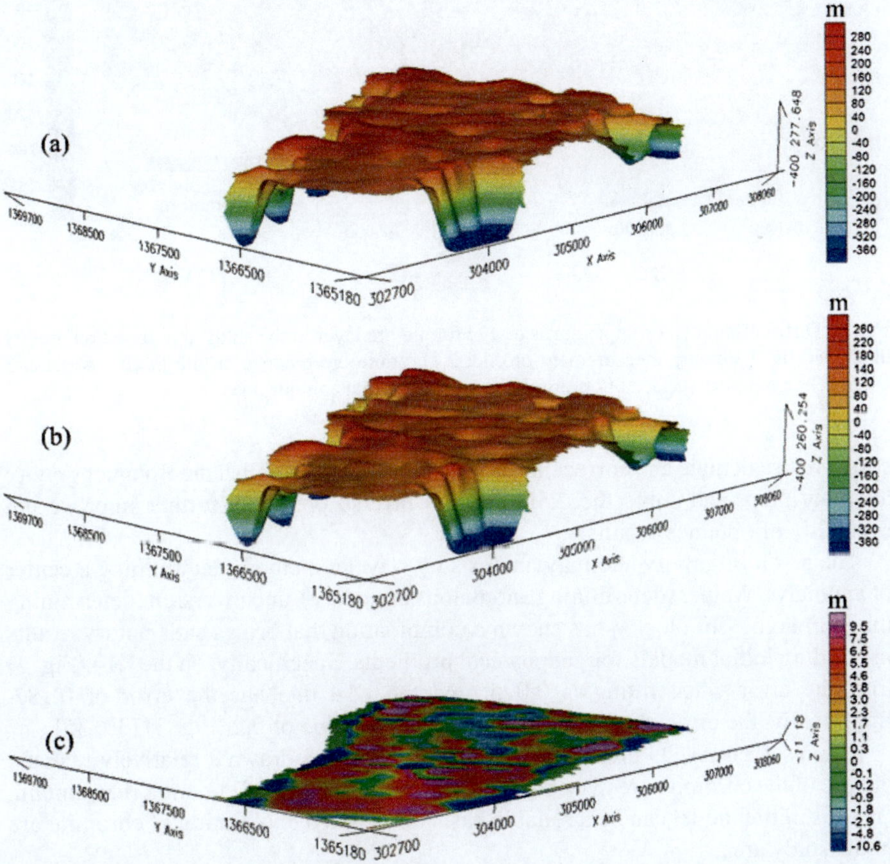

Fig. 4 3D structure of chromite ore of the study area: **a** Upper surface of chromite ore layer (input for solving 3D model, established based on 2.5D analysis results along 25 gravity profiles); **b** upper surface of chromite ore layer (the result of solving the 3D inverse problem); **c** error in determining depth to the upper surface of chromite ore layer (difference between the values before and after inversion)

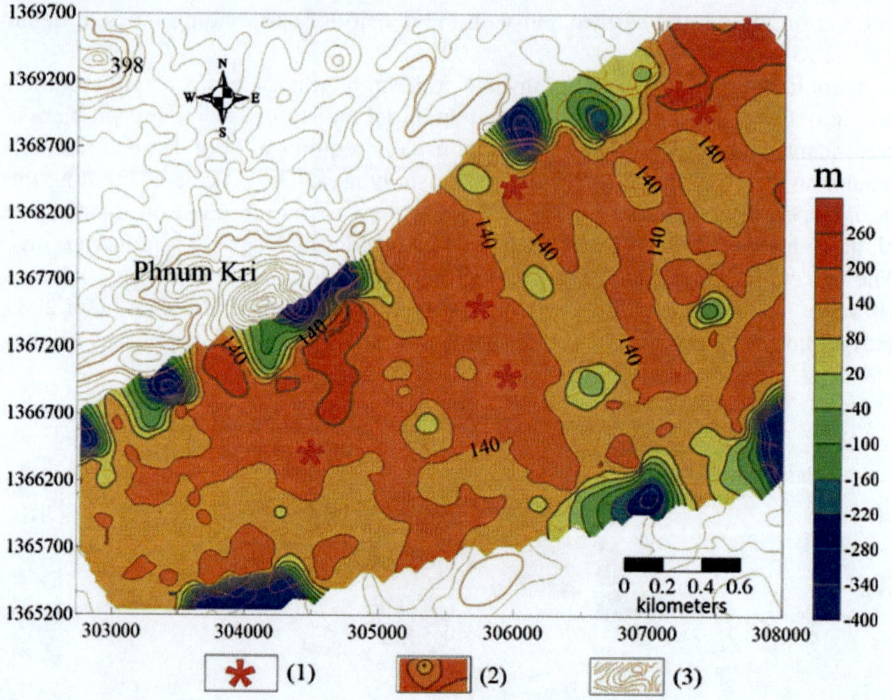

Fig. 5 Distribution map of upper surface of chromite ore layer in the study area based on gravity data analysis: 1 Gravity measurement profile; 2 chromite ore outcrop on the ground surface; 3 contour line of upper surface of chromite ore layer; 4 terrain contour line

software to calculate and correct the terrain parameters, establish the Bouguer gravity anomaly map, and solve the 2.5D and 3D inverse problems further improve the reliability of obtained results.

The result of gravity anomaly analysis by wavelet method (determining a center of anomaly), Weiner (determining anomalous boundary), and inversion (determining the morphology of blocks) has shown a combination that brings satisfactory results in building initial models for geophysical problems. Specifically, on the L14 (Fig. 3) line, the error when fitting is 0.095, and the L19 line has the error of 0.187. Importantly, the error are smaller than the relative value of 0.2 (Fig. 3) [30, 35].

The research results presented in Figs. 4 and 5 have drawn a relatively realistic picture of the chromite ore distribution in the northeast area of Phnum Kri Mountain, and this initial model can be a reliable basis for detailed exploration of chromite ore in the study area.

5 Conclusion

The commendable result in this study is the measurement of 1324 high-precision gravity points along 25 profiles. Detection of 7 ore outcrops and the difference in density value between surrounding rock ($g = 2.65$–2.75 g/cm^3) and chromite ore ($g = 3.90$–4.20 g/cm^3) is a great advantage by using gravity data to study the chromite ore distribution. The process of building the initial model for solving the gravity inverse problem to study chromite ore distribution along 25 highly detailed gravity measurement profiles in this paper is quite rigorous. In addition, the use of Oasis Montaj software in calculating and correcting the terrain parameters, establishing the Bouguer gravity anomaly map, and solving the 2.5D and 3D inverse problems further improves the reliability of obtained results.

The northeast area of Phnum Kri Mountain (Pursat province, Cambodia) has extensive chromite ore reserves. The chromite ore body has a fungal structure with the depth to the upper boundary of the ore layer from the outcrop on the ground surface to 200–300 m depth.

The combination of wavelet problem in determining the center of the object, the issue of determining the object boundary, and the inverse gravity problem in 2.5D linear analysis was first published by the authors to limit the multi-solution of the 2.5D linear decomposition problem. In addition, correcting the mine surface depth resulted from the structure inversion problem also helps to build the more accurate depth map of the mine surface, improving the ability to exploit in the future.

References

1. Sabah, L., Al-Rahim, A.M.: Constrained 3D gravity inversion for the mesopotamia basin—a case study. Int. J. Sci. Res. (2016)
2. Kanthiya, S., Mangkhemthong, N., Morley, C.K.: Structural Interpretation of Mae Suai Basin, Chiang Rai Province, Based on Gravity Data Analysis and Modelling. Published by Elsevier Ltd (2019). https://doi.org/10.1016/j.heliyon.2019.e01232
3. Ghazala, H.H., Ibraheem, I.M., Lamees, M., Haggag, M.: Structural study using 2D modeling of the potential field data and GIS technique in Sohag governorate and its surroundings. Upper Egypt (2018)
4. Kebede, H., Alemu, A., Fisseha, S.: Upward continuation and polynomial trend analysis as a gravity data decomposition, case study at Ziway-Shala basin. Central Main Ethiopian rift (2020)
5. Lenhart, A., Jackson, C.A.L., Bell, R.E., Duffy, O.B., Gawthorpe, R.L., Fossen, H.: Structural Architecture and Composition of Crystalline Basement Offshore West Norway (2019)
6. Trong, C.D., Bui, X.N., Hung, P.N., Tuan, T.A., Bach, M.X., Khai, P.C., Nghia, N.V., Trieu, C.D.: Seismic hazard assessment for Thuong Tan-Tan my quarries (Vietnam) (2020)
7. Dinh, T.C., Xuan, B.M., Nam, H.P., Anh, T.T., Kha, V.T., Dinh, T.C.: Seismic hazard assessment for South-Central region, Vietnam, pp. 167–191. Springer Nature Switzerland AG (2021)
8. Phong, N.T. (eds.): Report on ministerial-level scientific and technological research project. Research, evaluate the current status and propose solutions to improve the level of geophysical technology in the fields of operation under the Ministry of Natural Resources and Environment (2015)

9. Gozlan, E., Cull, J.: Micro-gravity for mineral exploration. Geophysical Conference and Exhibition. Brisbane (Aug 2001)
10. Ercan, Ö.A., Şeren, A., Elmas, A.: Gold and silver prospection using magnetic, radiometry and microgravity methods in the K ışladağ province of western Turkey. Resour. Geol. **64**(1), 25–34 (2014)
11. Phnom penh securities plc, Mining sector in Cambodia (2011)
12. Truong, V.H.: Survey results of chromite ore in Phnum Saram-Phnum Kri area, Phnum Kra Vanh district, Pursat province, Cambodia. Final report, archived at Geosimco Company, p. 90. Ho Chi Minh City (2007)
13. Cao D.T., Pham H.L.: Fault Tectonics in Vietnam, p. 208. Science and Technics Publishing House, Hanoi (2002)
14. Cao, D.T., Pham, H.L., Dao, V.C., Dang, T.H., Le, V.D., Cao, D.T., Thai, A.T., Pham, N.H., Mai, X.B. (eds.): Tectonic deformation in the East Vietnam Sea and adjacent areas in the Cenozoic, p. 308. Publishing House for Science and Technology, Hanoi (2019)
15. Cao, D.T., Pham, N.H., Mai, X.B., Le, V.D., Thai, A.T., Nguyen, H.T.: Application of high-precision gravity method in predicting the chromite prospects in Pursat, Cambodia. Final Report of Science and Technology Task, p. 70. Archived at the Institute of Geophysics (2008)
16. Sailhac, P., Galdeano, A., Giberd, D., Moreau, F., Delor, C.: Identification of sources of potential fields with the continuous wavelet transform: complex wavelets and applications to magnetic profile in French Guiana. J. Geophys. Res. **105**(B8), 19455–19475 (2000)
17. Li, Y., Braitenberg, C., Yang, Y.: Interpretation of gravity dats by the cotinuous wavelet transpormation: the case study of Chad lineament (North Central Africa). J. Appl. Geophys. **90**, 62–70 (2013)
18. Duong, Q.C.T., Duong, H.D., Tran, X.V.: Interpretation of gravity anomaly data using the wavelet transform modulus maxima. J. Mar. Sci. Technol. **17**(4B), 151–160 (2017)
19. Werner, S.: Interpretation of magnetic anomalies at sheet-like bodies. Sveriges Geologiska undersokning, ser. C. Arsbok 43, N. 06. Stockholm (1949)
20. Klingele E., Marson I. and Kahle H.G., Automatic interpretation of gravity gradiometric data in two dimensions: vertical gradient. Geophys. Prospect. (1991)
21. Hartman, D.R., Teskey, D.J., Friedberg, J.: A system for rapid digital aeromagnetic interpretation. Geophysics **36**, 891–918 (1971)
22. Hendra, G., Darharta, D.: Constrained two-dimensional inversion of gravity data. J. Math. Fundam. Sci. **46**(1), 1–13 (2014)
23. Guillen, A., Menichetti, V.: Gravity and magnetic inversion with minimization of a specific functional. Geophysics **49**(8), 1354–1360 (1984). https://doi.org/10.1190/1.1441761
24. Barbosa, V.C.F., Silva, J.B.C.: Generalized compact gravity inversion. Geophysics **59**(1), 57–68 (1994). https://doi.org/10.1190/1.1443534
25. Mendonca, C.A., Silva, J.B.C.: The equivalent data concept applied to the interpolation of potential field data. Geophysics **59**(5), 722–732 (1994). https://doi.org/10.1190/1.1443630
26. Mendonca, C.A., Silva, J.B.C.: Interpolation of potential field data by equivalent layer and minimum curvature: a comparative analysis. Geophysics **60**(2), 399–407 (1995). https://doi.org/10.1190/1.1443776
27. Blakely, R.J.: Potential Theory in Gravity and Magnetic Applications. Cambridge University Press, Cambridge (1995). https://doi.org/10.1017/CBO9780511549816
28. Silva, J.B.C., Barbosa, V.C.F.: Interactive gravity inversion. Geophysics **71**(1), J1–J9 (2006). https://doi.org/10.1190/1.2168010
29. ZondGM2D.: Program for 2D Interpretation of Magnetic and Gravity Data, p. 80. Saint-Petersburg (2013). https://geodevice.co/local/download/?id=5934
30. Grandis, H., Darharta, D.: Constrained two-dimensional inversion of gravity data. J. Math. Fund. Sci. **46**(1), 1–13 (2014). ITB Journal Publisher, ISSN: 2337-5760. https://doi.org/10.5614/j.math.fund.sci.2014.46.1.1
31. Amin, R.K., Ali, N.K., Farshard, S.: Interpretation of gravity data using 2-D continuous transformation and 3-D inverse modeling. J. Appl. Geophys. **12**, 54–62 (2015)

32. Laith, S., Ali, M.A.: Constrained 3D gravity inversion for the mesopotamia basin—a case study. Int. J. Sci. Res. 339–345 (2018)
33. Cordell, L., Henderson, R.G.: Iterative three-dimensional solution of gravity anomaly data using a digital computer. Geophysics 596–601 (1968)
34. Oldenburg, D.W.: The inversion and interpretation of gravity anomalies. Geophysics 526–536 (1974)
35. GMSYS.: Gravity and Magnetic Modeling, vol. 7.0.1 (2008). http://www.geosoft/resources/technotes/3DViewsOasismontaj.pdf
36. Cao, D.T.: Gravity and Gravimetric Methods, p. 276. Science and Technics Publishing House, Hanoi (2000)
37. Cao, D.T., Le, V.D., Pham, N.H.: Application of high-precision gravity method in studying shallow geological structure in Vietnam. J. Min. Earth Sci. No. 14/4-2006, Hanoi 61–66 (2006)

Relationship Between Shear Wave Velocity and Soil Depth and Evaluation of Soil Liquefaction in Quaternary Sedimentary Layer

Yu Zhou, Xuedong Li, Yi Zhang, Yibin Li, Xiaoyong Zhang, and Qiang Tang

Abstract Earthquake can cause severe damage to major projects of Quaternary sedimentary layer, and the vibration liquefaction by shear wave velocity has been widely used due to clear physical significance, high prediction reliability, and good economy. Taking Suzhou as a typical Quaternary sedimentary area, the shear wave velocity data of 40 borehole profiles are fitted and analyzed. The results show that there is an obvious correlation between the shear wave velocity and the buried depth of shallow soil layer. By testing the distribution of buried depth points, the distribution does not meet the normal distribution. The shear wave velocity of different soils varies greatly due to different depth or weathering degree, and the interval difference is between 60 and 190 m/s. Based on liquefaction discrimination methods, the liquefaction of soil layer is distinguished. The results show that the liquefaction probability of Quaternary sedimentary layer is limited. This paper can provide basic data for evaluating seismic impact based on Quaternary sedimentary layer.

Keywords Quaternary sedimentary layer · Shear wave velocity · Data fitting · Soil depth

1 Introduction

The Quaternary is the most recent geological unit in the history of Earth's development, and the division of its lower limit is still highly controversial [1]. Currently, most people prefer 1.8–2.6 Ma, i.e., the Quaternary usually refers to the history since

Y. Zhou · Q. Tang (✉)
School of Rail Transportation, Soochow University, Suzhou 215131, China
e-mail: tangiang@suda.edu.cn

Y. Zhou
e-mail: ZhouYu345@outlook.com

X. Li · Y. Zhang · Y. Li · X. Zhang
Jiangsu Suzhou Geological Engineering Survey Institute, Suzhou 215011, China

The Fourth Geological Brigade of Jiangsu Geological and Mineral Bureau, Suzhou 215004, China

L. Q. Nguyen et al. (eds.), *Advances in Geospatial Technology in Mining and Earth Sciences*, Environmental Science and Engineering,
https://doi.org/10.1007/978-3-031-20463-0_17

2.6 Ma before present, and although the Quaternary continues for a brief period of time, it has a very close relationship with human beings [2]. Many important geological events occurred during the Quaternary period, such as climate change, sea level fluctuations, vegetation turnover, volcanic eruptions, and uplift of the Qinghai-Tibet Plateau, all of which had a significant impact on the environment and resources on which we depend for our survival [3]. Therefore, the study of the characteristics of Quaternary sedimentation, earthquakes, and the effects caused by post-earthquake can better serve human beings, which is of great significance for solving practical problems, mitigating, and preventing disasters, accelerating engineering construction and improving people's living environment.

Soil shear wave velocity is a major physical quantity that distinguishes geodynamics from soil statics, which reflects the ability of a site to propagate seismic waves and is an important parameter to characterize the dynamic properties of the site soil [4]. Soil shear wave velocity is the main criterion for site classification in most of the current building seismic design codes, and it has very important applications in distinguishing saturated sandy soil and pulverized soil liquefaction, calculating the microseismic excellence period and analyzing the seismic effect of the site [5]. In geotechnical seismic engineering, the shear wave velocity of soil is widely used and researched in the following aspects: (1) as a parameter to classify the soil parameters, foundation stiffness and damping ratio of construction sites, (2) as one of the bases to discriminate the liquefaction of saturated sand and chalk, (3) in the analysis of seismic response of geotechnical buildings and foundations, and (4) as a controlling parameter to reshape the properties of in situ saturated sand configuration with disturbed sand in the laboratory [6]. In the same area of the city, to carry out many intensive engineering construction, repetition of a number of deep site shear wave velocity test not only affects the progress of the project but also increase the investment in the project [7]. If the existing borehole data, combined with the local geotechnical genesis and properties, to give a reasonable site shear wave velocity with the depth of the soil depth changes in the empirical relationship, will be conducive to speed up the construction process of major projects in the area, which has significant social and economic benefits [8].

Due to the unique accumulation environment of the Yangtze River Delta, Suzhou is in an area covered by loose Quaternary sediments and has a more complex urban environment [9]. Suzhou has a deep soft soil layer typical of the Quaternary period, and the long-period ground shaking of distant and large earthquakes may cause serious seismic hazards to major projects in Suzhou [10]. As one of the economically developed areas in the downstream plain of Yangtze River, Suzhou plays a key role in the coastal, riverine, and coastal opening strategy of Jiangsu and the urban integration strategy of Yangtze River Delta. In recent years, the construction of major projects in Suzhou has developed rapidly, such as the construction of lifeline projects for rail transportation, electric power facilities, and regional water supply, which have become the arteries driving the economic development of the region. Therefore, this paper chooses to use Suzhou as a typical area to provide a basis for seismic impact evaluation of Quaternary sedimentary layers. The shear wave velocity values of various soil types at different depths in the region are calculated, and the empirical relationship of shear wave velocity with depth is obtained by fitting, and the measured shear wave velocity is used to evaluate the soil liquefaction situation.

2 Geomorphology and Quaternary Stratigraphy of Suzhou

2.1 Landform Type and Main Features

Suzhou is located at the front edge of the Yangtze River Delta and the northeastern shore of Taihu Lake [11], the specific location and regional distribution are shown in Fig. 1. The terrain is generally high in the southwest and low in the northeast, and the Quaternary loose sediments are thick and complex in spatial distribution. In the west of Suzhou, there are low hill remnants and intermountain depressions. The low hill remnants are formed by tectonic denudation, while the intermountain depressions are composed of alluvial and alluvial lacustrine deposits. In the east, there are vast alluvial and tidal plains with many lakes and river ponds.

2.2 Quaternary Strata

Based on the characteristics of the sedimentary structure, the soil can be divided into 3 main layers at depth in the case of engineering applications. The main stratigraphic distribution in Suzhou is shown in Table 1.

(1) Fill layer (Q_4ml)

Brownish gray to gray, local brownish yellow, the upper part is loose miscellaneous fill. The lower part is mainly filled with vegetation, cultivated soil, which is soft to plastic, mainly accumulated in the old city with a thickness of 1.5 ~ 5.0 m.

(2) Late Pleistocene $\left(Q_3^{2-3}\right)$ alluvial and lacustrine sedimentary facies

③ ~ 1 Powdery clay ~ clay: yellowish brown, grayish yellow, can be ~ hard plastic state, containing ferromanganese nodules. The top of the layer is buried 0.8 ~ 3.6 m; the layer thickness is 1.2 ~ 4.8 m, with missing.

③ ~ 2 Powdered clay: gray–yellow, the lower part of the gradient to gray, can be ~ soft plastic, layer top burial depth 3.0 ~ 7.5 m layer thickness 0.4 ~ 5.0 m.

(3) Late Pleistocene $\left(Q_3^{2-2}\right)$ shallow marine phase, land–sea interactions deposited into

④ ~ 1 Powdered soil: grayish yellow to gray, slightly dense, rapid response to shaking and vibration, containing mica flakes, occasionally see shell fragments layer top burial depth 5.4 ~ 9.2 m, layer thickness 0.8 ~ 5.7 m.

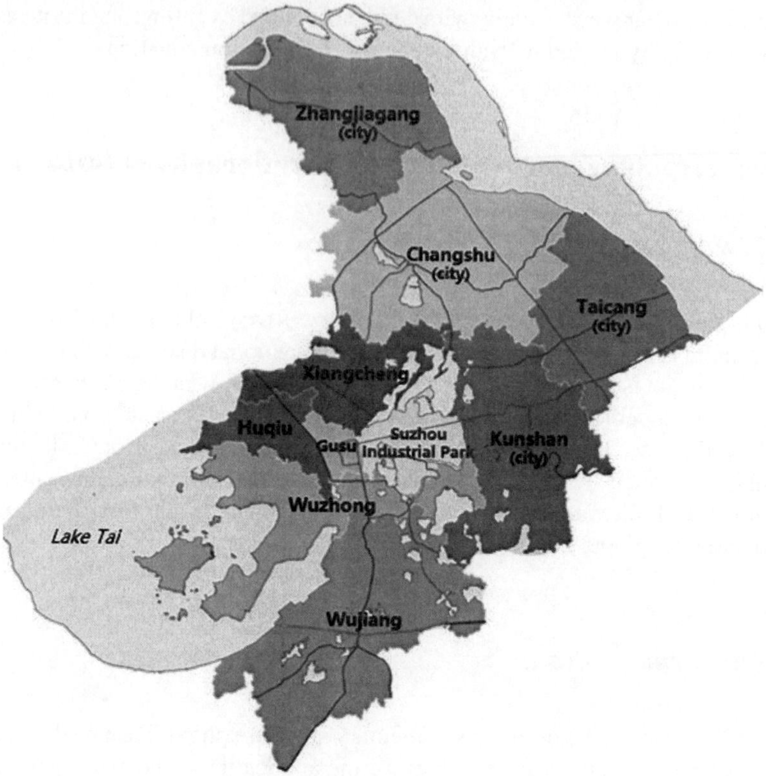

Fig. 1 Location and regional distribution of Suzhou City

④ ~ 1a Powdered clay sandwiched with powder: grayish yellow to gray, containing mica flakes, occasionally shell fragments, locally sandwiched with silty powdered clay, interlayer with powdered clay. Powdered clay is soft flow plastic; powdered clay is slightly dense, medium shaking response, poor homogeneity, layer top burial depth 5.5 to 9.7 m, layer thickness 0.8 to 6.0 m, irregular distribution in lenticular shape.

④ ~ 2 Powdered soil-powdered sand: gray, with a thin layer of clay, good permeability and water-rich, slightly dense to medium dense state, see a small amount of mica flakes and shell fragments, the top of the layer buried 6.5 ~ 15.9 m, layer thickness 2.7 ~ 14.1 m.

⑤ Powdery clay: gray, locally interspersed with silty powdery clay and thin laminated powdery clay and plastic powdery clay, mainly soft plastic to flow plastic. The top of the layer is 7.3 ~ 20.9 m deep, and the layer thickness is 0.5 ~ 17.4 m.

In terms of the Late Quaternary paleogeography, the general trend of changes in the southern plain of the Yangtze River Delta, where the urban area of Suzhou is located, is consistent with other areas along the eastern coast of China over the past 40,000 years [12]. The region has experienced two major land and sea changes.

Table 1 Main stratigraphic distribution in Suzhou

Chronostratigraphic		Code name	Depth (m)	Lithology
Holocene series		Qh	2	Filled soil
			5.3	Bluish gray mucky clay
Pleistocene	Upper Pleistocene	Qp_3	19.8	Bluish gray, yellowish gray silty clay
			24.55	Bluish gray, brown clay
			25.95	Bluish gray mucky clay
			31.53	Grayish yellow, cyan silty clay
			37	Cyan silty clay
			41.2	Gray powder sand
	Middle Pleistocene	Qp_2	50.9	Silty clay of bluish gray and taupe
			57.2	Gray, gray-green silty clay
			68.5	Polychromatic silty clay, silty sand
			74.35	Gray-green, gray-brown silty clay
			87.7	Gray, light gray silt

During the period of sea invasion, the Taihu Lake bay and shallow marine or lagoonal deposits were formed, mostly powdery clay or interbedded with chalk, silt and sub-clay, grayish yellow or gray, during the period of sea retreat, dark green clay or yellowish brown clay, and sub-clay layers were formed. Deep soft soil accumulation of Quaternary period is formed in Suzhou city, and loose soft soil layers are developed in most areas. Especially, in the coastal plain and ancient lake and marsh area, there are mostly soft clayey soils and saturated sandy soils. Under the action of strong ground shaking, soft soil seismic trap or sandy soil liquefaction may occur, which seriously affects the safety of major engineering structures as well as the economic and social benefits.

3 Statistical Information and Analysis Methods

3.1 Statistical Information

The information selected for this work is derived from the measured shear wave velocity data. To ensure the reliability of the test results, the testers did data calibration before the test work was carried out. Several sets of instruments were used for field testing and comparison. Supervision checks were strengthened during the work. The data were re-read and re-checked to minimize random factors and to ensure the consistency and reliability of the test results.

The information used in this paper was obtained from 57 boreholes at major engineering sites and along the rail line in urban Suzhou. This paper adopts the single-hole geophone method, and the test equipment includes vibration source, downhole geophone, trigger, and recorder. The ground excitation is used; the upper part of the wooden board pressed with weights is struck with a hammer; the SJ-5 trigger is installed on the wooden board; the downhole geophone adopts JK-55A three-component geophone to accept seismic waves, and the recording instrument adopts WAVE2000 site vibration tester. The vertical spacing of the measurement points was 1.0 m, tested from the bottom up and partially retested according to the specification to determine the shear wave velocity of the soil layers at different depths.

3.2 Analysis Method

The shear wave velocity of soil layer is mainly related to the soil type and depth of burial. At present, the relationship between shear velocity is mostly used in the following forms when calculating the relationship with depth variation. (1) linear function $Vs = a + bH$, (2) quadratic polynomial function $Vs = a + bH + cH^2$, and (3) power function $Vs = aH^b$ where Vs is the soil shear wave velocity (m/s), H is the burial depth of soil (m), and a, b, and c are the fitted parameters [13]. In this paper, R^2 is used to test the fit of the regression equation to the observed data. The value of R^2 ranges from 0 to 1, and the closer it is to 1, the better the fitting effect is.

The results of established studies show that not all soils have a significant correlation between shear wave velocity and burial depth [7], and the correlation between them is related to the degree of soil consolidation. The correlation between shear wave velocity and depth of burial may not exist for underconsolidated soils, while it exists for normally consolidated soils [14]. Based on this view, the empirical relationship between shear wave velocity and depth of burial for soils was calculated in this paper by excluding fill soils (miscellaneous fill, plain fill), silt and silty soils.

4 Analysis of Statistical Results

4.1 Empirical Relationship Between Soil Shear Wave Velocity and Soil Depth

The field measurements of site soil shear wave velocity for 57 boreholes are plotted in Fig. 2. It can be visualized that the site soil shear wave velocity increases essentially linearly with increasing soil layer and depth. Two mathematical models (1) and (2) are

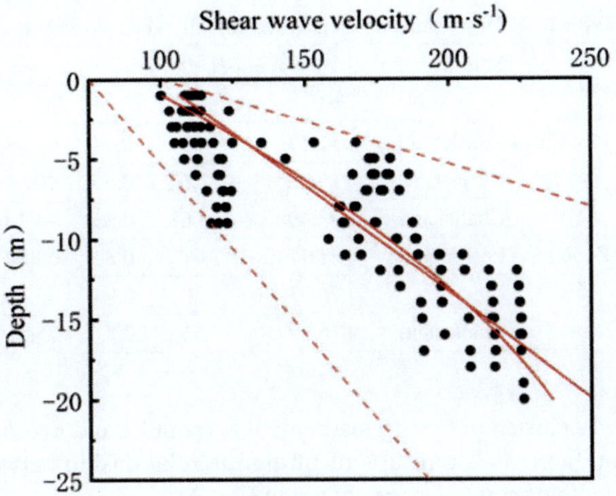

Fig. 2 Fitting effect of in situ soil shear wave velocity *Vs* and soil depth *H*

Table 2 Values of empirical relationship parameters for the 2 model fits

Models	Relationship formula	*a*	*b*	*c*	R^2
1	$V_s = a + bH$	99.20	−7.58		0.79
2	$V_s = a + bH + cH^2$	90.29	−10.37	−0.15	0.79

selected for regression analysis. The values of the empirical relationship parameters fitted using the two models are listed in Table 2. The fitting effect of site soil shear wave velocity *Vs* and soil depth H is shown in Fig. 2. It can be found that the fitting effect of linear functional relationship model 1 is not satisfactory compared with model 2, and the fitting effect of quadratic functional relationship model 2 is more satisfactory.

4.2 Variation of Shear Wave Velocity Along the Burial Depth for Different Soil Types in Each Unit Partition

In the western part of Suzhou, low hill remnants and intermountain depressions are formed by tectonic denudation and are mostly between 100 and 200 m in elevation. The eastern part of Suzhou is a vast alluvial and tidal plain. There are many lakes and rivers and ponds, and it is a typical water networked plain. The terrain is flat, with elevations ranging from 2 to 4 m, and tilts slightly from west to east. Therefore, the variation of shear wave velocity with depth in two areas of Suzhou is studied separately [15].

Considering the influence of soil type on the relationship between shear wave velocity Vs and soil depth H, the soils in the urban area of Suzhou were classified into four categories: clay, powdery clay and chalk sand, and the regression analysis

Table 3 Fitting results of shear wave velocity versus depth in soil layer

	Soil layer	$V_s = a + bH$			$V_s = a + bH + cH^2$			
		a	b	R^2	a	b	c	R^2
Zone II	Powdery clay	61.17	−7.53	0.84	61.95	−7.38	0.006	0.84
	Clay	134.25	−12.67	0.55	107.19	−26.73	−1.73	0.55
	Chalk sand	92.49	−10.68	0.85	−91.68	−42.40	−1.35	0.86
Zone I	Powdery clay	61.77	−7.41	0.86	65.80	−6.63	0.03	0.86
	Clay	76.33	−7.30	0.94	80.42	−6.20	0.04	0.93
	Chalk sand	76.42	−7.51	0.81	75.87	−7.63	−0.005	0.81

was carried out using mathematical model 1 and model 2, and the results are shown in Table 3. The results of fitting the relationship between shear wave velocity and depth of soil layer are shown in Fig. 3.

From the regression results, the fitted R^2 of model 2 is basically larger than that of model 1. Therefore, in the statistics of the empirical relationship between soil shear wave velocity and depth of burial in Suzhou City, model 2 is recommended. The best correlation coefficient is 0.94, and the worst correlation coefficient is 0.55. Except for clay in zone 2, which is lower than 0.7, the correlation coefficients of all other soil layers are greater than 0.7, indicating that the correlation of statistical relationships is good and statistically significant.

5 Shear Wave Velocity Discrimination of Soil Liquefaction

Among the seismic disasters occurring worldwide, liquefaction disasters, as the main secondary disasters caused by earthquakes, have been increasingly taken seriously by people [16]. The severity of earthquakes can often be reflected by the severity of liquefaction occurring because of the failure of several types of building foundations, dam damage, water bubbling from farmland, silting of rivers and canals, road collapse, soil cracks, residual deformation, etc., causing serious damage to a large number of houses and infrastructure [17]. The major earthquakes that occurred in history all had large areas of liquefaction [18, 19].

The calculation is based on the formula listed in the National Standard *Code for Seismic Design of Buildings* (GB 50,011–2010).

$$N_{\mathrm{cr}} = N_0\beta[\ln(0.6d_{\mathrm{s}} + 1.5) - 0.1d_{\mathrm{w}}]\sqrt{3/\rho_{\mathrm{c}}} \tag{1}$$

where N_{cr} represents the critical value of the standard penetration hammering number for liquefaction discrimination, N_0 represents the base value of the standard penetration hammering number for liquefaction discrimination, d_s represents the depth of the standard penetration point of saturated soil (m), d_w represents the water table (m), ρ_c represents the percentage of clay content, when it is less than 3 or sandy soil,

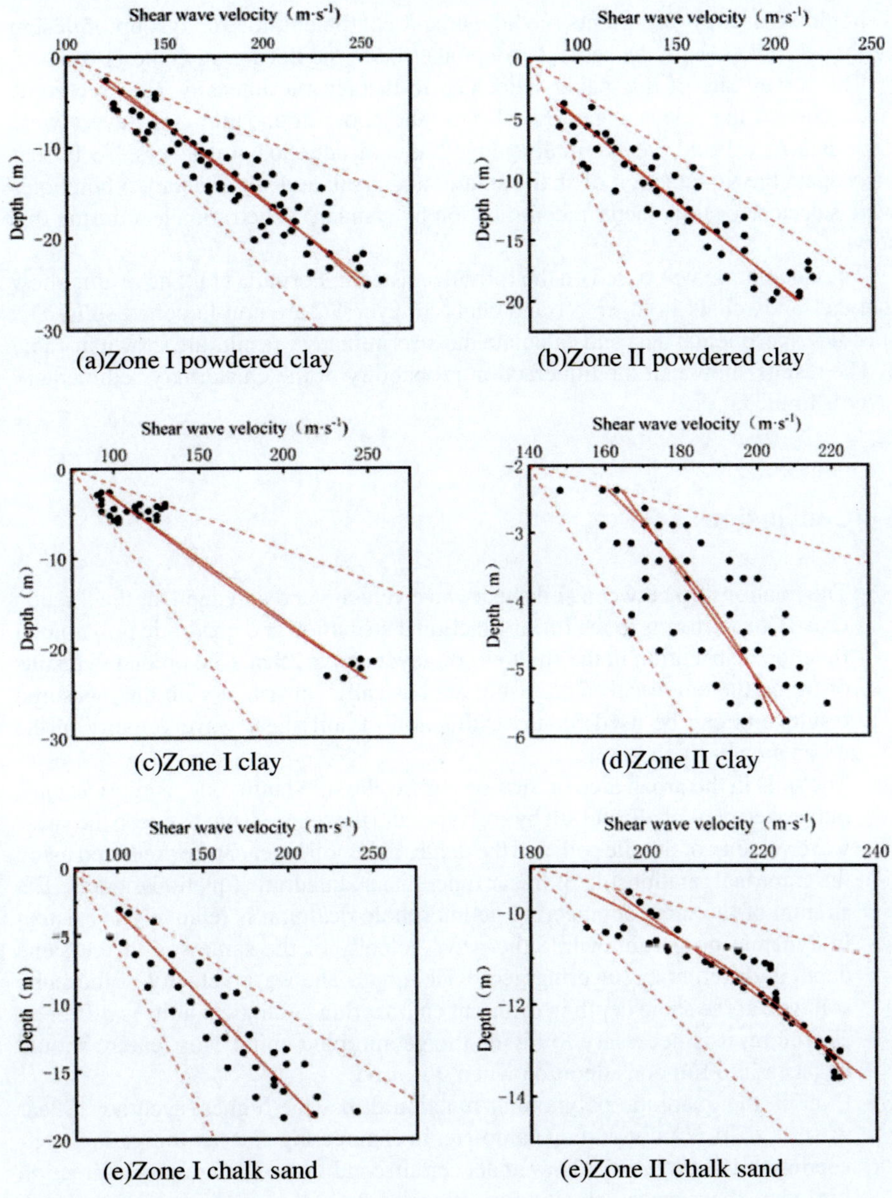

Fig. 3 Fitting curves of shear wave velocity with depth for different soils in various geomorphological divisions of Suzhou

3 should be used; β represents the adjustment coefficient, the first group of design earthquake takes 0.80, the second group takes 0.95, the third group takes 1.05.

The survey site of this paper is located in the seismic intensity 7-degree zone. According to the relevant geological data, the above foundation soil layers were deposited in $Q3$ and were initially judged as non-liquefied soil layers. To further investigate the liquefaction of shallow saturated chalk and silt in detail, 6 boreholes were selected for liquefaction identification by standard penetration test during this survey.

The calculation was based on the formulae listed in Formula (1). The results show that the sandy chalk in layer ⑤1 and chalk in layer ⑤2 are non-liquefied soils. The specific experimental data and calculate the discriminatory results are shown in Table 4. The results show that the liquefaction probability of the Quaternary sedimentary layer is limited.

6 Conclusion

(1) The relationship between soil shear wave velocity and soil depth at the Suzhou urban site conforms to the linear function distribution and quadratic polynomial function distribution in the shallow soil layer above 25 m. The predicted results of fitting the empirical relationship are basically consistent with the measured results and can be used for the estimation of soil shear wave velocity in the urban area of Suzhou.

(2) The soils in the urban area of Suzhou can be classified into four categories: clay, pulverized clay, chalk and silt by soil type, and the relationship between the shear wave velocity of the site soil, and the depth of the soil layer can be predicted using the empirical relationship of linear function and quadratic function models. The stratum of the same engineering geomorphological unit is relatively consistent in dynamic properties, while the wave velocity of the same soil at the same depth in different engineering geological units. The wave velocity of the same soil type at the same depth in different engineering geological units is different. Therefore, it is necessary to divide the geomorphic units. This feature should be taken into full consideration when it is used.

(3) Overall, the quadratic polynomial function fit has the highest accuracy. Shear wave velocity is an important parameter in engineering disaster mitigation engineering and has a statistical law under certain conditions. It is a research direction for future shear wave velocity projections to study the conditions under which shear wave velocity has a statistical law, to continue to explore the factors affecting shear wave velocity, and to determine the law of change of shear wave velocity with these factors.

(4) Substituting the existing typical shear wave velocity liquefaction discrimination method into the actual situation of Suzhou project, the results show that the liquefaction probability of Quaternary sedimentary layer is limited.

Table 4 Liquefaction of soil layers

Hole number	Layer number	Soil type	Standard penetration depth d_s (m)	Groundwater table depth d_w (m)	Viscous particle content P_c (%)	Standard penetration	N_{cr}	Liquefaction judgment
1	⑤1	Powdered clay clay	9.3	0.50	9.3	18.0	6.07	No
	⑤2	Chalk sand	11.8	0.50	3.0	19.0	11.76	No
	⑤2	Chalk sand	13.3	0.50	3.0	19.0	12.32	No
2	⑤1	Powdered clay	9.8	0.50	8.6	13.0	6.45	No
	⑤2	Powdered clay	11.3	0.50	9.6	15.0	6.46	No
	⑤2	Chalk sand	12.3	0.50	3.0	18.0	12.14	No
	⑤2	Chalk sand	14.3	0.50	3.0	18.0	12.66	No
	⑤2	Chalk sand	15.8	0.50	3.0	19.0	13.14	No
3	⑤1	Powdered clay	9.8	0.50	11.9	9.0	5.48	No
	⑤1	Powdered clay	11.3	0.50	8.5	10.0	6.87	No
	⑤2	Powdered clay	12.3	0.50	6.5	15.0	8.24	No
	⑤2	Chalk sand	14.3	0.50	3.0	16.0	12.66	No
4	⑤1	Powdered clay	9.8	0.50	9.3	14.0	6.20	No
	⑤1	Powdered clay	11.3	0.50	7.0	13.0	7.57	No
	⑤2	Chalk sand	12.3	0.50	3.0	17.0	12.14	No
	⑤2	Chalk sand	14.3	0.50	3.0	18.0	12.66	No
	⑤2	Chalk sand	15.3	0.50	3.0	16.0	12.98	No
5	⑤1	Powdered clay	12.3	0.50	6.5	18.0	8.24	No
	⑤1	Chalk sand	14.3	0.50	3.0	21.0	12.66	No

(continued)

Table 4 (continued)

Hole number	Layer number	Soil type	Standard penetration depth d_s (m)	Groundwater table depth d_w (m)	Viscous particle content P_c (%)	Standard penetration	N_{cr}	Liquefaction judgment
	⑤1	Chalk sand	15.8	0.50	3.0	23.0	13.14	No
	⑤2	Powdered clay	18.6	0.50	9.3	22.0	7.91	No
	⑤2	Powdered clay	19.8	0.50	5.9	25.0	10.16	No
6	⑤1	Powdered clay	8.3	0.50	13.3	11.0	4.84	No
	⑤1	Powdered clay	9.8	0.50	7.5	13.0	6.90	No
	⑤1	Powdered clay	11.3	0.50	9.3	14.0	6.56	No
	⑤2	Chalk sand	12.3	0.50	3.0	18.0	12.14	No
	⑤2	Chalk sand	14.3	0.50	3.0	19.0	12.66	No
	⑤2	Powdered clay	16.8	0.50	-	15.0	-	No
	⑤2	Chalk sand	17.3	0.50	3.0	18.0	13.58	No

Acknowledgements The research presented here is supported by the National Natural Science Foundation of China (52078317), Natural Science Foundation of Jiangsu Province for Excellent Young Scholars (BK20211597), project from Bureau of Housing and Urban-Rural Development of Suzhou (2021–25; 2021ZD02; 2021ZD30), Bureau of Geology and Mineral Exploration of Jiangsu (2021KY06), China Tiesiju Civil Engineering Group (2021–19), CCCC First Highway Engineering Group Company Limited (KJYF-2021-B-19) and CCCC Tunnel Engineering Company Limited (8gs-2021-04).

References

1. Finney, S.C.: The 'Anthropocene'as a ratified unit in the ICS international chronostratigraphic chart: fundamental issues that must be addressed by the task group. Geol. Soc., Lond, Spec. Publ. **395**(1), 23–28 (2014)
2. Barnosky, A.D.: Megafauna biomass tradeoff as a driver of Quaternary and future extinctions. Proc. Natl. Acad. Sci. **105**(1), 11543–11548 (2008)
3. Qin, D., Tao, S., Dong, S., et al.: Climate, Environmental, and Socioeconomic Characteristics of China//Climate and Environmental Change in China: 1951–2012. Springer, Berlin, Heidelberg (2016)
4. Bian, X., Li, W., Hu, J., et al.: Geodynamics of high-speed railway. Transp. Geotech. **17**, 69–76 (2018)
5. Kovacs, W.D., Leyendecker, E.V., Leiss, J.S., et al.: Data requirements for the seismic review of LNG facilities. US National Bureau of Standards (NEL), Prepared for Federal Energy Regulatory Commission. Washington DC, USA (1984)
6. Chaudhary, M.T.A.: Influence of site conditions on seismic design parameters for foundations as determined via nonlinear site response analysis. Front. Struct. Civ. Eng. **15**(1), 275–303 (2021)
7. Louie, J.N.: Faster, better: shear-wave velocity to 100 meters depth from refraction microtremor arrays. Bull. Seismol. Soc. Am. **91**(2), 347–364 (2001)
8. Fan, X., Scaringi, G., Korup, O., et al.: Earthquake-induced chains of geologic hazards: patterns, mechanisms, and impacts. Rev. Geophys. **57**(2), 421–503 (2019)
9. Wu, Z., Zhou, C., Huang, X., et al.: Main active faults and seismic activity along the Yangtze River economic belt: based on remote sensing geological survey. China Geol. **3**(2), 314–338 (2020)
10. Liu, W.: A geostatistical approach toward shear wave velocity modeling and uncertainty quantification in seismic hazard (2018)
11. Wang, Y., Wu, W., Boelens, L.: City profile: Suzhou, China—the interaction of water and city. Cities **112**, 103119 (2021)
12. Wu, L., Li, F., Zhu, C., et al.: Holocene environmental change and archaeology, Yangtze River Valley, China: review and prospects. Geosci. Front. **3**(6), 875–892 (2012)
13. Feng, X.U.E., Qing-liang, W.A.N.G., Ya-xuan, H.U., et al.: Study on the relationship between soil depth and shear wave velocity of rock and soil mass in Xian City. China Earthq. Eng. J. **36**(2), 249–255 (2014)
14. Hu, X., Zhou, Z., Chen, H., et al.: Seismic fragility analysis of tunnels with different buried depths in a soft soil. Sustainability **12**(3), 892 (2020)
15. Hongtao, J., Fubao, W., Dayuan, Y.: Study of late quaternary paleogeography and engineering geological conditions in Suzhou City. Geoscience **23**(1), 82–86 (2003)
16. Youd, T.L.: Geologic effects-liquefaction and associated ground failure. In: Proceedings of the Geologic and Hydraulic Hazards Training Program, pp. 210–232 (1984)
17. Tokimatsu, K., Tamura, S., Suzuki, H., et al.: Building damage associated with geotechnical problems in the 2011 Tohoku Pacific earthquake. Soils Found. **52**(5), 956–974 (2012)

18. Seed, H.B., Idriss, I.M.: Analysis of soil liquefaction: Niigata earthquake. J. Soil Mech. Found. Div. **93**(3), 83–108 (1967)
19. Sun, R., Yuan, X.M.: Shear wave velocity discriminant formulae suitable for liquefaction of soils at different depths. Chin. J. Geotech. Eng. **41**(3), 439–447 (2019)

Characterization of the Natural Dolomite from Thanh Liem Area, Vietnam, and Its Applications

Nguyen Thi Thanh Thao, Le Thi Duyen, and Phenglilern Sensousit

Abstract Natural dolomite is a common mineral in Thanh Liem area, Vietnam. Currently, many dolomite mines in the area have been searched, explored and put into operation, meeting raw materials for different fields of use. However, the literature evaluating the quality characteristics of limestone is limited. This paper aims to assess the quality of the natural dolomite in the study area by combining previous data and some new analytical results such as X-ray fluorescence (XRF), X-ray diffraction (XRD), the scanning electron microscope with energy-dispersive X-ray spectroscopy (SEM–EDS), inductively coupled plasma mass spectrometry (ICP-MS) and differential thermal analysis-thermogravimetry (DTA-TG). The results show that the regional dolomite has good quality and can be used in many different fields. Dolomite ($CaMg(CO_3)_2$) was the predominant mineral in the samples, followed by calcite ($CaCO_3$) and other trace minerals (quartz,..). The main element oxide contents of CaO and MgO are 32.5–42.1%, and 12.7–19.6%, respectively. The content of other oxides such as Al_2O_3, T.Fe, SiO_2, MnO and K_2O is not significant. The mechanical and physical properties of the stone completely meet the fields of civil construction. In addition, to improve the quality of raw materials used for each field, some mineral processing charts of dolomite and their main applications are also presented in the paper. It will be useful information for planning, exploiting and using this dolomite effectively.

Keywords Dolomite · Dong Giao formation · Thanh Liem area

N. T. T. Thao (✉) · L. T. Duyen · P. Sensousit
Hanoi University of Mining and Geology, Hanoi, Vietnam
e-mail: nguyenthithanhthao@humg.edu.vn

L. T. Duyen
e-mail: lethiduyen@humg.edu.vn

N. T. T. Thao · N. T. T. Thao · L. T. Duyen
Research and Advanced Technology Applications in Environmental, Material and Earth Sciences, Hanoi University of Mining and Geology, Hanoi 100000, Vietnam

L. Q. Nguyen et al. (eds.), *Advances in Geospatial Technology in Mining and Earth Sciences*, Environmental Science and Engineering,
https://doi.org/10.1007/978-3-031-20463-0_18

1 Introduction

Dolomite is the name of a mineral with the chemical formula $CaMg(CO_3)_2$ and is also the name of a carbonate sedimentary rock. Dolomite rocks are made up mostly of the mineral dolomite. Dolomite was first described in 1791 by the French naturalist and geologist, Déodat Gratet de Dolomieu (1750–1801) [1]. The oxide composition in dolomite rock includes CaO of 30.41%, MgO of 21.86% and CO_2 of 47.73%. With the characteristics of the composition, physical and mechanical properties, as well as low cost, dolomite is selected to be used in many different fields such as road base material, a feed additive for livestock, a sintering agent and flux in metal processing, high-grade bricks, ceramics, for post-treatment of desalinated water [2–4].

In Vietnam, the potential of dolomite is quite large and they are distributed in many provinces such as Ninh Binh, Ha Nam, Yen Bai, Bac Kan, Thai Nguyen, Thanh Hoa, Nghe An, Lam Dong, Dak Lak... [2]. The origin of dolomite in Vietnam is mainly divided into two forms: primary sedimentary and metamorphic sedimentary. The carbonate sedimentary formations are mainly from Triassic to Devonian in the Dong Giao Formation (T_{2a} *dg*), the Bac Son Formation (CP *bs*), the Dai Thi Formation (D_1 *dd*), the Duong Dong Formation (D_{1-2} *dđ*)... The dolomite from metamorphic sedimentary mainly belongs to the Nui Vu Formation ($Є_1$ *nv*), the Dac Uy Formation ($Є_2$ *du*), the Cam Duong Formation ($Є_2$ *cđ*) and the Sinh Vinh Formation (O_3-S_1 *sv*)... [5]. Depending on the origin and formation conditions, characteristics of quality and physicochemical properties of dolomite are different, and therefore, the field of use for each type of dolomite is also different. Thus, it is necessary to evaluate the quality of dolomite rock in each area.

Thanh Liem area, Vietnam, is an area with great potential for dolomite materials. Although many small mines are being exploited, there are limited studies on the quality of the dolomite systematically in the area. This paper aims to assess the quality of the natural dolomite in the study area by combining previous data and some new analytical results. Some mineral processing charts of dolomite and their main applications also are presented in the paper. It will be useful information for planning, exploiting and using this dolomite effectively.

2 Geological Characteristics

In the study area, the most common are the Tan Lac, Dong Giao formations and the undivided Quaternary sediments. The Tan Lac Formation has black clay sediments that meet the criteria of being used as a cement additive. The Tan Lac Formation has a transitional relationship with the Dong Giao Formation. The sediments of the Dong Giao Formation are exposed to a strip extending from the northwest to the southeast, concentrating more in the southwest and the garbage strip in the central area of the study area. Based on the petrographic composition, the formation is divided into lower and upper strata. The lower sub-formation (T_{2a} dg_1) consists of dolomite lime interspersed with a thin layer of clay lime. The rock is gray, light gray, medium to thickly layered structure with a thickness of 400–800 m. The upper sub-formation (T_{2a} dg_2) is limestone layered thinly to gray–green mass, interspersed with lime clay and dolomite lime with a thickness of 500–900 m. undivided Quaternary (Q) is distributed over a large area of the province. They are mainly composed of dark gray clay and sand, gray ash of marine origin, containing plant remains, and lenticular peat (Fig. 1).

3 Materials and Analytical Methods

Natural dolomite samples were taken at the typical outcrops in Thanh Liem area, Nam Ha province (Fig. 1). The samples were ground to a fine powder and homogenized using an agate mortar. A range of analytical methods, including petrographic microscopy, X-ray diffraction analysis (XRD), X-ray fluorescence (XRF), scanning electron microscopy-energy dispersive X-ray spectroscopy (SEM–EDS), inductively coupled plasma mass spectrometry (ICP-MS) and differential thermal analysis-thermogravimetry (DTA-TG), were used to assess the quality of the natural dolomite from the study area.

4 Results and Discussion

4.1 Characteristics of the Natural Dolomite from the Study Area

4.1.1 Mineral Composition

The results of the petrographic analysis show that in the studied sample, there are granular and rhombic dolomite aggregates with sizes from 0.05 to 0.8 mm (Fig. 2). Mixed with dolomite are calcite aggregates accounting for 2–20%, granular variation. The surface of dolomite is opaque and agglomerated, without polymorphic crystallization. The rock-forming components are uniformly distributed; they have a weakly oriented structure and a variable crystalline grain structure.

The results of mineral composition analysis by *X*-ray diffraction (XRD) method show that the dolomite samples in the study area have mainly dolomite minerals (79–98%) and calcite from few to 20% (Fig. 3). The XRD pattern shows that dolomite mineral (Do) is the main mineral in the sample with typical XRD peaks at 24.0°, 30.9°, 37.3°, 41.1°, 44.9°, 51.1°, 60.0° and 67.4° (Fig. 3a) [6]. The FT-IR pattern in Fig. 3b indicates that dolomite mineral (Ca, $Mg(CO_3)_2$) is the significant mineral in the sample with the presence of the bands at 3020 cm^{-1}, 2627 cm^{-1} and 728 cm^{-1} (Fig. 3b). The FT-IR bands detected at 3020 cm^{-1} and 2627 cm^{-1} are combination

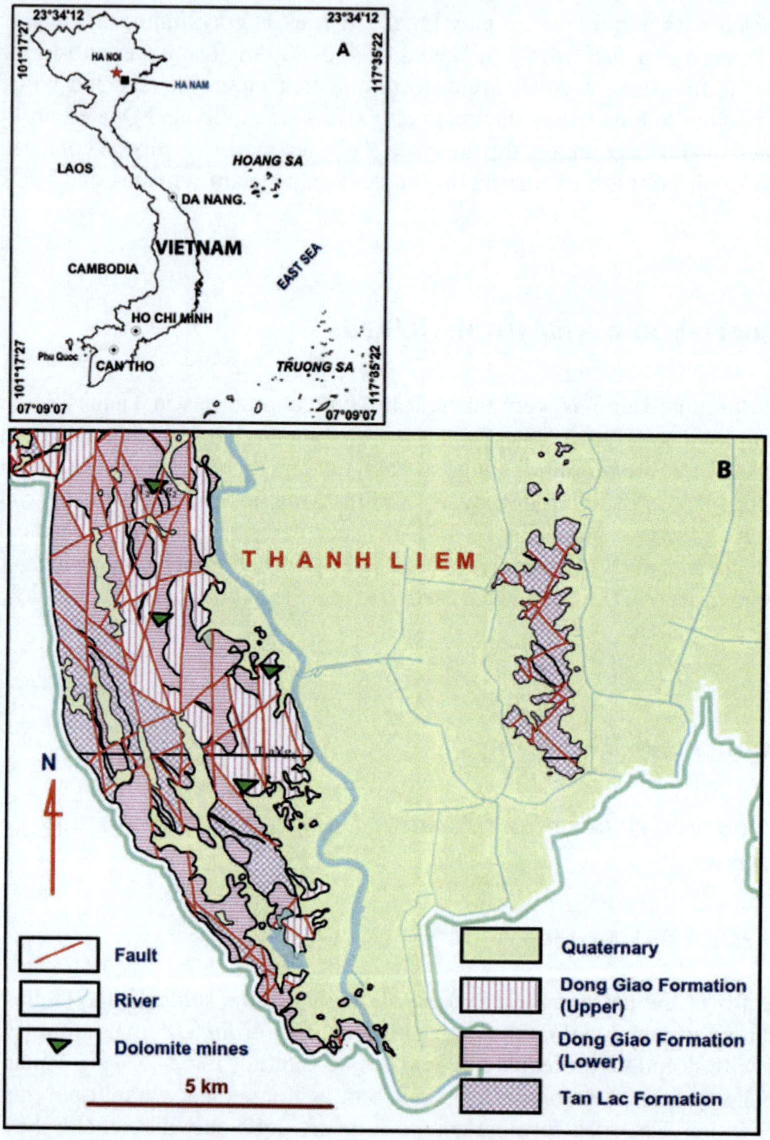

Fig. 1 Location of the study area **a** and geological map of Thanh Liem area **b**

frequencies [7], and the band at 728 cm^{-1} is assigned to the in-plane bending mode of $CO_3{}^{2-}$ in the dolomite structure [8]. The appearance of both the 2627 cm^{-1} and 728 cm^{-1} absorption bands in a sample is especially useful for indicating the presence of dolomite.

Fig. 2 Petrographic images of the natural dolomite from Thanh Liem area (*dl—Dolomite; Q–Quartz*)

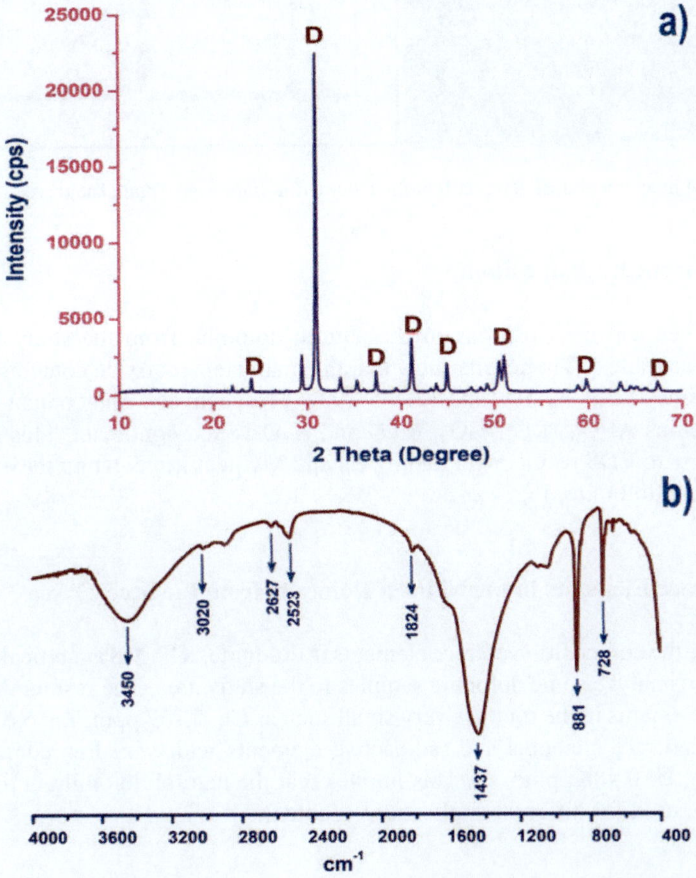

Fig. 3 XRD **a** and FT-IR **b** patterns of the natural dolomite from Thanh Liem area (*D—Dolomite*)

Table 1 Chemical composition of the natural dolomite from the study area

Chemical composition $\frac{\text{min}\div\text{max}}{\text{average}}$ (%)			
CaO	MgO	Al_2O_3	T.Fe[a]
$\frac{32.5\div42.1}{35.9}$	$\frac{12.7\div19.6}{16.6}$	$\frac{0.01\div0.03}{0.02}$	$\frac{0.02\div0.09}{0.06}$
SiO_2	MnO	K_2O	LOI[b]
$\frac{0.25\div0.28}{0.26}$	$\frac{0.01\div0.01}{0.01}$	$\frac{0.01\div0.02}{0.02}$	$\frac{46.7\div47.8}{47.2}$

Note [a]Total iron content; [b]Loss on ignition

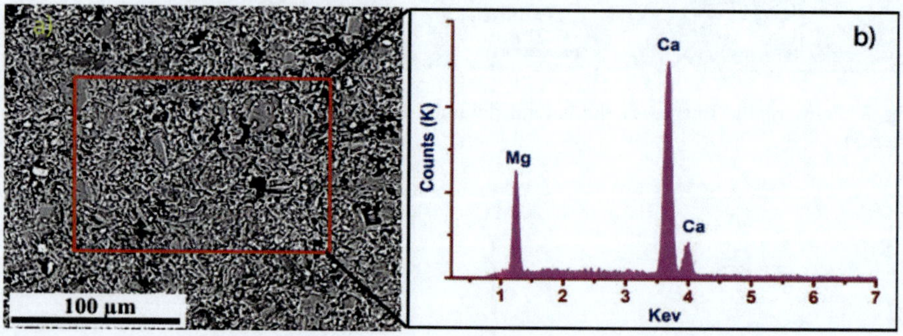

Fig. 4 SEM image: **a** and EDS result **b** of the natural dolomite from Thanh Liem area

4.1.2 Chemical Composition

Average chemical compositions of the natural dolomite from the study area are presented in Table 1. The results show that the main element oxide contents of CaO and MgO are 32.5 $\div$ 42.1%, and 12.7 $\div$ 19.6%, respectively. The content of other oxides such as Al_2O_3, T.Fe, SiO_2, MnO and K_2O is not significant. This result is consistent with EDS results with mainly Ca and Mg peaks, indicating the existence of quite pure dolomite (Fig. 4).

4.1.3 Trace Elements in the Natural Dolomite from the Study Area

To evaluate the composition of trace elements in dolomite, ICP-MS analytical method was used to analyze some dolomite samples in the study area. The results show that the metal elements in the rock are very small such as Cu: 5.257 ppm; Zn: 6.623 ppm. Precious and rare elements and radioactive elements with very low content (Th: 0.528 ppm; U: 0.488 ppm; …). This implies that the natural dolomite in the study area can be used for different applications (Table 2).

Table 2 Average trace elements in of the natural dolomite from the study area

Elem	(ppm)	Elem	(ppm)	Elem	(ppm)	Elem	(ppm)
Li	2.969	Cu	5.257	Nb	5.014	Ta	0.056
Be	0.113	Zn	6.622	Mo	0.152	W	9.339
P	0.062	Ga	0.153	Ag	0.046	Au	0.773
K	87.251	As	16.141	Ba	5.450	Pb	1.845
Ti	8.384	Se	5.147	La	1.581	Bi	0.093
V	9.878	Rb	0.548	Ce	2.946	Th	0.528
Co	0.867	Sr	122.108	Nd	1.468	U	0.488
Ni	2.224	Y	1.759	Hf	0.404	–	–

4.1.4 Physicomechanical Properties

The results of mechanical and physical parameters of dolomite rock samples in Thanh Liem area, Ha Nam province are shown in Table 3. The results show that dolomite has good physical and mechanical properties, meeting the fields of civil construction.

Table 3 Summary of mechanical and physical parameters of dolomite rock samples in Thanh Liem area, Ha Nam province

Parameters	Unit	Max	Min	Average
Natural humidity (W)	%	0.24	0.13	0.18
Water absorption (W_{hn})	%	0.48	0.46	0.47
Natural volumetric mass (γ_0)	g/cm^3	2.69	2.67	2.685
Saturated volumetric mass (γ_{bh})	g/cm^3	2.70	2.69	2.695
Dry volumetric weight (γ_c)	g/cm^3	2.69	2.66	2.67
Density (ρ)	g/cm^3	2.74	2.73	2.735
Porosity (n)	%	2.02	2.00	2.01
Shear strength	kG/cm^2	88	86	87
Natural compressive strength (δ_{tn})	kG/cm^2	703	400	540
Dry compressive strength (δ_k)	kG/cm^2	693	637	665
Saturated compressive strength (δ_{bh})	kG/cm^2	673	387	537
Dry tensile strength (δ_k)	kG/cm^2	69	64	66
Saturated tensile strength (δ_{bh})	kG/cm^2	64	59	61
Dry strength coefficient (f_k)	kG/cm^2	6.93	6.37	6.65
Saturated strength coefficient (f_{bh})	kG/cm^2	6.45	5.85	6.15

4.1.5 Thermal Behavior of Natural Thanh Liem Dolomite

The typical DTA-TG curve of the dolomite sample from Thanh Liem area is presented in Fig. 5. It can be seen that the observed weight loss was 46.29% between 600 °C and 850 °C. The weight loss detected in the temperature range of 100–120 °C, was followed by a weight loss attributed to the decomposition of carbonates. The weight loss in this temperature range can be attributed to the chemically bound water [9, 10].

4.2 *Mineral Processing of the Natural Dolomite of the Study Area*

With good quality, Thanh Liem area dolomite can be used for different applications. Depending on each application, steps in mineral processing are designed differently. Figure 6 presents two dolomite processing schemes: dolomite as raw material for ferrous metallurgy and high-quality dolomite (Fig. 6).

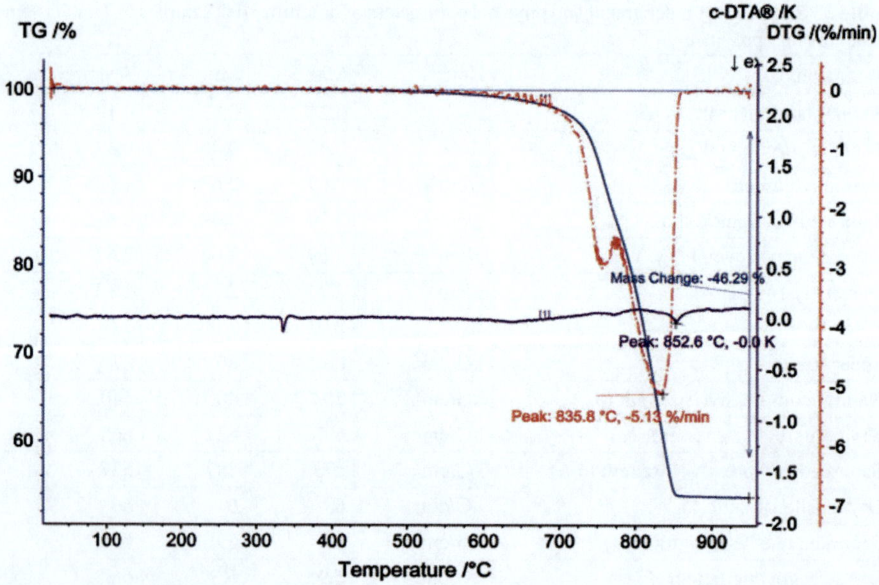

Fig. 5 DTA-TGA curves of the natural dolomite from Thanh Liem area

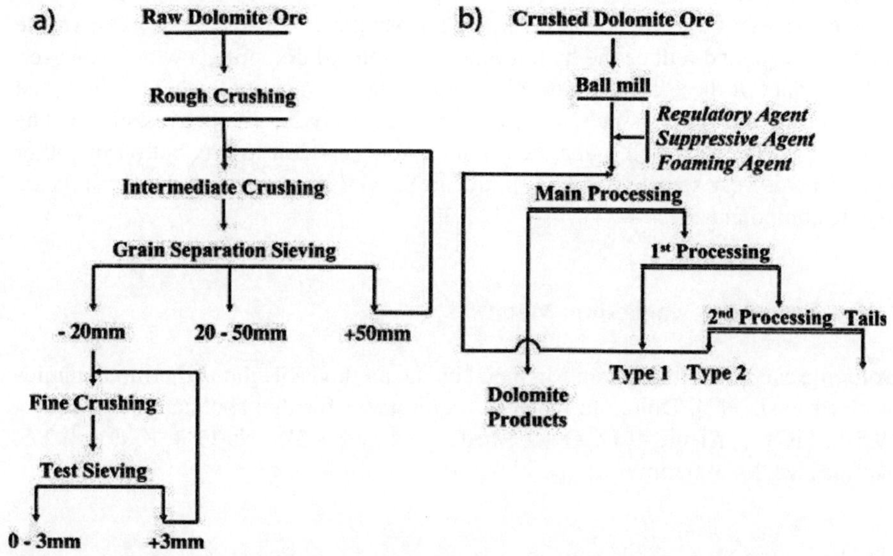

Fig. 6 Flowsheets for the natural raw dolomite processing in the study area. **a**—For ferrous metallurgy; **b**—For high-quality dolomite)

4.3 *Some Applications of Dolomite Products*

4.3.1 Producing Refractory Materials in Ferrous Metallurgy

Dolomite can be used as a flux or slag disintegration aid as well as in the preparation of sintering magnesium ores [11]. Dolomite used as flux in metallurgy has the following standards: MgO > 17–19%, SiO_2 < 6%, R_2O_3 + MnO < 5%, not mixed with S, P, particle size <25 mm below 8%, strength instantaneous compressive resistance >300 kg/cm^2. Therefore, dolomite used in sintering of metallurgical magnesium needs to meet the following requirements: CaO + MgO > 53%, MgO > 16%, insoluble residue < 2.5%, clay < 3%, particle size 5–75 mm accounting for 80%.

4.3.2 Producing Heat-Resistant Brick

Martensite is a self-adhesive powder made up of a mixture of magnetite and dolomite (containing 30–50% dolomite). Dolomite concrete bricks have high fire resistance, so they are used to cover electric furnaces and steel rolling furnaces [12]. Here, dolomite is required to be good at resisting fire and slag in the sintering state, that is, it does not decompose into mixtures of silicates and alumosilicates Mg, Ca, Fe and Mn which are easy to melt and easily corrode.

The oxides of Fe, Al, Ti and Mn are all beneficial because they improve the sintering stage and reduce the hydration of the sintered dolomite powder. However, if the amount of these compounds is too high, a large amount of calcium aluminate will appear, causing the sintered dolomite powder to reduce its fire resistance. The presence of free CaO will make the product more resistant to fire, but on the other hand, CaO is very sensitive to CO_2 in the air as well as reacts with Al, Fe to create fusible compounds.

4.3.3 Producing Magnesium Metal

Dolomite can be used to obtain Mg metal by the methods of silicon thermodynamics or electrolysis [13]. Dolomite material requirement for this application is MgO > 19.5%, $SiO_2 + Al_2O_3 + Fe_2O_3 + Mn_3O_4 < 2$ and > 5%, $Na_2O + K_2O < 0.2\%$, particle size 20–300 mm.

4.3.4 In the Chemical and Pharmaceutical Industry

In addition to the optimal materials such as magnesium, carnalite and saltwater, people also use dolomite to produce MgO. Since then, a series of important materials have been produced such as "white magnesium" (also known as "light magnesium" or "technical carbon magnesium" with the formula ($5MgO.4CO_2.6H_2O$), calcined magnesium (which is very light and absorbs gas). strong) used to make medicinal herbs, as heat insulators, catalysts, dye preparations, paints, aromatics, … [14].

4.3.5 In Environmental Treatment

Recently, the materials that are available, low cost and non-toxic materials are interested in many domestic scientists in the treatment of polluted water environment. Clay minerals which are often used to treat water pollution are mainly bentonite, vermiculite, kaolin and halloysite [15, 16]. However, natural clay materials have been used in many other fields with a high economic value, so their reserves are decreasing day by day. Therefore, finding alternative sources of raw materials is a necessary task, and dolomite is a potential source of raw materials for this application [17–19].

5 Conclusions

In conclusion, to assess the quality of the natural dolomite, samples were taken from the typical outcrops in Thanh Liem area. Analysis methods of XRD, XRF, SEM–EDS, ICP-MS and DTA-TG were used to characterize the dolomites. The analytical results indicate that dolomite ($CaMg(CO_3)_2$) was the predominant mineral in the samples, followed by calcite ($CaCO_3$) and other trace minerals (quartz,..). The main element oxide contents of CaO and MgO are 32.5–42.1%, and 12.7–19.6%, respectively. The content of other oxides such as Al_2O_3, T.Fe, SiO_2, MnO and K_2O is not significant. The mechanical and physical properties of the stone completely meet the fields of civil construction. The metal elements in the rock are very small such as Cu: 5.257 ppm; Zn: 6.623 ppm. Precious and rare elements and radioactive elements with very low content (Th: 0.528 ppm; U: 0.488 ppm; ...). Two flowsheets for the natural raw dolomite processing in the study area were set up with many technological solutions such as selective crushing and chemical processing to increase the quality of dolomite. In general, the quality of dolomite in the study area is of good quality, which can be used for many different application areas. It will be useful information for planning, exploiting and using this dolomite effectively.

References

1. Linnaeus, C.: Systema Naturae per Regnum Tria Naturae, Secundum Classes, Ordines, Genera, Species cum Characteribus and Differentiis, 236 pp. Tomus III. Laurentii Salvii, Holmiae (1978)
2. Nguyen, A.D., Kieu, Q.N., Phan, V.H., Tran, T.L.: Active dolomite in Ha Nam area and technical characteristics of unburnt bricks made from this material. Environ. Resour. **2**, 34–36 (2015). (Vietnamese)
3. Tsirambides, A.: Industrial applications of the dolomite from Potamia Thassos Island, N. Aegean Sea, Greece. Mater. Struct. **34**, 110–113 (2001)
4. Jerzy, T., Emilia, W., Mateusz, M., Paweł, L, Marek, K., Stanisław, S.: Petrographic and geotechnical characteristics of carbonate aggregates from Poland and their correlation with the design of road surface structures. Materials **14**, 2034 (2020)
5. Luong, Q.K., Bui, H.B., Truong, H.M.: Features on the quality and resource potential of dolomite in Ninh Binh province. In: The 2nd International Conference on Advances in Mining and Tunneling, pp. 43–46 (2012)
6. Said, S.A.L., Anwar, U.H., Abdul, R.I.M., Salih, S.: Use of X-ray powder diffraction for quantitative analysis of carbonate rock reservoir samples. Powder Technol. **175**(3), 115–121 (2007)
7. Nguyen, T.T., Janik, L.J., Raupach, M.: Diffuse reflectance infrared Fourier transform (DRIFT) spectroscopy in soil studies. Aust. J. Soil Res. **29**, 49–67 (1991)
8. Farmer, V.C.: Infrared spectra of minerals. In: Farmer, V.C. (Ed.), Mineralogical Society Monograph No. 4, p. 399. Mineralogical Society, London (1974)
9. Gunasekaran, S., Anbalagan, G.: Thermal decomposition of natural dolomite. Bull. Mater. Sci. **30**(4), 339–344 (2007)
10. Magdalena, O.H., Maciej, J.: Thermal behavior of natural dolomite. J. Therm. Anal. Calorim. **119**, 2239–2248 (2015)
11. Sen, P.C., Ramakrisana Rao, M., Bhaskar Rao, H.V.: Dolomite as a refractory raw material for lining L.D converters. Trans.—Indian Ceram. Soc. **23**(1), 131–143 (2014)
12. Mahmoud, R., Emad, M.M.E.: Multi-impregnating pitch-bonded Egyptian dolomite refractory brick for application in ladle furnaces. Ceram. Int. **35**(2), 813–819 (2009)
13. Demiray, Y., Yücel, O.: Production and refining of magnesium metal from Turkey originating dolomite. High Temp. Mater. Proc. **31**, 251–257 (2012)

14. Lihong, S., Chunlin, Y., Qingfeng, W., Zhaohui, L., Weibin, Z., Hanlie, H.: Photocatalytic degradation of diphenhydramine in aqueous solution by natural dolomite. RSC Adv. **10**(63), 38663–38671 (2020)
15. Tran, L.T.: Research on modification of vermiculite clay minerals to improve heavy metal adsorption capacity in water. VNU Sci. J.: Earth Environ. Sci. **32**(3), 82–91 (2016) (Vietnamese)
16. Bui, H.B., Hoang, N., Nguyen, T.T.T., Le, T.D., Vo, T.H., Nguyen, T.D., Luong, Q.K., Do, M.A.: Performance evaluation of nanotubular halloysites from weathered pegmatites in removing heavy metals from water through novel artificial intelligence-based models and human-based optimization algorithm. Chemosphere **282**, 131012 (2021)
17. Yamkate, N., Chotpantarat, S., Sutthirat, C.: Removal of Cd^{2+}, Pb^{2+}, and Zn^{2+} from contaminated water using dolomite powder. Hum. Ecol. Risk Assess.: Int. J. **23**(5), 1178–1192 (2017)
18. Huang, H., Zhang, D., Guo, G., Jiang, Y., Wang, M., Zhang, P., Li, J.: Dolomite application for the removal of nutrients from synthetic swine wastewater by a novel combined electrochemical process. Chem. Eng. J. **2018**(335), 665–675 (2018)
19. Liat, B., Noga, F., Ori, L.: Potential applications of quarry dolomite for post treatment of desalinated water. Desalin. Water Treat. **1**(1–3), 58–67 (2009)

A Mine Production Tracking Platform and Its Initial Application in the Digital Transformation for a Vietnamese Coal Exploitation Company

Dinh-Van Nguyen, Trung-Kien Dao, Viet-Tung Nguyen, Cong-Dinh Dinh, Trung-Kien Nguyen, Nguyen Quynh Nga, and Chu Thi Khanh Ly

Abstract Promoting the application of information technology in production management is essential to improve the efficiency. Mine production data usually come in multiple categories such as rough and detailed planning, execution, assessment, examination, inspection and accreditation. These data need to be frequently exchanged between related people and groups. In current Vietnamese coal companies, this is generally achieved by using a common shared folder on a local network, which is highly vulnerable to multiple risks, e.g., data loss or defect, lack of access permission control, file simultaneous access problem, difficulties in synthesizing data for reports… and overall, will result in efficiency decline in production administration. In this study, a data management platform for the mine production data tracking which aims to solve the above problems by using a centralized server is introduced. The solution is then deployed in a coal exploitation company in Vietnam. With support from Viettech Company Ltd, an initial evaluation of the system shows a 10% human errors reduction.

D.-V. Nguyen (✉) · D.-V. Nguyen (✉) · T.-K. Dao · V.-T. Nguyen
MICA Institute, Hanoi University of Science and Technology, Hanoi, Vietnam
e-mail: dinh-van.nguyen@mica.edu.vn

T.-K. Dao
e-mail: trung-kien.dao@mica.edu.vn

V.-T. Nguyen
e-mail: viet-tung.nguyen@mica.edu.vn

C.-D. Dinh
VietTech Company Ltd, Hanoi, Vietnam

T.-K. Nguyen
National Economic University, Hanoi, Vietnam

N. Q. Nga · C. T. K. Ly
National Academy of Public Administration, Hanoi, Vietnam
e-mail: nganq@napa.vn

C. T. K. Ly
e-mail: khanhly@napa.vn

L. Q. Nguyen et al. (eds.), *Advances in Geospatial Technology in Mining and Earth Sciences*, Environmental Science and Engineering,
https://doi.org/10.1007/978-3-031-20463-0_19

Keywords Mine production data tracking · Digital transformation · Centralized service · Data management

1 Introduction

Although many different definitions have been proposed, digital transformation in a broad sense can be understood as the process of applying novel technologies in the field of digital communication and data processing to solve problems of life [1]. This is a process closely related to the Industrial Revolution 4.0, promoting the application of science and technology to help change the way people live and work. For businesses, this is the process of integrating advanced technologies into their activities, helping to fundamentally and comprehensively change the way businesses operate and create value. Thus, this is not only a process of applying technology alone, but it also requires a change for people in terms of perception, culture, as well as the way of thinking, interacting, communicating, and working. Because of that deeply integrated nature, the digital transformation process will be very different between each industry, as well as within each business, and there is no universal formula.

In addition, digital technologies also enhance the production of data from the use process itself, and are a potential source of information for analysis, forecasting, decision-making, helping to solve problems in optimizing the operating processes of the organization [2]. The study in [3] shows that the digital transformation process for an organization is not only a reorganization of the operating structure, but also requires planners to build a model that integrates the strategy of the organization with that for digital technology [4], and proposed the concept of a digital business strategy (DBS), which is an organization's strategy created and implemented based on digital resources to create differentiated value. However, in [5], Kane pointed out that digital technologies themselves do not create much value for the organization, but applying them to the right circumstances allows to find new ways to create value.

In addition to new technologies for data processing, new system models have been introduced that bring high efficiency to applications, such as cloud computing, edge computing, Internet of things (IoT) [6]. Cloud computing is essentially a computing model that allows the allocation and sharing of storage, processing, and computing resources at the request of users through a network environment. This model has the outstanding advantage of separating the responsibilities for administration and maintenance of data and system resources from the process of using those resources and allows turning those responsibilities into the form of a service to provide to users, thereby helping to reduce costs in the stages of system development, distribution, operation, upgrade and maintenance. This model has now become especially popular for data storage services for individual users, but it also has great potential to improve the operational efficiency of businesses [7].

Although Vietnam's coal reserves are still very large, the coal seams are getting deeper and deeper, coal mining conditions are increasingly difficult, causing production costs to increase and reduce the competitiveness of domestic coal compared to imported coal. The mechanization and automation in the design, operation and supervision of the exploitation process have been interested by Vietnam National Coal and Mineral Industries Group (Vinacomin) for a long time, but the results achieved are still limited. Output of open-pit mines after decades of exploitation has gradually declined, while on the other hand, mining by opencast method greatly affects the composition of natural resources and the environment, so the proportion of opencast coal mining from 2013 to 2019 decreased from 49 to 40% [8]. Most of the open-pit coal mines are increasingly exploited to the depths, so the danger level, the coefficient of rock removal and the transport speed increase. From 1995 to 2019, the coefficient of rock removal increased from 3.41 m^3/ton to 11 m^3/ton, the transport distance increased from 1.0 km to 4.2 km. In underground coal mining, due to the complex geological and structural conditions of the coal seam, the high risk of gas explosion, the transportation supply is also increasing, causing the investment rate from 2000 to 2019 to increase from 50 to 180 USD/ton. Faced with these facts, since 2015, Vietnam has transformed from a coal exporter to a net coal importer, and the rate of trade deficit is increasing sharply [9].

In recent years, Vinacomin has stepped up investment in the development of software related to management operations in the group as well as in its member companies, with the goal of standardizing the management computerization problem [10–12]. The basic factors that help increase labor productivity, reduce manual labor and production costs, and reduce the risk of pit accidents. However, in general, the management in the group is still highly manual, only focusing on basic operations, not meeting the actual needs. With the targeted deadline for digital transformation which is set to be year 2025 by Vinacomin, a lot of work is still needed to be done [13]. Therefore, continue to further improve the capacity, human resource management and development are identified as a key task determining success. In this paper, a software system proposed for the production data management in a coal mining company is introduced.

2 Software Design and Implementation

Each stage of the mining process may use the data presented in datasheets as well as other attached information. Each stage needs to be performed by individuals, groups, or units. In addition, the data in the previous stage also need to be transferred to the individual or group performing the following stage to be able to continue the work. Therefore, the need for information exchange within an enterprise is huge, as shown in Fig. 1, for example. However, at present, the companies do not have any tools to support the sharing and exchange of data and coordinated group work. Meanwhile, the information being updated into the datasheets is happening very often at different stages, making the information exchange flow and the reception of updated information even more complicated.

The solution proposed in this study aims to solve the aforementioned problem by implementing a data cloud in order to gather all the production data, so that

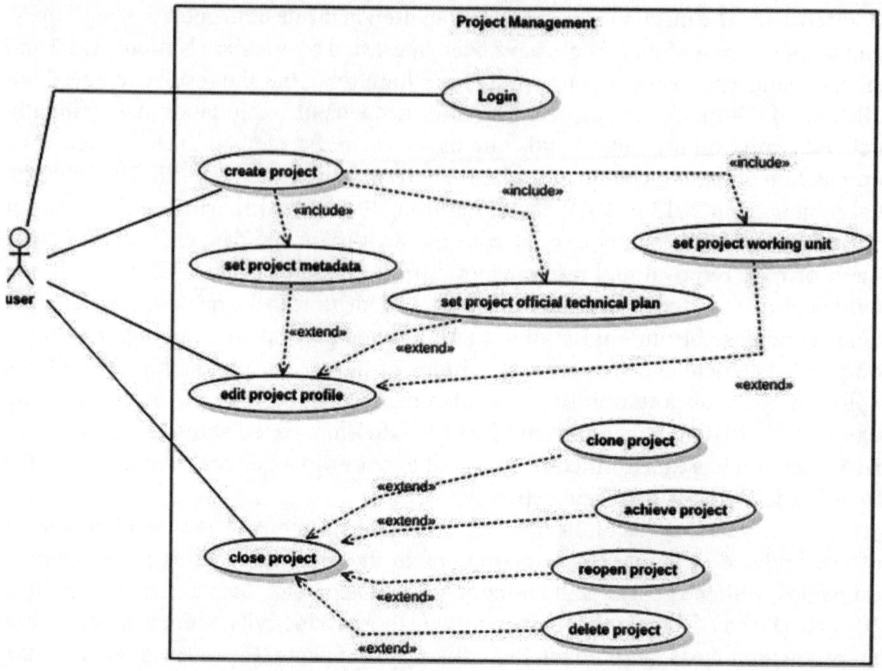

Fig. 1 Project management use case

individuals and groups are able to synchronize their data in real-time. The following functional requirements (FRs) are identified in Table 1.

Table 1 Functional requirements analysis

#FR	Description
FR1	*Users' role & permission management*
FR1.1	Manage user's role based on division unit
FR1.2	Manage user's data access permission based on project group
FR2	*Project management*
FR2.1	Manage project profile (create, update, close, achieve)
FR2.2	Manage project data phases (inspection, planning, design, blasting, exploitation)
FR2.3	Allow user of the same permission group to manage, edit project data collaboratively
FR2.4	Allow versioning and conflict resolving for shared documents
FR3	*Report generation*
FR3.1	Allow generating company working progress report
FR3.2	Allow generating customized project report
FR3.3	Allow generating official project report

Given those requirements, we will discuss some important use cases, database design, sequence diagram of the system. Since the system is built using modular approaches with event-driven NodeJS framework, there will be no class analysis. The diagrams follow IBM UML standards [14].

2.1 Use Case Analysis

There are 3 major use cases of the system: (1) The user initializes and manages different stages of a project; (2) The user edits shared mining data collaboratively via excel add-ins or web application; and (3) The user generate report from mining data. The first use case is presented in Fig. 1. In this use case, user can perform project management actions such as: create, update and close project. The second use case shows user can edit shared mining data from center server (see Fig. 2). Multiple users can perform edit and update actions at the same time. There will be a synchronization process running in order to ensure data synchronization between users. In addition, user can perform conflict resolve in case a conflict is raised. The last major use case is report generation where the systems automatically generate company mining statistic report or user's defined customized report (see Fig. 3). User can also pick from a list of pre-defined official reports for mining data within the company.

2.2 Database Design and Permission Management

The database schema, as shown in Fig. 4, includes two main groups, i.e., user information and production data. The software provides a permission control mechanism to manage user's administrative right and data access possibility based on the group that the user takes part in [15]. The list of account information fields is shown in Table 2.

For administrative right, depending on user account group, user can perform several actions on account management such as: add, remove, edit account's information. The root account has full administrative right on all user accounts. For data access permission, user can only access data within his/her division data views. In any project, one division should be involved in different phases of mining production. The data access permission should also be aligned with these project phases.

In addition to user management, the software allows administrators to manage division and its involved projects. Two types of division are defined: administrative division and on-site working division. The administrative divisions are responsible for setting up technical plan, quality control and accounting. On-site working divisions, in contrast, are unit of workers who directly works at mining tunnel. This, in turn, allows managing data access permission of users. User access permission to certain data view of a project is inherited from his/her division. Since a user can

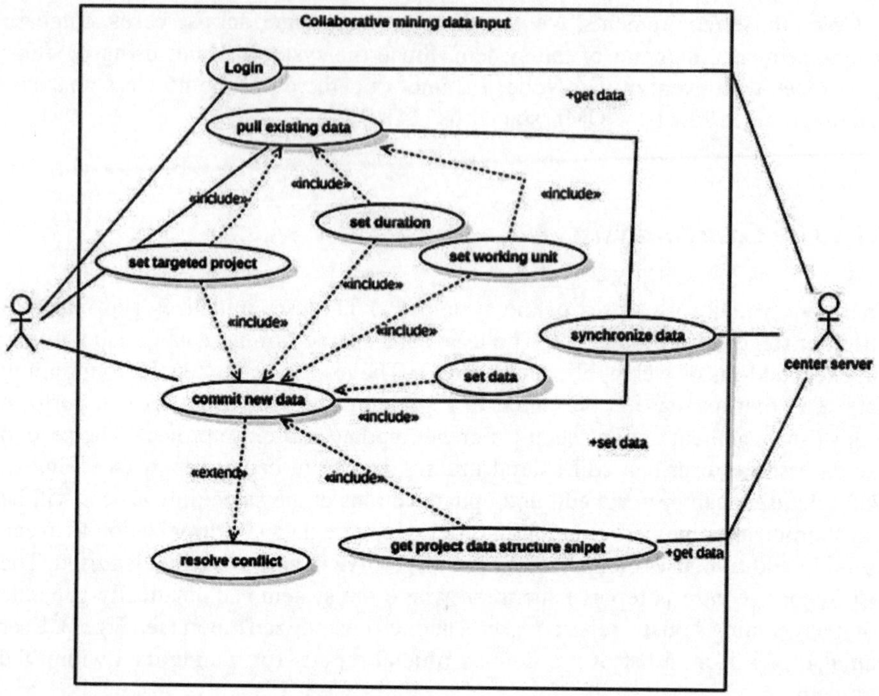

Fig. 2 Collaborative mining data input use case

be in multiple divisions at the same time, the user's data access permission is the cumulative sum of his/her inherited data access permission.

2.3 Mining Data Management

To track mining production data, the platform allows multiple users from different divisions to work simultaneously with mining project data. This includes creation of a new project, setting up mining annual plan for the entire company and for the project specifically, update daily mining production data, on-site quality control and checking data, etc. Here, we introduce the concept of global plan which is a collection of mining plans that should be met. For each plan, a structure consist of multiple production targets are formed.

Mining project is another concept introduced to manage the mining production data. Each project starts with a plan to mine at several locations for a selection of divisions. Data of one project includes: a global mining plan, daily geodetic data, technical assurance, on-site division mining plan and accreditation data. Data from different users and groups once committed will be synchronized to a centralized data cloud, so that other users with right permissions can update to their datasheets. This

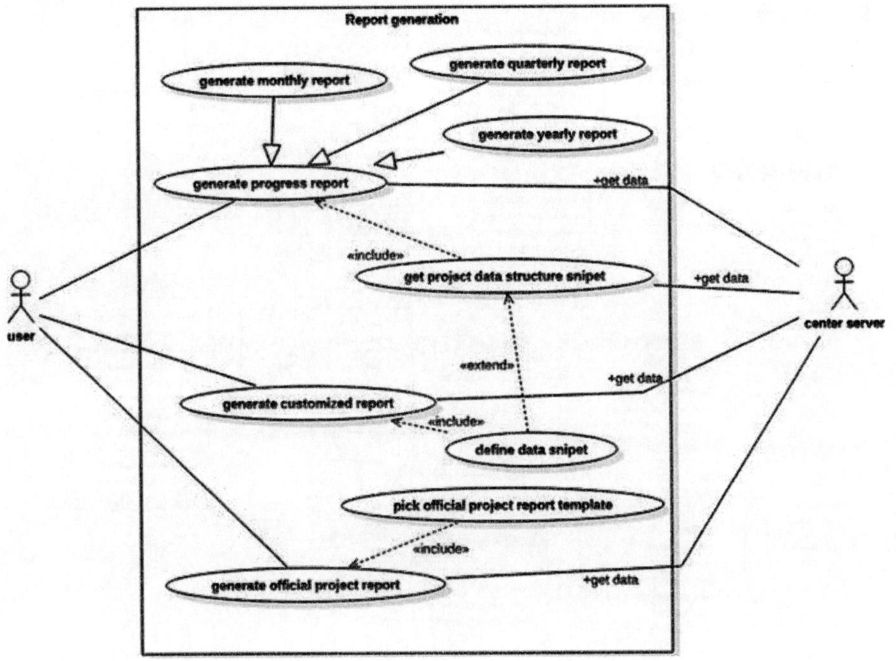

Fig. 3 Report generation use case

helps users to work seamlessly on the same data without need to exchange data files through emails or sharing networks, which causes bottleneck in the workflow, and is also exposed to many security threats. The process of committing new data to the center server is explained in the sequence diagram Fig. 5.

Mining production data then can be viewed with multiple dimensions, as detailed in Table 3. Any combination of these dimensions results in a different view of the centralized data. By defining multiple data dimensions, user can pick different view of the centralized data to have a good understanding about the current production progress. Furthermore, this allows the platform to perform automatic report generation as well as data analysis of the production flow in the future. In the future, a big data management in the mining industry should be considered [16].

2.4 *Conflict Resolution Strategy*

The goal of this platform is to allow multiple users to work with the centralized database at the same time. Achieving this will speed up the data input process as well as enhance error detection chance. Data should be seamlessly synchronized between divisions hence allows a silent and efficient data acquisition process for the higher level of management.

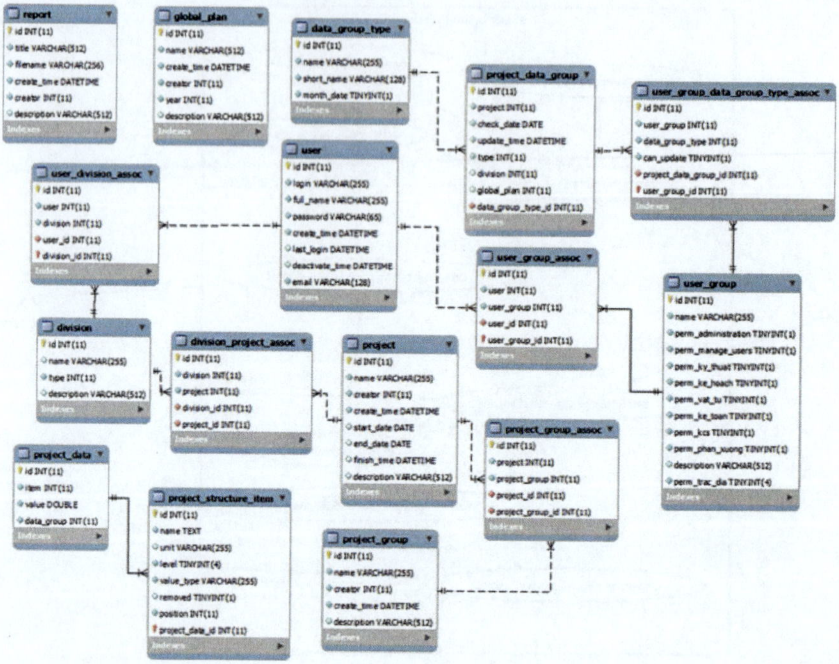

Fig. 4 Database schema

Table 2 User's profile information

Data	Type	Description
Username	Text	Username for login
Fullname	Text	User's full name
Email	Email	User's email address
Division	Enum	User's division unit
Account group	Enum	Define user's permission (data access and administrative right). A user can belong to multiple account groups
Account permission	Enum	Define the administrative right of the user based on user's account groups
Data permission	Enum	Define user data permission (read/write of different data views)
Date of creation	Date	Time of creation
Last access	Date	Time of the last logged in session

In this platform, user can perform data update and synchronization using two different methods: via web interface or Excel add-in. Both methods allow user to view current production data and input in new data given sufficient permission. The data will then be recorded along with user account information to form a new version of the centralized database.

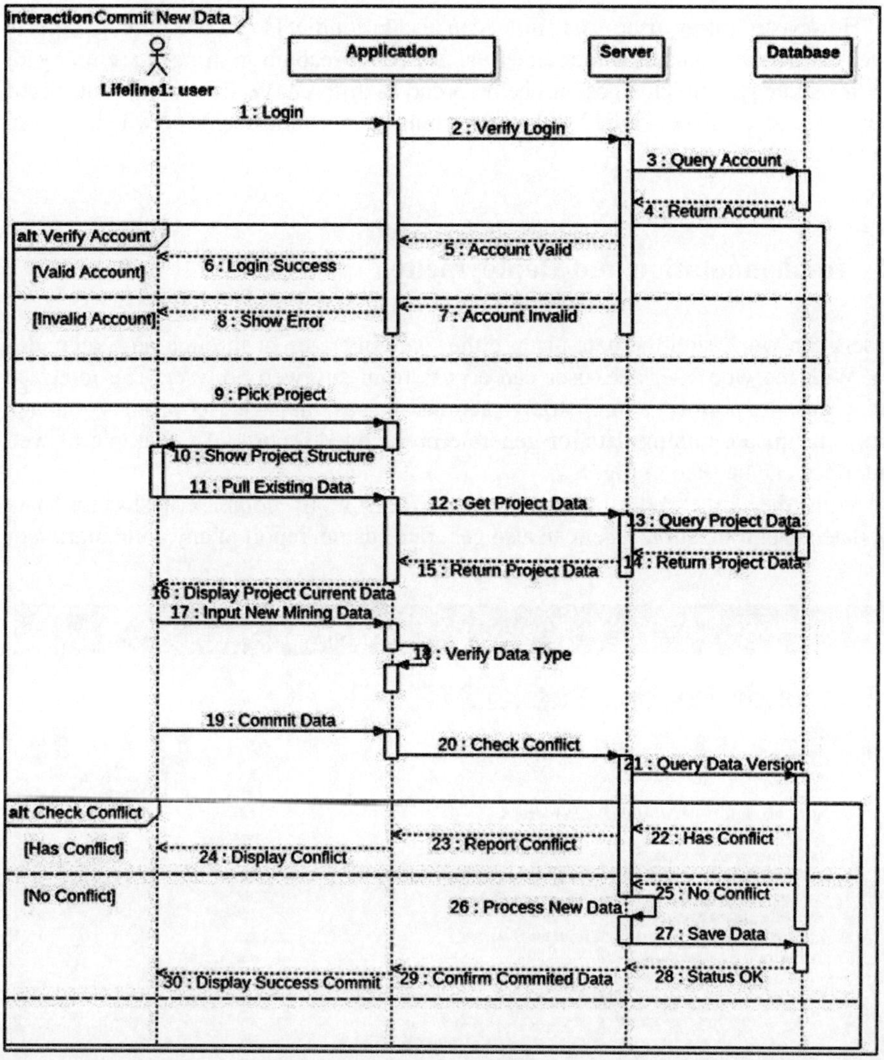

Fig. 5 Data commit and synchronization sequence diagram

Table 3 Mining project data dimensions

Dimension	Type	Description
Project	ID	The project ID number
Division	Enum	Involved divisions in the project
Date	Date	Time dimension for the project
Location	Enum	A project can have different mining locations
Target	Number	Different mining targets such as: total extracted coal, mining distance

However, multiple users' data input can lead to conflict [17] if one mining target is set by two different users at the same time. A passive resolution strategy is employed. In this strategy, both changes will be recorded as different version in the centralized database. User will be notified with new coming changes and decide on which version to be kept.

3 Implementation and Deployment

Users can work with the data using either web interface or through an Excel add-in. With the web interface, user can access from any web browser. The interface allows user to perform administrative actions such as: manage user profiles, manage projects, update mining data, or generate pre-defined reports. An example of web interface can be seen in Fig. 6.

With the Excel add-in, user can perform more complex tasks for data update/synchronization. User can also generate custom report of any combination of

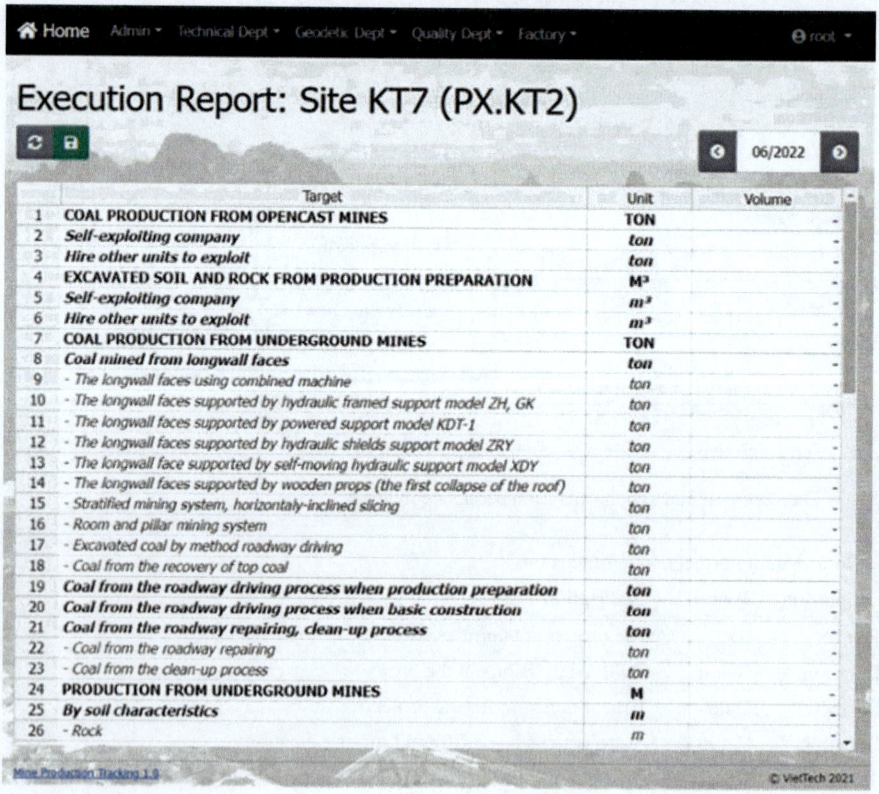

Fig. 6 Platform web interface

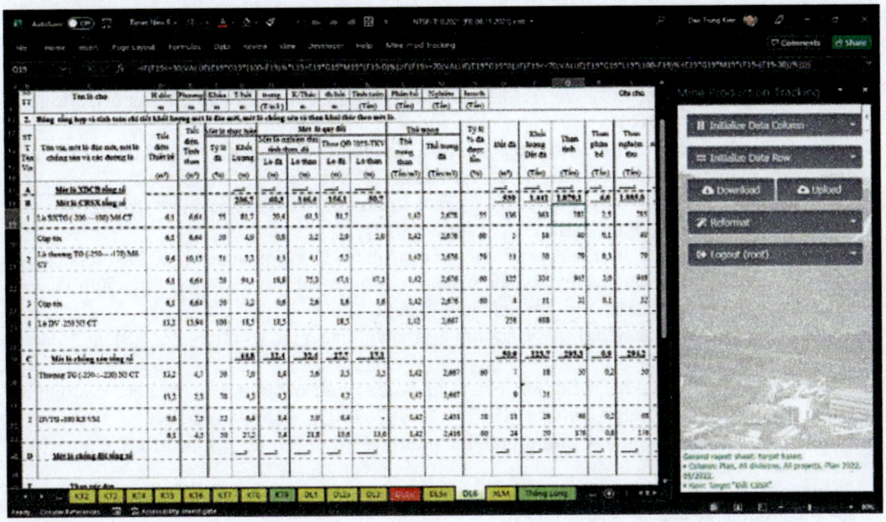

Fig. 7 Excel add-in interface

data dimensions. An example of the Excel add-in working interface is demonstrated in Fig. 7. The centralized server acts as a web server that provides centralized data and services such as account login, report generation and data backup.

4 Conclusion

In this paper, a design and implementation of a mine production tracking platform is presented. Using a combination of an online web app and an Excel add-in, the platform allows multiple users to collaboratively working with a centralized data of mine production. It enhances not only working efficiency but also allows real-time production data report and conflict resolution. Initial application in a coal company shows a sight of improvement especially in eliminating a vast majority of human error within the process. An initial evaluation of the system carried out by Viettech Company Ltd. shows an average 10% of human errors reduction across all departments in the company. However, deploying such platform in real-case scenario demands a throughout transformation in workflow. Hence, having the platform, the focus is now on investing on human resources to adapt with the new technology.

Acknowledgements This work was supported by the Vietnam Ministry of Science and Technology (MOST) by the project entitled "Digital transformation solution based on geospatial data to improve operational efficiency and management of underground coal mining in Vietnam".

References

1. Demirkan, H., Spohrer, J.C., Welser, J.J.: Digital innovation and strategic transformation. IT Prof. **18**(6), 14–18 (2016)
2. Günther, W.A., Mehrizi, M.H.R., Huysman, M., Feldberg, F.: Debating big data: a literature review on realizing value from big data. J. Strateg. Inf. Syst. **26**(3), 191–209 (2017)
3. Bharadwaj, A., El Sawy, O.A., Pavlou, P.A., Venkatraman, N.: Digital business strategy: toward a next generation of insights. MIS Q. 471–482 (2013)
4. Kahre, C., Hoffmann, D., Ahlemann, F.: Beyond business-IT alignment-digital business strategies as a paradigmatic shift: a review and research agenda. In: Proceedings of the 50th Hawaii International Conference on System Sciences (Jan 2017)
5. Kane, G.C.: The American red cross: adding digital volunteers to Its ranks. MIT Sloan Manag. Rev. **55**(4), 1 (2014)
6. De Donno, M., Tange, K., Dragoni, N.: Foundations and evolution of modern computing paradigms: cloud, IoT, edge, and fog. IEEE Access **7**, 150936–150948 (2019)
7. Shroff, G.: Enterprise Cloud Computing: Technology, Architecture, Applications. Cambridge University Press (2010)
8. Chinh, N.T.: The current situation, supply—demand, coal imports: challenges and development policies. Vietnam Energy Mag. (2020)
9. Vietnam National Coal and Mineral Industries Group (Vinacomin): report of the action program implementation conference on focusing on promoting the application of informatization and automation in production and management to improve production and business efficiency in the period 2017–2020 (2017)
10. East Sea Energy Environment.: https://esec.vn/en/du-an-ecostruxure-power-scada-cho-khach-hang-cao-son-coal-vietnam-vinacomin. (May 2022)
11. Vietnam National Coal and Mineral Industries Group (Vinacomin).: http://www.vinacomin.vn/dang-uy/tuyen-than-cua-ong-buoc-dot-pha-cong-nghe-tu-dong-hoa-202110051816520468.htm
12. Vietnam National Coal and Mineral Industries Group (Vinacomin).: http://www.vinacomin.vn/tin-tuc-vinacomin/nganh-than-chuyen-doi-theo-xu-the-cach-mang-40-202007171532165145.htm
13. Quang Ninh newspaper.: https://baoquangninh.com.vn/chuyen-doi-so-trong-cac-doanh-nghiep-nganh-than-3182638.html
14. IBM Developer.: https://developer.ibm.com/articles/an-introduction-to-uml/
15. Triartono, Z., Negara, R.M., Sussi.: Implementation of role-based access control on OAuth 2.0 as authentication and authorization system. In: 6th International Conference on Electrical Engineering, Computer Science and Informatics (EECSI), pp. 259–263 (2019)
16. Qi, C.-C.: Big data management in the mining industry. Int. J. Miner. Metall. Mater. **27**, 131–139 (2020)
17. Ji, T., Chen, L., Yi, X., Mao, X.: Understanding merge conflicts and resolutions in Git rebases. In: IEEE 31st International Symposium on Software Reliability Engineering (ISSRE) (2020)

Shear Strength of Poorly Graded Granular Material in Multi-Stage Direct Shear Test

Sung-Sik Park, Tan-No Nguyen, Dong-Kiem-Lam Tran, Keum-Bee Hwang, and Hee-Young Sung

Abstract Shear strength of soils plays an important role in geotechnical stability design. A growing number of studies have been carried out on the assessment of the shear strength of granular material using direct shear tests. However, for the same soil sample, different testing procedures may create different shear strengths causing difficulties in selecting the suitable design parameters. In this study, single-stage and multi-stage reversal direct shear tests on granular material were investigated under the drained condition as the dried sample state. Tested samples were glass beads. The drained shear strength of granular material was investigated under various normal stresses of 50, 100, and 150 kPa and three shearing rates of 0.01, 0.1, and 1.0 mm/min corresponding to slow, intermediate, and rapid speeds. The study results indicated that the shear stress of samples obtained from the different testing approaches tended to increase as the reversals increased in both the single-stage method and the multi-stage method. The strength envelopes revealed a good agreement between the fourth reversal in the single-stage test and the multi-stage direct shear test. In addition, the friction angle of glass beads was independent of shear rates under drained conditions.

Keywords Multi-stage method · Direct shear test · Granular material · Shear rate

S.-S. Park · T.-N. Nguyen (✉) · D.-K.-L. Tran · K.-B. Hwang · H.-Y. Sung
Department of Civil Engineering, Kyungpook National University, 80, Daehak-Ro, Buk-Gu, Daegu 41566, Republic of Korea
e-mail: tannonguyen.ce@gmail.com

S.-S. Park
e-mail: sungpark@knu.ac.kr

D.-K.-L. Tran
e-mail: lam.trandongkiem@uah.edu.vn

L. Q. Nguyen et al. (eds.), *Advances in Geospatial Technology in Mining and Earth Sciences*, Environmental Science and Engineering,
https://doi.org/10.1007/978-3-031-20463-0_20

1 Introduction

Granular materials are increasingly used as fill materials for embankment dams, highway embankments or earth-retaining structures due to the high strength, the rapidly drained water, or the compaction characteristic [1, 2]. The shear strength parameters are critical factors for fill or base materials in order to ensure the stability of geotechnical structures. Hence, it is necessary to determine the shear strength parameter of granular material. The shear strength of soils is commonly determined by experimental tests using the direct shear device [3–5]. Besides, using direct shear apparatus is less cost and less complicated for the sample preparation and operation. There are a range of factors that impact on the strength of soils such as grain size, particle shape, relative density, uniformity, stress level, specimen size, laboratory testing method, or shear rate [3, 4, 6–9]. Among these factors, testing procedures, namely single-stage reversal and multi-stage reversal direct shear methods, and shear rates are taken into account for estimating the shear strength of granular material in this study.

The reversal direct shear tests have been commonly used to evaluate the shear strength of soils when the sample preparations are less time consuming and less expensive in comparison with the non-reversal direct shear tests [10, 11]. As the different testing procedures were applied, the different shear strengths of soils were found. Therefore, the single-stage and multi-stage reversal direct shear tests were carried out to determine the shear strength of granular material. Generally, the single-stage reversal tests are conducted on one soil sample at a constant normal stress for several reversals, while the multi-stage reversal tests are carried out on one soil sample at different normal stresses for different reversals [12].

Regarding shear rate effects, based on the previous studies [3, 13–15], some revealed that the shear strength went up or fluctuated by increasing strain rates and normal stresses, while others concluded that the shear strength was independent of the shear speed. Thus, for a better understanding of the effects of the shear rates on the shear strength, this study employed three deformation rates of 0.01, 0.1, and 1.0 mm/min that were representative of low speed, medium speed, and high speed.

The contributions of this study are to: (1) evaluate the impacts of single-stage and multi-stage reversal direct shear approaches on the shear strength of granular material; (2) determine the shear strength envelopes under various testing programs; (3) investigate the influences of shearing rates on the shear strength parameters.

2 Materials and Methodology

2.1 Materials

To evaluate the shear behavior of granular material, glass beads are commonly used as references for soil tests [16, 17]. Glass bead used to prepare samples consists

Table 1 Physical properties of glass bead

D_{50}, mm	$\gamma_{d,min}$ (kN/m^3)	$\gamma_{d,max}$ (kN/m^3)	C_u	C_c
0.64	1.49	1.61	1.45	0.96

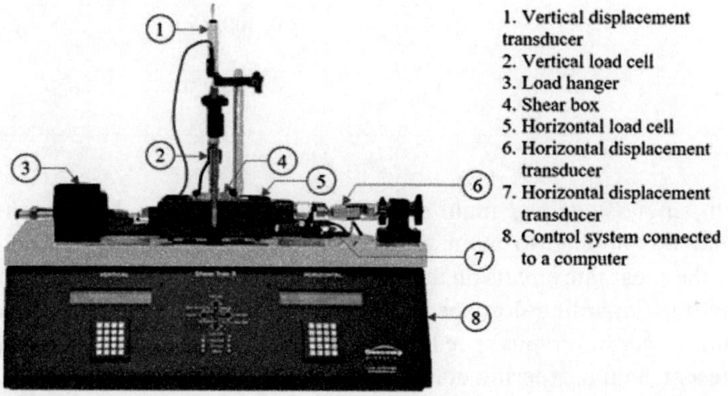

Fig. 1 Direct shear test made by Geocomp corporation (Massachusetts, USA)

of round shapes and smooth surfaces. The physical properties of glass bead were summarized in Table 1. It is classified as a poorly graded granular material. All samples were tested under dry condition.

2.2 Direct Shear Apparatus

The direct shear approach is one of the most common tests in geotechnical engineering practice to measure the strength properties of soils. For generating the reversal and re-shearing processes of the samples, direct shear apparatus manufactured by Geocomp (Acton, Massachusetts) is adopted in this study. The testing machine is shown in Fig. 1. The direct shear box is 63.5 mm in diameter and 25.0 mm in height. The shearing test is operated by accuracy and precision of a micro-stepper motor and gear box arrangement.

2.3 Testing Procedure

Testing procedure consists of three steps: (1) sample preparation using dry deposition method for a target density of 15.36 kN/m^3, (2) consolidation under nominal normal stresses of 50, 100, and 150 kPa, (3) shearing and re-shearing stage for the single-stage and multi-stage tests.

Table 2 Testing program

Testing method	No. of reversal	Shear rate (mm/min)	Nominal normal stress (kPa)	Shearing displacement (mm)
Single-stage	0	0.01; 0.1; 1.0	50, 100, 150	6.5
Single-stage reversal	4	0.1	50, 100, 150	6.5
Multi-stage reversal	4	0.1	50, 100, 150	6.5

For both single-stage and multi-stage reversal techniques, the shearing rate of 0.1 mm/min was applied to maintain a drained testing condition. In addition, to investigate the shear rate effects on the shear strength of granular material, three rates of 0.01 mm/min (regarding slow speed), 0.1 mm/min (regarding intermediate speed), 1.0 mm/min (regarding rapid speed) was conducted on the single-stage approach. Table 2 presents initial experimental test parameters for direct shear test.

For the single-stage reversal direct shear method, the procedure is explained in detail by the following steps:

(1) Applying a nominal normal stress, consolidating a sample, and then shearing a sample at a constant shearing rate to 6.5 mm of shearing displacement;
(2) Removing the nominal applied normal stress, re-winding and re-applying the same normal stress as step 1, and then re-shearing at the same shearing rate and shearing displacement as step 1;
(3) Repeating the step 2.

It is noteworthy that samples tested in the single-stage method without reversal processes (see Table 2) are solely applied to the first step as the single-stage reversal method.

For the multi-stage reversal direct shear test, the procedure is described as below:

(1) Applying the first normal stress, consolidating a sample and then shearing a sample at a constant shearing rate to 6.5 mm of shearing displacement;
(2) Removing the first applied normal stress, re-winding and re-applying the second nominal normal stress, and then re-shearing at the same shearing rate and shearing displacement as step 1;
(3) Repeating the step 2 until the target normal stress is applied.

3 Results and Discussion

The experimental test results are demonstrated in the following sections that are highlighted by the findings of testing procedure effects, shear strength envelopes, and shearing rate effects. Due to a gradual change of shearing area of sample, corrected test values of nominal normal and shear stresses were calculated in this study.

3.1 *Effect of Shear Reversals on Shear Strength*

The effects of single-stage approach and multi-stage approach using reversal direct shear tests were demonstrated in Fig. 2 and Fig. 3, respectively. As shown in Figs. 2a and 3a, the test results observed from both methods were no obvious peak shear stress until shearing stopped at 6.5 mm of shearing displacement for each step. Therefore, the peak shear strength at 6.5 mm of shearing displacement at each step was served as the shear stress at the failure.

Fig. 2 Single-stage reversal direct shear test results

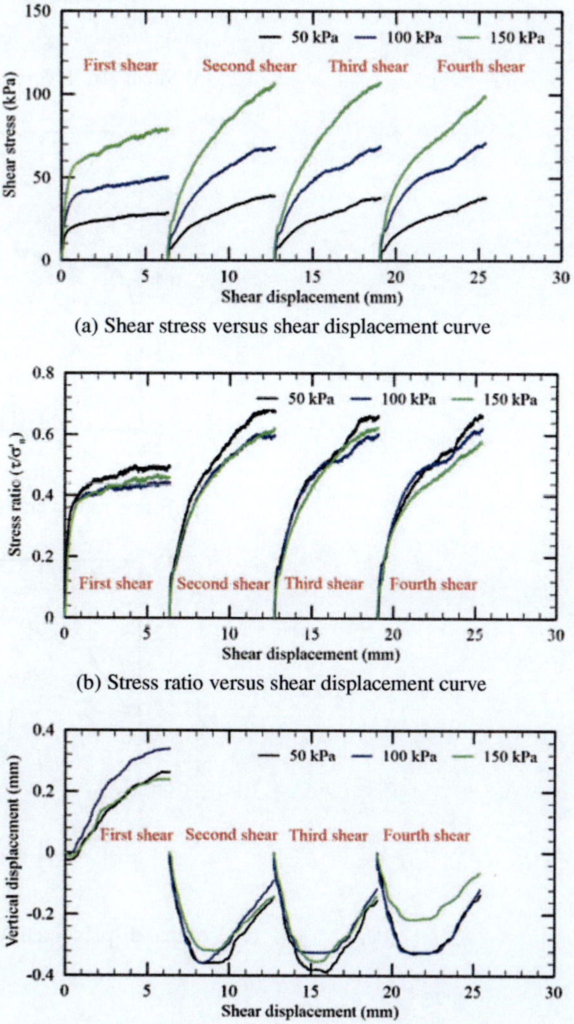

(a) Shear stress versus shear displacement curve

(b) Stress ratio versus shear displacement curve

(c) Vertical displacement versus shear displacement curve

Fig. 3 Multi-stage reversal direct shear test results

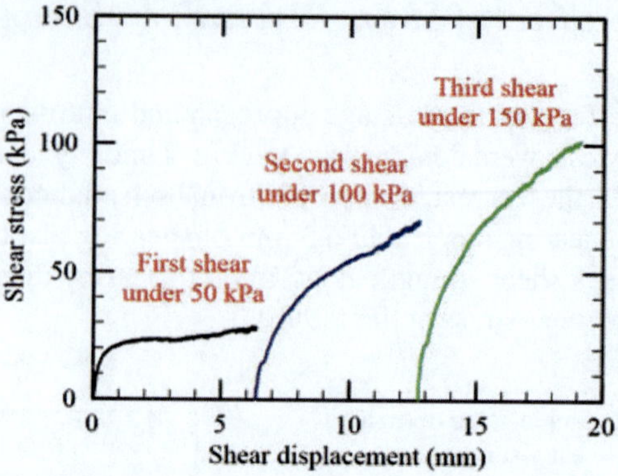

(a) Shear stress versus shear displacement curve

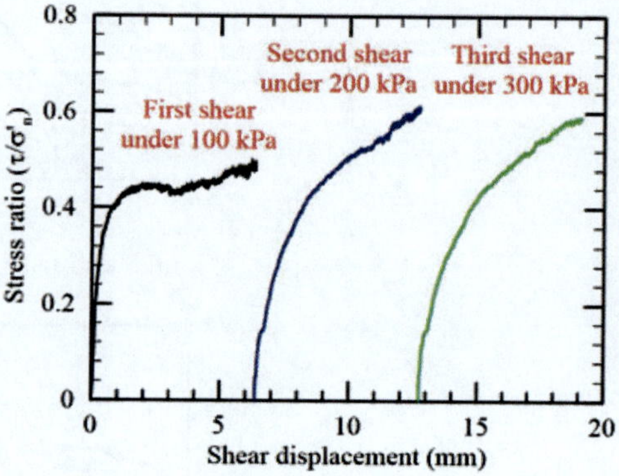

(b) Stress ratio versus shear displacement curve

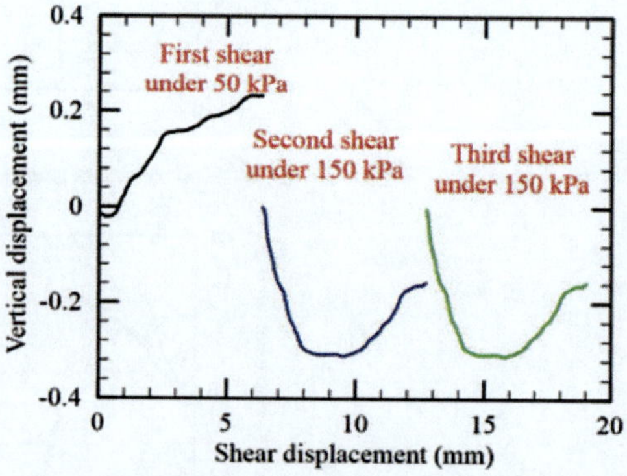

(c) Vertical displacement versus shear displacement curve

For the single-stage reversal procedure, the shear strength increased significantly from the first shearing to the second shearing and subsequently remained stable for the third shearing stage as plotted in Fig. 3a. However, a slight decrease in shear stress was obtained by the fourth shearing stage. For the relationship between the stress ratio and shear displacement presented in Fig. 3b, the stress ratios, the corrected shear stress divided by the corrected normal stress, were about 0.42 and 0.62 for the first shearing and the following shear reversals, respectively. A minor contraction and a major dilation were found for the first shearing from all the nominal normal stress of 50, 100, and 150 kPa, whereas the samples were firstly observed with the contraction and followed by the dilation for the latter reversals (see in Fig. 2c).

For multi-stage reversal procedure, while the shear strengths increased from about 30 kPa to over 100 kPa under increasing the nominal normal stresses from 50 to 150 kPa, respectively (see in Fig. 3a), the stress ratios fluctuated with increasing the nominal normal stresses (see in Fig. 3b). For the first shearing by applying 50 kPa nominal normal stress, a small contraction was initially found and rapidly accompanied by rising dilation (see in Fig. 3c). For the second and third shearing under 100 kPa, and 150 kPa, respectively, the contraction and dilation behaviors of samples were the same trend as that of samples observed from the single-stage reversal procedure.

Based on the above findings, it is highlighted that the shear strength results of samples under the nominal normal stress of 100 kPa at the fourth reversal of single-stage method were similar to that of the second reversal of multi-stage method. The findings were also found the same trend when applying the nominal normal stress of 150 kPa under these procedures.

3.2 Shear Strength Envelopes

Figure 4 depicts the shear strength envelopes of samples recorded from the single-stage reversal procedure (see in Fig. 4a) and the multi-stage reversal procedure (see in Fig. 4b). The findings indicated that all envelopes were a linearly approximate straight line.

For the single-stage reversal method, the lowest friction angle of 24.7° was obtained from the first shearing under different normal stresses, while the highest friction angle of 31.9° was observed from the third shearing. For the second shearing, the friction angle was similar to that of the third shearing. A slight decrease of 1° in the friction angle was found from the fourth shearing compared to the third shearing.

For the multi-stage reversal method, the friction angle of glass beads was 30.5° that was quite fitted with the friction angle obtained from the fourth shearing of the single-stage reversal method. It means that the various shear strength parameters should be taken into account for the stability of structures. Although the various angles of friction were obtained from the sing-stage and multi-stage approaches, these values were in a range of previous study [18].

Fig. 4 Shear strength envelopes under different testing procedures

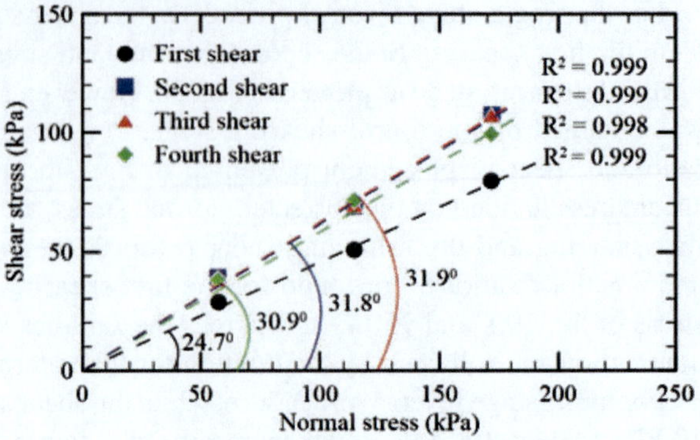

(a) Single-stage reversal method

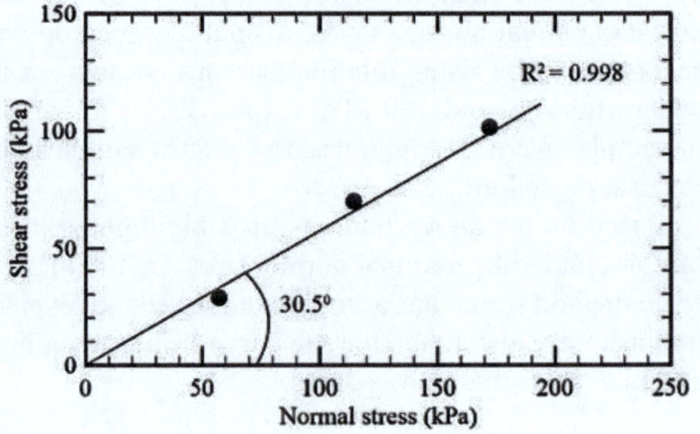

(b) Multi-stage reversal method

3.3 *Effect of Shearing Rates on Shear Strength*

Figure 5 illustrates the shear stress versus the normal stress curve. The influence of shearing rates on the shear strength of granular material were investigated. The testing results on glass beads indicated no variation in the friction angle for the speeds between 0.01 mm/min and 1.0 mm/min under the different nominal normal stresses. The friction angle of samples was 26° with the straight best-fit line for the shear strength envelope. In other words, the shear strength of granular materials was uninfluenced by the rate. The same result was established from other studies [14, 15].

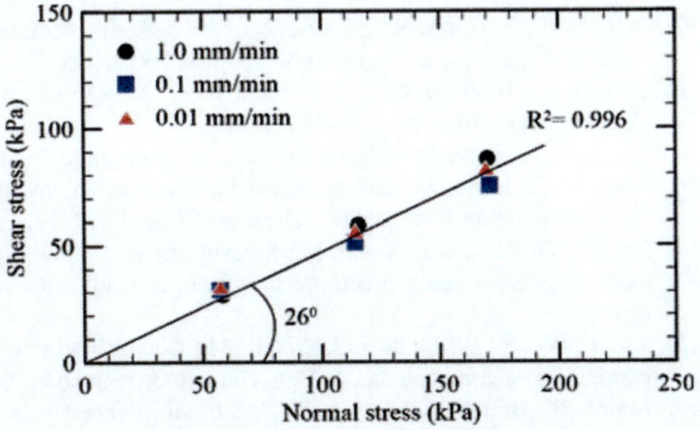

Fig. 5 Shear rate effects on shear strength envelope

4 Conclusion

The goal of this paper was to study the shear strength of granular material. The experimental tests were prepared under the drained condition as the dried samples prepared. The influences of different testing approaches and shear rates on the shear strength of granular material were discussed in this study. It is worth noting that the single-stage or multi-stage reversal procedures showed a range of shear strength parameters. In other words, selecting the appropriate testing methods for the different purposes of design parameters was crucial in geotechnical engineering. Additionally, the experimental test results presented that no effects of shearing speeds of 0.01, 0.1, and 1.0 mm/min and the nominal normal stresses of 50, 100, and 150 kPa on the friction angle were achieved.

Acknowledgements This study was supported by the National Research Foundation of Korea (NRF) grant funded by the Korea Government (MSIT) (Nos. NRF-2021R1I1A3059731 and NRF-2018R1A5A1025137).

References

1. Alias, R., Kasa, A., Taha, M.R.: Particle size effect on shear strength of Granular materials in direct shear test. Int. J. Civil, Environ., Struct., Constr. Architectural Eng. **8**, 1144–1147 (2014)
2. Lee, D.-S., Kim, K.-Y., Oh, G.-D., Jeong, S.-S.: Shear characteristics of coarse aggregates sourced from quarries. Int. J. Rock Mech. Min. Sci. **46**, 210–218 (2009)
3. Moon, H.D., Kim, J.S., Woo, S.-W., Tran, D.-K.-L., Park, S.-S.: Effect of shear rate on strength of non-cemented and cemented sand in laboratory testing. J. Korean Geotech. Soc. **37**, 23–36 (2021)
4. Kouakou, N.M., Cuisinier, O., Masrouri, F.: Estimation of the shear strength of coarse-grained soils with fine particles. Transp. Geotech. **25**, 100407 (2020)
5. Xu, Y.: Shear strength of granular materials based on fractal fragmentation of particles. Powder Technol. **333**, 1–8 (2018)
6. Asadzadeh, M., Soroush, A.: Direct shear testing on a rockfill material. Arab. J. Sci. Eng. **34**, 379 (2009)

7. Park, S.-S., Jeong, S.W.: Effect of specimen size on undrained and drained shear strength of sand. Mar. Georesour. Geotechnol. **33**, 361–366 (2015)
8. Simoni, A., Houlsby, G.T.: The direct shear strength and dilatancy of sand–gravel mixtures. Geotech. Geol. Eng. **24**, 523–549 (2006)
9. Jewell, R.A.: Direct shear tests on sand. Geotechnique **39**, 309–322 (1989)
10. Nam, S., Gutierrez, M., Diplas, P., Petrie, J.: Determination of the shear strength of unsaturated soils using the multistage direct shear test. Eng. Geol. **122**, 272–280 (2011)
11. Xu, Y., Williams, D.J., Serati, M.: Measurement of shear strength and interface parameters by multi-stage large-scale direct/interface shear and pull-out tests. Meas. Sci. Technol. **29**, 085601 (2018)
12. Xu, Y., Wu, S., Williams, D.J., Serati, M.: Determination of peak and ultimate shear strength parameters of compacted clay. Eng. Geol. **243**, 160–167 (2018)
13. Mamo, B., Banoth, K., Dey, A.: Effect of strain rate on shear strength parameter of sand. In: Proceedings of the 50th Indian Geotechnical Conference (IGC-2015), Pune, India (2015)
14. Pincus, H., Yamamuro, J., Lade, P.: Effects of strain rate on instability of Granular soils. Geotech. Test. J. **16**, 304–313 (1993)
15. Feng, D., Zhang, J., Deng, L.: Three-dimensional monotonic and cyclic behavior of a gravel–steel interface from large-scale simple-shear tests. Can. Geotech. J. **55**, 1657–1667 (2018)
16. Li, Y.R., Aydin, A.: Behavior of rounded granular materials in direct shear: mechanisms and quantification of fluctuations. Eng. Geol. **115**, 96–104 (2010)
17. Li, Y., Aydin, A., Xu, Q., Chen, J.: Constitutive behavior of binary mixtures of kaolin and glass beads in direct shear. KSCE J. Civ. Eng. **16**, 1152–1159 (2012)
18. Afzali-Nejad, A., Lashkari, A., Shourijeh, P.T.: Influence of particle shape on the shear strength and dilation of sand-woven geotextile interfaces. Geotext. Geomembr. **45**, 54–66 (2017)

High-Resolution Seismic Reflection Survey of Young Sediment at Can Gio Coast, Ho Chi Minh City, Vietnam

Thuan Van Nguyen, Cuong Van Anh Le, and Man Ba Duong

Abstract Young river sediment layers and seabed can be investigated by application of high-resolution seismic method supplied with a sub-bottom profiler. Contrast between different acoustic impedance leads to reflection of seismic waveform propagation that can image the interest geological targets. Soai Rap River is one of the two most important water gates in Ho Chi Minh city, Vietnam to the outside world. Studying its seabed and young sedimentology plays a vital in developing transportation and sea-ward economic Vietnamese policy. We have collected, processed, and interpreted ten of 2D high-resolution seismic profiles and two prior drill holes in the survey area, Can Gio district, Ho Chi Minh city. Its 3D seabed and 3D Holocene shallow sediment representations are interpolated different 2D boundaries which can be achieved by analysis of conventionally processed seismic data and seismic textural attributes (i.e., energy and entropy). Four resulted layers consists of water and young Holocene sediments and one trough channel in the middle of the river can be interpreted.

Keywords High-resolution seismic · Holocene · Bathymetry

1 Introduction

High frequency reflection seismic data has been used for researching formations of young river sediment, lowland, and seabed visualization [1–3]. Seismic waves

T. Van Nguyen · C. Van Anh Le (✉)
University of Science, Ho Chi Minh City, Vietnam
e-mail: lvacuong@hcmus.edu.vn

T. Van Nguyen
e-mail: nvthuan@hcmus.edu.vn

Vietnam National University Ho Chi Minh City, Ho Chi Minh City, Vietnam

M. B. Duong
Ho Chi Minh City Institute of Resources Geography, VAST, Ho Chi Minh City, Vietnam
e-mail: dbman@hcmig.vast.vn

L. Q. Nguyen et al. (eds.), *Advances in Geospatial Technology in Mining and Earth Sciences*, Environmental Science and Engineering,
https://doi.org/10.1007/978-3-031-20463-0_21

emitted from a sub-bottom profiler can propagate through its surrounding environment and reflect to its receiver after hitting seismic boundaries of different acoustic impedance zones [4]. With its smaller energy rather than other explosion traditional sources (i.e., air guns), the interest depth of this high-resolution seismic method can only go to hundreds of meters below the water surface. Data processing/analyzing stages can convert the raw measured data into interpretable data [5].

Our work focusses on collecting, processing, and interpreting ten 2D high-resolution seismic profiles to image seabed and young sediment at the fork area of Soai Rap River, Can Gio coast, Ho Chi Minh City, Vietnam. We have applied 3D visualization of their processed data in supporting analyzing stages. Interpreted seismic results are supported with information of two nearby drill holes. The drill holes' locations are in Thanh An island, Can Gio District [4].

2 Study Area

The acquisition area is the fork of Soai Rap River, Can Gio district, Ho Chi Minh city, Vietnam shown in Fig. 1. Known as a coastal area, Can Gio district plays a vital role in the economic development of the city when it has many important rivers connecting to the outside world. Its Soai Rap river is the second maritime route (the other is Long Tau river) for domestic and international large ships to transport imported and imported products to seaports in Ho Chi Minh city [6].

Recognized by UNESCO in 2000 as the first Mangrove Biosphere Reserve in Vietnam for having the highest diversity of mangrove living species, Can Gio also plays as "green lungs" of the city. Can Gio formed around 7000–6000 cal BP [8, 9] thanks to the Saigon-Dong Nai river basin. The factors such as semi-diurnal regime tidal from the estuary and average water discharge from Soai Rap and Long Tau rivers in the rainy season can form sand bars locating in front of their estuaries and Can Gio foreshore [4, 10].

Soai Rap river, an outlet of the Sai Gon–Dong Nai river, received a much smaller discharge of sediment compared to the Mekong River basin [11]. It is also close to the Mekong River delta. Therefore, knowledge of sediment of research area can be beneficial from geology of the Mekong delta basin and Ho Chi Minh city [4, 10, 12–17]. In Ho Chi Minh City, its northern part gets the higher elevation which are occupied by the pre–Holocene Cenozoic materials [18, 19]. Holocene sediment formations are seen in its southern areas as Nha Be and Can Gio districts [18, 19]. Investigations show that Holocene sediment patterns are linked to Sai Gon–Dong Nai river systems [18, 19]. Different acoustics impedances of types of young sediments can inspire application of high-resolution seismic method to image seabed and lowland topography [4, 5].

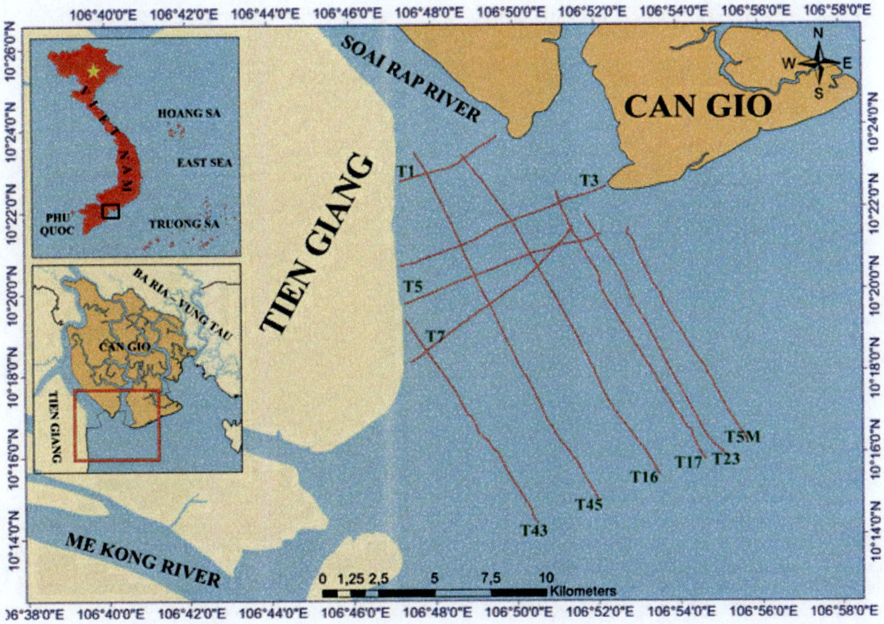

Fig. 1 Location of the survey area in Soai Rap river, Ho Chi Minh City, Vietnam [7]

3 Methodology

High-resolution seismic method is applied in searching subsurface image of maritime areas and young sediment structures thanks to wave reflections from boundaries of different acoustic impedance layers [2, 4, 20, 21].

In our research, the sub-bottom profiler, namely, SB-216S [22], typed with fixed transmitter–receiver offset is carried in a ship for data collection. Being laid 1.5 m underwater, it emits signals of the frequency range from 2 to 16 kHz in each 20 ms length which propagate through water and sediment environments [22]. During data collection stage, the ship can be detected by the GPS locator equipment while the professional software, Edge Tech Discover [23], helps to visualize the reflection seismic waves seen as electric signals in the computer monitor. In 2017, ten 2D profiles were collected in the research area, ranging from 5500 m to around 12000 m. In Fig. 2, one example shows that how the equipment is laid in the ship and seismic waves propagate through surrounding environment.

Data formats depend on data analysis/processing stages. For data collection, its seismic data format is binary with the extension *.jsf [25]. For processing/ analyzing the uninterpretable raw data to interpretable ones, we have used three professional software, Reflexw [26], OpendTect [27], and MATLAB codes [4].

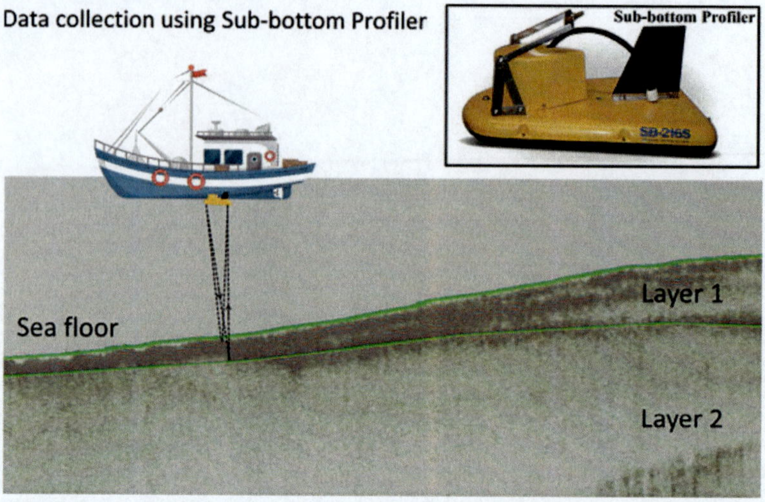

Fig. 2 Deployment of the sub-bottom profiler (SBS-216S) [4, 22, 24]. Boundary of two different acoustic impedance layers (i.e., water and sand) can be interpreted as green lines

3.1 Data Processing

Interpretable seismic data can be achieved via processing stages (Fig. 3). The first stage focuses on using processing tools in Reflexw for having interpretable seismic data in 2D view while the other stage utilizes the application of OpendTect for interpreting all the 2D data in 3D visualization.

Fig. 3 Processing steps for the raw seismic data (top image) in the profile T1 [4]. Suitable amplitude gains in Divergent Compensation applied to the Step 1 processed data (middle image) can make the Step 2 processed data (bottom image) visible in the larger travel time

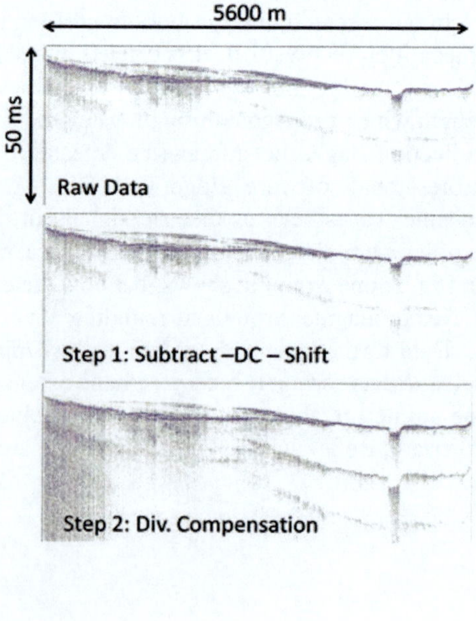

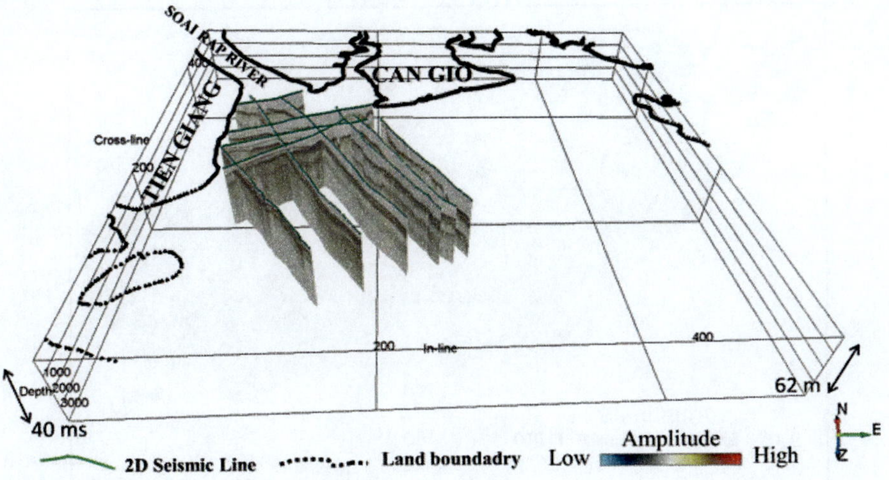

Fig. 4 Survey setting for the area. All the ten seismic profiles are put in 3D visualization. They will be input for further seismic attributes calculation

In the first stage, we applied two steps as Substract-DC-shift and Divergent (div.) Compensation Gain for all the ten seismic lines [4, 26]. A time constant shift is removed for each trace in Substract-DC-Shift filter. The Compensation Gain filter plays a vital role in reimbursing spherical energy loss while the waves propagate deeper [26].

Figure 3 illustrates one example of our first seismic processing stage for the profile T1. In the deeper part, a visible amplitude difference can be seen between two images, the raw data (top image) and the final processed image (bottom image).

In the second stage, with the help of OpendTech software, three main works are listed:

(i) 3D visualization of all the 2D processed seismic lines resulted from the first stage (Fig. 4)
(ii) validating the data quality (Fig. 5)
(iii) computation of their different seismic attributes, 3D interpolated horizons, and the Holocene layer thickness.

Coordinates of all the seismic data profiles are used to form survey settings in OpendTect platform (Fig. 4). Validating the seismic quality could be done using the technique of checking a meeting point of any two different seismic profiles [4]. In Fig. 5, all the meeting points defined from couples of two cross different seismic lines are illustrated. That is, their similar depths of their meeting points can prove high reliability of the seismic quality in terms of data measurement and analysis.

Seismic Textural Attributes: Reflectors can be recognized from extreme values of processed seismic data and its seismic attributes [4, 5, 28, 29]. We have used application of entropy and energy textural attributes to help interpretation of the main seismic reflection.

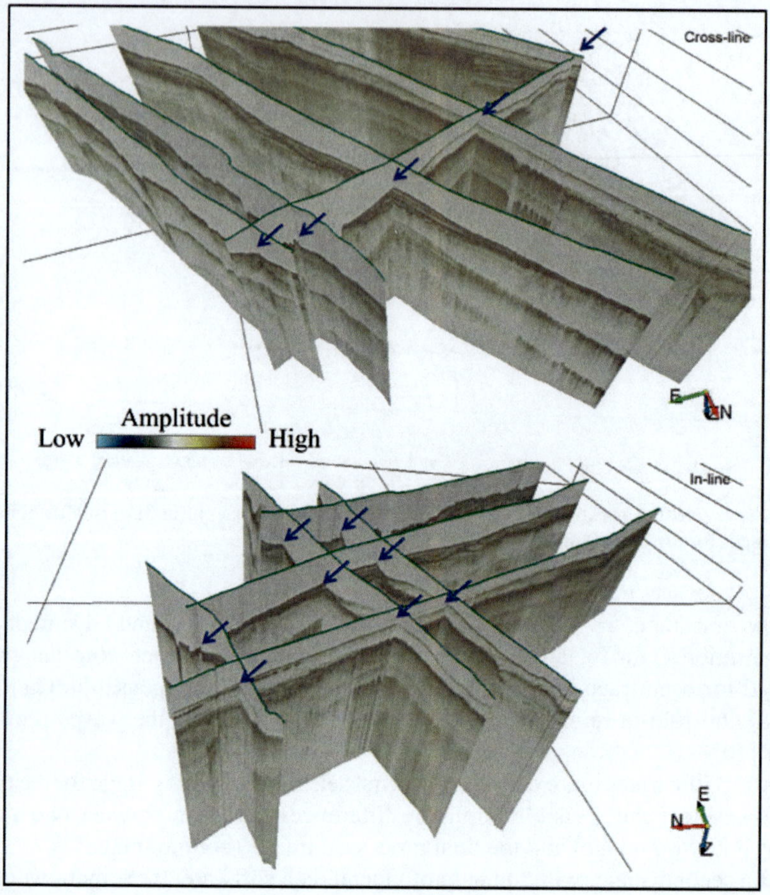

Fig. 5 Validity check of seismic data in depth analysis of the meeting points in their seismic profiles. In the meeting points, their depths of any two cross profiles should be similar. Two images (top and bottom) show the checking in different 3D view. Blue arrows refer locations of meeting points

Definition of textural attributes as energy, entropy, contrast, and homogeneity are well-known relating to detecting "zones of common signal character" [30]. The gray-level co-occurrence Matrix (GLCM) [31, 32] is established from seismic processed input data, then seismic textural attribute output such as energy or entropy are calculated from the GLCM data by the equations below[5]:

$$\text{Energy} = \sqrt{\sum_{i,j=0}^{N-1} P_{i,j}^2} \tag{1}$$

$$\text{Entropy} = \sum_{i,j=0}^{N-1} P_{i,j}\left(-\ln\left(P_{i,j}\right)\right) \tag{2}$$

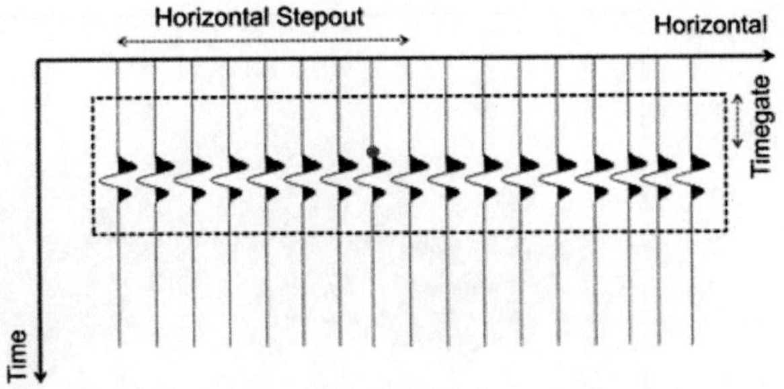

Fig. 6 Representation of 2D sliding window for computing seismic textural attributes [4, 27]. The sliding window is seen as a dashed rectangular characterizing horizontal width and time length

where $P_{i,j}$, shows the *i*th row and *j*th column of the GLCM matrix *P*.

The energy one brightens areas of high textural stability in each calculation zone which can be continuous features. The term, "Entropy" express how large randomness in the calculation zone is.

A calculation zone, namely a sliding window is characteristic through horizontal width and time length (see Fig. 6). For our research, we use time length as 2×27 ms and horizontal width as 2×9 traces.

4 Results and Discussion

We have interpreted some reflectors in the interest area. Using distinguished waveforms of processed seismic data and its seismic textural attributes (i.e., energy and entropy), three manually extracted boundaries can form three shallow layers as (i) water body, (ii) mixture of mud, sand, and silk clays layers (Present), (iii) sand silk clay sediments (Holocene), and hard gray sediment (Pleistocene).

The water body is easily recognized from strong seismic data and extreme values of its textural seismic attributes (see Fig. 7). In the Fig. 7, processed seismic data and its seismic attributes are shown for providing different view of a geology object. For example, water body can be seen as the space bounded by the surface to strong reflection boundary (see wheat color line in Fig. 7) while zone of low value for energy texture and zone of high value for entropy are represented for the water body.

Thanks to the seismic data attributes, we have interpreted different 2D seismic boundaries to form the three shallow layers. For illustrating all the 2D boundaries in 3D visualization, Fig. 8 refers seabed, top and bottom of Holocene layer as green, wheat, and pink colors, respectively. The well match in depth between of any two intersect lines can prove high correctness in seismic interpretation.

3D surfaces are interpolated from the 2D seismic boundaries (Figs. 9 and 10). We have used the Matlab built-in function, scatteredInterpolant.m [33] to build up the surfaces. 3D seabed is firstly interpreted and interpolated from strong visible reflection seismic amplitude events (Fig. 9). For the Top and Bottom interfaces of the Holocene layer, it is not consistent for strong reflection to interpret. Then, we used the visible patterns of the Holocene layer in textural seismic attributes (see

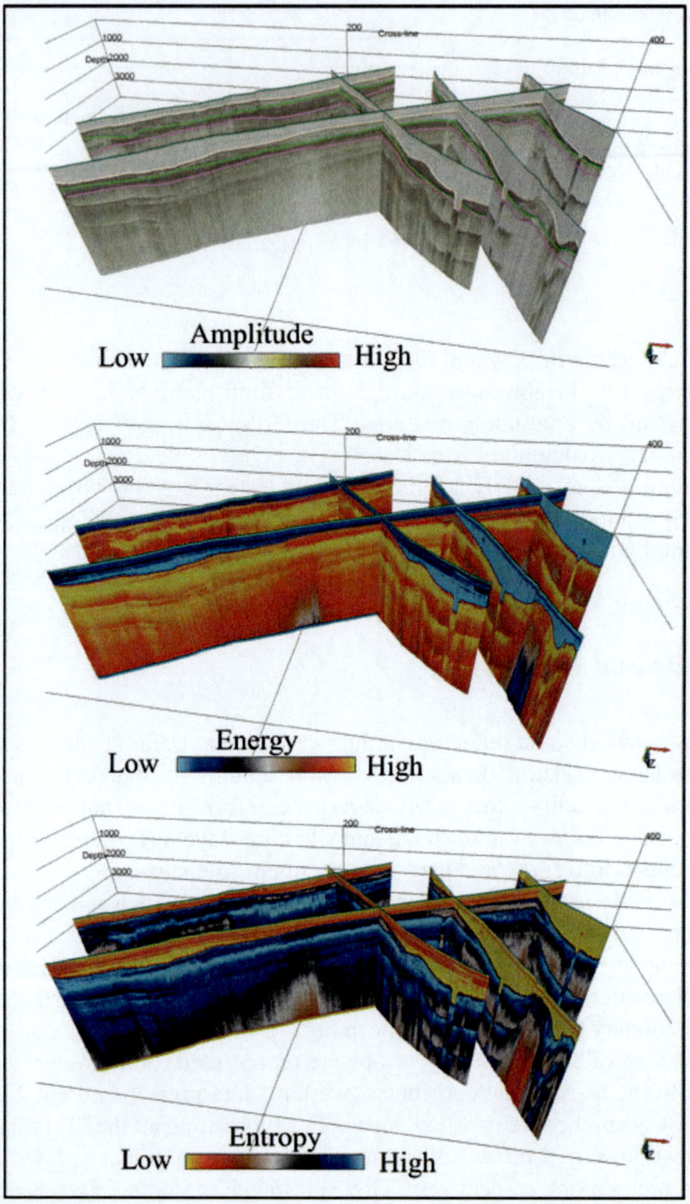

Fig. 7 Processed seismic data (top image) and its texture attributes, energy (middle image) and entropy (below image). Wheat, green, and pink lines are interpreted as seabed, top, and bottom of Holocene layer, respectively

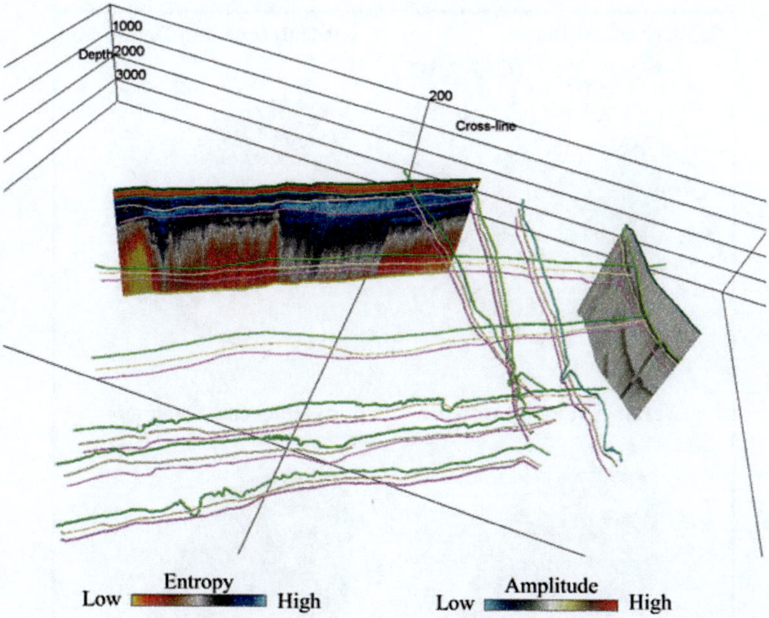

Fig. 8 Representation of all 2D interpreted seismic boundary. Processed seismic data and its textural attribute, entropy. Green, wheat color, and pink lines are interpreted as seabed, top and bottom of Holocene layer, respectively

Figs. 7 and 8). The boundaries between Holocene and Pleistocene layers are hardly seen through the processed seismic data but visible in their attributes helping its interpretation in different seismic profiles.

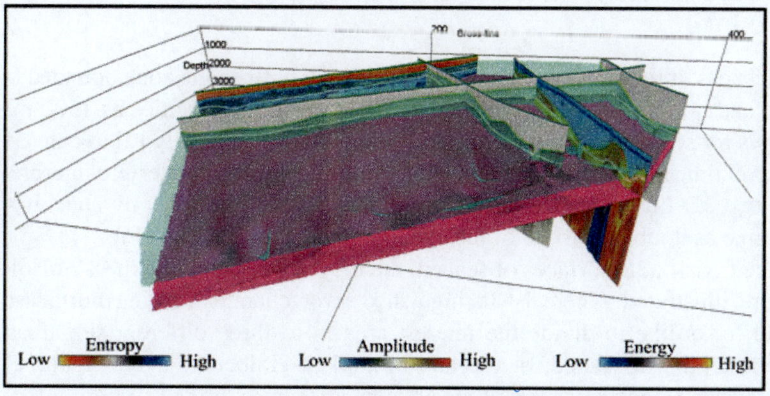

Fig. 9 3D seabed and Bottom Holocene made by 10 seismic profiles. Linear interpolation image with green surface for Seabed and pink one for Bottom Holocene

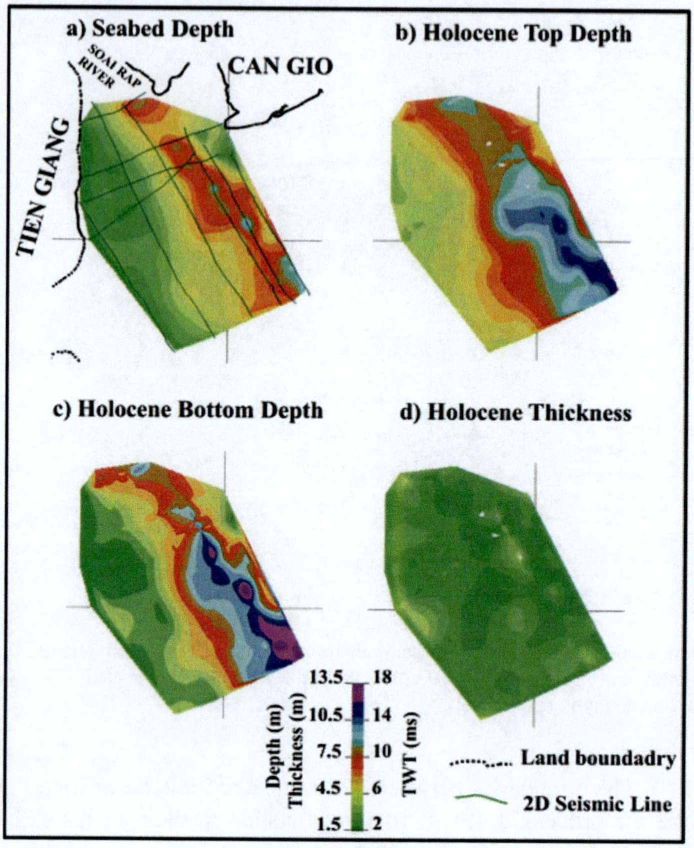

Fig. 10 3D seabed, top, bottom and thickness of Holocene layer made by ten seismic profiles. Holocene layer thickness can range from 1.5 m to around 4.5 m

Thickness of the Holocene layer can be calculated from the zone bounded by two 3D top and bottom surfaces. We have used propagation velocities as 1500 m/s and 1550 m/s for sea water and underground sediments, respectively [2] to convert two-way travel time (TWT) to depth. Note that multiple noise can interfere interpretation of the two 3D horizons. We have recognized seismic multiples by checking their travel time as double or triple of the original seismic events (See Fig. 7).

In the Fig. 10, 3D surfaces of seabed, the top, bottom, and thickness of Holocene layer are illustrated. There is distinguished river channel flowing northeast. The channel looks like to divide the interest area in to three different sub-areas with different depths in seabed, or top and bottom of Holocene layer. Figure 11 also shows a clear 3D visualization of the channel as seen as the red stripe.

According to the drill holes information [4] and the seismic interpretation result, we can determine three layers as follow:

(i) the first layer having mud, silk clay, sand, organic matter from Present time,

Fig. 11 3D view of small the Soai Rap river channel

(ii) the Holocene layer having clay sediment, and
(iii) the hard clay Pleistocene matters.

5 Conclusion

3D image of seabed and sedimentation in Soai Rap River fork can be presented using interpretation of 2D high-resolution seismic data. Suitable interpolation of different 2D boundaries which express zones of contrast of acoustic impedance can bring the 3D visualization come true with the help of useful professional software Matlab and OpendTect. Better interpreting the boundaries needs combination of different textural seismic attributes rather than just utilizing only the seismic processed data itself. Great match between seismic results and drill hole information as lithology can prove existence of Holocene layer and Pleistocene matters which helps to government officials in deepening the riverbed and canals for large ship transportation, especially, in Can Gio, Ho Chi Minh city. The distinguished Soai Rap channel appears in the seismic data can prove the fact that the high-resolution seismic data can detect small but meaningful geology features (Fig. 12).

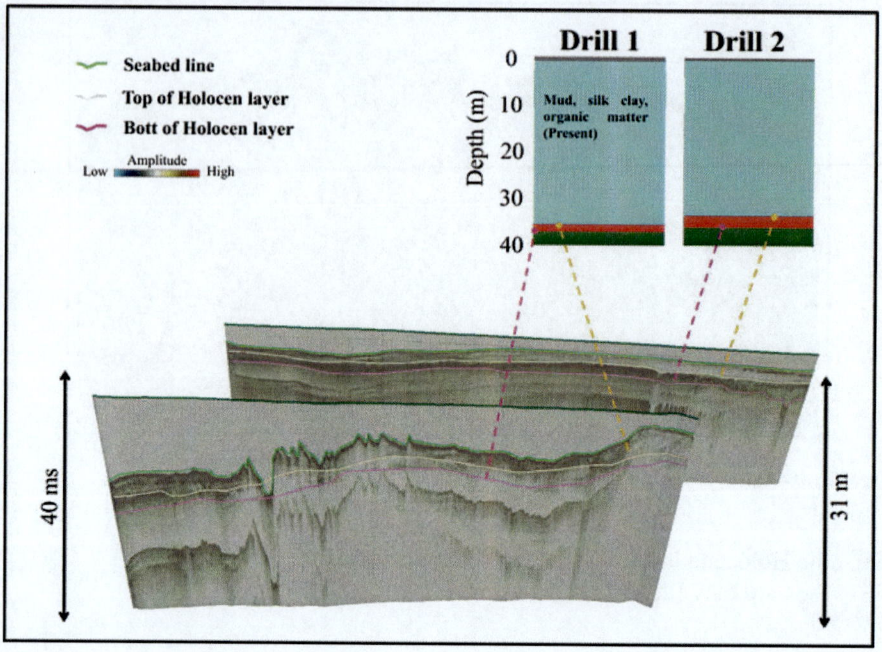

Fig. 12 Representations of two drill holes [4] with their two nearby seismic data profiles

Acknowledgements This research is funded by Vietnam National University Ho Chi Minh City (VNU-HCM) under grant number C2022-18-17. We would like to thank Mr. Nguyen Quang Dung for his help.

Conflicts of Interest The authors declare no conflict of interest.

Author Contribution All the authors contributed to the data analysis, interpretation, and manuscript. C.V.A.L mainly wrote the manuscript.

References

1. Ianniruberto, M., Campos, J.E., Araújo, V.: Application of shallow seismic profiling to study riverbed architectural facies: A case study of the Tocantins river (Pará-Brazil). An. Acad. Bras. Ciênc. **84**, 645–654 (2012)
2. Bui Viet, D., Stattegger, K., Unverricht, D., Van Phung, P., Nguyen Trung, T.: Late Pleistocene-Holocene seismic stratigraphy of the Southeast Vietnam Shelf. Global Planet. Change **110**, 156–169 (2013)
3. Novak, B., Björck, S.: Late Pleistocene–early Holocene fluvial facies and depositional processes in the Fehmarn Belt, between Germany and Denmark, revealed by high-resolution seismic and lithofacies analysis. Sedimentology **49**, 451–465 (2002)
4. Le, C.V.A., Duong, M.B., Kieu, T.D.: High–resolution seismic reflection survey of Holocene Sediment distribution at Thi Vai River, Ho Chi Minh City, Vietnam. Lecture Notes in Civil Engineering, vol. 2, pp. 290–304. Springer (2020)

5. Le, C.V.A., Harris, B.D., Pethick, A.M.: New perspectives on solid earth geology from seismic texture to cooperative inversion. Sci. Rep. **9**, 14737 (2019)
6. https://tuoitre.vn/luong-soai-rap-chinh-thuc-cho-tau-bien-30-000-tan-ve-tp-hcm-long-an-20200826115103097.htm
7. https://www.bandovn.vn/vi/page/mau-ban-do-hanh-chinh-nuoc-cong-hoa-xa-hoi-chu-nghia-viet-nam-181
8. David, F., Meziane, T., Tran-Thi, N.-T., Van, V.T., Thanh-Nho, N., Taillardat, P., Marchand, C.: Carbon biogeochemistry and CO 2 emissions in a human impacted and mangrove dominated tropical estuary (Can Gio, Vietnam). Biogeochem. **138**, 261–275 (2018)
9. Fujimoto, K., Umitsu, M., Nguyen, V.L., Ta, T.K.O., Kawase, K., Huynh, D.H., Nakamura, T.: Geomorphological evolution and mangrove habitat dynamics related to Holocene sea-level changes in the northern Mekong river delta and the Dong Nai river delta, southern Vietnam. In: Paul E. Schmidt (ed.) River Deltas: types, structures and ecology. Nova Science Publishers, Inc. (2011)
10. Bui, V.T., Huynh, T.T., Le, T.N.D., Ly, H.M., Le, P.T.: Monitoring and predicting the shoreline change in Can Gio area in condition of the sea level rise. Sci. Tech. Development J. **17**, 45–53 (2014)
11. Le, X.T., Vo, Q., Reyns, J., Pham, V.S., Tran, A.D., Dang, T.D., Roelvink, D.: Sediment transport and morphodynamical modeling on the estuaries and coastal zone of the Vietnamese Mekong Delta. Cont. Shelf Res. **186**, 64–76 (2019)
12. Stattegger, K., Tjallingii, R., Saito, Y., Michelli, M., Thanh, N.T., Wetzel, A.: Mid to late Holocene sea-level reconstruction of Southeast Vietnam using beachrock and beach-ridge deposits. Global Planet. Change **110**, 214–222 (2013)
13. Thoang, T., Giao, P.: Subsurface characterization and prediction of land subsidence for HCM City, Vietnam. Eng. Geol. **199**, 107–124 (2015)
14. Ta, T.K.O., Nguyen, V.L., Tateishi, M., Kobayashi, I., Tanabe, S., Saito, Y.: Holocene delta evolution and sediment discharge of the Mekong River, southern Vietnam. Quatern. Sci. Rev. **21**, 1807–1819 (2002)
15. Ta, T.K.O., Nguyen, V.L., Tateishi, M., Kobayashi, I., Saito, Y.: Holocene delta evolution and depositional models of the Mekong River Delta, southern Vietnam. River Deltas–Concepts, Models, and Examples, vol. 83. SEPM (Society for Sedimentary Geology) (2005)
16. Ta, T.K.O., Nguyen, V.L., Tateishi, M., Kobayashi, I., Saito, Y.: Sedimentary facies, diatom and foraminifer assemblages in a late Pleistocene-Holocene incised-valley sequence from the Mekong River Delta, Bentre Province, Southern Vietnam: the BT2 core. J. Asian Earth Sci. **20**, 83–94 (2001)
17. Nguyen, V.L., Ta, T.K.O., Saito, Y.: Early Holocene initiation of the Mekong River delta, Vietnam, and the response to Holocene sea-level changes detected from DT1 core analyses. Sed. Geol. **230**, 146–155 (2010)
18. Kitazawa, T.: Pleistocene macrotidal tide-dominated estuary–delta succession, along the Dong Nai River, southern Vietnam. Sed. Geol. **194**, 115–140 (2007)
19. Kitazawa, T., Nakagawa, T., Hashimoto, T., Tateishi, M.: Stratigraphy and optically stimulated luminescence (OSL) dating of a Quaternary sequence along the Dong Nai River, southern Vietnam. J. Asian Earth Sci. **27**, 788–804 (2006)
20. Yutsis, V., Krivosheya, K., Levchenko, O., Lowag, J., de León Gómez, H., Ponce, A.T.: Bottom topography, recent sedimentation and water volume of the Cerro Prieto Dam. NE Mexico. Geofísica internacional **53**, 27–38 (2014)
21. Aiello, G., Di Fiore, V., Marsella, E., Passaro, S.: High resolution seismic data coupled to Multibeam bathymetry of Stromboli island collected in the frame of the Stromboli geophysical experiment: implications with the marine geophysics and volcanology of the Aeolian Arc volcanic complex (Sicily, Southern Tyrrhenian sea, Italy). Springerplus **3**, 1–27 (2014)
22. https://www.edgetech.com/edgetech-introduces-new-portable-sub-bottom-profiling-system/
23. https://www.edgetech.com/resource-center/
24. VectorStock: Ship Image
25. EdgeTech: JSF Data File Description (2016)

26. https://www.sandmeier-geo.de/reflexw.html
27. dGB Earth Sciences: OpendTect dGB Plugins User Documentation version 4.6. (2015)
28. Yilmaz, O.: Seismic data analysis: processing, inversion, and interpretation of seismic data. Soc. Explor. Geophys., United States of America (2001)
29. Le, C.V.A., Harris, B.D., Pethick, A.M., Takam Takougang, E.M., Howe, B.: Semiautomatic and automatic cooperative inversion of seismic and magnetotelluric data. Surveys Geophys. **37**, 845–896 (2016)
30. Chopra, S., Marfurt, K.J.: Seismic attributes for prospect identification and reservoir characterization. Tulsa, Okla. (8801 South Yale St., Tulsa OK 74137–3175): Society of Exploration Geophysicists, United States of America (2007)
31. http://www.fp.ucalgary.ca/mhallbey/tutorial.htm
32. Haralick, R.M., Shanmugam, K., Dinstein, I.: Textural features for image classification. IEEE Trans. Syst. Man, Cybernetics SMC-3, 610–621 (1973)
33. https://www.mathworks.com/help/matlab/ref/scatteredinterpolant.html

Analysis of Geological Structures by 2D Magnetotelluric Inversion in Bang Hot Spring Area, Quang Binh Province

Cuong Van Anh Le, Duy Thong Kieu, Ngoc Dat Pham, and Hop Phong Lai

Abstract The magnetotelluric method can reveal geology structures information through conductivity distribution from a few hundred meters to kilometers depending on electromagnetic data frequency measurement. Although the raw magnetotelluric data does not clearly illustrate faults or geothermal reservoirs, they can be more visible in the conductivity models after inversion. However, the 2D inverted conductivity image is often distorted by geoelectrical strike as fault zones. Hence, in this work, we have integrated geology strike analysis to 2D magnetotelluric inversion to have better data quality interpretation in the Bang hot spring area, Quang Binh Province, Vietnam. The inverted conductivity model from the 2D integrated magnetotelluric inversion can show the fault system and possible hot source zone of Bang area. Great compatibility between the magnetotelluric inversion result with other geophysical data such as seismic and prior geology information can inspire its application in geothermal research.

Keywords Magnetotelluric · 2D inversion · Impedance tensor · Strike analysis · Hot spring

---

Contribution: C.V.A.L processed and inverted the MT data. All the authors contributed to the data analysis, interpretation, and manuscript. C.V.A.L mainly wrote the manuscript.

C. Van Anh Le (✉)
University of Science, Ho Chi Minh City, Vietnam
e-mail: lvacuong@hcmus.edu.vn

Vietnam National University Ho Chi Minh City, Ho Chi Minh City, Vietnam

D. T. Kieu
Hanoi University of Mining and Geology, Hanoi, Vietnam
e-mail: kieuduythong@humg.edu.vn

N. D. Pham · H. P. Lai
Institute of Geological Sciences, Vietnam Academy of Science and Technology, Hanoi, Vietnam

L. Q. Nguyen et al. (eds.), *Advances in Geospatial Technology in Mining and Earth Sciences*, Environmental Science and Engineering,
https://doi.org/10.1007/978-3-031-20463-0_22

1 Introduction

Electromagnetic wave propagation ignited from the interaction between charged flows from the sun and the Earth magnetic field is the origin of the magnetotelluric method [1]. When the waves travel down to the ground, inductions with geological structures with different resistivities appear leading to formation of electric and magnetic fields orthogonally recorded on the surface.

The impedance tensor Z can vary depending on electromagnetic frequency and resistivity of the survey area. Its diagonal components express the lateral effects while its off-diagonal components consist of basic information of the vertical resistivity profile [2]. For 2D interpretation in the vertical direction, the two modes Transversal Electric (TE) and Transversal Magnetic (TM) can describe the behaviors of the electromagnetic wave. That is, TE is more effective for deep anomalies while TM mode is more responsive to the horizontal anomalies [1, 2]. Then, their apparent resistivities can be calculated for brief interpretation of the research area.

Common preliminary way for researching the impedance is to investigate the strike direction [1, 3–5]. Assumption of environments as 2D or quasi-symmetric 3D models is important to detect strike angle. Theories have taken account of impedance tensor interpretation. For the approaches of Sims and Bostick [3] and Yee and Paulson [6], sum of diagonal components of the impedance tensor is minimized or sum of the off-diagonal ones maximized. Meanwhile, the other approaches direct to the theories that some meaningful relationship between the electric and magnetic fields when polarized at an angle to the original measured axis [4, 5]. For example, in the polarized angle, the electric field is orthogonal to the magnetic field [5] while there exists states in which eigen orthogonal electric fields or eigen orthogonal magnetic fields [4].

Inversion, a powerful tool to build a resistivity model, can illustrate geological structures. This process minimizes the difference between the recorded data and synthetic data of the resistivity model respect to several constraints such as smoothness, structural and petrophysical constraints [7–11].

Vietnam has a high potential for hot springs that can attract tourism. Thus, using geophysical techniques such as MT to define the structure related to hot springs is needed [12–15]. In Quang Binh province, Vietnam, a geophysical project was launched for investigating the geology of Bang geothermal reservoir. For researching the geology structures in the Bang hot spring area, we have conducted a workflow as some steps: (i) strike analysis of the MT data and (ii) running 2D inversion for different lines in the data with and without applying strike analysis.

2 Methods

2.1 Magnetotelluric Impedance Tensor

An impedance tensor represents relationship between the horizontal parts of the two electric fields and magnetics fields at a field site through the equation [1].

$$\begin{aligned} E_x &= Z_{xx} H_x + Z_{xy} H_y \\ E_y &= Z_{yx} H_x + Z_{yy} H_y \end{aligned} \tag{1}$$

where E is electric field, H is the magnetic intensity, Z is impedance tensor, subscript x and y mark the direction, x points toward North and y points toward East (Fig. 1).

The relationship between electric and magnetic fields recorded in the field trip can be shown by their tensor $\hat{Z} = \begin{bmatrix} Z_{xx} & Z_{xy} \\ Z_{yx} & Z_{yy} \end{bmatrix}$ [1]. When the axis of a new survey setup forms an angle θ with the axis of the old survey (see Fig. 1), then its new tensor $Z'(\theta)$ can be calculated via the following equations [1].

$$Z'(\theta) = \begin{bmatrix} Z'_{xx}(\theta) & Z'_{xy}(\theta) \\ Z'_{yx}(\theta) & Z'_{yy}(\theta) \end{bmatrix} \tag{2}$$

$$Z'_{xx}(\theta) = Z_2 + Z_3 \sin 2\theta + Z_4 \cos 2\theta \tag{3}$$

$$Z'_{xy}(0) - Z_1 + Z_3 \cos 20 - Z_4 \sin 20 \tag{4}$$

$$Z'_{yx}(\theta) = -Z_1 + Z_3 \cos 2\theta - Z_4 \sin 2\theta \tag{5}$$

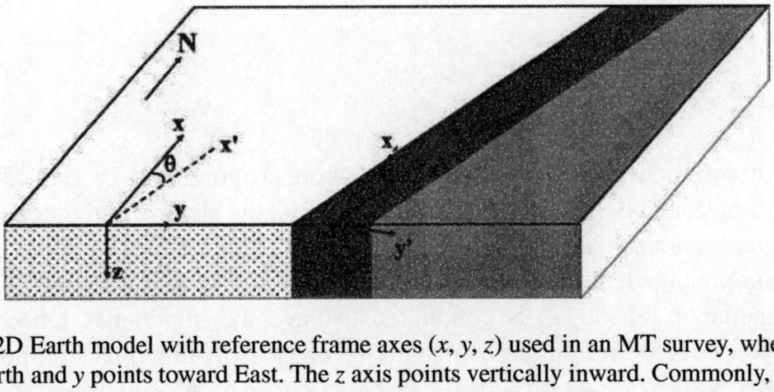

Fig. 1 A 2D Earth model with reference frame axes (*x*, *y*, *z*) used in an MT survey, where *x* points toward North and *y* points toward East. The *z* axis points vertically inward. Commonly, a reference frame rotated around the z axis (*x*', *y*', *z*) is also used. θ indicates the angle between the *x*' and *x* axes

$$Z'_{yy}(\theta) = Z_2 - Z_3 \sin 2\theta - Z_4 \cos 2\theta \tag{6}$$

where,

$$Z_1 = \frac{1}{2}(Z_{xy} - Z_{yx});\ Z_3 = \frac{1}{2}(Z_{xy} + Z_{yx}); \tag{7}$$

$$Z_2 = \frac{1}{2}(Z_{xx} + Z_{yy});\ Z_4 = \frac{1}{2}(Z_{xx} - Z_{yy}) \tag{8}$$

Some complex invariants which are independent in the angle θ are listed as Z_1, Z_2, and det $\hat{Z}$ with

$$\det \hat{Z} = Z_{xy}Z_{yy} - Z_{xy}Z_{yx} \tag{9}$$

In a 2D model, conductivity can vary in two directions (i.e., *y*'- and *z*- axes) and be constant in the geoelectrical strike (i.e., *x*' axis) (Fig. 1). In these 2D cases, Maxwell's equations can be decoupled into two modes [1].

i. Mode xy (E_x, H_y), also known as Transversal Electric (TE) mode, with currents (electric fields) parallel to the strike direction.
ii. Mode yx (H_x, E_y), or called as Transversal Magnetic (TM) mode, with currents (electric fields) perpendicular to the strike.

The magnetotelluric impedance tensor $\underline{Z}$ in 2D models is non-diagonal and expressed as:

$$\underline{Z}_{2D} = \begin{pmatrix} 0 & Z_{xy} \\ Z_{yx} & 0 \end{pmatrix} = \begin{pmatrix} 0 & Z_{TE} \\ Z_{TM} & 0 \end{pmatrix} \tag{10}$$

where Z_{xy} is equal E_x/H_y and Z_{yx} equal E_y/H_x coming from TE and TM sets of equations, respectively. They usually have opposite signs.

2.2 Strike Analysis

To strike direction, we have applied the method proposed by Eggers [5]. When measured direction matches with the strike direction, the electric and magnetic fields can be perpendicular. Polarization of electric field is the ratio between the two electric field components leading to the computation of the strike angle [5]. Then, rotation of impedance tensor [1] is conducted to have the impedance tensor for the inversion procedure. The apparent resistivity of two off-diagonal components as $\rho_Z = 0.2 * \text{period} * |Z|^2$ can provide the brief geophysical information of the interest environment.

The Eggers's idea is to define special eigenstates in which electric and magnetic fields (H_τ, E_τ) are perpendicular.

$$\vec{E}_\tau \vec{H}_\tau = 0 \Leftrightarrow E_x H_x + E_y H_y = 0 \tag{11}$$

Then, a scalar ζ will exist:

$$\zeta = \frac{E_x}{H_y} = -\frac{E_y}{H_x} \tag{12}$$

So: $E_\tau = \zeta\left[\vec{H}_\tau \times \vec{1}_z\right]$; $\vec{1}_z$ is the wave propagation direction. Then, $\hat{Z}\vec{H}_\tau = \zeta\left[\vec{H}_\tau \times \vec{1}_z\right]$.

Expanding the Eq. (1) and using the Eq. (11), we have:

$$\begin{cases} Z_{xx} H_x + Z_{xy} H_y = \zeta H_y \\ Z_{yx} H_x + Z_{yy} H_y = -\zeta H_x \end{cases} \tag{13}$$

Or

$$\begin{cases} Z_{xx} H_x + (Z_{xy} - \zeta) H_y = 0 \\ (Z_{yx} + \zeta) H_x + Z_{yy} H_y = 0 \end{cases} \tag{14}$$

Finding magnetic field $\overrightarrow{H_\tau}$ and solving the system of the Eq. (13) can lead to the following equation:

$$Z_{xx} Z_{yy} - (Z_{xy} - \zeta)(Z_{yx} + \zeta) = 0 \Leftrightarrow \zeta^2 - 2Z_1\zeta + \det \hat{Z} = 0 \tag{15}$$

which has solutions ζ of the quadratic equation:

$$\zeta_{1,2} = Z_1 \pm \sqrt{Z_1^2 - \det \hat{Z}} \tag{16}$$

The invariants ζ_1 and ζ_2 are the two eigenvalues of the fields the fields $\left(\vec{E}_{1\tau}, \vec{H}_{1\tau}\right)$ and $\left(\vec{E}_{2\tau}, \vec{H}_{2\tau}\right)$.

$$\vec{E}_{1\tau} = \zeta_1\left[\vec{H}_{1\tau} \times \vec{1}_z\right] \tag{17}$$

$$\vec{E}_{2\tau} = \zeta_2\left[\vec{H}_{2\tau} \times \vec{1}_z\right] \tag{18}$$

The field couples $\left(\vec{E}_{1\tau}, \vec{H}_{1\tau}\right)$ and $\left(\vec{E}_{2\tau}, \vec{H}_{2\tau}\right)$ are perpendicular vector pairs. The major axes of the polarization electric and magnetic ellipses are mutually perpendicular.

Establishing polarization electric and magnetic ellipses needs some important parameters as polarization ratio, ellipticity and an angle between the axis and its major polarization electric axis [1, 5, 7]. Note that after the angle rotation, the electric and magnetic fields belong to an eigen state (Fig. 2).

$$P_{E_{1,2}} = \frac{E_{1,2y}}{E_{1,2x}} = \tan\theta_{E_{1,2}} e^{i\phi_{E_{1,2}}} = \frac{\zeta_{1,2} - Z_{xy}}{Z_{xx}} = \frac{Z_{yy}}{\zeta_{1,2} + Z_{yx}} \tag{19}$$

$$\varepsilon_{E_{1,2}} = tg\left[\frac{1}{2}\arcsin(\sin 2\theta_{E_{1,2}} \sin\phi_{E_{1,2}})\right] \tag{20}$$

$$\tan 2\alpha_{E_{1,2}} = \tan 2\theta_{E_{1,2}} \cos\phi_{E_{1,2}} \begin{cases} \alpha_E \in [0, \pi/2] \text{ when } \cos\phi_E > 0 \\ \alpha_E = 0 \text{ when } \cos\phi_E = 0 \\ \alpha_E \in [-\pi/2, 0] \text{ when } \cos\phi_E < 0 \end{cases} \tag{21}$$

To identify degree of the three dimensionality of the structure, skew can be used:

$$A_E = \left| \left| \alpha_{E_1} - \alpha_{E_2} \right| - 90^\circ \right| \tag{22}$$

For evaluate dimensionality of a research area, some notices are suggested:

(i) 1D structure: $\underline{Z}_{1D} = \begin{pmatrix} 0 & Z_{xy} \\ -Z_{xy} & 0 \end{pmatrix}$, then it infers $\zeta_1 = \zeta_2 = Z_{xy} = Z$, $P_{E_{1,2}} = 0/0$, $\alpha_{E_{1,2}} = 0/0$. The major value of Z looks like to the Tikhonov-Cagniard tensor while the major orientation cannot be defined.

(ii) 2D structure: $\underline{Z}_{2D} = \begin{pmatrix} 0 & Z_{xy} \\ Z_{yx} & 0 \end{pmatrix}$, α -rotated impedance tensors have major ellipse axes as $\zeta_{1,2} = Z^{//,\perp}$ and $\alpha_{E_{1,2}} = -\alpha, -\alpha + \pi/2$. The ellipticity values $\varepsilon_{1,2} = 0$. This is a linear polarized case.

(iii) 3D structure: $\underline{Z}_{3D} = \begin{bmatrix} Z_{xx} & Z_{xy} \\ Z_{yx} & Z_{yy} \end{bmatrix}$, then the ellipticity values $\varepsilon_{1,2} \neq 0$ express the inhomogeneity of the research environment.

2.3 Inversion

Smooth approach is integrated in inverting MT data for establishing its subsurface electrical resistivity model [8]. The penalty function has two main parts; (i) misfit of the field ψ_d and synthetic data and (ii) resistivity model constraint ψ_m (i.e., smooth condition) as described in the Eq. (2).

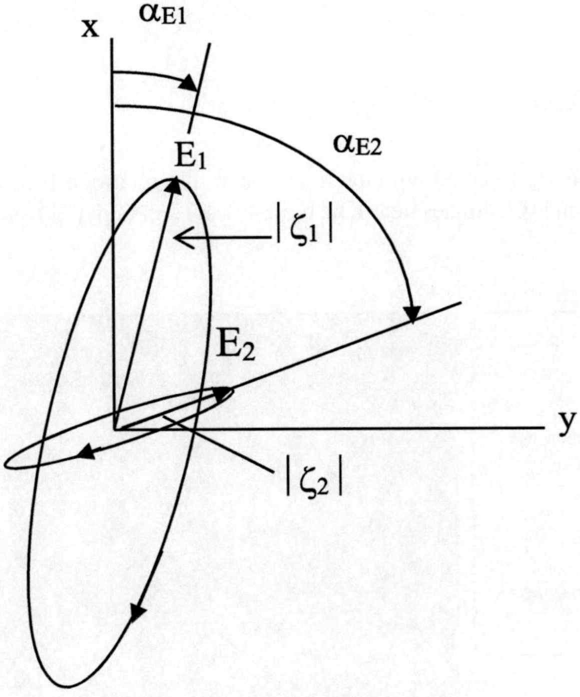

Fig. 2 Representation of polarization electric fields in eigen states [5]. *x*-axis points toward North and *y* points toward East. The *z* axis points vertically inward. After applying Eggers' method, a reference frame rotated around the *z* axis (*x*', *y*', *z*) is also used via an angle $\alpha_E \cdot \alpha_{E_1}$ or α_{E_2} indicates an angle between the *x*' (an eigenstate) and *x* axes (fieldtrip measurement setup)

$$\psi = \psi_d(m) + \lambda\psi_m(m) = d - F(m)^2 + \lambda\Delta m^2 \tag{23}$$

where *d* is the measured data, F the modeling operator, Δm^2 the difference between of the adjacent resistivity model parameters, and λ Lagrange multiplier. The penalty function is minimized to get the resistivity model in which the Lagrange multiplier plays a key role in balancing the two parts. We have applied the code modEM2D [8] in which the nonlinear conjugate gradient approach is used to minimize the penalty function.

3 Results

3.1 Field Data

The hot spring Bang location with the hot temperature recorded 100 °C degree in its exposed footprint [16] situates near Khe Giua–Vinh Linh fault (F1) and other smaller faults (Fig. 3).

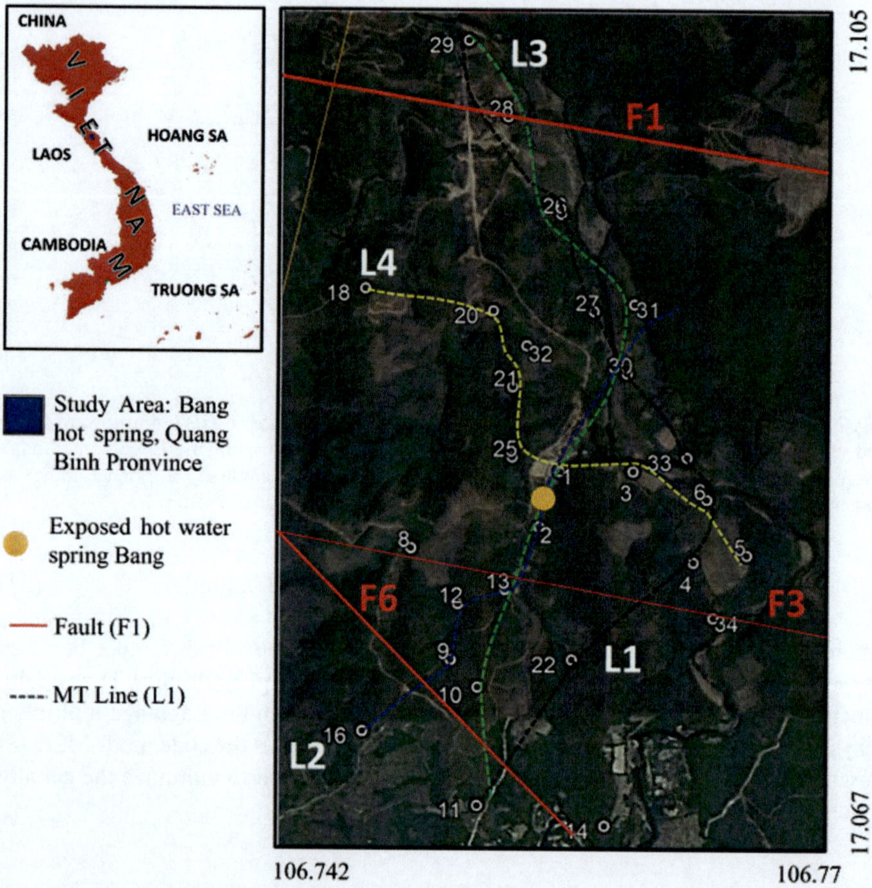

Fig. 3 The map expresses the Bang hot spring area in Quang Binh Province, Vietnam (blue square in left image) and the MT measurement locations (white circles) are set up nearby geological faults (right image). There are four 2D MT lines for inversion as Lines L1, L2, L3, and L4. Line L1 and Line L3 start from MT location "11" and end at MT location "29". Line L2 starts at MT location "16" and end at MT location "31" while Line L4 begins at MT location "18" and stops at MT location "5"

Noise filter and strike direction analysis by Eggers [5] are applied to the full impedance tensor prior 2D inversion. The MT data for 2D inversion include off-diagonal components (TE and TM) and have the frequency band ranging from 320 to 0.35 Hz within 28 stations. Note that we have used modEM2D to invert all the four MT lines as Lines L1, L2, L3, and L4. Line L1 and Line L3 start from MT location "11" and end at MT location "29". Line L2 starts at MT location "16" and ends at MT location "31" while Line L4 begins at MT location "18" and stops at MT location "5" (Fig. 3). According to the skin depth equation, the research depth can be around 8.5 km with the background resistivity 100 Ω.m and the minimum frequency 0.35 Hz.

3.2 Strike Analysis Result

After computing eight invariants of MT stations in the lines L3 and L4 from the Eggers' method (See Sect. 2.2 Strike analysis), some remarks are presented from Figs. 4, 5, 6, 7, 8, 9, 10 and 11.

(i) Low ellipticity values of the two eigenstates E1 and E2 can refer to 2D dimensionality or symmetrical 3D characteristics for the research area (Figs. 4, 5, 8 and 9). The zones defined by mostly the low ellipticity values is prior known as the zones having less fractures or the footprint of Bang hot spring. Low skew values can have the same meaning ideas with the low ellipticity when it explains the 2D or symmetrical 3D characteristics.

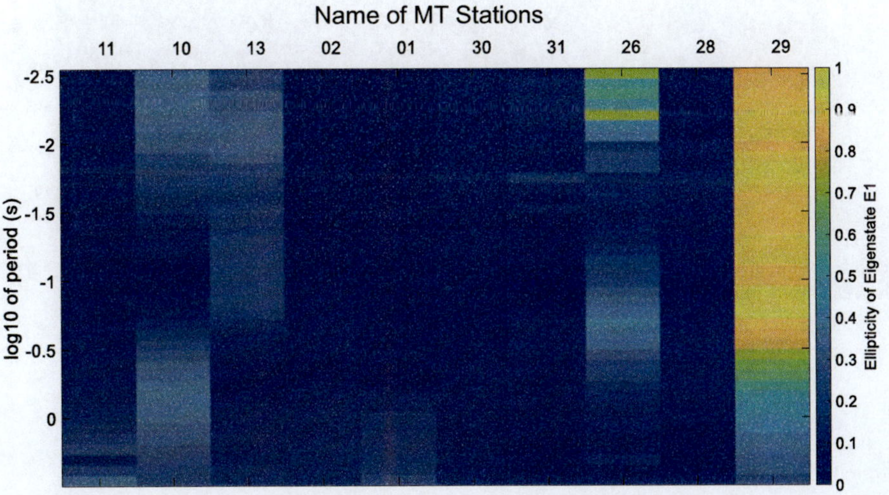

Fig. 4 The map expresses the ellipticity of E1 of Line L3. The exposed hot spring in the middle of the MT stations "01" and "02" are presented with the low ellipticity of eigenstate E1

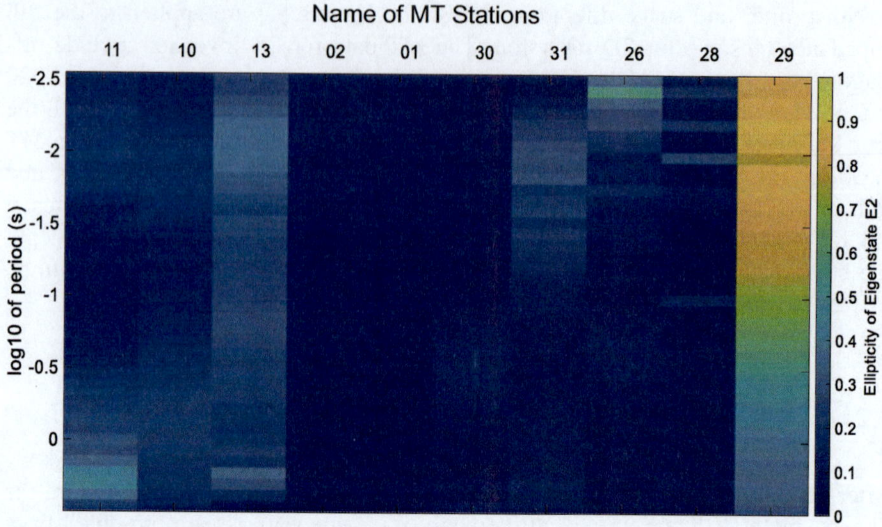

Fig. 5 The map expresses the ellipticity of E2 of Line L3. The exposed hot spring in the middle of the MT stations "01" and "02" is presented with the low ellipticity of eigenstate E2

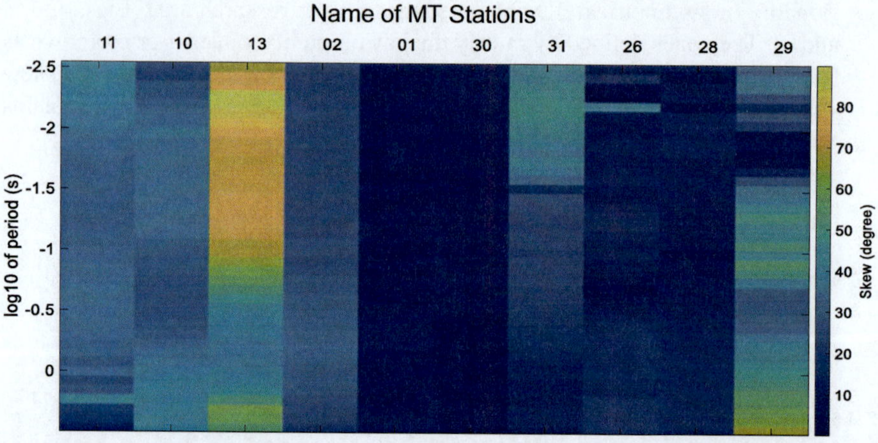

Fig. 6 Skew of Line L3. The exposed hot spring in the middle of the MT stations "01" and "02" is presented with the low Skew values

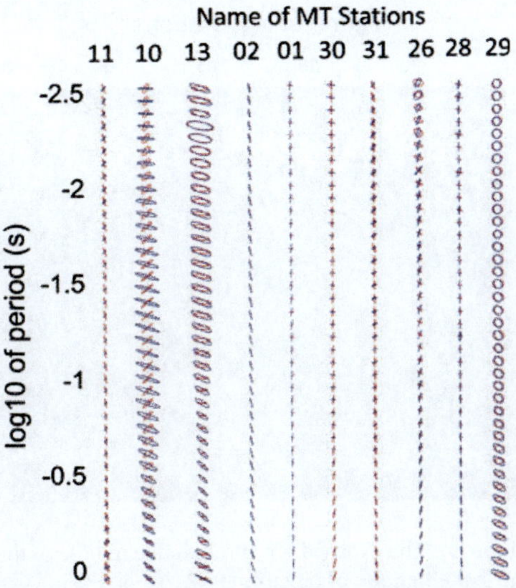

Fig. 7 The map expresses the ellipses of E1 and E2 of Line L3. Two small linear ellipses having orthogonal tendency in each period can show 2D or symmetrical 3D characteristics in MT stations "01" and "02"

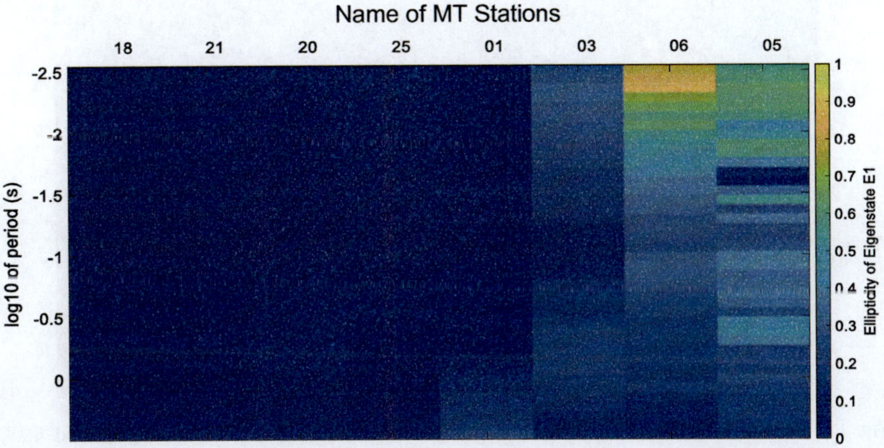

Fig. 8 Ellipticity E1 of Line L4. The exposed hot spring in the middle of the MT stations "01" and "02" is presented with the low ellipticity of eigenstate E1

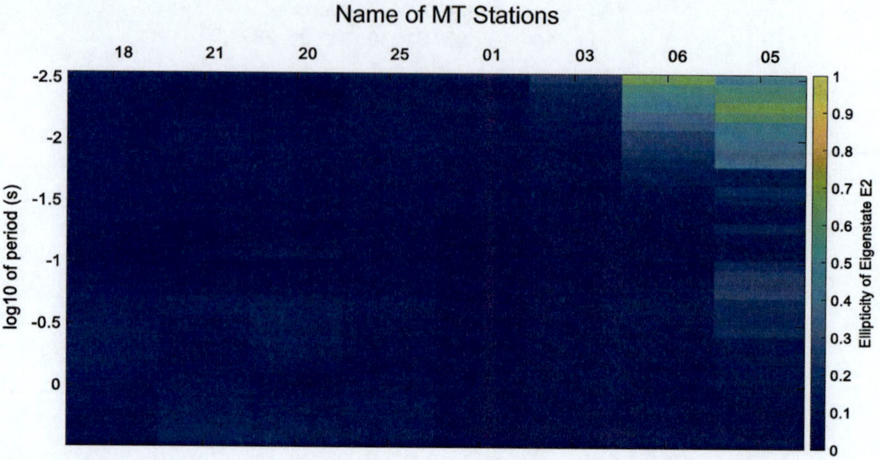

Fig. 9 Ellipticity E2 of Line L4. The exposed hot spring in the middle of the MT stations "01" and "02" is presented with the low ellipticity of eigenstate E2

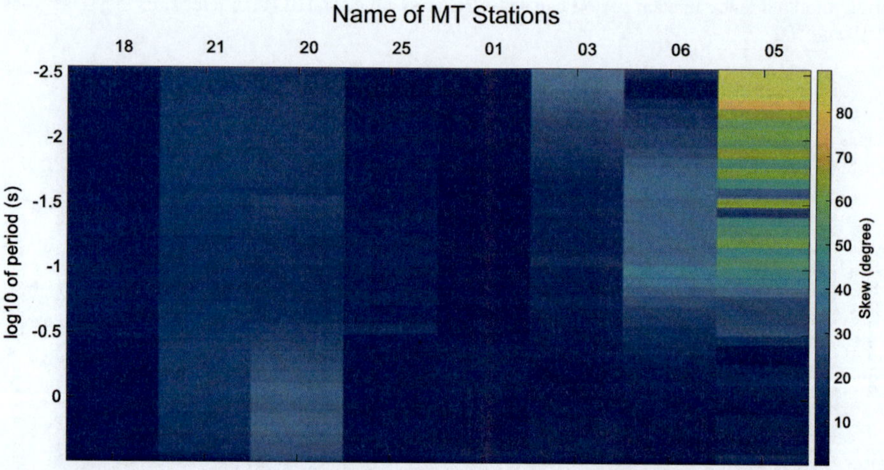

Fig. 10 Skew of Line L4. The exposed hot spring in the middle of the MT stations "01" and "02" is presented with the low Skew values

(ii) High ellipticity and high skew values show 3D homogeneity of the environment when the areas are affected by big faults or chaos of different geological objects.

Ellipses of the MT stations are shown in Figs. 7 and 11 to expresses full 8 parameters of each tensor impedance. They can help interpreters quickly recognizes characteristics of the environment. In Figs. 7 and 11, for some stations as "11", "02", and "01", two eigenstates' ellipses have a tendency in being orthogonal each other and showing 2D or symmetrical 3D characteristics.

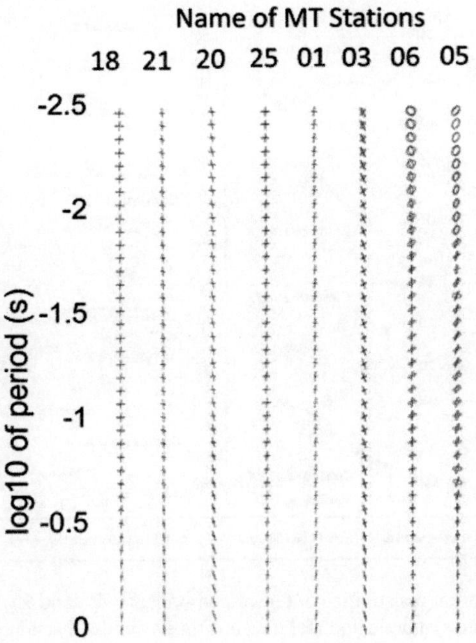

Fig. 11 Ellipses of Line L4. Two small linear ellipses having orthogonal tendency in each period can show 2D or symmetrical 3D characteristics in MT stations "01" and "02"

3.3 Inversion Result

Running 2D MT inversion is conducted on the Cray Cascade system at the Pawsey supercomputing system in Perth city, Western Australia. We set up the prior model as 100 Ω.m for all the 2D lines. We have run several 2D inversions for both real data with and without strike angle analysis.

In Fig. 12, the term "TETM measured." expresses the real data points measured in the field trip while the term "TETM syn." refers the synthetic data points modeled after 2D inversion. The term "E1 measured." expresses the real data points rotated by the strike analysis approach by Eggers [5] while the term "E1 syn." refers the synthetic data points modeled after 2D inversion. The rotation transformation equations are shown in the Eqs. (2–5). The rotation angle is defined as medium values of α_{E_1} from the eigenstate case E1. Then, we figure out that running 2D inversion with strike angle analysis provides the smaller misfit values. After inversion with the real data of the "E1 measured." type, the synthetic and real data have good similarity level and the root mean square misfits reduce by half after at least 15 iterations.

OpendTect software [17] is used for 3D visualizing the four 2D MT resistivity results from inverting the data with strike analysis (Figs. 13 and 14).

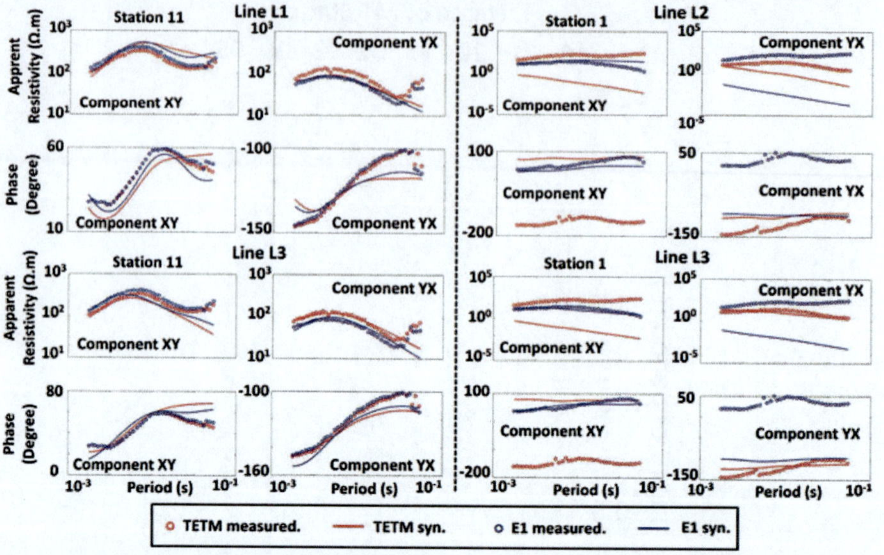

Fig. 12 Display of apparent resistivities of tensor components Z_{xy} and Z_{yx} for some MT stations in different MT lines. The synthetic and real data are similar in the distribution of data points. The term "TETM measured." expresses the real data points measured in the field trip while the term "TETM syn." refers the synthetic data points modeled after 2D inversion. The term "E1 measured." expresses the real data points rotated by the strike analysis approach by Eggers [5] while the term "E1 syn." refers the synthetic data points modeled after 2D inversion

The 2D inversion result of the Line L3 shows the great match with the seismic result [18] (Fig. 13). Positions of the faults (F1 and F3) can be seen as highly conductive zones in Line L3 (Fig. 13). Meanwhile, two separated zones as high and low velocity are appropriate with the two zones of high and low resistivity, respectively. The exposed Bang spring footprint and the fault F3 are shown in the biggest area of the low resistivity zone. For the smaller fault F6, it looks like that only MT resistivity results in Lines L3 and L1 can reflect its existence with highly conductive zone in the near surface (around 100 m below the surface) (See Fig. 13 and the bottom image in Fig. 14).

In Fig. 14 (see top image for Line L1 and Line L2), a conductive zone runs at depth around 2 km and their start locations (MT location "11") to the location (MT location "1") nearby the position of the exposed Bang hot spring. Figure 14 also shows a conductive layer in near surface close to the Bang footprint (see middle image). Line L4 recognizes a resistive layer dip at its end profile that matches greatly with the result of Line L1 although they share different resistivity values (see bottom image in Fig. 14).

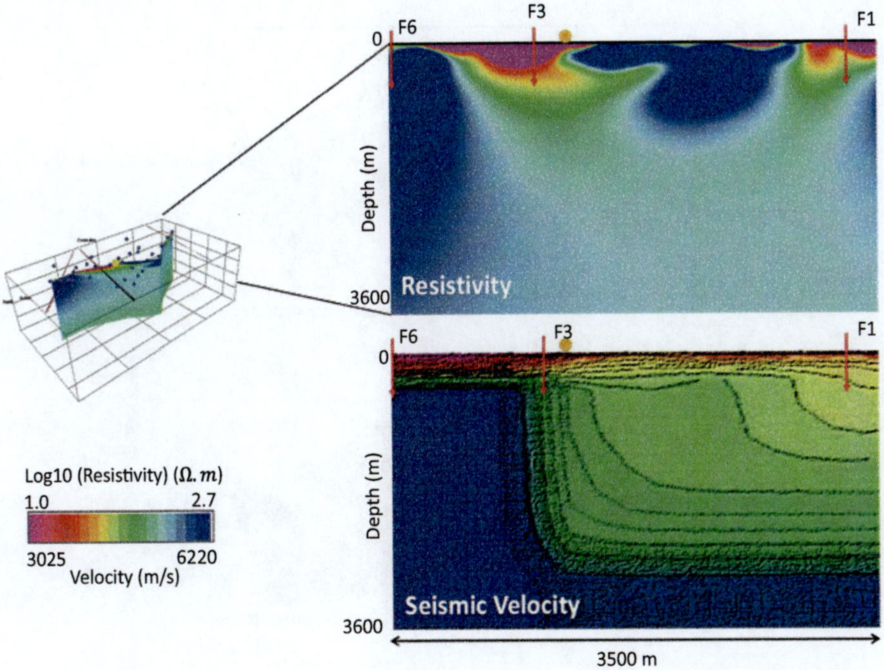

Fig. 13 Display of inverted resistivity in the Line L3 and seismic velocity model. Agreement between the two models is demonstrated by indications of faults, exposed hot spring footprint and geophysical parameter shapes. Note the seismic velocity section is extracted from the works of Tran et al. [18]

4 Conclusion

2D MT inversion could provide information of geology structures from distribution of resistivity. Analysis of dimensionality is important to give first impression about the research area. The low values of ellipticity and skew parameters can show 2D or symmetrical 3D characteristics in the zones with no existence of major geological faults (uncomplex environments) and vice versa. The real data rotated after strike angle analysis are necessary for better misfit analysis in MT inversion, leading to high validity of resistivity interpretation. For the Quang Binh case, the resistivity image is consistent with the seismic velocity in geophysical interpretation. Moreover, geological faults and hot spring exposed areas can be seen by conductive zones. Great compatibility between the inverted electrically resistivity distribution of the 2D lines gives good insight for geology structures.

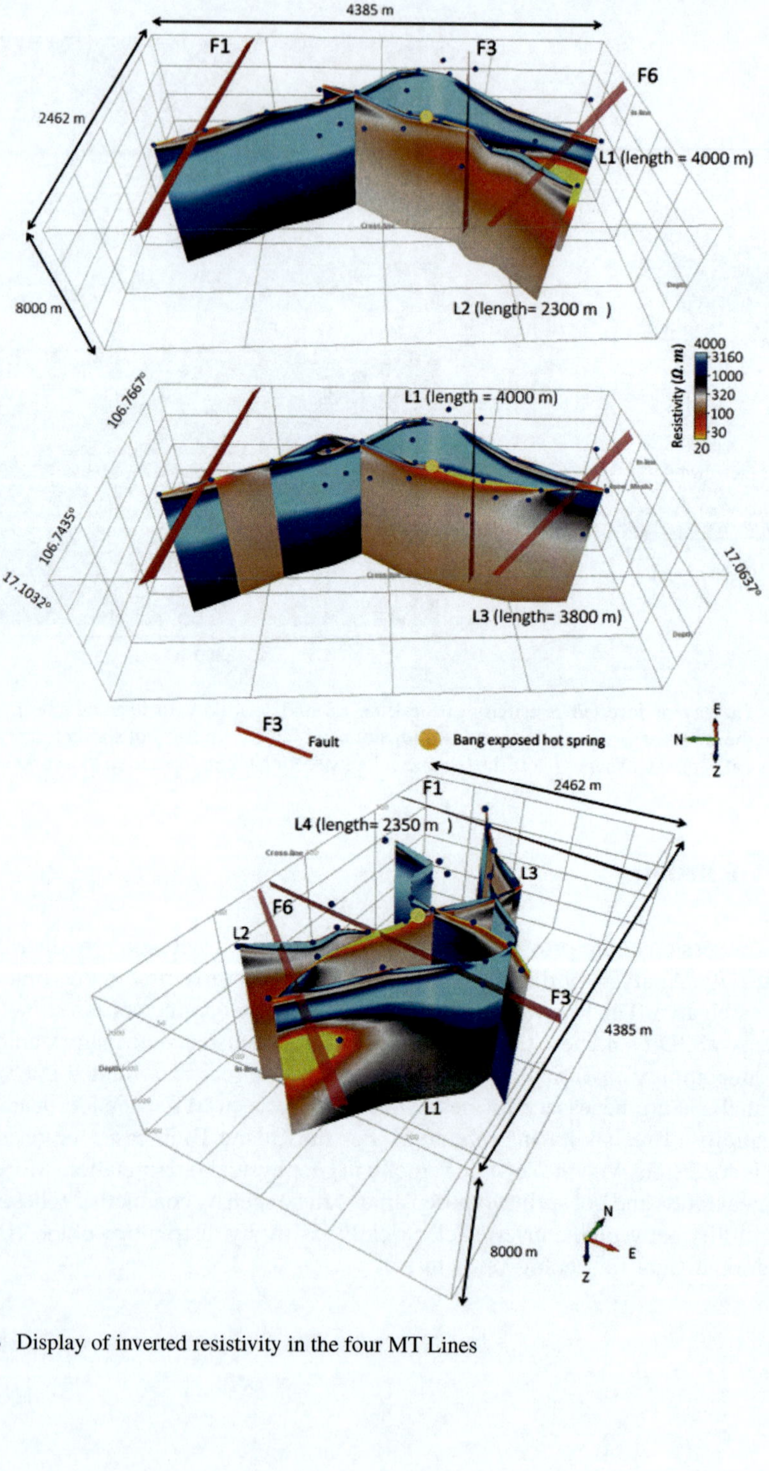

Fig. 14 Display of inverted resistivity in the four MT Lines

Acknowledgements This research is funded by Vietnam National University, Ho Chi Minh City (VNU-HCM) under grant number C2022-18-06. This work was supported by resources provided by the Pawsey Supercomputing Centre with funding from the Australian Government and the Government of Western Australia. We would like to thank Gary Egbert, Anna Kelbert and Naser Meqbel for granting us access to the 2D MT inversion code ModEM2D. We appreciate dGB Earth Science for providing software OpendTect. C.V.A.L. would like to give very special thanks to Mr. Can Van Le and Ms. Loan Thi Bui for their constant supports and guidance.

References

1. Berdichevsky, M.N., Dmitriev, V.I.: Models and methods of Magnetotellurics. Springer (2008)
2. Berdichevsky, M.N.: Marginal notes on magnetotellurics. Surv. Geophys. **20**, 341–375 (1999)
3. Sims, W.E., Bostick Jr, F.: Methods of magnetotelluric analysis. Texas Univ at Austin Electronics Research Center (1969)
4. LaTorraca, G., Madden, T., Korringa, J.: An analysis of the magnetotelluric impedance for three-dimensional conductivity structures. Geophysics **51**, 1819–1829 (1986)
5. Eggers, D.E.: An eigenstate formulation of the magnetotelluric impedance tensor. Geophysics **47**, 1204–1214 (1982)
6. Yee, E., Paulson, K.: The canonical decomposition and its relationship to other forms of magnetotelluric impedance tensor analysis. J. Geophys. **61**, 173–189 (1987)
7. Le, C.V.A., Harris, B.D., Pethick, A.M., Takam Takougang, E.M., Howe, B.: Semiautomatic and automatic cooperative inversion of seismic and magnetotelluric data. Surveys Geophys. **37**, 845–896 (2016)
8. Kelbert, A., Meqbel, N., Egbert, G.D., Tandon, K.: ModEM: a modular system for inversion of electromagnetic geophysical data. Comput. Geosci. **66**, 40–53 (2014)
9. Zhdanov, M.S.: Geophysical electromagnetic theory and methods. Amsterdam, The Netherlands (2009)
10. Kieu, D.T., Kepic, A., Le, V.: Fuzzy clustering constrained magnetelluric inversion-Case study over the Kevitsa ultramafic intusion, Northern Finland. Near Surface Geoscience 2016-First Conference on Geophysics for Mineral Exploration and Mining, vol. 2016, pp. 1–5. European Association of Geoscientists & Engineers (2016)
11. Le, C.V.A., Harris, B.D., Pethick, A.M.: New perspectives on solid earth geology from seismic texture to cooperative inversion. Sci. Rep. **9**, 14737 (2019)
12. Le, C.V.A., Kieu, T., Pham, D., Lai, P.: Inversion of 3D magnetotelluric data for geothermal exploration at Quang Binh province, Vietnam. EAGE-GSM 2nd Asia Pacific Meeting on Near Surface Geoscience and Engineering, vol. 2019, pp. 1–5. European Association of Geoscientists & Engineers, Malaysia (2019)
13. Bartel, L., Jacobson, R.: Results of a controlled-source audiofrequency magnetotelluric survey at the Puhimau thermal area, Kilauea Volcano, Hawaii. Geophysics **52**, 665–677 (1987)
14. Blake, S., Henry, T., Muller, M.R., Jones, A.G., Moore, J.P., Murray, J., Campanyà, J., Vozar, J., Walsh, J., Rath, V.: Understanding hydrothermal circulation patterns at a low-enthalpy thermal spring using audio-magnetotelluric data: A case study from Ireland. J. Appl. Geophys. **132**, 1–16 (2016)
15. Wannamaker, P.E., Stodt, J.A., Pellerin, L., Olsen, S.L., Hall, D.B.: Structure and thermal regime beneath the South Pole region, East Antarctica, from magnetotelluric measurements. Geophys. J. Int. **157**, 36–54 (2004)
16. Doan, T., Tran, V., Lai, P., Le, S., Pham, D., Duong, N., Dinh, T., Nguyen, Q.: Results of magnetotelluric survey for studying geothermal system in the Bang area, Quang Binh province. Vietnam J. Earth Sci. **37**, 48–56 (2015)

17. Huck, H.: The road to open source: Sharing a ten years' experience in building OpendTect, the open source seismic interpretation software. In: 74th EAGE Conference and Exhibition, Copenhagen, Denmark (2012)
18. Tran, V., Dinh, T., Doan, T., Lai, P., Duong, N., Nguyen, Q., Pham, D.: Utilization of seismic refraction data for the study of structure of Bang hot-water source, Le Thuy, Quang Binh. Vietnam J. Earth Sci. **38**, 393–408 (2016)

Physicochemical Characteristics of the Middle Triassic Limestone in Ha Nam Province, Vietnam and the Ability of Adsorption of Heavy Metal Ions from Aqueous Environments

Bui Hoang Bac, Le Thi Duyen, Nguyen Thi Thanh Thao, and Nguyen Huu Tho

Abstract Calcium carbonate rocks of middle Triassic age formations account for a large amount, up to 25% of Vietnam's limestone potential. Ha Nam is one of the areas with large reserves of middle Triassic limestone of the Dong Giao Formation, up to billions of m^3. Dozens of limestone quarries have been licensed to exploit the area for different applications such as cement, lime, light powder, and common building materials. The research on limestone as a raw material in the treatment of contaminated water is still limited. In this study, the middle Triassic limestone sample (Dong Giao formation) from Ha Nam area is used to determine physicochemical characteristics and then to investigate the ability to remove Pb^{2+} ions from an aqueous solution. The analysis results show that the limestone in the study area is of good quality. The mineral composition is mainly calcite and the CaO content is up to 92–100%. Other properties of limestone such as physical and radioactive properties indicate that the limestone here can be used safely for different fields. For environmental treatment, different conditions such as contact time, solution pH, adsorbent

B. H. Bac (✉) · L. T. Duyen · N. T. T. Thao
Hanoi University of Mining and Geology, Hanoi, Vietnam
e-mail: buihoangbac@humg.edu.vn

L. T. Duyen
e-mail: lethiduyen@humg.edu.vn

N. T. T. Thao
e-mail: nguyenthithanhthao@humg.edu.vn

B. H. Bac
Innovations for Sustainable and Responsible Mining (ISRM) Research Group, Hanoi University of Mining and Geology, Hanoi 100000, Vietnam

B. H. Bac · L. T. Duyen · N. T. T. Thao
Research and Advanced Technology Applications in Environmental, Material and Earth Sciences, Hanoi University of Mining and Geology, Hanoi 100000, Vietnam

N. H. Tho
Ministry of Construction of Vietnam, Hanoi, Vietnam

L. Q. Nguyen et al. (eds.), *Advances in Geospatial Technology in Mining and Earth Sciences*, Environmental Science and Engineering,
https://doi.org/10.1007/978-3-031-20463-0_23

weight and initial concentration of Pb^{2+} were tested. The results show that experimental conditions of $pH_0 = 6.2$ and a temperature of 30 °C, with 0.1 g of limestone material and an initial concentration of Pb^{2+} of 50 mg/L, the Pb^{2+} adsorption can reach equilibrium after about 40 min and efficiency reached 98.88%. The adsorption process follows the Langmuir adsorption isotherm model with the maximum monolayer adsorption capacity of 76.22 mg/g and follows the pseudo-second-order kinetic equation. The result indicate that unmodified the Triassic limestone powder has a significant potential for the removal of heavy ions from an aqueous solution.

Keywords Triassic limestone · Aqueous environments · Adsorption · Vietnam

1 Introduction

Vietnam is a country with great potential for calcium carbonate rocks, including limestone (1754 billion tons, of which the defined reserve is about 2.21 billion tons, accounting for 0.13%), mainly distributed in the Northeast (46.22%), Northwest (25.92%), North Central (22.64%) of Vietnam [1]. In this country, limestone rocks have been used in many fields such as building materials, tiling, calcining lime, metallurgical flux, processing earthenware, light powder, heavy powder, fertilizer, and cement. However, with the government's policy of using natural resources effectively, the expansion of limestone applications is an important and necessary task.

Dong Giao Formation ($T_2a\text{đ}g$) was defined by Jamoida and Pham Van Quang in 1965 in the geological mapping of 1:500,000 scale of the northern part [2]. This formation is widely distributed in the provinces of Son La, Lai Chau, Hoa Binh, Ha Tay, Ninh Binh, Ha Nam, and Thanh Hoa. This formation's limestone rocks are light gray and dark gray with medium layers to blocks, with a thickness of over 500–900 m. This limestone is considered to be of good quality and currently, many limestone mines have been explored and exploited by some enterprises for producing cement materials. According to Dr. Doan Huy Cam, the potential of calcium carbonate rocks of the Middle Triassic age formations (T_2), which are mainly formations of the Dong Giao Formation ($T_2a\text{đ}g$), accounts for 25% of the limestone potential of the whole country (about 435 billion tons) [1]. In which, Thanh Liem area, Ha Nam province is one of the regions with large limestone reserves of the Dong Giao formation, up to billions of m^3. Dozens of limestone quarries have been licensed to exploit in the area. Limestone is here mainly used for some industries such as cement, lime, light powder, and common building materials [3].

Recently, the materials that are available, low cost and non-toxic materials are interested in many domestic scientists in the treatment of polluted water environment. Due to having high adsorption capacity, clay materials are widely used in water pollution treatment. In our country, clay minerals are often used to treat water pollution such as bentonite [4–6], vermiculite [7], kaolin [8], halloysite [9, 10]... However, natural clay materials have been used in many other fields with a high economic value, so their reserves are decreasing day by day. Therefore, finding alternative sources of raw materials is a necessary task. Research on Dong Giao limestone as a raw material to treat contaminated water with heavy metal ions is still limited in Vietnam. Thus, with great potential, along with good quality, adequate mineral exploitation, and processing conditions, the research and application of limestone of the Dong Giao formation in Ha Nam areas in the field of polluted water treatment will meet many advantages and be feasible.

In this paper, the natural Triassic limestone of Dong Giao formation from Ha Nam province is characterized and used to study the ability of removal of Pb^{2+} in an aqueous solution. The achieved results will open an effective application for this abundant limestone material.

2 Experimental and Analytical Methods

Natural Triassic limestone samples of Dong Giao formation are taken at the typical outcrops in Ha Nam province. The sample is crushed and then ground with a mill, and then finely ground with an onyx mortar. Then, the sample is used for further experiments and analysis. For petrographic and SEM–EDS analysis, the sample is cut and polished fully.

2.1 Experimental Methods

2.1.1 Experimental Study of the Effect of Some Parameters on the Adsorption Efficiency of Pb^{2+} by Limestone Material

The experiments are conducted by adding a quantity of Ha Nam limestone powder to 50 ml of Pb^{2+} solution. The influence of different physicochemical parameters on the adsorption process is examined. Pb^{2+} solutions with initial concentrations of 20 ÷ 150 mg/L are prepared. The contact time is varied between 10 and 80 min. The pH of the solutions is adjusted in the range of 2.1 ÷ 7.1 and the dose of limestone material changed from 0.01 to 0.2 g. The mixture is then stirred continuously at 100 rpm using a mechanical shaker at room temperature. After filtration to remove the solid, the remaining concentration of Pb^{2+} is determined by using the inductively inducing plasma-mass spectrometric method (ICP-MS).

The adsorption capacity and the adsorption efficiency are determined by Eqs. (1) and (2) [11].

$$Q = (C_0 - C) \cdot V/m. \tag{1}$$

$$H = (C_0 - C) \cdot 100/C_0. \tag{2}$$

where Q (mg/g) and H (%) represent the adsorption capacity and removal efficiency of Pb^{2+}, respectively. C_0 (mg/L) and C (mg/L) are the initial and equilibrium concentrations of Pb^{2+} in solution. V (l) is the volume of the adsorbent solution and m is the mass of Ha Nam limestone powder (g), respectively.

2.1.2 Adsorption Isotherm

The Pb^{2+} adsorption capacity of the limestone material from the study area is calculated based on the isothermal adsorption of Langmuir and Freundlich [12].

Langmuir linear equation:

$$\frac{C_e}{Q} = \frac{C_e}{Q_m} + \frac{1}{K_L \cdot Q_m}. \quad (3)$$

Freundlich linear equation:

$$\ln q = \ln K_F + \frac{1}{n} \cdot \ln C_e \quad (4)$$

where C_e (mg/L) is the equilibrium concentration of Pb^{2+}, Q (mg/g) is the adsorption capacity at equilibrium. Q_m (mg/g) is the maximum adsorption capacity, K_L is the Langmuir coefficient related to the adsorption energy. K_F and n are the constants of the Freundlich model.

2.1.3 Adsorption Kinetics

The adsorption kinetics is studied by two kinetic models: the pseudo-first-order model (Eq. 5) and the pseudo-second-order model (Eq. 6).

$$\ln(Q_e - Q_t) = \ln Q_e - k_1 t \quad (5)$$

$$t/Q_t = t/Q_e + 1/(k_2 \cdot Q_e^2). \quad (6)$$

where Q_e is the adsorption capacity at equilibrium (mg/g), Q_t is the adsorption capacity at time t (mg/g). k_1 and k_2 are pseudo-first-order (min^{-1}) and pseudo-second-order (g/mg/min) rate constants, respectively.

2.2 *Analytical Methods*

The analytical methods used in this study include X-ray Diffraction analysis (XRD), X-ray Fluorescence (XRF), Scanning Electron Microscopy-Energy Dispersive X-Ray Spectroscopy (SEM–EDS) and Inductively Coupled Plasma Mass Spectrometry (ICP-MS) and other analyses.

3 Results and Discussion

3.1 *Characteristics of the Limestone from the Study Area*

3.1.1 Mineral Composition

The limestone in Ha Nam area has a light gray, greenish-gray color. The rock has a thick layer or block structure, with a micro-grain architecture. According to the results of petrographic analysis, the mineral composition of the rock is mainly calcite 92–100%, dolomite (from 0 to 5%), and quartz (from 0 to 3%) and, little organic matter. The characteristics of minerals under the petrographic microscopy are as follows: calcite minerals have the form of crystalline granules, micro-granular polymorphs. The size of calcite is mainly from 0.001 to 0.05 mm, sometimes the limestone is recrystallized to create polymorphic granules with the size of 0.05–0.3 mm. Under glass, calcite is colorless, well-absorbed, completely cleavage, and has high-order white interference. Dolomite is automorphic and, semi-automorphic granules with sizes of 0.05–0.3 mm, colorless, clearly pseudo-adsorbent, completely cleavage, high-order white interference, alternately scattered growth. Quartz has the form of crystalline grains, polymorphic microparticles, size 0.01–0.05 mm, colorless, not cleavage, first-order white light interference, replacing scattered irregular diffusion patterns. Organic matter was found in most samples. They exist in the form of dark, non-luminescent, unevenly dispersed dust.

The results of mineral composition analysis by X-ray diffraction (XRD) and FT-IR methods are presented in Fig. 1. The XRD pattern shows that calcite (Cal) is the main mineral in the sample with typical XRD peaks at 23.1°, 29.4°, 36.0°, 39.4°, 43.2°, and 48.5° (Fig. 1a) [13]. The FT-IR pattern in Fig. 1b indicates that calcite ($CaCO_3$) is the significant mineral in the sample. The characteristic absorption peaks of calcite are stretching vibrations of the C-O at 1427 cm^{-1} and bending vibrations of the C-O at 876 and 712 cm^{-1} (Fig. 1b) [14, 15].

3.1.2 Chemical Composition

Average chemical compositions of the limestone from the Ha Nam area are presented in Table 1. The results show that the element oxide contents of CaO and MgO are

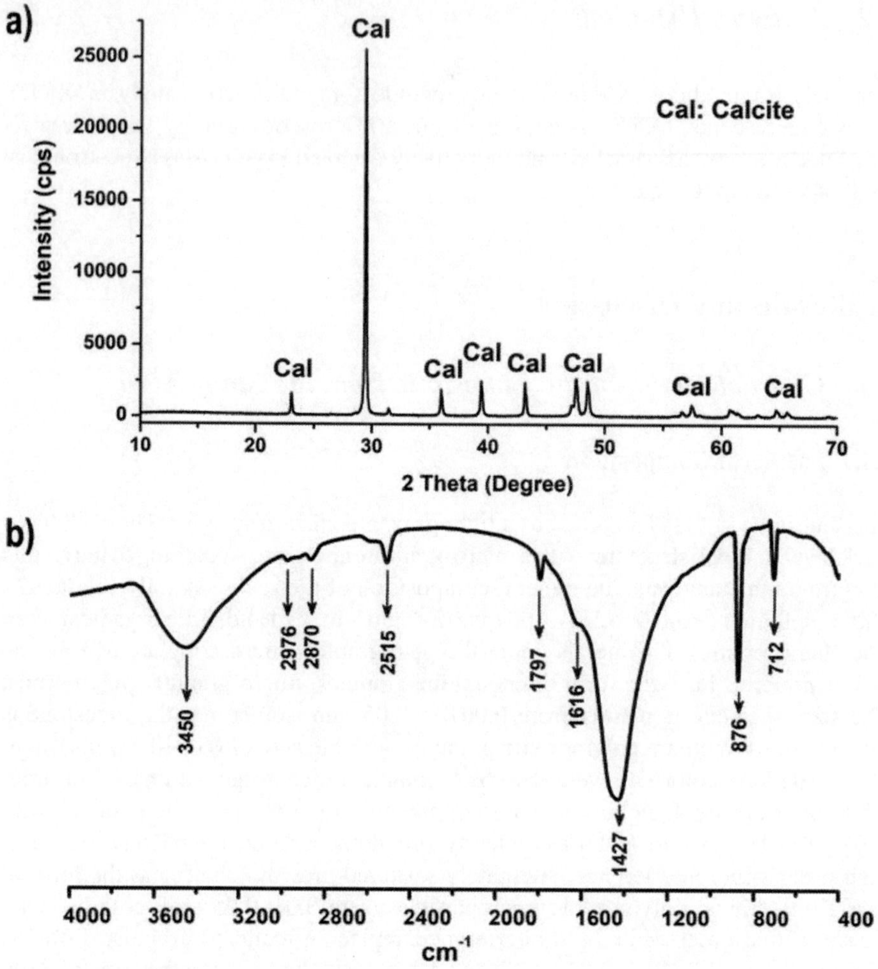

Fig. 1 XRD (**a**) and FT-IR (**b**) patterns of the limestone from the Ha Nam area

53.05 ÷ 54.71%, and 0.06 ÷ 1.40%, respectively. The content of other oxides such as Al_2O_3, T.Fe, SiO_2, MnO, and K_2O is not significant. This result is consistent with EDS results with mainly Ca peak, indicating the existence of quite pure calcite mineral (Fig. 2).

3.1.3 Radioactive Properties

The results show that the limestone in the study area has radioactive intensity varying from 6.3 R/h to 7.5 R/h, K content: 0.3–0.4%, U: 2.3–2.8 ppm, Th: 5.6–7.5 ppm. With such results, it can be confirmed that limestone in the study area has very low

Table 1 Chemical composition of the limestone from Ha Nam area

Chemical composition $\frac{\text{min}\div\text{max}}{\text{average}}$ (%)			
CaO	MgO	Al_2O_3	T.Fe[a]
$\frac{53.05\div54.71}{53.86}$	$\frac{0.06\div1.40}{0.94}$	$\frac{0.001\div0.035}{0.016}$	$\frac{0.014\div0.064}{0.036}$
SiO_2	MnO	K_2O	LOI[b]
$\frac{0.20\div1.12}{0.50}$	$\frac{0.012\div0.062}{0.034}$	$\frac{0.005\div0.020}{0.013}$	$\frac{42.77\div43.53}{43.22}$

Note [a] Total iron; [b] Loss on ignition

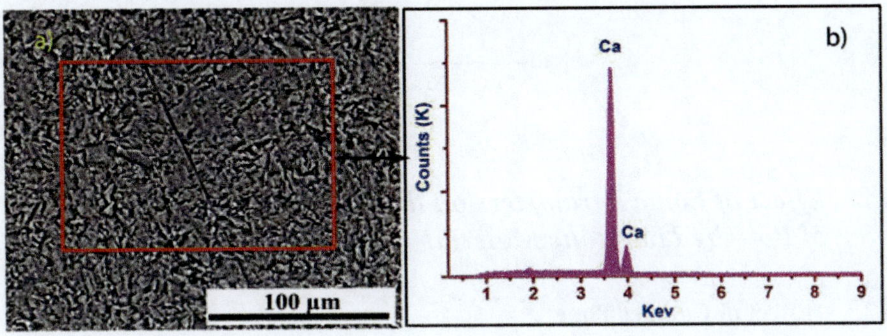

Fig. 2 SEM image (**a**) and EDS (**b**) results of the natural limestone from Ha Nam area

radioactivity, so its exploitation and use will not affect the environment and human health.

3.1.4 Physico-Mechanical Properties

The results of mechanical and physical parameters of the natural limestone from the Ha Nam area are shown in Table 2. The results show that the limestone has good physical and mechanical properties, meeting the fields of civil construction.

Table 2 Summary of mechanical and physical parameters of natural limestone samples in Ha Nam province

Parameters	Unit	Max	Min	Average
Natural Humidity (W)	%	0.09	0.16	0.13
Natural Volumetric Mass (γ_0)	g/cm^3	2.67	2.70	2.68
Saturated Volumetric Mass (γ_{bh})	g/cm^3	2.68	2.71	2.69
Natural Compressive Strength (δ_{tn})	kG/cm^2	632	715	685
Saturated Compressive Strength (δ_{bh})	kG/cm^2	569	682	646
Natural Tensile Strength (δ_k)	kG/cm^2	64.4	72.8	69.7
Saturated Tensile Strength (δ_{bh})	kG/cm^2	59.4	71.3	65.6

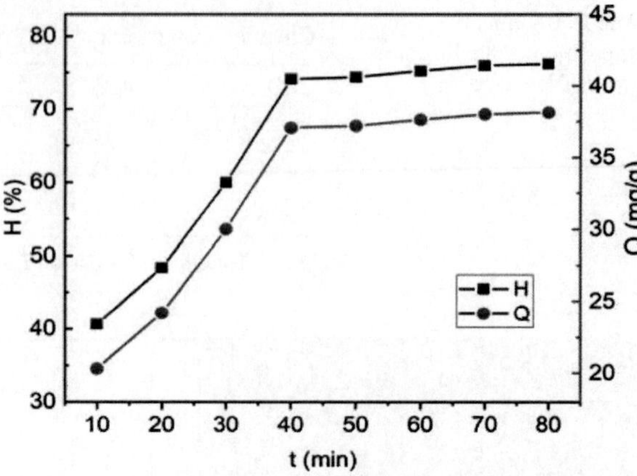

Fig. 3 The variation of Pb^{2+} adsorption capacity Q (mg/g) and the adsorption efficiency H (%) according to contact time (t); $m_{\text{limestone}} = 0.05$ g; $C_0 = 50$ mg/L; $pH_o = 5$; $T = 30$ °C

3.2 *Effect of Some Parameters on the Adsorption Efficiency of Pb^{2+} by Limestone Material*

3.2.1 Effect of Contact Time

The change in adsorption efficiency and capacity of the Triassic Ha Nam limestone by the contact time is shown in Fig. 5. During the experimental period, the adsorption capacity and efficiency increased rapidly in the initial 40 min. After the contact time of 40 min, the adsorption capacity and efficiency increased slowly and steadily as the adsorption reached equilibrium. To obtain high adsorption capacity and efficiency (37.06 mg/g and 74.13%, respectively), a contact time of 40 min is selected for further experiments (Fig. 3).

3.2.2 Effect of pH

The ability to remove Pb^{2+} ions is highly dependent on the pH of the solution because pH changes the surface properties of the adsorbent, the adsorption media, and the mechanism of Pb^{2+} ion removal in the aqueous solution. To avoid precipitation of $Pb(OH)_2$, the pH values are adjusted in the range of 2.1–7.1. The change in adsorption capacity and efficiency of limestone by the change in pH are presented in Fig. 4. It can be seen that, the efficiency and adsorption capacity increase as the pH increases. This result is explained because in an acidic media, limestone is protonated and then the surface of the limestone particles will be positively charged, thus reducing the number of adsorption centers of the material. On the other hand, in an acidic environment with low pH values, some limestone particles will be dissolved, thus reducing the adsorption capacity and efficiency. In the pH range from 5 to 6.2, the adsorption capacity and efficiency increased rapidly (29.3 mg/g and 58.6% to 47.5 mg/g and 95%, respectively). However, when increasing the pH from 6.2 to 7.1, the adsorption

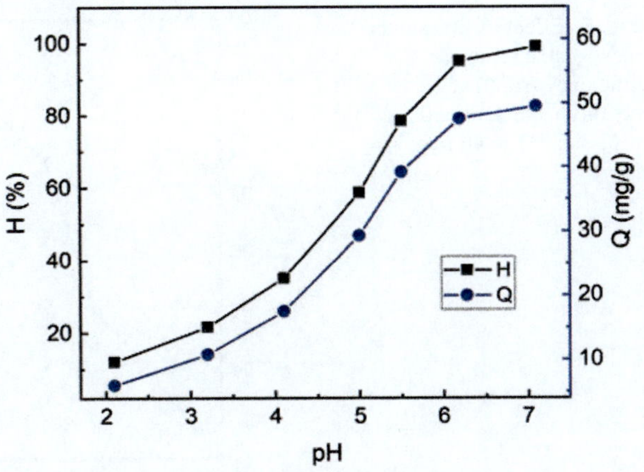

Fig. 4 Variation of Pb^{2+} adsorption capacity Q (mg/g) and the adsorption efficiency H (%) according to pH; $m_{\text{limestone}} = 0.05$ g; $C_o =$ 50 mg/L; $t = 40$ min; $T =$ 30 °C

capacity and efficiency increased not much (49.5 mg/g and 99%). Therefore, it is possible to choose pH = 6.2 close to the natural pH of the Pb^{2+} ion medium (5.7) for further experiments.

3.2.3 Effect of Adsorbent Mass

The effect of the limestone mass on Pb^{2+} adsorption efficiency is presented in Fig. 5. It indicates that when the limestone mass increases from 0.01 to 0.1 g, the adsorption efficiency increases rapidly from 92.96 to 98.88%. When the adsorbent mass increases from 0.1 to 0.2 g, the adsorption efficiency increases insignificantly because the adsorption process reached equilibrium. To achieve high adsorption efficiency (98.88%) with the minimum amount of adsorbent used, the mass of limestone material of 0.1 g is selected for treatment of Pb^{2+} from the aqueous solution.

3.3 Adsorption Isotherm

The appropriate experimental conditions for removal of Pb^{2+} from aqueous solution are 0.1 g of limestone powder, a contact time of 40 min, a pH value of 6.2 at room temperature (30 °C), and differential initial Pb^{2+} concentrations. Then, the remaining Pb^{2+} concentration at equilibrium (C_e) is determined and the values of $\ln C_e$, $\ln Q$, and the C_e/Q ratio can be calculated (Table 3) and the isotherm equations of Langmuir (Fig. 6a) and Freundlich (Fig. 6b) adsorption are established.

Based on these adsorption isotherm curves, Langmuir and Freundlich constants can be calculated, respectively and the result is shown in Table 4.

From the results, the Pb^{2+} adsorption on Triassic Ha Nam limestone material can be described by both Langmuir and Freundlich's isothermal adsorption models.

Fig. 5 Effect of limestone mass on the removal efficiency *H* (%) at conditions of $C_0 = 50$ mg/L; $pH_0 = 6.2$; $t = 40$ min; $T = 30$ °C

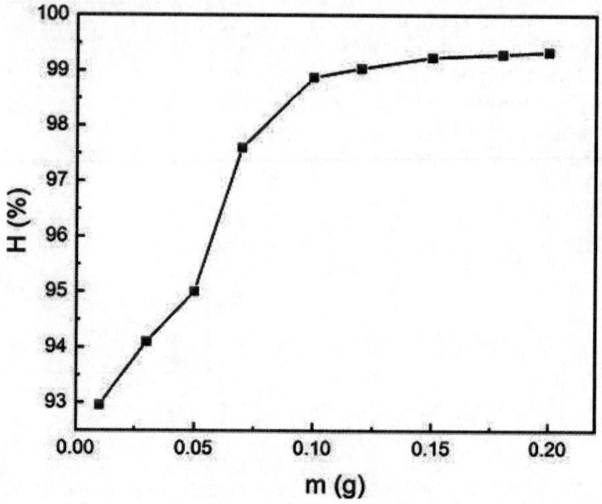

Table 3 Values of $\ln C_e$, $\ln Q$, C_e/Q at different initial concentrations of Pb^{2+}

C_o (mg/L)	C_e (mg/L)	*H* (%)	*Q* (mg/g)	C_e/Q (g/L)	$\ln C_e$	$\ln Q$
20	0.21	99.00	9.90	0.0212	−1.561	2.292
30	0.31	98.96	14.84	0.0210	−1.168	2.698
50	0.60	98.79	24.70	0.0244	−0.505	3.207
70	1.40	98.00	34.30	0.0408	0.336	3.535
100	3.23	96.70	48.39	0.0667	1.172	3.879
120	5.16	95.70	57.42	0.0898	1.641	4.050
150	6.5	95.67	71.75	0.0906	1.872	4.273

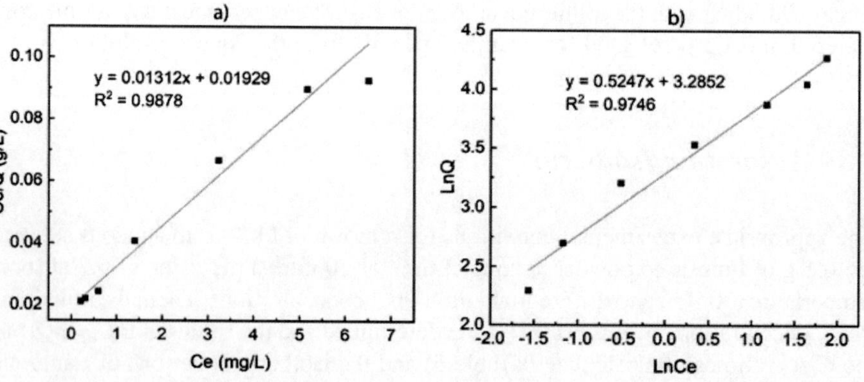

Fig. 6 Adsorption isotherm curves at 30 °C follow **a** Langmuir and **b** Freundlich models

Table 4 Experimental constants Q_m, K_L, K_F, and n in Langmuir and Freundlich equations

Langmuir			Freundlich		
Q_m (mg/g)	K_L	R^2	n	K_F	R^2
76.22	0.68	0.9878	1.906	26.71	0.9746

However, it can be seen that the R^2 value of the Langmuir isothermal model ($R^2 = 0.9878$) is higher than that of the Freundlich model ($R^2 = 0.9746$) with the maximum adsorption capacity of 76.22 mg/g. It indicates that the Langmuir isotherm is more appropriate than the Freundlich isotherm. This result indicates adsorption ability of Pb^{2+} from aqueous solution by the limestone in the study area is equivalent to limestone in some areas of the world [16, 17] and better than some other natural adsorbents such as kaolin, and sericite [18, 19].

3.4 Adsorption Kinetics

Pb^{2+} adsorption is studied under experimental conditions of 50 mL of Pb^{2+} 50 mg/L solution, 0.05 g limestone mass, $pH_o = 5.0$, $T = 30$ °C, and different contact time. Then, the adsorption capacity at adsorption equilibrium (experimental Q_e) is determined and the graphs of pseudo-first-order (Fig. 7a) and pseudo-second-order kinetic equations (Fig. 7b) are established.

From Fig. 7, the adsorption rate constants (k) and the equilibrium adsorption capacity (Q_e) are defined. The results are presented in Table 5. It can be seen that the Q_e value calculated from the pseudo-first-order kinetic equation (58.14 mg/g) is higher than the experimental Q_e value (38.13 mg/g). Meanwhile, the Q_e calculated from the pseudo-second-order kinetic equation (47.39 mg/g) is not much different

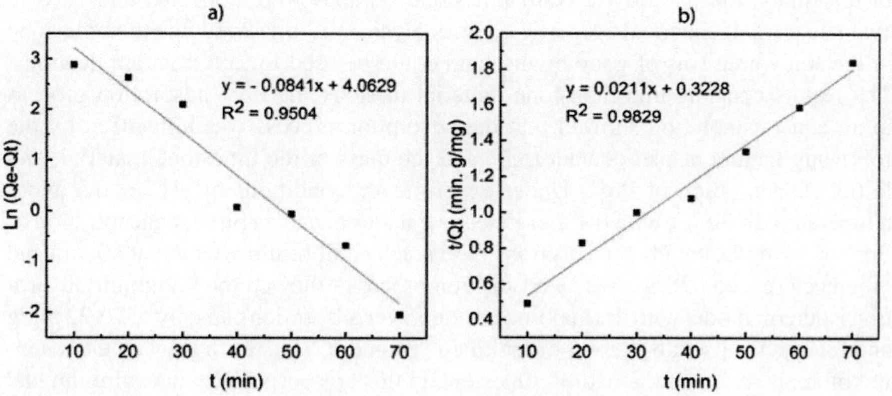

Fig. 7 Description of the experimental data by **a** pseudo-first-order and **b** pseudo-second-order kinetic equations

Table 5 Values of k and Q_e calculated from pseudo-first-order and pseudo-second-order kinetic equations

Pseudo-first-order kinetic equation			Pseudo-second-order kinetic equation			Experimental Q_e (mg/g)
Q_e (mg/g)	k_1 (min^{-1})	R^2	Q_e (mg/g)	k_2 (g/mg/min)	R^2	
58.14	0.08410	0.9504	47.39	0.00138	0.9829	38.13

from the experimental Q_e value (38.103 mg/g). In this case, the regression coefficient $R^2 = 0.9829$, approximately equal to 1. These results prove that the Pb^{2+} adsorption process by Ha Nam limestone material following the pseudo-second-order kinetic equation is the best fit for the experimental data. Accordingly, the adsorption rate constant is determined to be 0.00138 g/mg/min.

3.5 *Mechanism of the Adsorption Process*

The process of removing heavy metals in the aqueous environment using natural limestone has been published by several authors. Depending on the pH values of the aqueous solutions and the nature of metal ions, the heavy metal removal mechanism can be physical adsorption, surface complexation, surface precipitation, ion exchange, and chemical precipitation [16, 17]. For Pb^{2+} adsorption by the natural limestone [20, 21], the mechanism of the adsorption process can be explained in two stages: surface complexation and ion exchange.

4 Conclusions

In this study, the natural Ha Nam limestone samples were collected and used to evaluate physicochemical properties. The analysis results showed that the limestone in the study area was of good quality and could be used for different applications. The results of using this limestone material to study the Pb^{2+} adsorption process in an aqueous solution showed that the adsorption process was influenced by the following factors of pH, contact time, and the mass of the limestone material, and initial concentration of Pb^{2+}. Under experimental conditions of $pH_o = 6.2$ and a temperature of 30 °C, with 0.1 g of limestone material and an initial concentration of Pb^{2+} of 50 mg/L, the Pb^{2+} adsorption could reach equilibrium after about 40 min and efficiency reached 98.88%. The adsorption process followed the Langmuir adsorption isotherm model with the maximum monolayer adsorption capacity of 76.22 mg/g and follows the pseudo-second-order kinetic equation. This result showed the potential of applications of the natural limestone in this area, especially in environmental treatment.

References

1. Cam, D.H.: Doctoral thesis "Assessment of calcium carbonate rock resources and economic use of them in socio-economic development, especially in the development strategy of Vietnam's cement manufacturing industry". Hanoi University of Mining and geology (2003) (Vietnamese)
2. Khuc, V., et al.: The Vietnamese geological quartile reference book. Department of Geology and Minerals of Vietnam (2000) (Vietnamese)
3. Report on exploration results of limestone as a cement material in Thanh Nghi 3 area, Thanh Nghi commune, Thanh Liem district, Ha Nam province. Department of Geology and Minerals of Vietnam (2017) (Vietnamese)
4. Ha, T.H.H., Dang, K.C., Nguyen, H.P.: Study on the adsorption capacity of Cu^{+2} in water on bentonite Tuy Phong, Binh Thuan. J. Chem. **49**(5AB), 62–67 (2011) (Vietnamese)
5. Ha, T.H.H., Nguyen, H.P., Dang, K.C.: Study on adsorption capacity of heavy metals on bentonite Tuy Phong, Binh Thuan. J. Chem. **49**(5AB), 68–73 (2011) (Vietnamese)
6. Le, T.H., Phan, C.: Study on the process of modified Thuan Hai bentonite and the application of Mn^{2+} ion adsorption in water. J. Sci. Tech. Univ. Danang **3**(26), 112–117 (2008) (Vietnamese)
7. Tran, L.T.: Research on modification of vermiculite clay minerals to improve heavy metal adsorption capacity in water. VNU Sci. J. Earth Environ. Sci. **32**(3), 82–91 (2016) (Vietnamese)
8. Tu, T.C.L., Hoang, T.T.T., Ha T.T.T.: Research on the application of kaolin clay and rice husk as a filter material for groundwater contaminated with manganese. Hydrometeorological J. **4**, 17–22 (2020) (Vietnamese)
9. Bui, H.B., Hoang, N., Nguyen, T.T.T., Le, T.D., Vo, T.H., Nguyen, T.D., Luong, Q.K., Do, M.A.: Performance evaluation of nanotubular halloysites from weathered pegmatites in removing heavy metals from water through novel artificial intelligence-based models and human-based optimization algorithm. Chemosphere **282**, 131012 (2021)
10. Bui, H.B., Hoang, N., Nguyen, T.T.T., Vo, T.H., Le, T.D., Nguyen, T.D., Nguyen, K.D., Nguyen, H.H.: Estimating heavy metals absorption efficiency in an aqueous solution using nanotube-type halloysite from weathered pegmatites and a novel Harris hawk optimization-based multiple layers perceptron neural network. Eng. Comp. (2021). https://doi.org/10.1007/s00366-021-01459-8
11. Neha, G., Atul, K., et al.: Adsorption studies of cationic dyes onto Ashoka (Saraca asoca) leaf powder. J. Taiwan Inst. Chem. Eng. **43**(1), 125–131 (2012)
12. Mobasherpour, I., Salahi, E.: Comparative of the removal of Pb^{2+}, Cd^{2+} and Ni^{2+} by nano crystallite hydroxyapatite from aqueous solutions: adsorption isotherm study. Arab. J. Chem. **5**(4), 439–446 (2012)
13. Jelena, M., Ágata, A.D., Ana, I.J., Manuel, F.C.P., Sofia, M., Sven, P., Fernando, J.A.S.B.: XRD Identification of Ore minerals during cruises: refinement of extraction procedure with Sodium Acetate buffer. Minerals **10**, 160 (2020)
14. Katarzyna, J.S.: The importance of Fourier-Transform infrared spectroscopy in the identification of carbonate phases differentiated in magnesium content. Spectroscopy **34**(6), 32–42 (2019)
15. Barhoum, A., El-Sheikh, S.M., Morsy, F., El-Sherbiny, S., Reniers, F., Dufour, T., Delplancke, M.P., Van Assche, G., Rahier, H.: Preparation and characterization of ultra-hydrophobic calcium carbonate nanoparticles, IOP Conf. Series: Mat. Sci. Eng. **64**, 012037 (2014)
16. Xuewen, S., Yuwei, C., Xianzhong, B., Xianping, L.: Porous vaterite and cubic calcite aggregated calcium carbonate obtained from steamed ammonia liquid waste for Cu^{2+} heavy metal ions removal by adsorption process. Appl. Surf. Sci. **536**, 147958 (2021)
17. Gao, Q., Qiaoqin, X., Haibo, L., Tianhu, C., Jingjing, X., Hongwei, L.: Removal of Cu(II) from aqueous solutions using dolomite–palygorskite clay: performance and mechanisms. Appl. Clay Sci. **118**, 107–115 (2015)
18. Yasmin, V., Matias, S.N., Éder, C.L., Ioannis, A., Marcos, L.S.O., Guilherme, L.D.: An overview of geological originated materials as a trend for adsorption in wastewater treatment. Geosci. Front. **13**, 101150 (2022)

19. Soliman, N.K., Moustafa, A.F.: Industrial solid waste for heavy metals adsorption features and challenges: a review. J. Mater. Res. Technol. **9**(5), 10235–10253 (2020)
20. Erol, P., Ali, M.Ö., Salih, D.: Serife Parlayici Adsorption of Cu^{2+} and Pb^{2+} ion on dolomite powder. J. Hazard. Mater. **167**, 1044–1049 (2009)
21. Sturchio, N.C., Chiarello, R.P., Cheng, L., Lyman, P.F., Bedzyk, M.J., Qian, Y., You, H., Yee, D., Geissbuhler, P., Sorensen, L.B., Liang, Y., Baer, D.R.: Lead adsorption at the calcite-water interface: synchrotron X-ray standing wave and X-ray reflectivity studies. Geochim. Cosmochim. Acta **61**, 251–263 (1997)

Local Mechanical Behaviors of Steel Box Girder During Skew Incremental Launching

Jiabao Du, Wen Niu, Yu Shi, Yongzhe Wu, Yuan Chen, and Qiang Tang

Abstract In the process of incremental launching construction, deviation error will pose a threat to the bridge safety. Two adverse conditions were selected for numerical analysis to investigate the steel box girder local stress during skew incremental launching and the effects of deviation error in the process. The maximum stress of the two working conditions is similar to that of the web when the skewed and cantilevered conditions are 275.2 MPa and 195.0 MPa, respectively. The top plates stress is 211.7 MPa under skewed condition and 73.2 MPa under cantilevered condition. There is a great difference between the two conditions. Attention should be paid to the change of top plates stress in the actual launching process. Among the three launching deviations, the longitudinal deviation presents the least influence on the girder. The vertical deviation depicts the greatest influence, which will lead to a significant increase in the local stress of the steel box girder. When the vertical displacement reaches 10 cm, the web stress reaches almost four times of the normal condition. Thus, for actual incremental launching, the synchronization of vertical launching shall be ensured as far as possible to reduce the risk during incremental launching.

Keywords Steel box girder · Numerical analysis · Stress analysis · Walking-type incremental launching

J. Du (✉) · Q. Tang
School of Rail Transportation, Soochow University, Suzhou 215131, China
e-mail: dujiabao1030@outlook.com

Q. Tang
e-mail: tangqiang@suda.edu.cn

W. Niu · Y. Shi · Y. Wu · Y. Chen
CCCC First Highway Engineering Group Co., Ltd, Beijing 100020, China

CCCC Tunnel Engineering Company Limited, Beijing 100020, China

L. Q. Nguyen et al. (eds.), *Advances in Geospatial Technology in Mining and Earth Sciences*, Environmental Science and Engineering,
https://doi.org/10.1007/978-3-031-20463-0_24

1 Introduction

With the continuous improvement of steel bridge construction technology, the incremental launching technology has also experienced the transition from single point launching to multi-point launching. Nowadays, walking-type incremental launching technology is widely used. The hydraulic synchronous control of walking-type incremental launching has the advantages of high precision, real-time adjustment and total and partial phase control [1, 2]. However, due to the influence of construction error, launching synchronization, operator proficiency, and other factors, it is difficult to ensure that the launching construction process is carried out completely according to the ideal state, which will have a great impact on the local stress state of the girder [3, 4]. Such launching construction deviation can be summarized as transverse deviation, longitudinal launching deviation, and vertical launching deviation [5]. During actual launching process, the asynchronous action of the launching equipment can be briefly summarized as follows: (1) Uneven sunshine and temperature [6, 7], (2) temporary equipment installation and beam splicing error [8], (3) deviation of performance of launching equipment [9], (4) error caused by non-standard operation of constructors in similar elevation measurement [10]. It can be seen that the deviation in the launching construction process is inevitable, but if it is not paid attention to in the construction process, it is likely to cause serious consequences, such as the damage to the launching equipment, the bridge installation error exceeding the allowable design range, the local plastic deformation, and even buckling failure of the beam.

Existing research about incremental launching of steel box girder mainly focuses on orthogonal launching process, and there is a lack of relevant research on oblique launching process. In this paper, the skew launching process of steel box girder was simulated by using finite element model. The maximum cantilever working condition and maximum reaction force working condition was selected to analyze the influence of launching deviation on local stress of the whole structure was studied. The sensitivity relationship between the stress of key components and launching deviation was also obtained.

2 Project Overview

A cable-stayed bridge is a 240 m long bridge with a span of (110 + 130) m. In this structure, 130 m span is the launching construction section, which is composed of 10 beam sections. Picture of the structure is shown in Fig. 1. The multi-point launching construction is carried out by using walking launching equipment. The main beam is a steel box girder, with a standard cross-sectional width of 30.5 m and a height of 3.0 m. As the project is affected by the load of existing roads, the 4# temporary pier used for launching construction is erected on the central divider of existing roads

Fig. 1 Picture of the structure

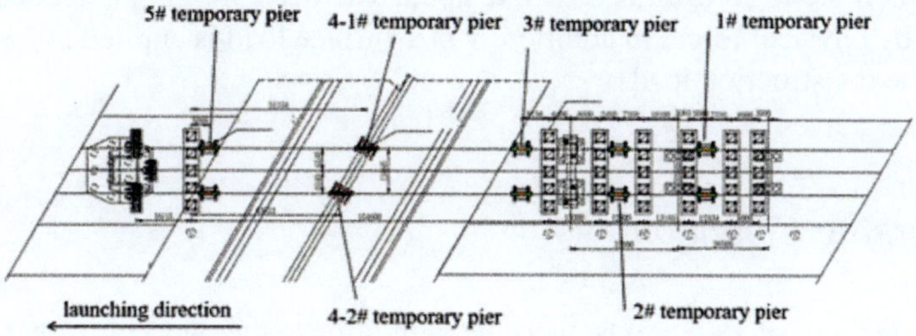

Fig. 2 Layout plan of launching construction pier

and obliquely intersects with the launching direction. The layout of the temporary pier for launching construction is shown in Fig. 2.

3 Numerical Analysis

3.1 Establishment of Numerical Model

ABAQUS is a general finite element software. Using finite element software ABAQUS establishes plate and shell element model of the whole structure. In order to analysis local stress, the hybrid element method is usually used for modeling [11, 12]. In order to obtain more accurate local stress of the whole girder and improve the analysis and calculation accuracy, the shell element model is used for the overall modeling of steel box girder under adverse working conditions. Using shell element simulates main beam and beam element simulate nose beam. The cushion block is simulated by solid element. In the finite element model, the physical parameters of steel box girder, cushion block, and steel nose beam are taken as density: 7850 kg/m^3, elastic modulus E: 2.06 GPa and Poisson's ratio μ: 0.3, grid size 0.2 m. The numerical model is shown in Fig. 3.

Fig. 3 Numerical model

Tie is used to bind the corresponding area of the beam bottom plate to the top of the cushion block, and between the main girder and the nose beam, and the bottom of the cushion block is restrained by fixed end. The dead weight of steel box girder is applied by physical force. In addition, 1 kPa surface load is applied to top plate to simulate the construction load.

3.2 Working Condition

For the selection of dangerous working conditions, the working conditions of maximum cantilever, maximum support reaction, and maximum internal force of launching construction are usually selected [4, 13, 14]. Beam element model was simulated by Midas/civil. In this paper, Midas/civil is used to establish the beam element model of the whole process of launching, to simulate the launching state of the girder in different construction stages. Based on the establishment of the whole launching process model, two working conditions A and B are selected from the calculation results as dangerous working conditions for analysis. The schematic of the numerical model of the two working conditions is shown in Fig. 4. Working condition A is the maximum cantilever condition. Working condition B is the maximum reaction working condition. The difference between the two conditions is that the nose beam has reached the 4–1# temporary pier under condition B.

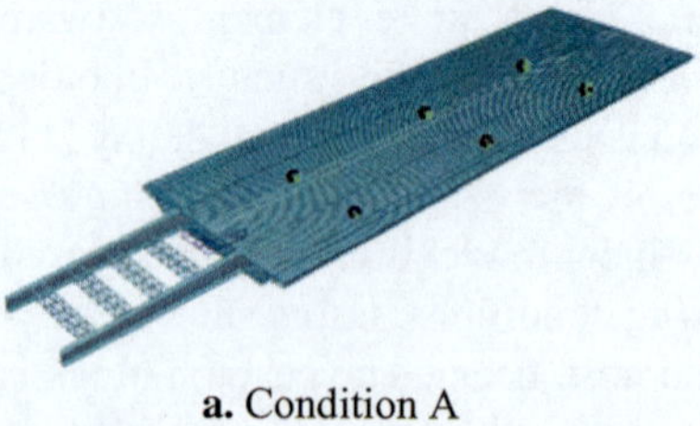

a. Condition A

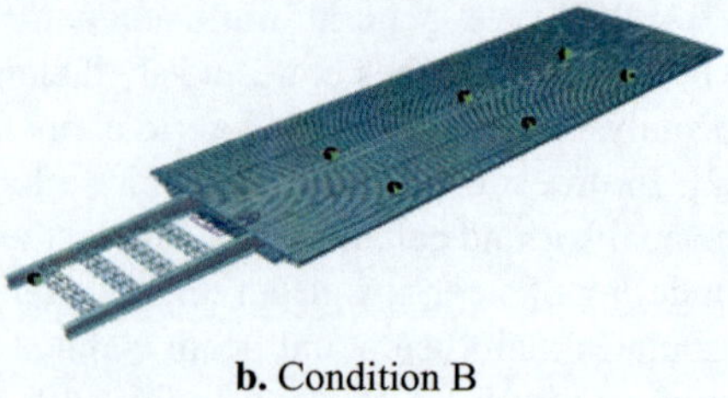

b. Condition B

Fig. 4 Analysis condition model diagram

The launching process of each working condition is 1 m. Add constraints at the corresponding position of the pier on the lower surface of the bottom plate to restrict the vertical displacement. The deformation of the pier and the temporary pier is not considered in the simulation process.

4 Results and Discussion

4.1 Stress Analysis

Whether orthogonal or oblique launching, the maximum cantilever working condition is an adverse working condition that must be considered for the launching construction method [15, 16]. The comparison of working conditions A and B selected in this paper can also reflect the difference between orthogonal launching and oblique launching. The maximum equivalent stress of main components of steel box girder under two adverse conditions was shown in Fig. 5, which shows that the stress difference between web, top plate, U-rib of top plate, and diaphragm under two different working conditions is larger.

As shown in Fig. 6, the maximum stress of the top plate under working condition A occurs at the connection between the top plate and the nose beam, and its value is 73.2 MPa, which is almost equal on the side of left and right. Under working condition B, the maximum equivalent stress of the top plate appears on the right side of the connection with the nose beam, and the maximum equivalent stress is 211.7 MPa, which is quite different from the stress at the connection between the left side of the top plate and the nose beam.

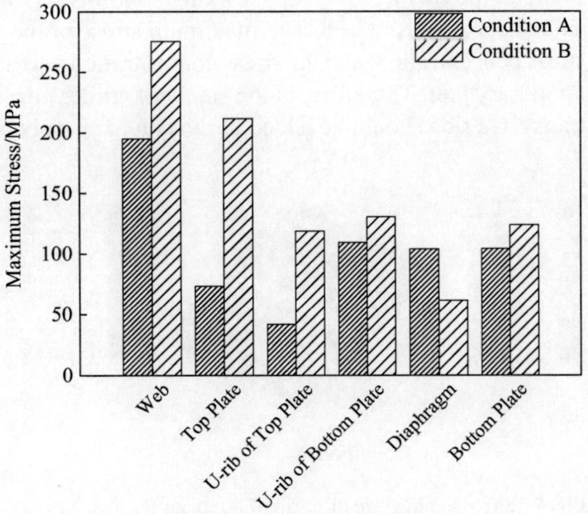

Fig. 5 Stress comparison under two working conditions

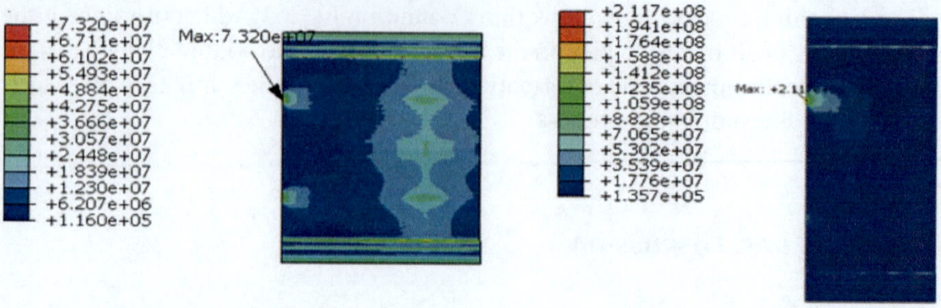

a. Condition A **b.** Condition B

Fig. 6 Stress nephogram of top plate/Pa

When only one side of the nose beam is launched onto the pier and the other side is suspended, the connection between the non-suspended side nose beam and the steel box girder is one of the key parts where stress concentration will occur, and the stress change of this part needs to be focused. Nevertheless, the maximum equivalent stress of the top plate is greater than that of the bottom plate under this working condition, which is significantly different from that of the bottom plate in the process of orthogonal launching.

As shown in Fig. 7, under working condition A, the stress distribution of U-rib of top plates is symmetrical and uniform, and the maximum equivalent stress value is 41.69 MPa. However, under working condition B, there is stress concentration in the U-rib of the top plate, and the maximum equivalent stress is 118.4 MPa, which occurs at the connection between the right side of the nose beam and the main beam.

As shown in Fig. 8, under working condition A, the maximum stress of the web is 195 MPa, and there is little difference in the stress size and distribution form between the two webs. Under working condition B, the maximum equivalent stress value appears in the web. The maximum stress of the web is 275.2 MPa, caused by stress concentration, and the stress concentration part appears at the left pier of the 3# temporary pier. Therefore, in the steel box girder constructed by oblique launching, transverse ribs should be added at the contact part between the web and the bottom

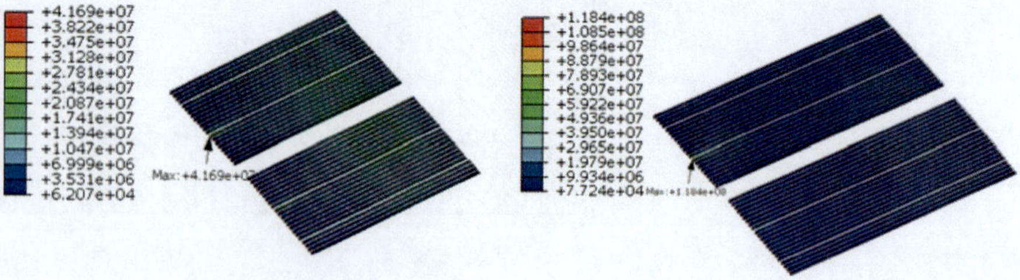

a. Condition A **b.** Condition B

Fig. 7 Stress nephogram of U-rib of top plate/Pa

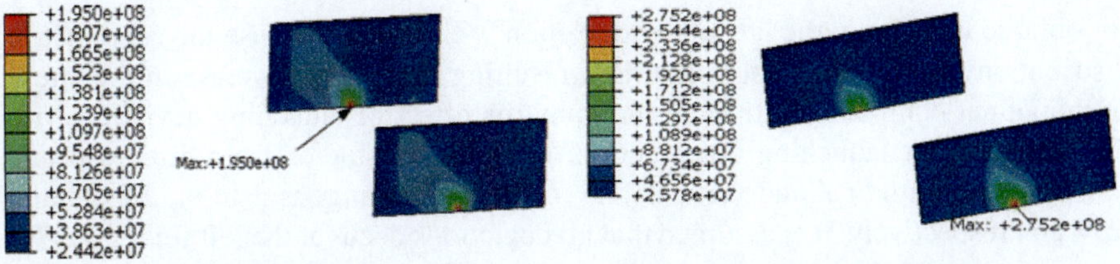

a. Condition A **b.** Condition B

Fig. 8 Stress nephogram of web/Pa

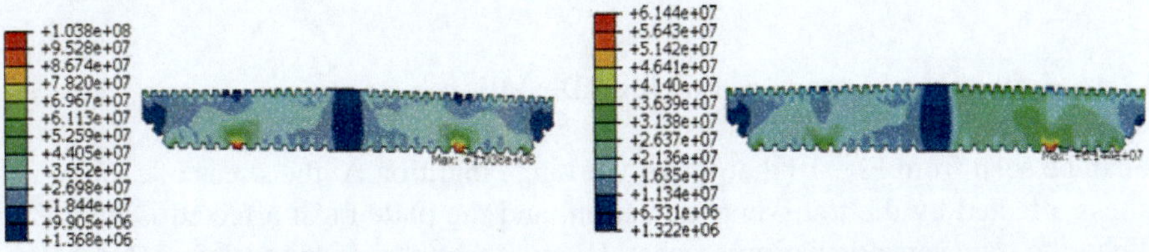

a. Condition A **b.** Condition B

Fig. 9 Stress nephogram of diaphragm/Pa

plate to improve the stress concentration, which is conducive to the performance of steel.

The maximum stress appears in the web, on the one hand, because the web, as the main stress-bearing component, bears the bending stress caused by the self-weight of the whole structure; on the other hand, because the pier under the analyzed working condition is located between the two diaphragm plates, the diaphragm plate does not directly participate in the stress at the pier, and more stress is borne by the web.

The main function of the diaphragm is to enhance the stiffness of the girder. By segmenting the steel girder, the slenderness ratio of the girder is reduced, so as to improve the local stability of the steel girder [16]. As shown in Fig. 9, under working condition A, the stress of the diaphragm is symmetrically distributed, and the maximum stress appears on the 3# temporary pier, with the maximum value of 103.8 MPa. Under working condition B, the maximum equivalent stress of diaphragm is 61.44 MPa, which is located near the left pier of 3# temporary pier.

4.2 *Sensitivity Analysis of Launching Deviation*

In the process of step-by-step launching construction, the launching deviation is inevitable in the construction process because it cannot perfectly meet the synchronization of each launching equipment [17]. The existence of launching deviation will cause the redistribution of internal stress in steel box girder [18], especially for oblique launching, the launching deviation will further increase the asymmetric distribution of stress in steel box girder, resulting in a sharp increase in the stress of individual components. In order to study the effect of launching deviation with the whole girder launching construction, this chapter studies from three aspects: transverse, longitudinal and vertical, and the deviation range is ±1 cm, ±5 cm and ±10 cm, respectively. It is assumed that all deviations occur at the left fulcrum of 3# temporary pier, the transverse deviation of the pushing equipment is positive to the left relative to the steel box girder, the longitudinal deviation of the pushing equipment is positive to the front relative to the steel box girder, and the vertical deviation of the pushing equipment is positive to the up.

4.2.1 Influence of Lateral Launching Deviation

It can be seen from Fig. 10 that under working condition A, the stress of each plate is less affected by the transverse deviation, and the plate most affected is the web. When the transverse deviation reaches 10 cm, the stress of the web is 216.4 MPa. Under condition B, the stress condition of the web with the largest stress in each plate is most affected by the transverse deviation, and other components are less affected by the transverse deviation. When the transverse deviation is within 5 cm, the stress change of each component is relatively small; when the lateral deviation exceeds 5 cm, the stress concentration of oblique launching is exacerbated, the stress value of the web rises sharply, and the structure is in an unsafe state.

The walking launcher has good lateral deviation correction function, so there will be no large lateral deviation in the actual construction process [19]. Combined with the data, it can be seen that the lateral deviation of launching construction has a great impact on the stress of the beam by the launching method. During oblique launching project of steel box girder, the lateral deviation within 5 cm under adverse working conditions has little change on the stress of each component of the girder, and the whole structure is safe. During the launching process, the beam axis can be monitored after each launching distance to ensure that the transverse deviation of the beam is

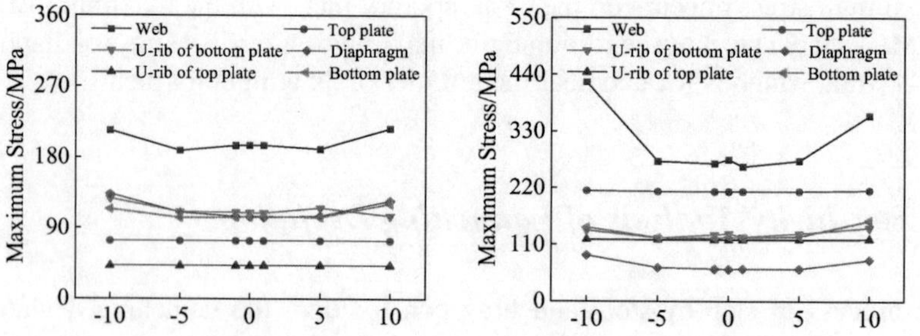

a. Condition A **b.** Condition B

Fig. 10 Lateral deviation/cm

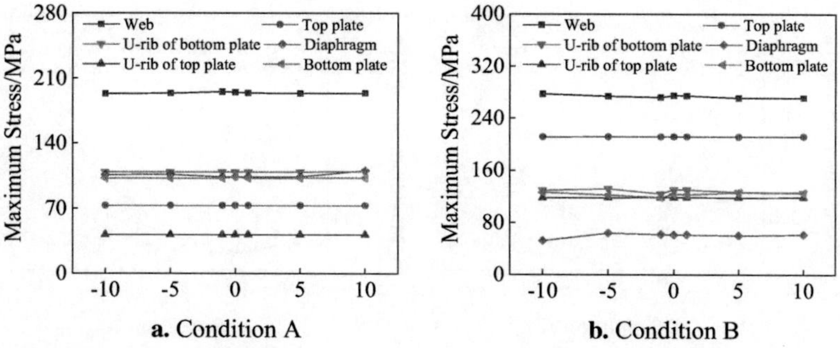

Fig. 11 Lengthways deviation/cm

within the safe range. If it exceeds the limit, it is easy to adjust in time to ensure the safety and efficiency of the construction process.

4.2.2 Influence of Longitudinal Launching Deviation

It can be seen from Fig. 11 that the longitudinal launching deviation has little effect on the stress of main components under working conditions A and B. Especially for working condition A, the stress state of each main component is hardly affected by the longitudinal launching deviation, which shows that for orthogonal launching, the longitudinal deviation is not the main factor of construction error. For working condition B, the longitudinal launching deviation will only have a small impact on each component of the steel box girder, and the longitudinal launching deviation within ±10 cm will hardly have a great impact on the construction process. The lateral deviation has no significant effect on the local stress of steel box girder.

4.2.3 Influence of Vertical Launching Deviation

The influence of vertical launching deviation can be seen from Fig. 12. The greater the vertical launching deviation under working condition A, the maximum equivalent stress value of each component increases. The sensitivity of asynchronous launching to the stress of each component of the main beam is different: the maximum equivalent stress of the bottom plate, web, U-rib of bottom plate and diaphragm of the main stressed components increases greatly with the increase of launching deviation, indicating that they are highly sensitive to asynchronous launching. The U-rib of top plate and top plate increases with the launching deviation, the stress level increases slowly, and the sensitivity is low. Among all the main components, the web is the most sensitive to the vertical launching deviation. Under working condition B, the greater the vertical launching deviation is, the greater the maximum equivalent stress of bottom plate, U-rib of bottom plate, diaphragm and the web will increase. Different

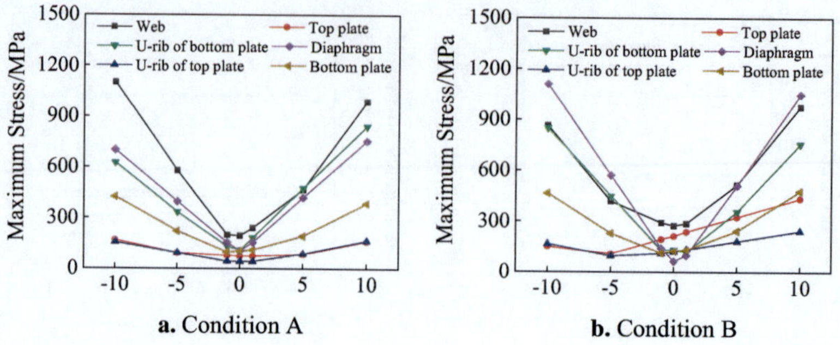

Fig. 12 Vertical deviation/cm

from condition A, the unsynchronized launching has the highest sensitivity to the stress of diaphragm. The influence of vertical deviation on local stress will far exceed the allowable stress of the material.

The vertical launching deviation will cause the beam body to have a lateral offset, resulting in a larger pier reaction force on the side with smaller launching height, and increase the local stress level of the side component. Especially for the oblique launching represented by working condition B, the phenomenon of eccentric load will be more significant. When the vertical deviation reaches 10 cm, the difference between the stress level of the diaphragm and the normal state is 1063.3 MPa. As a component to strengthen the transverse stiffness of the beam, the stress and deformation of the diaphragm is antisymmetric. The side with large launching height has a large stress level and is greatly affected by the launching deviation [20].

Under condition A and B, the vertical launching deviation will greatly improve the stress state of each component of the girder, which proves that the vertical launching deviation has the greatest impact on the stress state of the girder in the launching process. Therefore, the vertical launching deviation should be avoided as much as possible in the launching process. When the vertical launching deviation exists, the difference should be controlled within ±1 cm.

5 Conclusions

This paper simulates the skew incremental launching of steel box girder. Analyzes the influence of launching deviation on the local stress of steel box girder. The sensitivity relationship between local stress of steel box girder and launching deviation was also obtained. The main conclusions are as follows:

(1) Skew incremental launching will lead to local stress concentration and large change of component stress due to partial load. However, the maximum equivalent stress under skew condition appears in the web, and its value is 275.2 MPa.

The maximum equivalent stress under cantilever condition also appears in the web, and its value is 195.0 MPa.

(2) When the lateral launching deviation exceeds 5 cm, the maximum equivalent stress of the steel box girder web will increase greatly, which will bring hidden dangers to the construction safety. Therefore, the lateral launching deviation should be controlled within 5 cm.

(3) The vertical launching deviation is the most influential factor among all launching deviations. In the actual launching construction process, the synchronization of vertical launching shall be ensured as much as possible, and the deviation range should be controlled within 1 cm.

Acknowledgements The research presented here is supported by the National Natural Science Foundation of China (52078317), Natural Science Foundation of Jiangsu Province for Excellent Young Scholars (BK20211597), project from Bureau of Housing and Urban–Rural Development of Suzhou (2021-25; 2021ZD02; 2021ZD30), Bureau of Geology and Mineral Exploration of Jiangsu (2021KY06), China Tiesiju Civil Engineering Group (2021-19), CCCC First Highway Engineering Group Company Limited (KJYF-2021-B-19), and CCCC Tunnel Engineering Company Limited (8gs-2021-04).

References

1. Marchetti, M.E.: Specific design problems to bridges built using the incremental launching method. Engineering Struct., 185–210 (1984)
2. Zellner, W., Svensson, H.: Incremental launching of structures. J. Struct. Eng. **109**(2), 520–537 (1983)
3. Zhang, Y., Luo, R.: Patch loading and improved measures of incremental launching of steel box girder. Steel Const. **68**(1), 11–19 (2012)
4. Jung, K.H., Kim, K.S., Sim, C.W.: Verification of incremental launching construction safety for the Ilsun Bridge, the world's longest and widest prestressed concrete box girder with corrugated steel web section. J. Bridg. Eng. **16**(3), 453–460 (2011)
5. Li, C., He, J., Dong, C.: Control of self-adaptive unstressed configuration for incrementally launched girder bridges. J. Bridg. Eng. **20**(10), 04014105 (2015)
6. Hirmand, M., Shojaei, A., Riahi, H.T.: Finite element modeling of incremental bridge launching and study on behavior of the bridge during construction stages. Int. J. Civil Eng. **13**(1), 112–125 (2014)
7. Chang, M., Chang, H.S.: Temperature gradient and its effect on flat steel box girder of long-span suspension bridge. Sci. China Technol. Sci. **8**,11 (2013)
8. Shao, C.Y.: Innovative Structure Design and Construction-Jiubao Bridge. Struct. Eng. Int. (2015)
9. Granata, M.F., Margiotta, P., Arici, M.: A parametric study of curved incrementally launched bridges. Eng. Struct. **49**(Apr), 373–384 (2013)
10. Granata, M.F.: Adjustable prestressing for construction stages of incrementally launched bridges. European J. Environ. Civil Eng., 327–346 (2015)
11. Su, Q.T., Wu, C., Dong, B.: Analysis of flat steel-box-girder of cable-stayed bridge by finite mixed element method. Tongji Daxue Xuebao/Journal of Tongji University **33**(6), 742–746 (2005)
12. Hui, Z., Desroches, R.: Experimental and analytical studies on a streamlined steel box girder. J. Constructional Steel Res. **66**(7), 906–914 (2010)

13. Zhang, Z., Zhang, R., Hao, R.: Hangzhou Jiangdong Bridge designed as a spatial self-anchored suspension bridge. Struct. Eng. Int. **20**(3), 303–307 (2010)
14. Rosignoli, M.: Solution of the continuous beam in launched bridges. Struct. Build. **122**(4), 390–398 (2015)
15. Fasching, S., Huber, T., Rath, M.: Semi, recast segmental bridges: Development of a new construction method using thin walled prefabricated concrete elements. Structural Concrete (2021)
16. Granata, M.F.: Analysis of non-uniform torsion in curved incrementally launched bridges. Eng. Struct. **75**, 374–387, Sep 15 (2014)
17. Wang, J.F.: Incremental launching construction control of long multispan composite bridges. J. Bridge Eng. **20**(11), 4015006.1. (2015)
18. Tian, L., Yang, M., Chang, S.: Effects of a new method on stress amplitude and fatigue life of orthotropic steel box girder. J. Civ. Eng. **24**(4), 1–10 (2020)
19. Chacón, R., Uribe, N., Oller, S.: Numerical validation of the incremental launching method of a steel bridge through a small-scale experimental study. Experimental Tech. (2016)
20. Kang, Y.J.: Effect of load combinations on distortional behaviors of simple-span steel box girder bridges. Metals (2021)

GIS Applications in Land Adaptability Mapping for Perennial Industrial Crops in Nghe An Province, Vietnam

Hanh Thi Tong, Kien-Trinh Thi Bui, Cuong Manh Nguyen, and Yit Chanthol

Abstract Perennial industrial crops such as coffee, rubber, pepper, etc., are a group of key crops with high economic value, contributing an important role in promoting the socio-economic development of localities. Geographic information systems (GIS) and the analytical hierarchy process (AHP) are used to develop land adaptation maps for the perennial industrial crops based on land assessment methodology of Food and Agriculture Organization of the United Nations (FAO). This research is conducted in Nghe An province because it has the largest area of agricultural land in Vietnam and has the potential to develop perennial industrial crops. Ten impact factors including soil type, elements of soil mechanic, soil layer thickness, elevation, slope, distance to river, annual average rainfall, annual average air temperature, annual mean maximum temperature, and annual mean minimum temperature are selected for natural adaptation assessment of the perennial industrial crops. The results derived from weighted identification using AHP method in GIS indicate that the impact factors related to soil such as soil type, elements of soil mechanic, and soil layer thickness have the highest impact on perennial industrial crops adaptation. Validation of the adaptation map is verified with the current land use map of Nghe An province using the receiver operating characteristic (ROC) curve that comparing the actual land use and adaptation zones. Results from this research can assist managers, planners, as well as local communities in making rational decisions for the most effective land use.

Keywords AHP · GIS · Perennial industrial crops · Adaptability

H. T. Tong · Y. Chanthol
Le Quy Don Technical University, Hanoi, Vietnam
e-mail: tonghanh@lqdtu.edu.vn

K.-T. T. Bui (✉)
Thuyloi University, Hanoi, Vietnam
e-mail: bktrinh@tlu.edu.vn

C. M. Nguyen
Royal HaskoningDHV, Hanoi, Vietnam

L. Q. Nguyen et al. (eds.), *Advances in Geospatial Technology in Mining and Earth Sciences*, Environmental Science and Engineering,
https://doi.org/10.1007/978-3-031-20463-0_25

1 Introduction

In recent decades, Vietnam becomes one of the top five world's largest suppliers of coffee, rubber, pepper, and cashew nuts. Especially in 2021, each export turnover of coffee, cashew nuts, and rubber will reach 3 billion USD [1]. Hence, perennial industrial crops including coffee, rubber, tea, pepper, cashew, etc., are the main crops attributing to an important role in Vietnamese agricultural industry [2], which is considered as a strength of many localities which creates jobs and income for millions of workers.

Land adaptation evaluation which is a process of predicting land performance based on its attributes can help choose the most suitable cropping area to get the optimum benefit [3]. The land adaptation of perennial industrial crops was determined based on the multiple effects of the natural resources and socio-economic factors [2, 4, 5]. It is known that some efficient tools such as geographic information systems (GIS) and multi-criteria analysis (MCA) techniques are popularly used to identify adaptation land for agriculture [6–12]. GIS can improve the convenience and accuracy of spatial data, more productive analysis, and data access [13]. However, the impact factors of land adaptation are not equally important and their different important roles have not been taken into account in GIS [14]. Therefore, multiple-criteria decision-making tools are combined with GIS [15, 16]. The analytical hierarchy process (AHP) is one of the most widely used, it uses expert opinions to estimate the relative weight of criteria [17, 18]. AHP integrated with GIS allows the use of multilevel hierarchies comprising different criteria and restrictions and is one of the most promising methods for land suitability analysis [6–9, 19].

In Vietnam, there have been many land adaptation studies at commune and district level for native crop species or crop species that have been bringing high economic efficiency to the locality using the AHP method in GIS environment [20–27]. For instance, orange tree in Tan Phu commune (Tan Ky district—Nghe An province) [20], ramie in Ngoc Lac district—Thanh Hoa province [22], dragon tree in Bac Binh district—Binh Thuan province [24], rambutan in Long Khanh district—Dong Nai province [25], coconut tree in Mo Cay district—Ben Tre province [26], and local cinnamon tree in Tra Bong district—Quang Ngai province [27] are all special crops having high economic efficiency. Few other studies were conducted with the aim of transforming agricultural land use structure in Di Linh district—Lam Dong province [28], Kien Giang province [23]. However, all of these studies use some key influencing factors such as soil type, soil thickness, elevation, slope, and rainfall without taking into account other factors specified in [4].

The objective of this research is to assess the land adaptation of perennial industrial crops for agricultural planning purpose based on some attributes, also called impact factors. This study focuses on only the influences of natural resources, and ten main impact factors including (1) elevation, (2) slope, (3) distance to the rivers and streams, (4) soil type, (5) elements of soil mechanic, (6) soil layer thickness, (7) annual average air temperature, (8) annual mean maximum temperature, (9) annual mean minimum temperature, and (10) total average annual rainfall were used [11, 29]. In this paper,

GIS and AHP are applied in assessing the influence of these ten impact factors on the land adaptation of perennial industrial crops.

2 Data and Methods

2.1 Study Area

The study is conducted in Nghe An province, which is the largest province in Vietnam with total area being approximately 16,490 km^2 but the area planted with perennial crops only reach about 1% of the province's natural area [30]. Nghe An locates in the center of the North Central region, between 18°33′10″ N and 19°24′43″ N latitude; and 103°52′53″ E and 105°45′50″ E longitude. It has complex terrain with altitude ranging from 0.2 to 2711 m above sea level, and about 83% of the hilly area concentrated in the West of the province. The remaining Eastern parts consist of plains and coastal areas [31, 32].

The climate of Nghe An belongs to the tropical monsoon type, the temperature difference between months in one year is relatively high. The annual average, annual mean maximum, and annual mean minimum of air temperature range from 14 to 24 °C, from 20 to 34 °C in June and July, and from 4 to 16 °C in December to January, respectively. Average rainfall in the area varies from 1200 to 2000 mm/year [31] (Fig. 1).

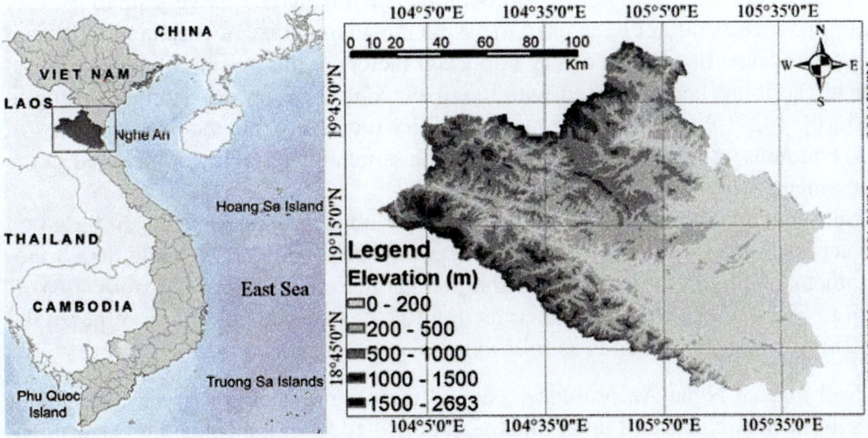

Fig. 1 Location of the study area

2.2 *Data Collection and Methods*

The overall methodology of land adaptation assessment process for the perennial industrial crops followed in this study is illustrated in Fig. 2. To summarize, (i) impact factors that could determine the growth and production of perennial industrial crops are identified and selected; (ii) these impact factors are mapped using collected data in a GIS environment; and (iii) the impact factors are compared to determine their degree of importance using AHP, then the pairwise comparison matrix is created. At last, all maps of impact factors are integrated and reclassified using weighted overlay in GIS.

2.2.1 Identification and Selection of Impact Factors

All impact factors are ranked according to their significance for the perennial industrial crops following expert experiences [5, 33], literatures [7–9, 12, 13], and/or government regulations [4, 34, 35]. According to [4], there are 12 common factors to assess adaptation for perennial industrial crops, including: average annual temperature; average annual minimum temperature; average maximum annual temperature; total annual rainfall; number of dry months/year (month), soil type; terrain slope; terrain slope; thickness of fine soil layer; agglomeration; mixed rock; mechanical components; and flooding.

Due to the difficulties in collecting data on the factors of agglomeration, mixed rock and flooding for each type of crop in Nghe An province, and according to expert experience, the weights of both agglomeration and mixed rock are very small. Hence, these two factors do not been used to assess the adaptability of perennial industrial crops. However, flooding is a very important factor for the perennial industrial crop growth, so it has been replaced with the distance to the river criterion according to studies [11, 29]. In addition, the number of dry months within the year in Nghe An is 5–6 months, at level N in the adaptation classification, so it is not included in the assessment.

Subsequently, a hierarchy composing of ten impact factors is chosen, based on the accessibility and relevance of three types of data (soil data, climate data, topographical, and hydrological data) within the natural conditions of the study area, as seen in Fig. 2. Accordingly, data collection for the assessment criteria of perennial industrial crops adaptation is as follows:

- Soil map of Nghe An province scale 1:100.000 includes soil type, elements of soil mechanic, and soil layer thickness, collected from the Nghe An Department of Natural Resources and Environment.
- Map of current land use in Nghe An province in 2015 scale 1:100,000, including data on land use types, serves as a basis for result verification of natural adaptation assessment of industrial perennial industrial crops and proposing planting zones in the research area. This map is inherited from Nghe An Department of Natural Resources and Environment.

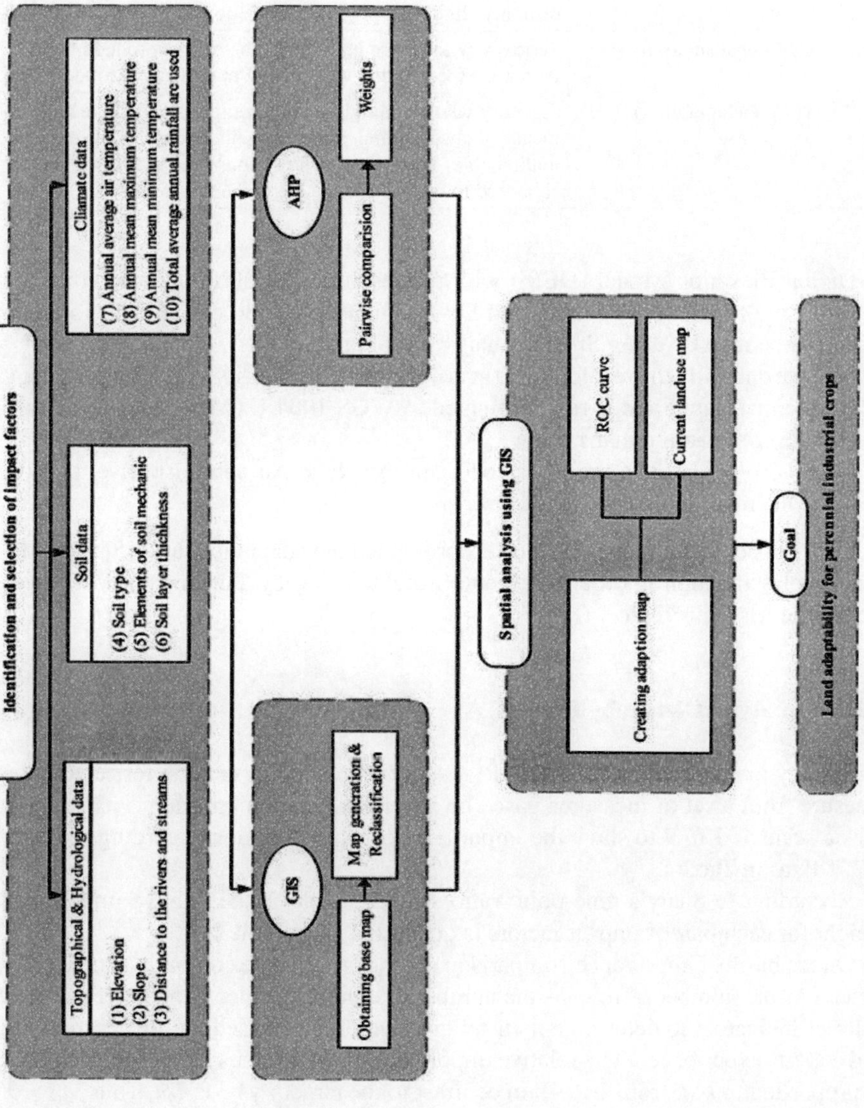

Fig. 2 Perennial industrial crop adaptation assessment process

Table 1 Classification of adaptation indicators according to FAO [5]

Score	Level of adaption	Explain
8–9	S1 (high adaptability)	Territory with minor limitations for agricultural use that do not affect productivity or significantly increase expenses
6–8	S2 (medium adaptability)	Territory with moderate limitations that reduce productivity or imply slight risks of soil degradation
4–6	S3 (low adaptability)	Territory with severe limitations that reduce productivity or increase expenses that will only be marginally justified
0–4	N (not adaptability)	Territory whose limitations can be eliminated with technical means or costs, although these modifications are currently unthinkable) or (territory with serious limitations that are supposed to be insurmountable in the long term

- Digital Elevation Model (DEM) with the resolution of 30 m adopted from the Ministry of Natural Resources and Environment is used to create elevation map and slope map by using Spatial Analysis Tool on ArcGIS.
- Climate data with the resolution 90 m is downloaded at https://worldclim.org/data/worldclim21.html and is re-coordinated to WGS 1984 UTM zone 48 to modify climate composition maps.
- The hydrological data are extracted from the Nghe An administrative map (to make the map of distance to the river).

Each impact factor map is divided according to land adaptation level [5] using the raster reclassification in order to execute weighted overlay. For this purpose, there are four levels as in Table 1 (Fig. 3).

2.2.2 Analytic Hierarchy Process

The Analytic hierarchy process (AHP) is a decision-support method for selecting a measure from a set of measures based on several evaluation criteria [19, 36]. AHP uses a scale of 1 to 9 to show the importance of expert opinion according to Saaty [17, 18], as in Table 2.

According to Saaty's nine-point ratio scale in Table 2, the relative importance weight for each pair of impact factors is quantified as in Table 3.

On the basis of importance comparison among criteria, the comparison matrix A_{ij} (where *i*—the number of row, *j*—the number of column) is then conducted between pairs of indicators to determine their relative importance based on logical thinking and experts experiences. The relative importance of *the i* versus *j* target is calculated in proportion to *k* (*k* from 1 to 9), in contrast to the target *j* vs. i is *1/k*. Thus, $a_{ij} > 0$, $a_{ij} = 1/a_{ji}$, *and* $a_{ii} = 1$.

The standardized method is chosen to calculate the weighs of ten impact factors in the study as in following equations. Table 7 shows the normalization of pairwise comparison matrix and the weights of ten impact factors.

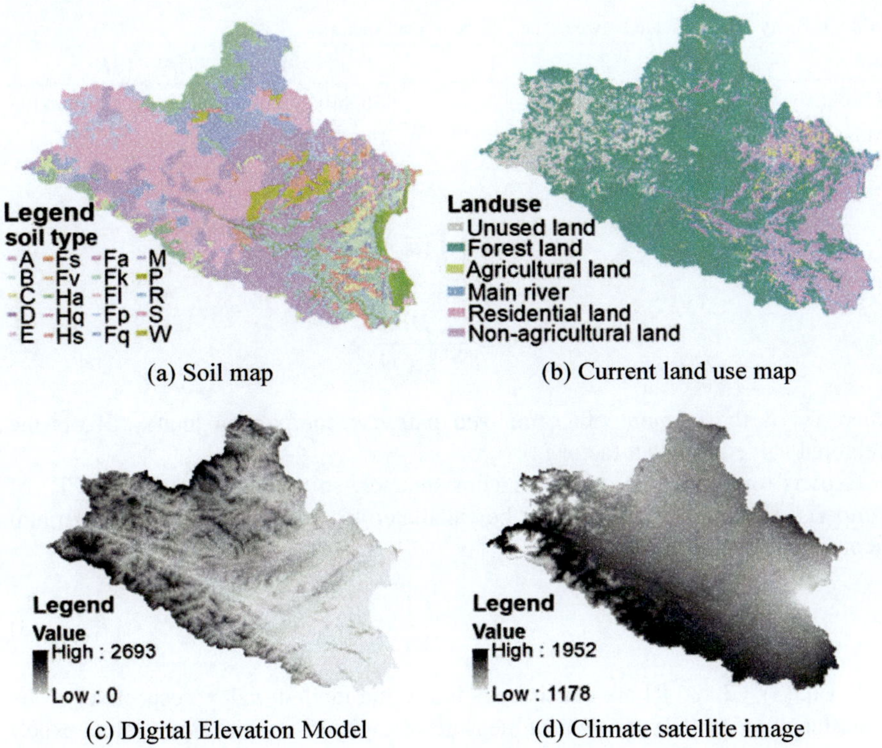

(a) Soil map (b) Current land use map

(c) Digital Elevation Model (d) Climate satellite image

Fig. 3 Collected data

Table 2 Scale comparing the importance of the criteria

Level of importance	Definition	Explain
1	Equal importance	Two components have equal properties
3	Slight importance	Experience and judgment are slightly inclined toward a criterion over the other
5	Strong importance	Experience and judgment are strongly inclined toward a criterion than the other
7	Very importance	A criterion takes precedence over the other and is demonstrated in practice
9	Absolutely importance	The importance of a criterion is higher than the other at the highest level
2, 4, 6, 8	Intermediate levels between upper levels	Need a compromise between two levels of perception

Table 3 Saaty relative scale between the pair of impact factors

9	7	5	3	1	1/3	1/5	1/7	1/9
Absolutely	Very	Strongly	Slightly	Equal important	Slightly	Strongly	Very	Absolutely
More important					Less important			

$$D_i = \prod_{j=1}^{n} \left(\overline{a_{ij}}\right)^{\frac{1}{n}} \tag{1}$$

$$W_i = \frac{D_i}{\sum_{i=1}^{n} D_i} \tag{2}$$

where $\overline{a_{ij}}$ is the element of normalized pairwise comparison matrix, W_i is the weighted value of impact factor i.

Experts' opinions may lead to inconsistencies, so a consistency ratio (CR) of comparison of impact factors must be calculated in order to establish a maximum inconsistency value [17, 18].

$$\mathrm{CR} = \frac{\mathrm{CI}}{\mathrm{RI}} \tag{3}$$

In Eq. (3), CI and RI are consistency index and random index, respectively. The consistency index (CI) serves as a measure of logical inconsistency against expert judgment during peer-to-peer comparisons [37], estimated by

$$\mathrm{CI} = \frac{\lambda_{\max}}{n-1} \tag{4}$$

in which $\lambda_{\max}$ is the maximum eigenvalue [17, 18] computed by using Eq. (5), RI is calculated for different "n" values (number of factors in the matrix) from a simulation of 100,000 matrixes, and simply obtained from Table 4.

$$\lambda_{\max} = \frac{1}{n} \sum_{i=1}^{n} \left(\sum a_{ij} \times W_i \right) \times \frac{1}{w_i} \tag{5}$$

Finally, CR must be < 0.1 (less than 10% inconsistency), then pairwise comparison matrix is automatically assumed to be consistent and acceptable.

Table 4 RI random metric questionnaire [17, 18]

n	1	2	3	4	5	6	7	8	9	10
RI	0.00	0.00	0.52	0.89	1.11	1.24	1.34	1.40	1.45	1.49

2.2.3 Spatial Analysis Using GIS

Spatial data of ten impact factors are generated in the raster format, georeferenced in the WGS84-Zone 48 N coordinate system, and each of them was reclassified into four adaptive classes using ArcGIS 10.5 software. Moreover, after weighting each impact factors using AHP, the aggregate adaptive map is made in the ArcGIS 10.5 software by Eq. (6)

$$S = \sum (W_i \times X_i) \quad (6)$$

where S is the adaptive index, W_i is the weighting of the impact factor i, and X_i is the value (quantimized) of the impact factor i, $i = (1, \ldots, n)$.

The ArcGIS 10.5 software is employed to drive raster layers for each impact factors using spatial interpolation (30 m cell size). Values in raster layers are reclassified to a common adaptability class ($S1$ for high adaptability, $S2$ for medium adaptability, $S3$ for low adaptability, and N for not adaptability).

At last, the overall adaptability raster is generated by applying the layer weighted overlay method. The reclassified raster layers are overlayed by multiplying each raster cell adaptability value by its layer weight and totaling the values to derive the map of overall adaptability for perennial industrial crop planting.

3 Results and Discussions

3.1 Adaptive Classification of Impact Factors

Ten impact factors are standardized according to suitability thresholds in [4] and listed in Table 5. Lastly, standardized thematic maps for each impact factor are generated as in Fig. 4 and corresponding values of adaptation levels are shown in Table 6.

Figures 4 and Table 6 indicate that, within three data types, topological and hydrological conditions have the best adaptability on perennial industrial crops, the criteria range for topographic suitability is relatively wide. In there, the elevation has great influence on the ecological adaptation of the crops. The altitude requirement for perennial industrial crops to grow varies from 30 to 400 m.

In general, perennial industrial crops are moderately adaptive with soil in Nghe An province. In terms of this, soil type hardly suits to perennial industrial crops, while element of soil mechanic has clear adaptability. The ideal area for planting these crops only accounts for a small proportion, about 3% of the total area, distributed mainly in Quy Hợp districts, Tan Ky, Nghia Dan, and Thai Hoa Town.

In climate data type, annual average rainfall in the region is extremely low, therefore it almost does not adapt to perennial industrial crops. The average and maximum temperature for moderately adaptable growth are 22–25 °C and 24–27 °C, respectively. The classification results indicate that no region in the province is highly suited

Table 5 Ecological requirements for perennial industrial crops [4]

Data type	Impact factor	Level of adaptation			
		S1	S2	S3	*N*
Topography and hydrology	Elevation (m)	30–400	400–600 3–30	600–800	>800 <3
	Slope (degree)	0–8	8–15	15–25	>25
	Distance to rivers and streams (m)	200–400	400–600	600–800	>800 <200
Soil	Soil type[a]	Fk, Fu, Fv, Fe	Fj, Fn, Fs, Fa, Fp, Fq, X, Xa, Fd, B, D	Ba, Fd, Fl	Other types of soil
	Elements of soil mechanic	Sandy loam	Clay loam	Loam	Loamy sand and other soil
	Soil layer thickness (cm)	>100	70–100	50–70	<50
Climate	Annual average air temperature	25–30 °C	22–25°C	15–22 °C >30 °C	<15 °C
	Annual mean maximum temperature (°C)	27–30	24–27 >30	22–24	<22
	Annual mean minimum temperature (°C)	>20	17–20	14–17	<14
	Total average annual rainfall (mm)	2000–2500	1700–2000	1300–1700 > 2500	<1300

[a] Soil types: A: Humus soil in the mountains; Fs: Red-yellow soil on clay and metamorphose; Ha: Reddish-yellow humus soil on acidic magma rocks; Fa: Yellow-red soil on acidic magma rocks; Fq: Pale yellow soil on sandstone; Py: A stream silt soil; B: Silvery gray soil on ancient silt; Ba: Silvery gray soil on acid ghoststones and sandstones; Hq: Pale yellow humus soil on sandstone; Hs: Red-yellow humus soil on clay rocks

for crop adaptation in terms of climate conditions. Only 1% of the area is medium suited for the crops, while the remaining areas (99%) is poorly suitable or not suitable for crop adaptation.

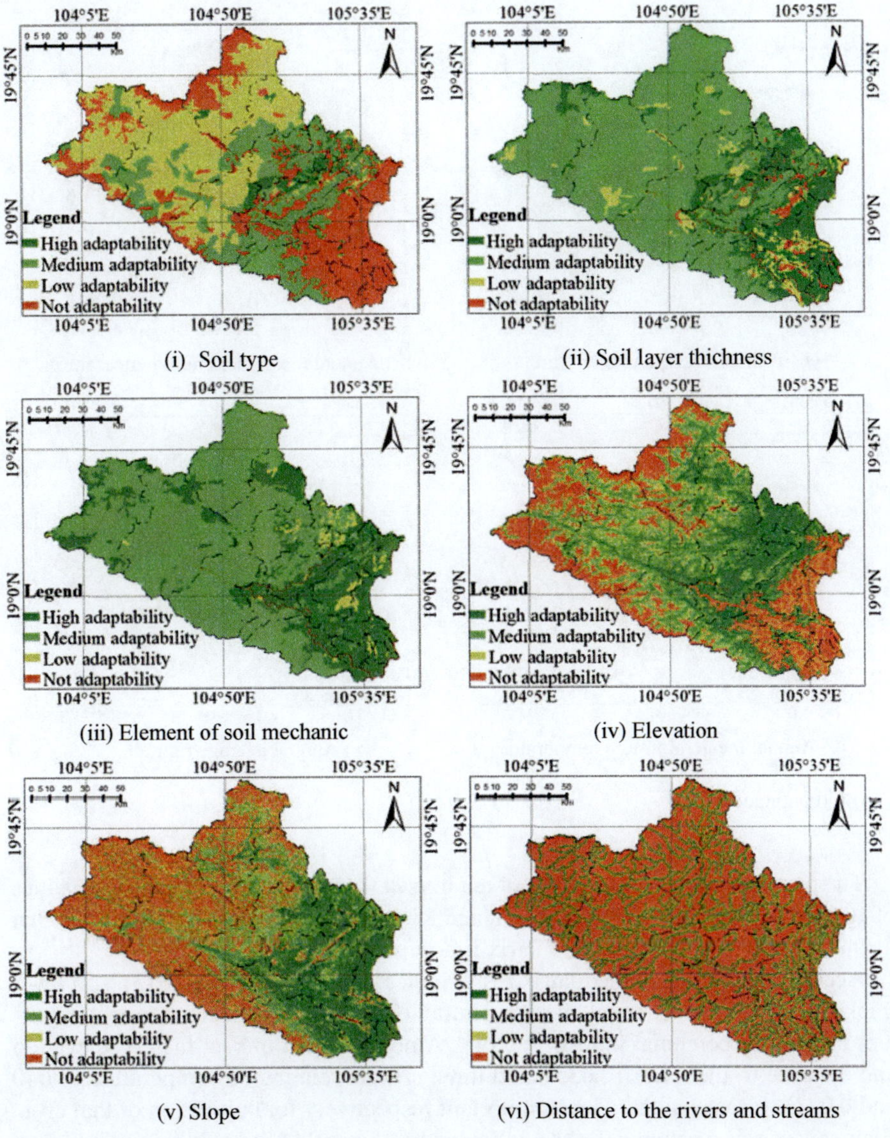

Fig. 4 Ten adaptation classification maps for perennials in Nghe An province

3.2 AHP Pairwise Comparison Matrix and Weights

The final pairwise comparison matrix of ten impact factor represents the relationship between all the impact factors in pair (Table 7).

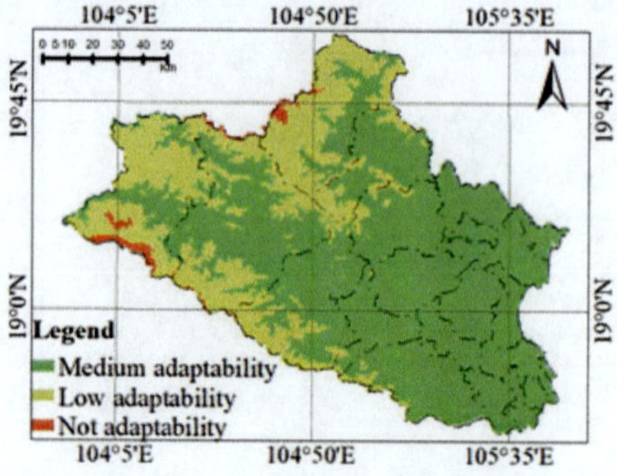

(vii) Annual average temperature

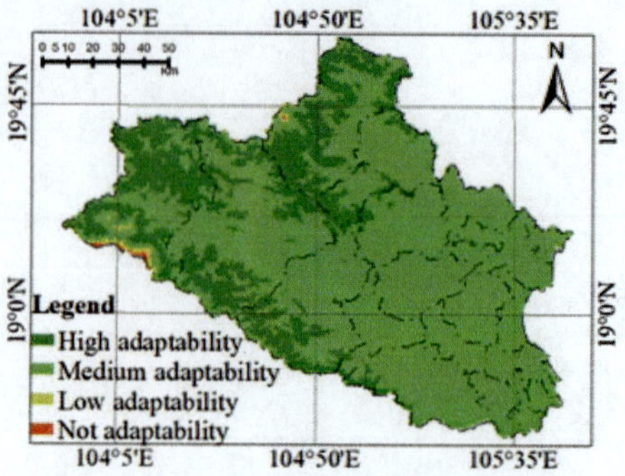

(viii) Annual mean maximum temperature

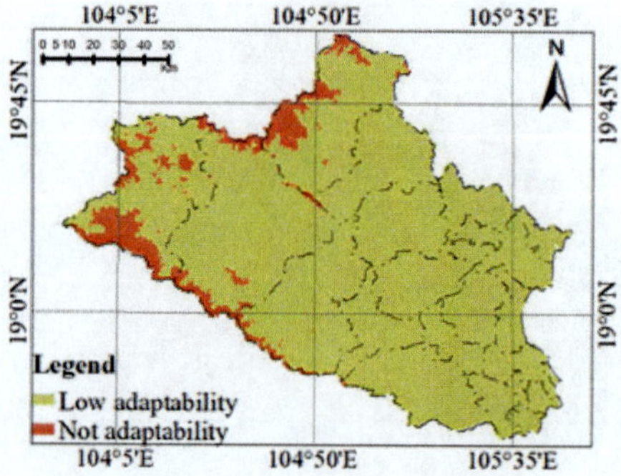

(ix) Annual mean minimum temperature

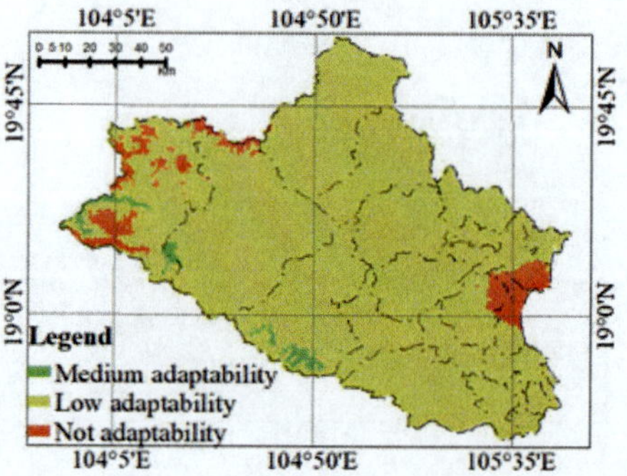

(x) Annual average rainfall

Fig. 4 (continued)

The pairwise comparison matrix of ten impact factors in Tables 7 and 8 has values of 10.009, 0.001, and 1.49 for λ_{max}, CI, and RI, respectively. Hence, CR is 0.001 that means the pairwise comparison matrix is acceptable.

According to Table 8, the matrix result indicated that soil type (0.315), soil layer thickness (0.174), and annual average rainfall (0.161) in order were the most important factors for perennial industrial crops. Among the ten impact factors, elevation and distance to the rivers (0.035), and three criteria relating air temperature (0.040 and 0.041) were considered least important respectively for cultivation of that crop. Elements of soil mechanic (0.096) followed by slope (0.065) got higher values than factors of air temperature.

Table 6 Adaptability classes of impact factors for perennial industrial crops

Data type	Impact factor	Level of adaptability	Area (ha)	Percent (%)
Topography and hydrology	Elevation	S1: High adaptability	7564.22	46
		S2: Medium adaptability	4641.85	28
		S3: Low adaptability	1780.40	11
		N: Not adaptability	2503.53	15
	Total		16,490.00	100
	Slope	S1: High adaptability	4239.72	26
		S2: Medium adaptability	2728.34	17
		S3: Low adaptability	3645.48	22
		N: Not adaptability	5876.46	36
	Total		16,490.00	100
	Distance to rivers and streams	S1: High adaptability	1827.01	11
		S2: Medium adaptability	1687.95	10
		S3: Low adaptability	1565.44	9
		N: Not adaptability	11,409.60	69
	Total		16,490.00	100
Soil	Soil type	S1: High adaptability	579.12	4
		S2: Medium adaptability	4381.15	27
		S3: Low adaptability	5990.96	36
		N: Not adaptability	5538.77	34
	Total		16,490.00	100
	Elements of soil mechanic	S1: High adaptability	4983.50	30
		S2: Medium adaptability	10,988.75	67

(continued)

Table 6 (continued)

Data type	Impact factor	Level of adaptability	Area (ha)	Percent (%)
		S3: Low adaptability	413.13	3
		N: Not adaptability	104.62	1
	Total		16,490.00	100
	Soil layer thickness	S1: High adaptability	3445.62	21
		S2: Medium adaptability	11,524.10	70
		S3: Low adaptability	1061.76	6
		N: Not adaptability	458.52	3
	Total		16,490.00	100
Climate	Annual average air temperature	S1: High adaptability	0.00	0
		S2: Medium adaptability	11,428.13	69
		S3: Low adaptability	4826.76	29
		N: Not adaptability	235.12	1
	Total		16,490.00	100
	Annual mean maximum temperature	S1: High adaptability	3973.68	24
		S2: Medium adaptability	12,404.10	75
		S3: Low adaptability	112.22	1
		N: Not adaptability	0.00	0
	Total		16,490.00	100
	Annual mean minimum temperature	S1: High adaptability	0.00	0
		S2: Medium adaptability	0.00	0
		S3: Low adaptability	15,174.19	92
		N: Not adaptability	1315.81	8

(continued)

Table 6 (continued)

Data type	Impact factor	Level of adaptability	Area (ha)	Percent (%)
	Total		16,490.00	100
	Total average annual rainfall	S1: High adaptability	0.00	0
		S2: Medium adaptability	221.91	1
		S3: Low adaptability	15,369.82	93
		N: Not adaptability	898.28	5
	Total		16,490.00	100

Table 7 Pairwise comparison matrix for evaluating relative importance of ten impact factors

Impact factor	(1)	(2)	(3)	(4)	(5)	(6)	(7)	(8)	(9)	(10)
(1) Soil type	1	3	3	1/5	1/3	1/3	3	1/7	1/3	5
(2) Elements of soil mechanic	1/3	1	3	1/7	1/5	1/3	3	1/7	1/7	3
(3) Soil layer thickness	1/3	1/3	1	1/9	1/7	1/5	1/3	1/9	1/7	3
(4) Annual average rainfall	5	7	9	1	3	5	7	1/3	3	9
(5) Annual average air temperature	3	5	7	1/3	1	3	5	1/3	3	7
(6) Annual mean maximum temperature	3	3	5	1/5	1/3	1	1	1/5	1/3	5
(7) Annual mean minimum temperature	1/3	1/3	3	1/7	1/5	1	1	1/7	1/5	3
(8) Elevation	7	7	9	3	3	5	7	1	3	1
(9) Slope	3	3	7	1/3	1/3	3	5	1/3	1	7
(10) Distance to the river	1/5	1/3	1/3	1/9	1/7	1/5	1/3	1	1/7	1
Sum	23.20	30.00	47.33	5.57	8.69	18.40	34.67	2.85	11.30	52.00

3.3 *Land Use Recommendations for Perennial Industrial Crops in Nghe An Province*

The natural adaptability map for the perennial industrial crops is established according to Eq. (6). The result of land suitability classification is presented in Fig. 5 and Table 9.

To examine the precision of the result map, receiver operating characteristic (ROC) curve analysis is used [38] because ROC analysis can yield better results than analyses

Table 8 Normalized pairwise comparison matrix and weights

Impact factor	(1)	(2)	(3)	(4)	(5)	(6)	(7)	(8)	(9)	(10)	*Weight*	Rank
(1)	0.05	0.08	0.05	0.07	0.21	0.14	0.22	0.04	0.02	0.07	0.315	1
(2)	0.01	0.03	0.05	0.11	0.02	0.11	0.07	0.04	0.18	0.07	0.096	4
(3)	0.04	0.01	0.02	0.02	0.01	0.01	0.01	0.13	0.01	0.04	0.174	2
(4)	0.16	0.18	0.14	0.06	0.25	0.20	0.16	0.10	0.20	0.13	0.161	3
(5)	0.10	0.13	0.11	0.05	0.08	0.12	0.11	0.10	0.20	0.10	0.041	6
(6)	0.10	0.18	0.08	0.13	0.03	0.04	0.07	0.13	0.02	0.27	0.040	7
(7)	0.13	0.10	0.05	0.02	0.02	0.01	0.04	0.04	0.01	0.04	0.040	7
(8)	0.22	0.18	0.14	0.47	0.25	0.24	0.16	0.29	0.20	0.13	0.035	9
(9)	0.18	0.08	0.11	0.05	0.03	0.12	0.11	0.10	0.07	0.10	0.065	5
(10)	0.01	0.03	0.25	0.02	0.10	0.01	0.05	0.03	0.09	0.05	0.035	9
Sum	1.00	1.00	1.00	1.00	1.00	1.00	1.00	1.00	1.00	1.00		

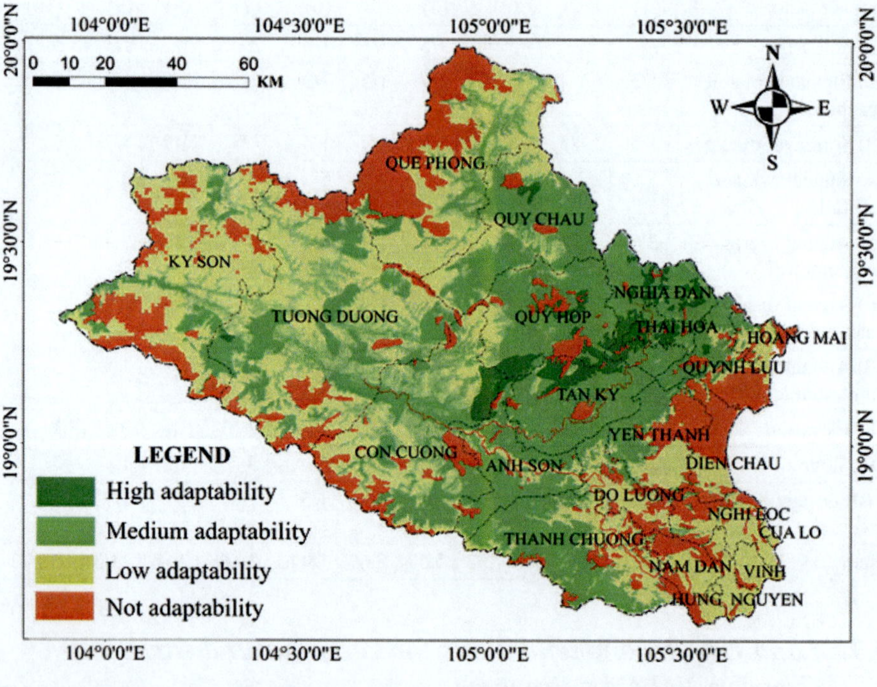

Fig. 5 Map of ecological adaptation classification for perennial industrial crops

Table 9 Summary of adaptable areas for perennial industrial crops in Nghe An province

Hierarchy of adaptation	Area (km^2)	Percentage (%)
High adaptability (S1)	544.21	3.30
Medium adaptability (S2)	6071.28	36.82
Low adaptability (S3)	6574.27	39.87
Not adaptability (N)	3300.25	20.01

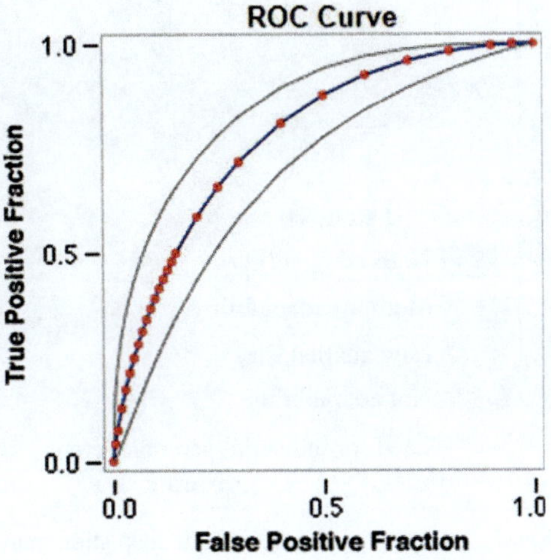

Fig. 6 ROC of land adaptability map for perennial industrial crops in Nghe An province

based on accuracy [39]. The area under the ROC curve (AUC) is important criteria to compare the performance of reclassify process [39–41]. As seen in Fig. 6, AUC = 0.9182 corresponds to the extremely good classifier [39].

The area is medium adaptability (6071.28 km^2 or 36.82% of it) concentrated in western, central, and partly southern locations of the area; low adaptability (6574.27 km^2/39.87%) dominating the northwestern and southeastern portions; not adaptability (3300.25 km^2/20.01%) scattered around the boundary of the area, and high adaptability (544.21km^2/ 3.30%) located near to the central of the area.

Due to some climate limitations, only 3% of the area provides high suitability for the crops, while 38% provides medium-crops suitability that mainly distributed in Tan Ky, Quy Hop, Nghia Dan, Thai Hoa town, and part of Con Cuong district. The remaining areas are low or non-adaptability for the perennial industrial crops.

The study has also shown that the area is potential for producing that crop. However, considerable attention should be given to crop selection that best fit the agroecology and proper management of the soils in order to get optimum yield.

At last, the current land use map is overlaid on the land adaptability map to validate the accuracy of potential areas for industrial perennial crop planting (Fig. 7).

The results indicate that the area with high crop adaptation coincides with the area of perennial industrial crops planting in the current land use map. The area with very high adaptation is 544.21 km^2. The total area of perennial industrial crops is

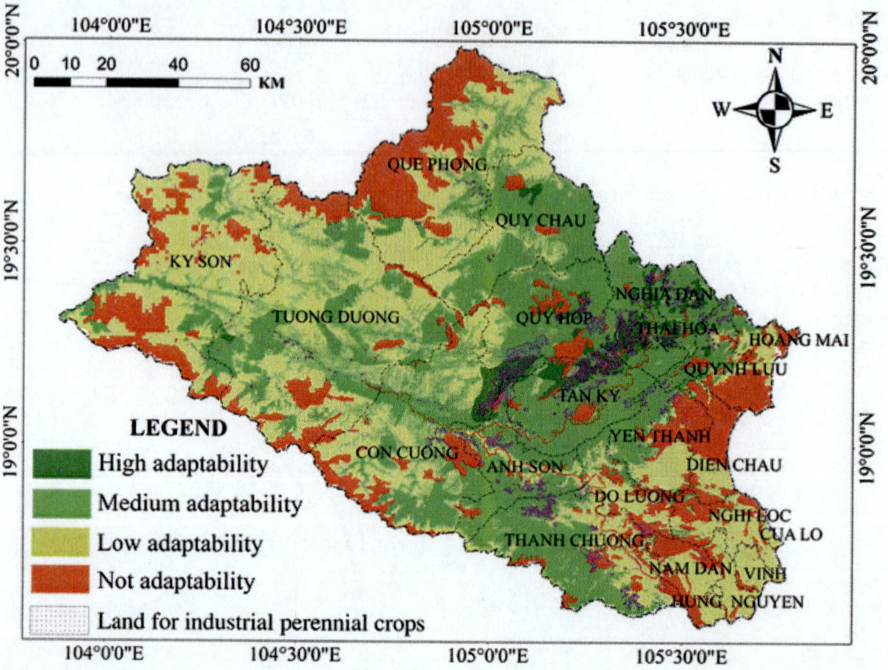

Fig. 7 Overlapping map of natural adaptation map and current status map of perennial industrial crops in Nghe An province

242.80 km^2 as per the current status map, while according to [31] the actual area of the crops is 12,401km^2. On this basis, we propose that the potential areas to plant perennial industrial crops are as follows: Quy Hop district, Thai Hoa town, Tan Ky, Nghia Dan, Anh Son, and to the west of Thanh Chuong district.

To enhance the level of perennial industrial crops adaptation, it is necessary to carry out soil reclamation by supplementing organic fertilizer to increase soil fertility and applying intercropping with the short-term industrial crops such as peanuts, soybeans, and chickpeas to stabilize protein to improve soil fertility. The intercropping also brings more income for local people. Temperature and rainfall also affect the level of crop adaptation. However, challenges caused by these factors can be solved by proactively providing water in the dry season by applying modern irrigation technology for water retention serving for irrigation. For the area that is not suitable for crop planting, reforestation or conversion to other purposes can be implemented to avoid wasting land.

4 Conclusions

Ten impact factors of soil, climate, and hydro-topographic characteristics were used to generate land adaptability evaluation for perennial industrial crops in Nghe An province, Vietnam. The GIS-based AHP model was employed in identifying adaptive areas for crop planting. According to the land adaptability map produced, high adaptability, medium adaptability, low adaptability, and not adaptability areas covered 544.21 km^2 (3.30%), 6071.28 km^2 (36.82%), 6574.27 km^2 (39.87%), and 3300.25 km^2 (20.01%), respectively.

Slope, air temperature, precipitation, and elements of soil mechanic were severely limiting the cultivation potential of the area for perennial industrial crops. It should be noted that growing crops which best fit natural conditions of the area should be taken into consideration in order to maintain soil health and accordingly improve the yield of the crop.

Because climate limitation was difficult to overcome, further studies can be made by incorporating socio-economic variables so as to improve the suitability results. This study also can be applied in different research areas to assess crop adaptation, providing information, and serving as references for planners to address challenges in enhancing crops productivity.

References

1. Ministry of Finance of the Socialist Republic of Vietnam. https://mof.gov.vn/webcenter/portal/btcen/pages_home
2. World Bank and IPSARD: Transforming Vietnamese agriculture: gaining more from less. Hanoi, Vietnam (2016)
3. Sonneveld, M.P.W., et al.: Thirty years of systematic land evaluation in the Netherlands. Geoderma **156**(3–4), 84–92 (2010)
4. National Institute of Agricultural Planning and Projection: Agricutural production land evaluation instruction for land use planning at district level, in TCVN 8409:2012. Ministry of Science and Technology (2012)
5. FAO: A framework for land evaluation. FAO Soils Bull. **32** (1976). http://www.fao.org/docrep/t0715e/t0715e06.htm
6. Trigoso, D.I., et al.: Land suitability analysis for potato crop in the Jucusbamba and Tincas Microwatersheds (Amazonas, NW Peru): AHP and RS–GIS approach. Agronomy (MDPI). **10**(12), 18 (2020)
7. Tadesse, M., Negese, A.: Land suitability evaluation for sorghum crop by using GIS and AHP techniques in Agamsa sub-watershed, Ethiopia. Cogent Food Agri. **6**(1), 18 (2020)
8. Singha, C., Swain, K.C., Swain, S.K.: Best crop rotation selection with GIS-AHP technique using soil nutrient variability. Agriculture (MDPI). **10**(6), 18 (2020)
9. Kahsay, A., et al.: Land suitability analysis for sorghum crop production in northern semi-arid Ethiopia: application of GIS-based fuzzy AHP approach. Cogent Food Agri. **4**(1), 24 (2018)
10. Yangouliba, I., Kwawuvi, D., Almoradie, A.: Suitable land assessment for rice crop in Burkina Faso using GIS, remote sensing and multi criteria analysis. J. Geographic Info. Syst. **12**(6), 683–696 (2020)

11. Pham, M.P., et al.: Modelling for Hevea brasiliensis and Manihot esculenta plantations responses to climate change in Song Hinh District of Phu Yen province, Vietnam. In: E3S Web of Conferences (2021)
12. Akıncı, H., Özalp, A.Y., Turgut, B.: Agricultural land use suitability analysis using GIS and AHP technique. Comput. Electron. Agric. **97**, 71–82 (2013)
13. Baroudy, A.A.E.: Mapping and evaluating land suitability using a GIS-based model. CATENA **140**, 96–104 (2016)
14. Kazemi, H., Sadeghi, S., Akinci, H.: Developing a land evaluation model for faba bean cultivation using geographic information system and multi-criteria analysis (A case study: Gonbad-Kavous region, Iran). Ecol. Ind. **63**, 37–47 (2016)
15. Torrieri, F., Batà, A.: Spatial multi-criteria decision support system and strategic environmental assessment: a case study. Buildings (MDPI). **7**(4) (2017)
16. Musakwa, W.: Identifying land suitable for agricultural land reform using GIS-MCDA in South Africa. Environ. Dev. Sustain. **20**, 2281–2299 (2018)
17. Saaty, T.L.: Fundamentals of decision making and priority theory with the analytic hierarchy process. RWS Publications (1994)
18. Saaty, T.L.: Decision making with the analytic hierarchy process. Int. J. Serv. Sci. **1**(1), 83–98 (2008)
19. Tran, T.V., et al.: Landslide susceptibility mapping for the Thao River catchment with high spatial resolution rainfall data. In: Proceedings of the 3rd International Conference on Sustainability in Civil Engineering. Springer (2021)
20. Tuyen, T.T.: Soil assessment for orange tree growth with the help of GIS and AHP. HNUE J. Sci. **63**(10), 116–127 (2018)
21. Kha, N.D.: Integrating GIS and AHP in rubber tree adaptation assessment in Chon Thanh district, Binh Phuoc province, Bachelor thesis, Nong Lam University–Ho Chi Minh City, Ho Chi Minh City (2014)
22. Hao, N.H., Loan, N.T.: Application of AHP method and GIS to evaluate the suitability of some soil characteristics for ramie plant in Ngoc Lac District, Thanh Hoa province. J. Sci.—Hong Duc University **50**, 21–29 (2020)
23. Trung, N.T., Tri, L.Q., Dinh, L.C.: GIS and AHP applications in land adaptation assessment for agricultural land use structure change in Kien Giang province. J. Sci.—Can Tho Univ. **53** (2018)
24. Luan, P.C., et al.: Application of GIS and analytical Hierarchy process in land suitability evaluation for dragon fruit: a case study of Bac Binh District, Binh Thuan Province. In: National Conference of GIS Application (2016)
25. Chau, L.M., Duc, T.T.: Using AHP and GIS techniques to evaluate suitable areas for rambutan growing by VietGAP standard in Long Khanh region, Dong Nai province. J. Vietnam Agri. Sci. Technol. **04**(101), 87–93 (2019)
26. Cuong, N.H.: Integration of GIS and decision tree in land evaluation for coconut trees in Mo Cay Nam district, Ben Tre province. VNU J. Sci. Earth Environ. Sci. **34**(1), 66–76 (2018)
27. Loi, N.V., Ngoc, T.K., Hung, P.D.: Application of GIS, AHP, and FAHP in assessing indigenous cinnamon species suitability in Tra Bong, Quang Ngai. Hue University J. Sci.: Agri. Rural Develop. **129**(3D), 5–19 (2020)
28. Loi, N.K., Tuan. V.M.: Integration of GIS and AHP techniques for analyzing land use suitability in Di Linh district, upstream Dong Nai watershed, Vietnam. In: FORTROP II: Tropical Forestry Change in a Changing World. Kasetsart University, Bangkok, Thailand (2018)
29. Pham, M.P., et al.: Evaluation of land suitability for Cunninghamia konishii Hayata (Cupressaceae) planting in Vietnam. Geograp. Environ. Sust. **14**(2) (2021)
30. General Statistics Office: Statistical yearbook of Vietnam 2020, G.S. Office, Editor. Statistical Publishing House (2020)
31. Nghe An province portal. https://www.nghean.gov.vn/
32. Ministry of planning and investment of the socialist Republic of Vietnam. https://www.mpi.gov.vn/Pages/tinhthanhchitiet.aspx?idTinhThanh=58

33. Smyth, A.J., Dumanski, J.: FESLM: an international framework for evaluating sustainable land management. In: World Soil Resources Report, F.a.A.O.o.t.U. Nations, Editor (1993)
34. Janowiak, M.K., et al.: Adaptation resources for agriculture: responding to climate variability and change in the Midwest and Northeast. United States Department of Agriculture (2016)
35. Farquharson, R., et al.: Everfarm®—Climate adapted perennial-based farming systems for dryland agriculture in southern Australia. In: National Climate Change Adaptation Research Facility, Gold Coast, p. 147 (2013)
36. Le, T.T.T., et al.: Developing a landslide susceptibility map using the analytic hierarchical process in Ta Van and Hau Thao Communes, Sapa, Vietnam. J. Disas. Res. **16**(4), 529–538 (2021)
37. Bozóki, S., Rapcsák, T.: On Saaty's and Koczkodaj's inconsistencies of pairwise comparison matrices. J. Global Optim. **42**, 157–175 (2008)
38. Fawcett, T., An introduction to ROC analysis. Pattern Recognition Lett. **27**(8), 861–874 (2006)
39. Hanczar, B., et al.: Small-sample precision of ROC-related estimates. Bioinformatics **26**(6), 822–830 (2010)
40. Flach, P., Hernández-Orallo, J., Ferri. C.: A coherent interpretation of AUC as a measure of aggregated classification performance. In: Proceedings of the 28th International Conference on Machine Learning (2011)
41. Calders, T., Jaroszewicz, S.: Efficient AUC optimization for classification, in knowledge discovery in databases: PKDD 2007, J.N. Kok, et al., Editors, 42–53 (2007)

Land-Use and Land-Cover Change Detection and Classification to Analyze Dynamics of Dragon Fruit Farming in Sand Dunes Area of Binh Thuan Province of Vietnam

Luan Hong Pham, Trong Dieu Hien Le, Lien T. H. Pham, Ho Nguyen, and Hong Quan Nguyen

Abstract Mapping of land use and land covers (LULCs) and their temporal changes are important subjects in remote sensing field. This study applies change detection and classification (CDC) in mapping and analyzing changes of dragon fruits area in sand dunes area of Binh Thuan province using MODIS NDVI time-series data from 2000 to 2020. Unlike the single-phase classification which assumes unchanged LULC in analyzed periods, the CDC divided time series into segments defined by change points for segment classification. The per-pixel change detection using Pelt method identified 18 wide-scaled change points in study area. Afterward, two-phased multi-class classification was conducted to delineate dragon fruit area using the Random Forest algorithm with statistical features of four time-series segments representing four periods 2000–2005, 2005–2010, 2010–2015, and 2015–2020. A robust paradigm combining validation of classification results with area estimation

L. H. Pham (✉) · H. Q. Nguyen
Vietnam National University–Ho Chi Minh City (VNU–HCM), 01 Marie Curie, Linh Trung Ward, Thu Duc City, Ho Chi Minh City, Vietnam
e-mail: hongluanosgeo@gmail.com

T. D. H. Le
Faculty of Resources and Environment, University of Thu Dau Mot, 06 Tran Van On Street, Thu Dau Mot City 820000, Binh Duong, Vietnam

L. T. H. Pham
Faculty of Environmental Sciences, University of Science, VNU–HCM, 227 Nguyen Van Cu, Ward 4, District 5, Ho Chi Minh City, Vietnam

H. Nguyen
Department of Land Management, Dong Thap University, Cao Lanh City 870000, Vietnam
e-mail: nguyenho@dthu.edu.vn

Institute of Landscape Ecology, University of Münster, 48149 Münster, Germany

H. Q. Nguyen
Center of Water Management and Climate Change, Institute for Environment and Resources (IER), VNU–HCM, 01 Marie Curie, Linh Trung Ward, Thu Duc City, Ho Chi Minh City, Vietnam

Institute for Circular Economy Development (ICED), VNU–HCM, 01 Marie Curie, Linh Trung Ward, Thu Duc City, Ho Chi Minh City, Vietnam

L. Q. Nguyen et al. (eds.), *Advances in Geospatial Technology in Mining and Earth Sciences*, Environmental Science and Engineering,
https://doi.org/10.1007/978-3-031-20463-0_26

was applied to analyze changing of dragon fruit areas. The result suggests a steady increase of dragon fruit areas in sand dune areas of Binh Thuan province from 2000 to 2020 especially in two districts Ham Thuan Bac and Ham Thuan Nam and recent conversions to dragon fruits in Bac Binh, Tuy Phong, and La Gi districts.

Keywords Land use and land cover · Change point detection · Classification · Time series · Dragon fruits · Binh Thuan

1 Introduction

Remote sensing is the science that efficiently supports learning dynamics of land use and land cover (LULC) [1, 2]. Analyzing history of LULC changes is greatly facilitated since the availability of Landsat, Moderate Resolution Image Spectroradiometer (MODIS), and Advanced Very High Resolution Radiometer (AVHRR) data with their long record of measurement. Despite their temporal comparability, it was noted that the MODIS and AVHRR data have been used frequently in regional and global researches on considering their coarse spatial resolution (250 m) demanding less computing resources [3, 4].

In analyzing LULC dynamics, the availability of remote sensing data demanded sophisticated methods to capture the full spectrum of LULC changes whether they are seasonal, abrupt, or trendy changes. However, in most LULC change studies, the simple single-phase classification combined with post-classification was applied to analyze LULC changes. The method includes picking the most cloud-free images to represent one period for LULC classification and post-classification change detection by overlaying classification results [1, 5, 6]. Shortcomings of the single-phase method arrived from its assumption of LULC consistency in the selected period (normally five or ten years) [7]. This assumption is hardly valid in dynamic landscapes, making LULC analysis subjected to biases. Multi-temporal classification [8, 9] was the improvement to the single-phased approach for its ability to capture phenological features of land by applying time-series classification [10]. However, the method still bases on the assumption about unchanged LULC in the considered period [11].

In this study, the Change Detection and Classification (CDC) method with multi-temporal data is proposed to address invalid assumption about LULC intra-consistency in analyzed periods [7, 10, 12]. The method as it was named includes two main steps: (1) change detection and (2) time-series classification. Change detection is the essential initial step in CDC in which land-cover change is realized by finding breakpoints in remote sensing time-series data [13]. It is a major topic in remote sensing science, taking into account multiple research and ongoing developing algorithms working on change detection, like Breaks For Additive Seasonal and Trend (BFAST) [14], Continuous Change Detection and Classification [10, 15], the Prophet [7, 16] to name a few. The core purpose of change detection is twofold, firstly, to identify time points of change and, secondly, to form temporal segments without LULC disturbances. The temporal-consistent segments are later input to time-series classification using temporal statistical features [10] or Dynamic Time Warping (DTW) [7, 17].

Dragon fruit is one of the key cash crops in Vietnam. Its area covered approximately 41,164.6 ha of land with yearly yield of 686,195 tons. The South East of Vietnam has the highest cultivating area of dragon fruit (26,964.7 ha), in which Binh Thuan province accounts for nearly 96% (26,026.4 ha) [18]. Binh Thuan province also has the highest area of dragon fruit in Vietnam, making it the main specialized area for dragon fruit cultivation.

Even though dragon fruit has a major position in socioeconomic development, farming of dragon fruits is not without trade-offs. Expansion and intensification of dragon fruit to meet demands for exportation brings environmental costs. Specifically, lighting techniques adopted by farmers at night to maintain dragon fruit production throughout a year would cause light pollution which has externalities on wildlife [19]. Among the most evident ecological impacts of light pollution are the disruption of birds' migration pattern and changing of sea turtle breeding behaviors [20, 21]. Locally, light pollution might hamper astronomic activities and esthetic value of night sky [19]. The increasing demand of lighting also poses pressure on the electricity supply system and indirectly contributes to greenhouse gases emission. Lastly, the non-competitiveness of Vietnamese dragon fruit on the international market and, therefore, its heavy dependence on the China market [22] imply financial risks to farmers. The local and global problems caused by dragon fruit farming raised concerns about the sustainability of land-use planning for dragon fruit in Vietnam. As Binh Thuan is Vietnam's key dragon fruit producer, land resources allocated to dragon fruit in the province should be monitored to negate negative impacts. A robust scientific evidence about the dynamics of spatial coverage and distribution of dragon fruit farms in the province would support better planning for the plant.

LULC dynamics in Binh Thuan province was the main topic in previous studies by [23–27]. However, [23] and [24] still applied the single-phase method to analyze general LULC dynamics. The research by [25–27] focused specifically on dragon fruits in Binh Thuan, using Visible Infrared Imaging Radiometer Suite (VIIRS) nighttime light (NTL) data. In this research, the CDC approach has been applied to analyze the dynamics of dragon fruit area in the sandy dune area of Binh Thuan province since 2000, using MODIS data. Two main research questions were addressed including 1) how is the capability of CDC to map spatial distribution of dragon fruit area since 2000? and 2) how did area of dragon fruit change over the last twenty years?

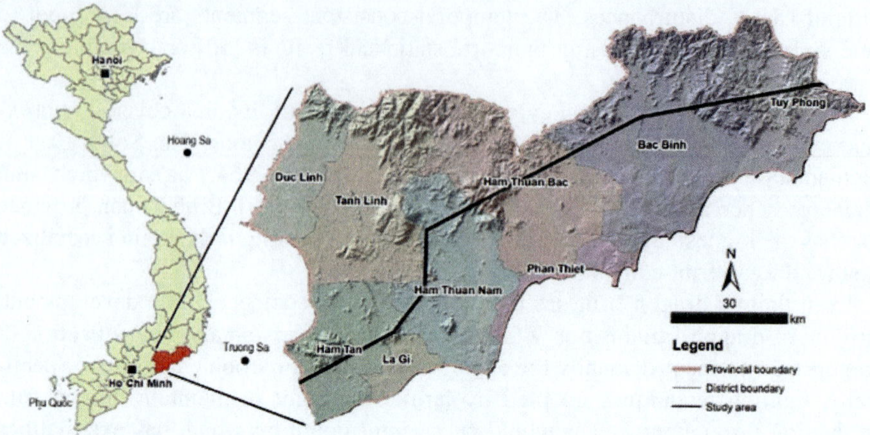

Fig. 1 Map of study area

2 Material and Methods

2.1 Study Area

The sand dunes area in Binh Thuan province (Fig. 1) is located at 10°34'N—11°24'N and 107°32'E—108°53'E. The area crosses six districts and one city of Binh Thuan province including Ham Tan, La Gi, Ham Thuan Nam, Ham Thuan Bac, Bac Binh, Tuy Phong, and Phan Thiet city. Topographically, the study area includes narrow coastal plain, coastal sand dunes, and mountainous areas with elevation ranges from 100 m, 100–200 m, and 500–800 m above mean sea level, respectively.

Land for agriculture accounted for nearly 89% of the whole province area, in which permanent crops took the highest ratio. Besides dragon fruit, which took nearly 26,000 ha in 2015, rubber tree and cashew were other main permanent crops with an area of 42,919 ha and 17,892 ha, respectively. Rice was the main annual crop, taking up approximately 53,849 ha [28].

2.2 MODIS NDVI Data

The research utilizes MODIS terra vegetation data (MOD13Q1 V6) [29] with a spatial resolution of 250 m and revisit time of 16 days. The MOD13Q1 V6 product gives global per-pixel vegetation values with two major indices including Normalized Difference Vegetation Index (NDVI) and Enhanced Vegetation Index (EVI). The two indices are computed from atmospherically corrected surface reflectances along with quality band for cloud masking. In this research, MODIS NDVI data from

2000 to 2020 were collected from the Google Earth Engine platform [30]. Preprocessing including reprojection, clipping to study area, and cloud masking were also conducted on the Google Earth Engine platform. Temporal interpolation is applied to fill masked data, afterward the Savitzky-Golay filter [31] with a filter window of five and polynomial order of two being conducted to smooth the NDVI time series for each pixel.

2.3 Time-Series Breakpoint Detection for LULC Change Analysis

Breakpoints on time-series data can be defined as points at which there are structural changes. In this research, the Prune Exact Linear Time (Pelt) method [32] is applied to search for change points for each pixel. Pelt is an offline method to efficiently find an unknown optimal number of change points on time series. The algorithm activates on the mechanism of finding segmentation that minimizes the cost given by cost function [33]. A test based on expert judgment was conducted to select two important parameters of the search method including the Radial Basis Function (RBF) cost function and minimum segment length of 23 time points. The rationale of picking minimum segment length of 23 was that with MODIS revisit time of 16 days, a segment of 23 time points could have above one-year time span to ensure that each pixel would not have two LULC changes in one year. Lastly, to remove breakpoints representing ephemeral LULC changes [10], breakpoints existed at less than 1% of study area (663 out of 58,685 pixels) are removed.

2.4 Time-Series Classification

Time-series classification is conducted on temporal NDVI segments identified by breakpoints. Feature-based Random Forest (RF) [34] is adopted on statistical features (Table 1) of NDVI temporal segment to build classifiers. The tenfold cross validation is applied to tune the number of trees and maximum features to build decision trees. The multi-class classification targeted delineation of dragon fruit in the study area. For that, permanent crops are firstly identified from built-up/bare lands, annual crops, forest, and water. Afterward, a second classifier is developed for the classified permanent crop areas to classify dragon fruits from other permanent crops like cashew, rubber tree, and coconut. Such a two-step method has been proved to succeed in identifying specific crops from general LULC types [35].

Training samples are collected separately for unchanged areas and by each identified temporal segment based on Landsat composite images and Google Earth. A two-phase sampling is also applied with respect to two-step multi-class classification. Figure 2 presents a sampling scheme to develop labeled data for classification.

Table 1 Time-series statistical features to build classifiers

Features	Formulas
Absolute sum of changes	$\sum_{i=1,\ldots,n-1} \lvert x_{i+1} - x_i \rvert$
Mean of absolute of changes	$\frac{1}{n-1} \sum_{i=1,\ldots,n-1} \lvert x_{i+1} - x_i \rvert$
Mean of changes	$\frac{1}{n-1} \sum_{i=1,\ldots,n-1} (x_{i+1} - x_i)$
Sum	$\sum_{i=1,\ldots,n} x_i$
Mean	$\frac{1}{n} \sum_{i=1,\ldots,n} x_i$
Median	$\begin{cases} x\left(\frac{n+1}{2}\right) \text{when } n\text{odd} \\ \frac{x\left(\frac{n}{2}\right)+x\left(\frac{n+2}{2}\right)}{2} \text{when } n\text{even} \end{cases}$
Maximum	$\max_{i=1,\ldots,n} (x_i)$
Minimum	$\min_{i=1,\ldots,n} (x_i)$
Kurtosis	$\frac{\sum_{i=1}^{n}(x_i-\overline{x})^4/n}{s^4}$
Skewness	$\frac{\sqrt{n(n-1)}}{n-2} \frac{\sum_{i=1}^{n}(x_i-\overline{x})^3/n}{s^3}$
Standard deviation	$\sqrt{\frac{\sum_{i=1}^{n}(x_i-\mu)^2}{n}}$
Variance	$\sqrt{\frac{\sum_{i=1}^{n}(x_i-\mu)^2}{n-1}}$
Complexity	$\sqrt{\sum_{i=1}^{n-1} (x_i - x_{i-1})^2}$
Numbers of peaks	$f\left(x_{1,\ldots,n}, m\right)$

Note x denotes time-point data, n denotes length of time series, s denotes standard deviation

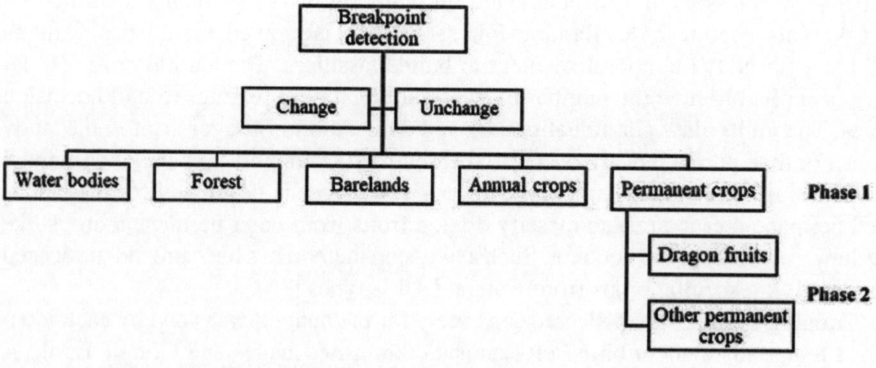

Fig. 2 Two-phase classification to delineate dragon fruits

2.5 Accuracy Assessment and Area Estimation

A robust paradigm that combines between validation and area estimation suggested in [36] is adopted in this research. The paradigm includes recommendations for sampling design for validation and area estimation based on reference data. Equation (1) proposed by [36, 37] presents a way to calculate sampling size (denoted as n) for each LULC class i based on its proportion in classification result and desired confidence level. Later, random stratified sampling is conducted to take samples for validation. Best available data including Landsat images and Google Earth imagery are used as reference layers for validation. Three accuracy metrics including user accuracy (UA), producer accuracy (PA), and overall accuracy (OA) are used to measure the classification performance [37, 38].

$$n = \frac{\left(\sum W_i S_i\right)^2}{\left[S\left(\hat{O}\right)\right]^2 + (1/N)\sum W_i S_i^2} \approx \left(\frac{\sum W_i S_i}{S\left(\hat{O}\right)}\right)^2 \tag{1}$$

in which: N is the total number of pixel in study area, $S\left(\hat{O}\right)$ is the standard error of wanted accuracy, W_i is the ratio of LULC class i in study area, and S_i is the standard deviation of class i.

Area of each LULC class k is estimated by multiplying its proportion on reference classification (denoted as $\hat{p}_{.k}$) to total area. $\hat{p}_{.k}$ is defined by Eq. (2)

$$\hat{p}_{.k} = \sum_{i=1}^{q} W_i \frac{n_{ik}}{n_{i.}} \tag{2}$$

in which: W_i is the ratio of LULC class i in study area, q is the number of LULC classes, n_{ik} is the number of samples of class k on reference data, classified as class i, and $n_{i.}$ is the number of samples of class i according to classification result.

Lastly, confidence interval of estimated area of dragon fruit is calculated by multiplying 1.96 by the standard error of percentage of estimated area (denoted as $S(\hat{p}_{.k})$). $S(\hat{p}_{.k})$ is defined by:

$$S(\hat{p}_{.k}) = \sqrt{\sum_i \frac{W_i \widehat{p_{ik}} - \widehat{p_{ik}}^2}{n_{i.} - 1}} \tag{3}$$

in which: $\widehat{p_{ik}}$ is the estimated percentage of pixels of class k on reference data, classified as class i.

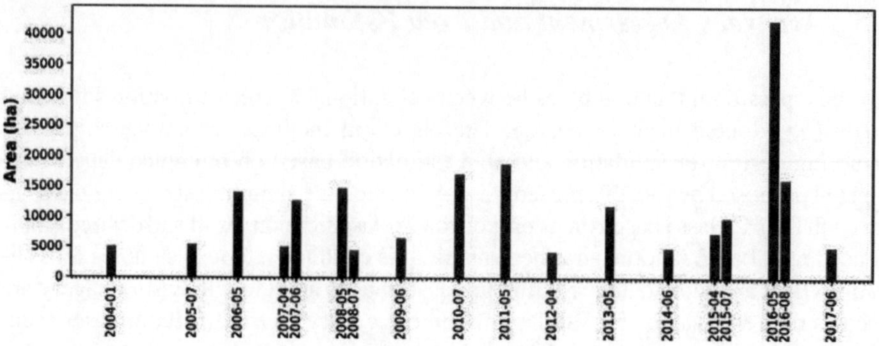

Fig. 3 Spatial popularity of detected breakpoints

3 Result

3.1 LULC Changes Detection

Application of the Pelt search method on NDVI time series identifies 18 significant breakpoints in the study area. Figure 3 demonstrates the popularity of these change points. It is noted that May 2016 was the point when there was large-scale LULC change reflected by its occurrence at nearly 41,000 ha of study area. Other major shifting times were May 2006, May 2008, July 2010, and May 2011. Besides, the search algorithm finds an area of approximately 1,600 km2 (Fig. 4) that shows no breakpoints in the period 2000–2020. It could be inferred that this area had stable LULC in the 20-year time span.

Figures 5a–c demonstrate the sensitivity of the Pelt search method in finding LULC change in the study area. Two main types of detected shifting in temporal features could be observed including changing of mean NDVI values happening as land surface became less vegetated (Fig. 5a) and changing of NDVI variance when land surface underwent changing from being (quite) stable to more dynamic or vice versa (Fig. 5b). Figure 5c shows robustness of the Pelt search method in avoiding commission error (i.e., identifying change points that did not exist). In pixels where there is conversion to dragon fruits (Fig. 5b), big difference in data variance can be observed.

3.2 LULC Classification to Delineate Dragon Fruits Area

Validating classification results of all temporal segments is very intensive, in which only four temporal segments are selected to represent four periods 2000–2005, 2005–2010, 2010–2015, and 2015–2020. Table 2 shows representative temporal segments

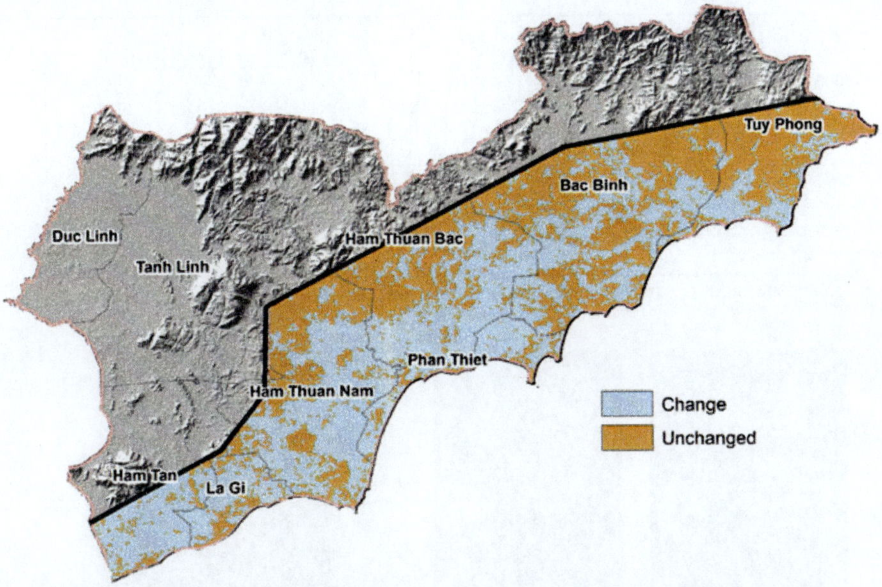

Fig. 4 Spatial distribution of change/unchanged area in twenty years (2000–2020)

and training sample sizes for each segment. However, we still recommend conducting classification for all derived segments identified in change detection.

Hyper-parameters of numbers of trees and features are optimized to build classifiers giving the best test scores. The hyper-parameters are tuned to develop classifiers with test scores higher than 0.90. The developed classifiers are adopted on temporal segments for multi-label classifications. Figures 6a–d present the classification results for four periods, respectively.

Figure 6a shows that the earliest planting of dragon fruits happened in Ham Thuan Bac, Ham Thuan Nam, and Phan Thiet. In the period 2005–2010 (Fig. 6b), dragon fruit area expanded more rapidly in Ham Thuan Nam, mostly in areas previously occupied by other permanent crops. In the next period (Fig. 6c), while dragon fruits kept expanding in Ham Thuan Nam, conversion to dragon fruits happened more rapidly in annual crop areas in Ham Thuan Bac. In the last period 2015–2020 (Fig. 6d) when dragon fruit was the major LULC type in Ham Thuan Nam and Ham Thuan Bac, its cultivation started to occur in Bac Binh district.

3.3 Accuracy Assessment and Area Estimation

Tables 3, 4, 5, and 6 present confusion matrices and accuracy metrics for classification results in four periods. In general, all developed classifiers achieve OA higher than 70%. Classification results of dragon fruits are acceptable where there are higher

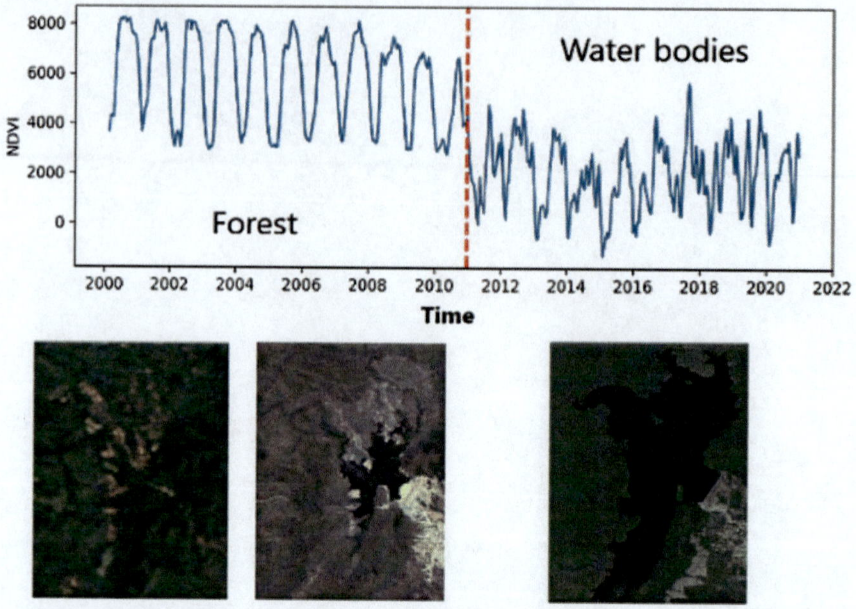

(a)

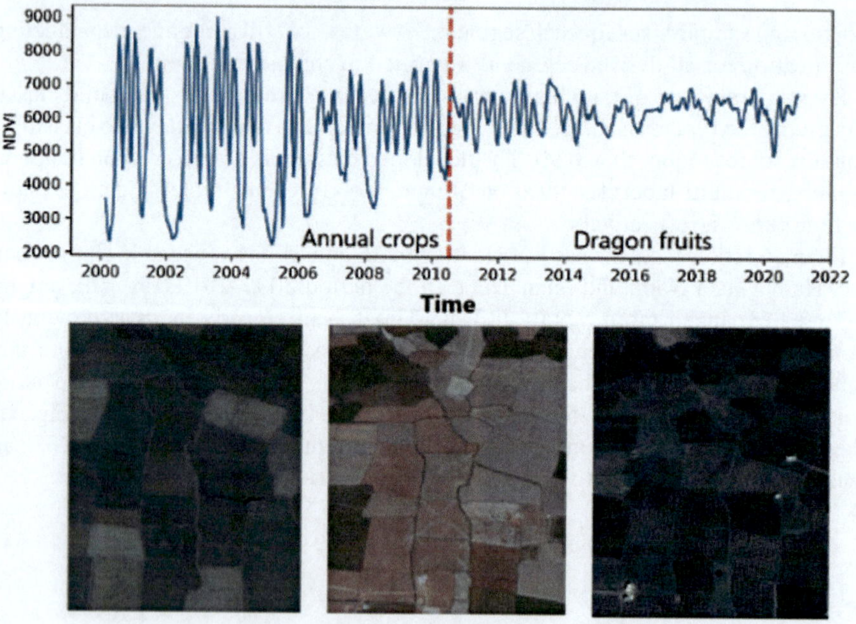

(b)

Fig. 5 **a** Sensitivity of the Pelt search method in detecting breakpoints in a) less vegetated area, **b** annual crop area converted to permanent crops, and **c** unchanged area

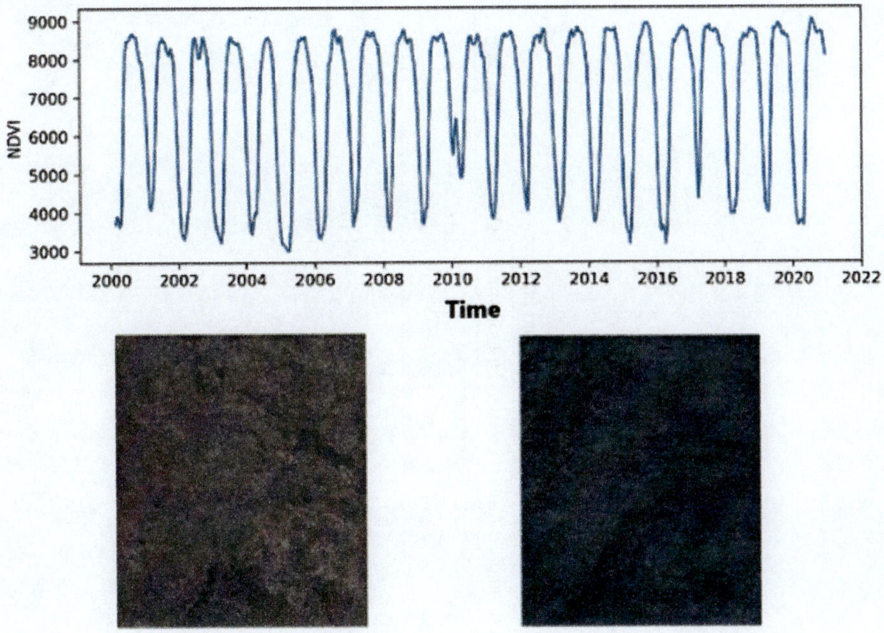

(c)

Fig. 5 (continued)

areas of dragon fruits. For example, UAs of dragon fruits in Ham Thuan Bac and Ham Thuan Nam were nearly 85% and 71%, respectively, in the period 2005–2010 (Table 4). Similarly, both UAs of dragon fruit at Ham Thuan Bac and Ham Thuan Nam in the next period (Table 5) are higher than 80%. High UA of dragon fruit implied that the classifiers were less likely to confuse other LULCs with dragon fruits; therefore, the results do not overestimate dragon fruit areas.

The PA of dragon fruit is also higher than 70% in most cases. Contrary to UA, the PA measured the probability that the classifiers missed to recognize dragon fruits. With PA of higher than 70%, especially in Ham Thuan Nam and Ham Thuan Bac, it is not likely that dragon fruits areas were underestimated. However, the developed classifiers perform poorly in areas with low areas of detected dragon fruit. Low performance of classification results was more common in the earliest period 2000–2005 (Table 3) where UAs of dragon fruits in Ham Thuan Bac and Ham Thuan Nam were only approximately 29 and 31%. In areas where dragon fruit was only newly cultivated like Bac Binh district in 2010–2015 (Table 4), UA was only 10%. Nevertheless, the low accuracies in lowly detected areas were expected under the random stratifying sampling scheme for validation, in which sampling size was proportional to detected population size.

Table 2 Representative temporal segments and total numbers of training sizes

Periods	Temporal segments	Training samples sizes						
		First phase					Second phase	
		Forest	Water bodies	Permanent crops	Bare lands	Annual crops	Dragon fruits	Other permanent crops
Unchanged	02/18/2000–12/18/2020	121	80	51	134	43	30	47
2000–2005	02/18/2000–01/01/2004	165	75	132	308	132	38	101
2005–2010	06/10/2009–07/12/2010	194	96	164	190	65	122	110
2010–2015	06/10/2014–04/23/2015	151	118	155	175	136	125	100
2015–2020	06/26/2017–12/18/2020	174	93	281	223	147	80	80

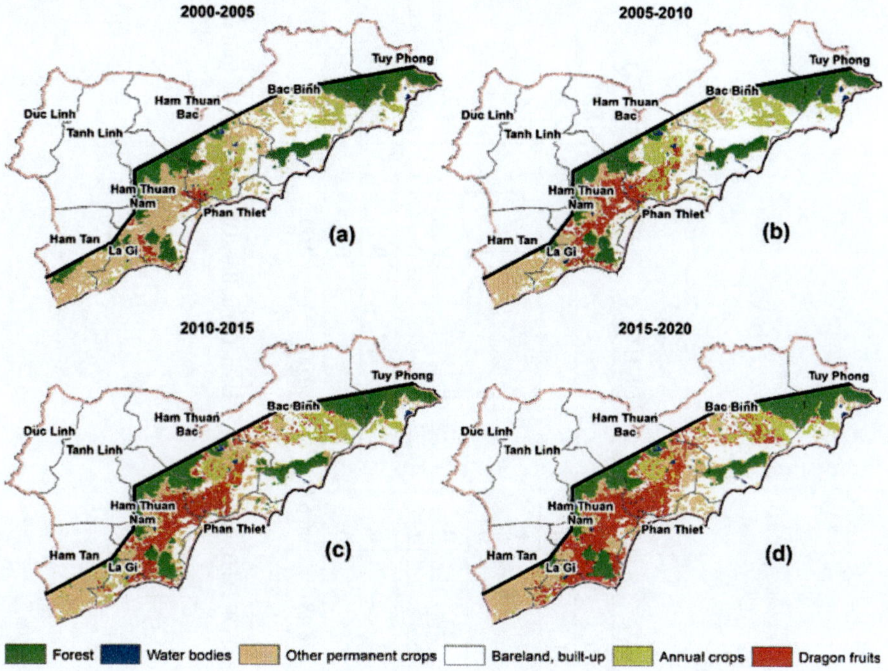

Fig. 6 LULCs classification results in four periods: **a** 2000–2005, **b** 2005–2010, **c** 2010–2015, and **d** 2015–2020

Analysis of confusion matrices show that dragon fruits are mostly confused with annual and other permanent crops and bare lands. For example, in the period 2015–2020 (Table 6), 29%, 46%, and 25% of samples misclassified as dragon fruits in Ham Thuan Bac were permanent crops, bare land crops, and annual crops, respectively. The numbers for Ham Thuan Nam in the same periods were 34%, 55%, and 4%, respectively.

Figure 7 shows developing areas of dragon fruits with a confidence level of 95%. Table 7 of Appendix presents numeric presentation of calculation. The area of dragon fruits in seven districts increases linearly over 20 years. The estimated dragon fruits areas in Ham Thuan Bac and Ham Thuan Nam districts in 2015–2020 were approximately 239km2. These areas were 20-fold the areas in the period 2000–2005. The increasing areas of dragon fruits in these two districts from 2000 to 2015 also show statistical significance. In specific, confidence intervals 95% of dragon fruits areas in Ham Thuan Nam in three periods 2000–2005, 2005–2010, and 2010–2015 are 4–35km^2, 146–219km^2, and 212–289km^2, respectively. In Ham Thuan Bac district, the fluctuations of dragon fruits areas in these three periods are 0–25km^2, 73–121km^2, and 164–235km^2, respectively.

Dragon fruit areas in districts other than Ham Thuan Nam and Ham Thuan Bac show an increasing trend but do not contribute much to total areas of dragon fruits. Dragon fruits appear in three districts Bac Binh, La Gi, and Phan Thiet in the period

Table 3 Confusion matrix and accuracy metrics of classification in period 2000–2005

		Reference data								
		FR	WB	PC	BA	AC	DF-HTB	DF-HTN	DF-PT	UA (%)
Classified data	FR	**180**	0	2	3	0	0	0	0	97.30
	WB	1	**16**	3	12	1	0	0	0	48.48
	PC	36	0	**156**	89	7	1	1	1	53.61
	BA	2	5	14	**272**	4	0	0	1	91.28
	AC	0	0	4	7	**62**	0	0	0	84.93
	DF-HTB	0	0	0	3	2	**2**	0	0	28.57
	DF-HTN	0	0	2	4	3	0	**4**	0	30.77
	DF-PT	0	0	0	0	0	0	0	**0**	
	PA (%)	82.19	76.19	86.19	69.74	78.48	66.67	80.00	0.00	
	OA (%)	76.89								

FR: forest, WB: water bodies, PC: permanent crops, BA: bare lands, AC: annual crops, DF-HTB: dragon fruit in Ham Thuan Bac, DF-HTN: dragon fruit in Ham Thuan Nam, DF-PT: dragon fruit in Phan Thiet

Table 4 Confusion matrix and accuracy metrics of classification in period 2005–2010

		Reference data										
		FR	WB	PC	BL	AC	DF-BB	DF-HTB	DF-HTN	DF-LG	DF-PT	UA (%)
Classified data	FR	**172**	0	1	1	0	0	0	0	0	0	98.85
	WB	2	**16**	3	14	0	0	0	2	1	0	42.11
	PC	15	0	**150**	38	3	0	0	3	0	0	71.77
	BL	1	5	30	**261**	13	1	2	6	0	0	81.82
	AC	1	0	5	5	**71**	0	5	0	0	0	80.68
	DF-BB	0	0	0	0	0	**0**	0	0	0	0	–
	DF-HTB	0	1	0	1	1	0	**17**	0	0	0	85.00
	DF-HTN	0	0	5	7	2	0	0	**34**	0	0	70.83
	DF-LG	0	0	1	0	0	0	0	0	**1**	0	50.00
	DF-PT	0	0	0	0	0	0	0	0	0	**1**	100.00
	PA (%)	90.05	72.73	76.53	79.82	78.89	0.00	70.83	75.56	50.00	100.00	
	OA (%)	80.33										

FR: forest, WB: water bodies, PC: permanent crops, BA: bare lands, AC: annual crops, DF-HTB: dragon fruit in Ham Thuan Bac, DF-HTN: dragon fruit in Ham Thuan Nam, DF-PT: dragon fruit in Phan Thiet, DF-LG: dragon fruit in La Gi, DF-BB: dragon fruits in Bac Binh, DF-TP: dragon fruits in Tuy Phong

Table 5 Confusion matrix and accuracy metrics of classification in period 2010–2015

		Reference data												
		FR	WB	PC	BL	AC	DF-BB	DF-HT	DF-HTB	DF-HTN	DF-LG	DF-PT	DF-TP	UA (%)
Classified data	FR	**168**	0	7	0	0	0	0	0	0	0	0	0	96.00
	WB	0	**17**	0	21	0	0	0	0	1	0	0	0	43.59
	PC	29	0	**132**	29	5	2	0	6	9	0	0	0	62.26
	BL	3	4	15	**233**	7	1	0	0	3	1	0	0	87.27
	AC	1	0	5	5	**61**	5	0	7	3	1	0	1	68.54
	DF-BB	0	0	2	5	2	**1**	0	0	0	0	0	0	10.00
	DF-HT	0	0	0	1	0	0	**0**	0	0	0	0	0	0.00
	DF-HTB	0	1	0	6	2	0	0	**36**	0	0	0	0	80.00
	DF-HTN	1	0	2	7	0	0	0	0	**46**	0	0	0	82.14
	DF-LG	0	0	0	1	1	0	0	0	0	**2**	0	0	50.00
	DF-PT	0	0	0	0	0	0	0	0	0	0	**1**	0	100.00
	DF-TP	0	0	0	0	0	0	0	0	0	0	0	**0**	–
	PA (%)	83.17	77.27	80.98	75.65	78.21	11.11	–	73.47	74.19	50.00	100.00	0.00	
	OA (%)	77.44												

FR: forest, WB: water bodies, PC: permanent crops, BA: bare lands, AC: annual crops, DF-HTB: dragon fruit in Ham Thuan Bac, DF-HTN: dragon fruit in Ham Thuan Nam, DF-PT: dragon fruit in Phan Thiet, DF-LG: dragon fruit in La Gi, DF-BB: dragon fruits in Bac Binh, DF-TP: dragon fruits in Tuy Phong

Table 6 Confusion matrix and accuracy metrics of classification in period 2015–2020

		Reference data												
		FR	WB	PC	BL	AC	DF-BB	DF-HT	DF-HTB	DF-HTN	DF-LG	DF-PT	DF-TP	UA (%)
Classified data	FR	**293**	0	1	14	0	0	0	0	0	0	0	0	95.13
	WB	1	**22**	1	41	0	0	0	0	0	0	0	0	33.85
	PC	24	0	**266**	79	6	2	0	16	7	2	1	0	66.00
	BL	5	2	28	**342**	4	3	0	2	1	0	0	2	87.92
	AN	3	1	4	8	**86**	7	0	10	2	0	0	4	68.80
	DF - BB	0	0	6	11	6	**13**	0	0	0	0	0	0	36.11
	DF - HT	0	0	0	1	0	0	**1**	0	0	0	0	0	50.00
	DF-HTB	0	0	7	11	6	0	0	**77**	0	0	0	0	76.24
	DF-HTN	2	1	16	26	2	0	0	0	**95**	0	0	0	66.90
	DF-LG	0	0	12	2	0	0	0	0	0	**7**	0	0	33.33
	DF-PT	1	0	1	1	0	0	0	0	0	0	**4**	0	57.14
	DF-TP	0	0	0	0	1	0	0	0	0	0	0	**1**	50.00
	PA (%)	89.06	84.62	77.78	63.81	77.48	52.00	100.00	73.33	90.48	77.78	80.00	14.29	
	OA (%)	75.44												

FR: forest, WB: water bodies, PC: permanent crops, BA: bare lands, AC: annual crops, DF-HTB: dragon fruit in Ham Thuan Bac, DF-HTN: dragon fruit in Ham Thuan Nam, DF-PT: dragon fruit in Phan Thiet, DF-LG: dragon fruit in La Gi, DF-BB: dragon fruits in Bac Binh, DF-TP: dragon fruits in Tuy Phong

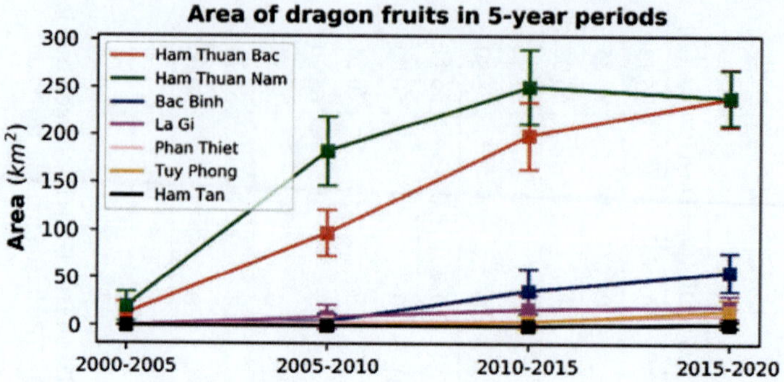

Fig. 7 Development of dragon fruit areas in four periods

2005–2010 with an estimated area below 10km^2. In the next periods, dragon fruits areas in these three districts increase gradually but did not reach over 60km^2. Bac Binh is the district with the highest area of dragon fruits among three with estimated areas of 57km^2 in the period 2015–2020. Lastly, Tuy Phong and Ham Tan are the two districts with the lowest areas of dragon fruits. Dragon fruits start to be cultivated in Tuy Phong in the period 2010–2015 with estimated areas of 4km^2, while in Ham Tan, dragon fruit farming started in the period 2015–2020 with an estimated area of 2 km^2.

4 Discussion

4.1 CDC in Studying Dynamics of Dragon Fruit Areas

Since the free access to remote sensing data archives like Landsat or MODIS and the appearance of Google Earth Engine as an open platform to process bulks of data, there have been multiple studies focusing on dynamics of LULCs. The availability of data demanded sophisticated methods without assumption about LULC consistency in analyzing LULC changes, which the "single-phase" paradigm could not guarantee. The CDC approach in this research showed that combining break detection with multi-temporal classification could provide robust insights about LULC dynamics with special focus on dragon fruits.

The Pelt search with spatial filter gave 18 breakpoints that were spatially common. The identification of breakpoints in NDVI time series helped define segments with consistent statistical features. Advantages of Pelt search were its explorative nature in finding unknown number of breakpoints and computation efficiency [33]. Finding significant breakpoints by filtering out those with low spatial coverage could address ephemeral changes in time series, regularly caused by abnormal values like cloud or

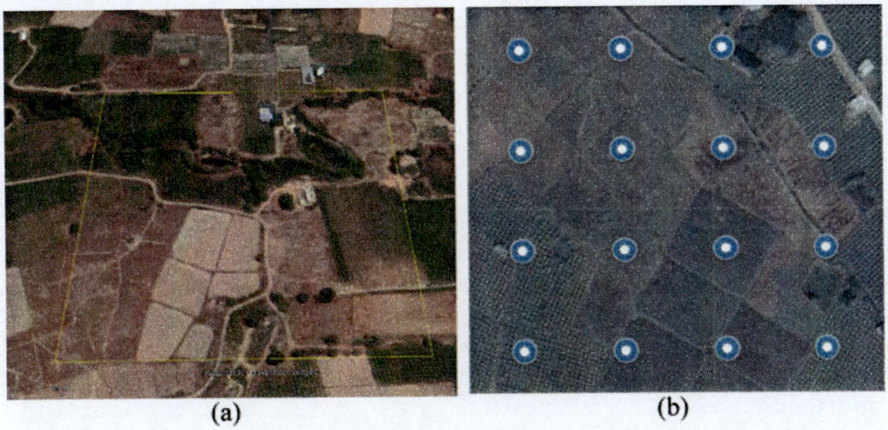

Fig. 8 **a** Dragon fruits and annual crops are mixed to create spatial heterogeneity within 250 × 250 m MODIS pixel, **b** Sub-pixel systematic sampling

cloud shadow. However, validation of breakpoints was largely dependent on the availability of historical satellite imaginaries which was not of high coverage everywhere on Earth.

The two-phase multi-temporal classifications delineated dragon fruits at acceptable levels of accuracy. The validation step revealed that confusion between dragon fruits and other LULC classes mostly originated from pixel heterogeneity (Fig. 8) caused by coarse spatial resolution of MODIS NDVI data. Spatial heterogeneity was also prevalent in areas where LULC conversion was happening. The expert judgment method applied in validating classification results might lead to underestimating or overestimating of the overall dragon fruit area because of ambiguity in levels of dragon fruits coverages for a pixel to be defined as dragon fruit fields. In such context, the statistically robust paradigm of validation and area estimation by [36] was adopted to address the uncertainties in area estimation. Based on reference classification, 95% confidence intervals of dragon fruits areas by districts were calculated to give statistically significant insights about dragon fruit areas in each period. It was noted that a random design for sampling with equal sampling size would be more suitable than the random stratified design as it treats low-detected dragon fruits in the early periods more fairly. The systematic sampling within pixel (Fig. 8) could also be integrated with random sampling for validation at sub-pixel level to address uncertainties from spatial heterogeneity.

4.2 Dynamics of Dragon Fruits Area

Application of CDC for four periods 2000–2005, 2005–2010, 2010–2015, and 2015–2020 showed steady increase of dragon fruit farming in the sandy dunes areas of Binh

Thuan province. Besides, the increasing trend showed signs of slowing down in recent years, especially at Ham Thuan Nam. This finding matched with [25] which revealed that dragon fruits area in the province expanded rapidly after 2014 but reached its peak after 2017. The increase happened rapidly in two districts Ham Thuan Bac and Ham Thuan Nam (12–199km^2 and 19–250km^2, respectively) from 2000 to 2015. Dragon fruit cultivation also started to expand to two districts Bac Binh and Tuy Phong in the period 2015–2020. With a total area of dragon fruit of approximately 584km^2 in 2015–2020 detected in this research (95% confidence interval of 474–699km^2), areas for dragon fruits have breached the ceiling of land planned for dragon fruit farming in the province. For continuous expansion of dragon fruits implies not only socioeconomic but also ecological cost, LULC planning in the province should focus on balancing between the benefits and externalities of conversion to dragon fruits farms.

5 Conclusion

In this research, the CDC method has been adopted to analyze the dynamics of dragon fruit cultivation in sandy dune areas of Binh Thuan province. Not relying on fragile assumption of intra-consistency, the application of CDC presented robust findings about spatial distribution and area change of dragon fruit from 2000 to 2020. The result showed steady increase of dragon fruit area in two districts Ham Thuan Bac and Ham Thuan Nam, and recent expansion of dragon fruit in Bac Binh, Tuy Phong, and La Gi.

The research also recommended validation of breakpoints in time series and conduction of segment-wise classification rather than the adopted period-wise classification. Lastly, the random sampling or systematic combined with random design instead of random stratified design for validating classification results can be applied to bring statistical significance to lowly detected dragon fruit areas.

Acknowledgements This study is funded by the Binh Thuan Department of Science and Technology, Vietnam

Appendix

See Table 7.

Table 7 Summary of dragon fruit areas and confidence intervals in four periods

	2000–2005			2005–2010			2010–2015			2015–2020		
	Area (km^2) with CI 95%	CI 95% (km^2)		Area (km^2) with CI 95%	CI 95% (km^2)		Area (km^2) with CI 95%	CI 95% (km^2)		Area (km^2) with CI 95%	CI 95% (km^2)	
		LB	UB		LB	UB		LB	UB		LB	UB
Ham Thuan Bac	12.03 ± 12.84	0	24.87	97.06 ± 24.39	72.67	121.45	199.3 ± 35.39	163.91	234.69	239.22 ± 30.01	209.21	269.23
Ham Thuan Nam	19.66 ± 15.44	4.22	35.1	182.59 ± 36.36	146.23	218.95	250.57 ± 38.8	211.77	289.37	239.58 ± 28.82	210.76	268.4
Bac Binh	0	0	0	4.05 ± 7.95	0	12	36.59 ± 23.59	13	60.18	56.88 ± 20.04	36.84	76.92
La Gi	0	0	0	8.84 ± 12.31	0	21.15	17.06 ± 15.13	1.93	32.19	20.17 ± 11.55	8.62	31.72
Phan Thiet	0	0	0	3.25 ± 0	3.25	3.25	5.12 ± 0	5.12	5.12	11.59 ± 7.85	3.74	19.44
Tuy Phong	0	0	0	0	0	0	4.07 ± 7.98	0	12.05	15.91 ± 11.7	4.21	27.61
Ham Tan	0	0	0	0	0	0	0	0	0	1.9 ± 3.73	0	5.63

Note CI stands for Confidence Interval, LB stands for Lower Bound, UP stands for Upper Bound)

References

1. Langat, P.K., et al.: Monitoring of land use/land-cover dynamics using remote sensing: a case of Tana River Basin, Kenya. Geocarto Int. **36**, 1470–1488 (2021)
2. Sarif, M.O., Gupta, R.D.: Spatiotemporal mapping of Land Use/Land Cover dynamics using Remote Sensing and GIS approach: a case study of Prayagraj City, India (1988–2018). Environ. Dev. Sustain. **24**, 888–920 (2022)
3. Klein, I., Gessner, U., Kuenzer, C.: Regional land cover mapping and change detection in Central Asia using MODIS time-series. Appl. Geogr. **35**, 219–234 (2012)
4. Lunetta, R.S., et al.: Land-cover change detection using multi-temporal MODIS NDVI data. Remote Sens. Environ. **105**, 142–154 (2006)
5. Tran, H., Tran, T., Kervyn, M.: Dynamics of land cover/land use changes in the Mekong Delta, 1973–2011: A Remote sensing analysis of the Tran Van Thoi District, Ca Mau Province, Vietnam. Remote Sens. **7**(3), 2899–2925 (2015)
6. Son, N.T., et al.: Assessment of urbanization and urban heat islands in Ho Chi Minh City, Vietnam using Landsat data Sustain. Cities Soc. **30**, 150–161 (2017)
7. Yan, J., et al.: A time-series classification approach based on change detection for rapid land cover mapping. ISPRS J. Photogramm. Remote. Sens. **158**, 249–262 (2019)
8. Guerschman, J.P., et al.: Land cover classification in the Argentine Pampas using multi-temporal Landsat TM data. Int. J. Remote Sens. **24**, 3381–3402 (2003)
9. Homer, C., et al.: Development of a 2001 national land-cover database for the United States. Photogramm. Eng. Remote. Sens. **70**, 829–840 (2004)
10. Zhu, Z., Woodcock, C.E.: Continuous change detection and classification of land cover using all available Landsat data. Remote Sens. Environ. **144**, 152–171 (2014)
11. Rogan, J., Franklin, J., Roberts, D.A.: A comparison of methods for monitoring multitemporal vegetation change using thematic mapper imagery. Remote Sens. Environ. **80**, 143–156 (2002)
12. Zhu, Z., et al.: Including land cover change in analysis of greenness trends using all available Landsat 5, 7, and 8 images: A case study from Guangzhou, China (2000–2014). Remote Sens. Environ. **185**, 243–257 (2016)
13. Zhu, Z.: Change detection using landsat time series: a review of frequencies, preprocessing, algorithms, and applications. ISPRS J. Photogramm. Remote. Sens. **130**, 370–384 (2017)
14. Verbesselt, J., et al.: Detecting trend and seasonal changes in satellite image time series. Remote Sens. Environ. **114**, 106–115 (2010)
15. Zhu, Z., et al.: Continuous monitoring of land disturbance based on Landsat time series. Remote Sens. Environ. **238**, 111116 (2020)
16. Taylor, S.J., Letham, B.: Forecasting at scale. Am. Stat. **72**, 37–45 (2018)
17. Lambert, J., et al.: Monitoring forest decline through remote sensing time series analysis. GIScience and Remote Sensing **50**, 437–457 (2013)
18. Dard, B.T.: Land-use planning to 2020 of Binh Thuan province (2016–2020). Binh Thuan Department of Agriculture and Rural Development (DARD) (2015)
19. Gallaway, T., Olsen, R.N., Mitchell, D.M.: The economics of global light pollution. Ecol. Econ. **69**, 658–665 (2010)
20. Verheijen, F.J., Photopollution: artificial light optic spatial control systems fail to cope with. Incidents, causation, remedies. Experimental Biol. **44**, 1–18 (1985)
21. Salmon, M., et al.: Behavior of Loggerhead Sea Turtles on an Urban Beach. II. Hatching Orientation. J. Herpetology **29**, 568 (1995)
22. Vuong, D.M., et al.: Analysis of value chain of dragon fruits in Cho Gao district, Tien Giang Province. Can Tho J. Sci. **36**, 10–22 (2015)
23. Hountondji, Y., Ozer, P.: Land use and land cover change analysis 1990–2002 in Binh Thuan Province, south central Vietnam. In: 1st International Conference on Energy, Environment and Climate Change (2011)
24. Tran, H.T., et al.: Drought and human impacts on land use and land cover change in a vietnamese coastal area. Remote Sens. **11**(3), 1–23 (2019)

25. Jia, S., et al.: Extreme development of dragon fruit agriculture with nighttime lighting in Southern Vietnam, in Remote sensing of agriculture and land cover/land use changes in South and Southeast Asian Countries. In: K. P. Vadrevu, et al. (ed.). Springer International Publishing, pp. 553–571 (2022)
26. Krauser, L.E., et al.: Shedding light on agricultural transitions, dragon fruit cultivation, and electrification in Southern Vietnam using mixed methods. Ann. Am. Assoc. Geogr. **112**(4), 1139–1158 (2022)
27. Wang, R., Shi, W., Dong, P.: "Mapping dragon fruit croplands from space using remote sensing of artificial light at night. Remote Sens. **12**(24), 1–17 (2020)
28. DONRE: Land-use planning to 2020 (2016–2020) of Binh Thuan province. 2017, Binh Thuan Department of Natural Resources and Environment.
29. Didan, K.: MOD13Q1 MODIS/terra vegetation indices 16-day L3 global 250m SIN Grid V006. NASA EOSDIS Land Processes DAAC (2015)
30. Gorelick, N., et al.: Google earth engine: planetary-scale geospatial analysis for everyone. Remote Sens. Environ. (2017)
31. Savitzky, A., Golay, M.J.E.: Smoothing and differentiation of data by simplified least squares procedures. Anal. Chem. **36**, 1627–1639 (1964)
32. Killick, R., Fearnhead, P., Eckley, I.A.: Optimal detection of changepoints with a linear computational cost. J. Am. Stat. Assoc. **107**, 1590–1598 (2012)
33. Truong, C., Oudre, L., Vayatis, N.: Selective review of offline change point detection methods. Signal Process. **167**, 107299 (2020)
34. Breiman, L.: Random forests. Mach. Learn. **45**(1), 5–32 (2001)
35. Van Tricht, K., et al.: Synergistic use of radar sentinel-1 and optical sentinel-2 imagery for crop mapping: a case study for Belgium. Remote Sens. **10**(10), 1–22 (2018)
36. Olofsson, P., et al.: Good practices for estimating area and assessing accuracy of land change. Remote Sens. Environ. **148**, 42–57 (2014)
37. Cochran, W.G.: Sampling techniques, 3rd edn. (1977)
38. Congalton, R.G.: A review of assessing the accuracy of classifications of remotely sensed data. Remote Sens. Environ. **37**(1), 35–46 (1991)

Random Forest Analysis of Land Use and Land Cover Change Using Sentinel-2 Data in Van Yen, Yen Bai Province, Vietnam

Xuan Quang Truong, Nguyen Hien Duong Dang, Thi Hang Do, Nhat Duong Tran, Thi Thu Nga Do, Van Anh Tran, Vasil Yordanov, Maria Antonia Brovelli, and Thanh Dong Khuc

Abstract Land use land cover (LULC) change has become a crucial topic that needs to be addressed when the studying global and local sustainable development. In this research, time-series of Sentinel-2 images from 2019 to 2020 are used to derive LULC change in Mu Cang Chai (MCC) and Van Yen (VY) districts, Yen Bai province, Vietnam. We identified seven main land cover types and collected reference data from visual interpretation using Google Earth. The random forest (RF) classification algorithm is applied to construct the classified LULC map in these regions of Yen

X. Q. Truong · T. T. N. Do
Hanoi University of Natural Resources and Environment, 41A Phu Dien Road, Hanoi 100000, Vietnam
e-mail: txquang@hunre.edu.vn

N. H. D. Dang · N. D. Tran
University of Science and Technology of Hanoi, 18 Hoang Quoc Viet, Hanoi 100000, Vietnam

T. H. Do (✉)
Osaka City University, 3 Chome-3-138 Sugimoto, Sumiyoshi Ward, Osaka 558-8585, Japan
e-mail: dohanghumg@gmail.com

V. A. Tran
Hanoi University of Mining and Geology, Hanoi 100000, Vietnam
e-mail: tranvananh@humg.edu.vn

V. Yordanov · M. A. Brovelli
Department of Civil and Environmental Engineering (DICA), Politecnico Di Milano, Piazza Leonardo da Vinci 32, Milan, Italy
e-mail: vasil.yordanov@polimi.it

M. A. Brovelli
e-mail: maria.brovelli@polimi.it

M. A. Brovelli
Istituto Per Il Rilevamento Elettromagnetico Dell'Ambiente, CNR-IREA, Via Bassini 15, 20133 Naples, Italy

T. D. Khuc
Hanoi University of Civil Engineering, 55 Giai Phong Street, Hanoi 100000, Vietnam
e-mail: dongkt@huce.edu.vn

L. Q. Nguyen et al. (eds.), *Advances in Geospatial Technology in Mining and Earth Sciences*, Environmental Science and Engineering,
https://doi.org/10.1007/978-3-031-20463-0_27

Bai province. The classification accuracy of the method is evaluated using producer's accuracy, user's accuracy, overall accuracy, and Kappa coefficient. We obtain a high overall accuracy (90.7%) with a corresponding Kappa coefficient of 0.85 for the classification in 2019. In the case of 2020, overall classification accuracy reaches about 91.1% and 0.87 of the Kappa coefficient. Then, the LULC change area in the period 2019–2020 of the study area is evaluated and discussed by using the transition matrix of LULC.

Keywords Remote sensing · Land use land cover · Random forest · Sentinel-2

1 Introduction

Land use land cover (LULC) change detection plays an important role in environmental modeling and Earth sciences applications. According to [1], land cover describes the (bio) physical characteristics such as the vegetation type and soil properties of the Earth's surface. Land use is defined by the actions of people on the land cover type, whether to production, development, maintenance, or mixed use. Land use and land cover change may have a strong impact on the environment and climate. It affects global and regional climate change by altering land biogeochemical and biophysical processes. LULC change is a major contributor to anthropogenic greenhouse gas (GHG) emissions [2]. Researchers demonstrated that from 1990 to 2010, the net flux of carbon caused by land use land cover change makes up 12.5% of carbon emissions. Over the 2009–2018 period, the global CO_2 emissions from land use and land cover change were 1.36 ± 0.42 PgC yr − 1 (1σ range) and reached a cumulative total of 206 ± 57 PgC in the 1750–2018 period [3].

Many previous studies have focused on applying remote sensing technology to monitor LULC change. In Vietnam, satellite images and geographical information system (GIS) techniques have been employed to produce valuable LULC maps for change detection [4]. However, in recent years, the population growth, the rapid socio-economic development, and the complexity of climate change have resulted in various changes in LULC. Therefore, it is needed to study a suitable and optimal methodology for monitoring changes and preparing LULC maps in Vietnam. This research analyzes LULC change in Mu Cang Chai (MCC) and Van Yen (VY) districts in Yen Bai province, which is in the Northwest region of Vietnam. Over the past few years, Yen Bai is focused on industrial development, which leads to lending land for the construction of road networks and residential areas; deforestation and mineral mining. As a result, LULC in the province has been changing rapidly.

Applying remote sensing to detect the long-term LULC variations is crucial in dealing with the effects of climate change and preparing plan for development. Remote sensing and GIS techniques provide ecologists and researchers a multi-temporal data sources to monitor the periodical variations over large regions. This research uses Sentinel-2 satellite data with high spatial and temporal resolutions along with its wide swath width and high revisit time. This study observes the potential of Sentinel-2 data for land cover classification and assess its suitability for land cover mapping with natural conditions in Vietnam. Various works have presented approaches for LULC classification with acceptable accuracy such as unsupervised algorithms (i.e., IsoData and K-means clustering). Recently, many machine learning algorithms have been studied and applied for land cover classification such as k-nearest neighbors (kNN), artificial neural networks (ANN), decision trees (DT), support vector machines (SVM), and random forest (RF). Researchers have evaluated the performance of three classification methods RF, kNN, and SVM classifiers applied to Sentinel-2 imagery with high overall accuracy (OA) ranging from 90 to 95% [5]. The research shown that the SVM classification outperforms the other algorithms (RF and kNN) with the highest OA and the least sensitivity to the size of training dataset. Random forest also proves its potential in many works on LULC in the world and in Vietnam [6–8]. Nguyen et al. [8] evaluated the performance of RF applied to Landsat 8 OLI image to classify ten different types of LULC and combined with spectral bands to obtain high OA of 90.32%. Piao et al. [6] improved the classification accuracy while applying RF to classify the LULC change between 2001 and 2018 and evaluate the forest change with very high OA (98.2% $\pm$ 1.6%).

The purpose of this research is to assess the LULC changes in the study area during the two years 2019 and 2020 using mainly the random forest classification algorithm.

2 Study Area and Data

2.1 Study Area

The research is conducted in Mu Cang Chai and Van Yen districts in Yen Bai Province, Vietnam. Mu Cang Chai is a rural district which is located between 21°39’ to 21°50’ latitude North and 103°56’ to 104°23’ longitude East. Mu Cang Chai has a total area of about 1,199 km^2 with an average elevation of 800 m above sea level. Van Yen district is located in 21°50’30″ to 22°12’ latitude North and 104°23’ to 104°30’ longitude East with a size of about 1389 km^2. The whole study area is located in the North-west of Yen Bai Province. A large natural forest area covers the study area, and as such it is also considered the strength of Mu Cang Chai. The climate and topography of this region are favorable for agriculture development with various kinds of land cover change. The terrain of Mu Cang Chai is strongly varied, with a steep slope and the average slope of the whole district is about 40°. Mu Cang Chai is also famous for the terrace fields. Residents take advantage of this structure to lead stream water from top of the mountain to the low-altitude fields. Mu Cang Chai has no rivers, however, there are dense stream systems here and high mountains to divide the terrain here.

Yen Bai province is located close to the Northern tropic in the tropical belt of the Northern hemisphere, resulting in tropical monsoon climate in this region. Yen Bai experiences two distinctive seasons: rainy and dry. In the dry season, the situation of frost and fog affects the production of citizens and is also an obstacle for the remotely-sensed data, especially in Mu Cang Chai. The average annual temperature is about 19.6 ÷ 23.3 °C. Due to the influence of topographical conditions, between different regions, the distribution of temperature is also different. Mu Cang Chai experiences the Northeast monsoon, therefore the temperature is even colder with the average temperature is just 14.3 °C (Fig. 1).

2.2 *Data*

The data used in this study are two scenes of Sentinel-2 satellite images collected in 2019 and 2020 from the Copernicus Programme. The images are provided from the onboard includes multi-spectral instrument (MSI) covering 13 spectral bands, in which nine bands blue, green, red, red-edge-1, -2, -3, NIR, and two SWIR with spatial resolutions of 10 m and 20 m are used in this research (Table 1). The satellite, combined with Sentinel-2B provides images in a short revisit time (5 days).

To collect reference datasets for 2020, firstly, four areas of 1 km^2 within the study area are randomly selected. Then, image objects of Sentinel-2 (2020) covering the areas are generated by image segmentation process using "region growing and merging" algorithm, threshold equals 0.1. Finally, a part of the objects is labeled into seven LULC classes: water, built-up, paddy field, rice terrace, none-forest vegetation, forest, and bare-land/others based on visual interpretation of Google Earth (2020) and the Sentinel-2 image with verification using forest map (2013). The reference data of 2019 is generated by modifying the reference data of 2020 based on Sentinel-2 images of 2019. The reference dataset is randomly divided into training and test dataset with the ratio 80% and 20%, respectively (Fig. 2).

3 Methodology

3.1 *Classification Algorithms and Random Forest*

3.1.1 LULC Classification Methods

There are two main approaches to classify remote sensing image, named pixel-based and object-based classification. Pixel-based method works at individual pixel and labels each pixel to a specified class. On the other hand, object-based method uses an image segmentation algorithm to group neighborhood pixels into an image object, then classifies the individual objects. The image object represents vector shapes with

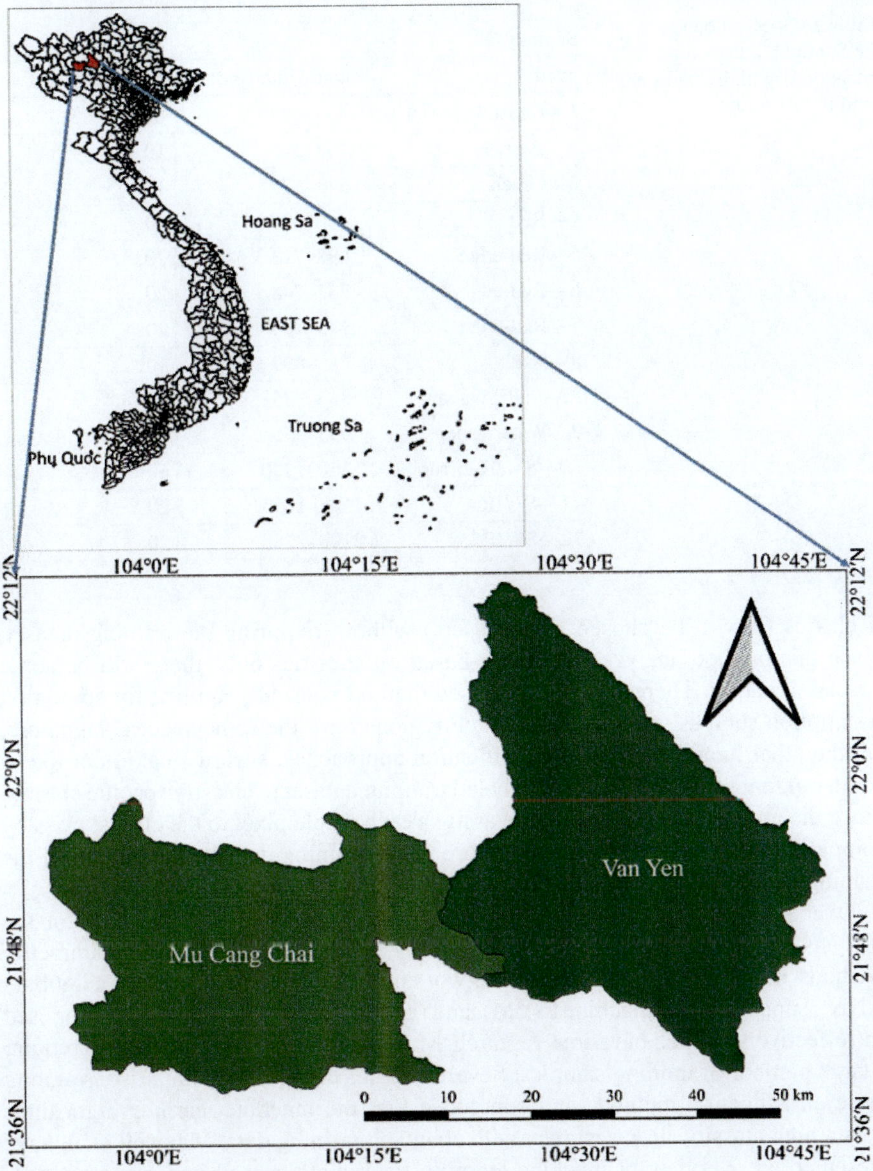

Fig. 1 Location of the study area in Yen Bai Province, Vietnam

size and geometry. Object-based classification was reported to perform better for LULC classification compared to the pixel-based approach [9].

Two major approaches for image classification are unsupervised and supervised techniques. Unsupervised classification automatically generates the clusters based on the similarity of spectral characteristics of pixels and then groups them into

Table 1 Spectral bands of the Sentinel-2 a satellite imagery. Highlight bands are used in this study

Sentinel-2 band	Bandwidth (nm)	Resolution (m)
1—Coastal aerosol	433–453	60
2—Blue	458–523	10
3—Green	543–578	10
4—Red	650–680	10
5—Red-edge 1	698–713	20
6—Red-edge 2	733–748	20
7—Red-edge 3	773–793	20
8—NIR	785–899	10
8A—NIR narrow	855–875	20
9—Water vapor	935–955	60
10—SWIR-cirrus	1360–1390	60
11—SWIR	1565–1655	20
12—SWIR	2100–2280	20

N classes (N will be chosen by the users) without requiring the training dataset. Since this method analyzes randomly based on statistics only, there will be some misclassification. Therefore, unsupervised method is usually applied for some first assumption such as to understand which regions have the same spectral signature. On the other hand, supervised classification approaches, such as maximum likelihood, random forest, is relying on labeled training dataset to classify specific classes. Since the image processing software assigns each single pixel to the closest class by comparing pixel's data with the signature of the training dataset, the quality of the training dataset affects classification result.

Over the past several decades, the use of machine learning algorithms for the analysis of remote sensing imagery is rapidly increasing [10]. The nonparametric methods including artificial neural networks, decision trees, k-nearest neighbors (kNN), support vector machine (SVM), and random forests (RF) have been proposed and effectively applied in various research. Most machine learning algorithms require a large number of training samples. Several studies carried out comparisons among these classification methods and concluded that the machine learning algorithms obtain high classification accuracy with abundant training data. Maxwell et al. have presented an overview of machine learning for remote sensed imagery classification by comparing some methods such as artificial neural networks, support vector machines, single decision trees (DTs), random forests, boosted DTs, and k-nearest neighbors (kNN) [11]. The SVM, RF, and boosted DTs have shown their potential with high overall accuracies. Abdulhakim has also compared four nonparametric algorithms such as SVM, RF, extreme gradient boosting (XGBoost), and deep learning (DL) in classification of LULC using Sentinel-2 satellite images [12]. The study proved that SVM, XGBoost, and RF produced similar classification accuracy.

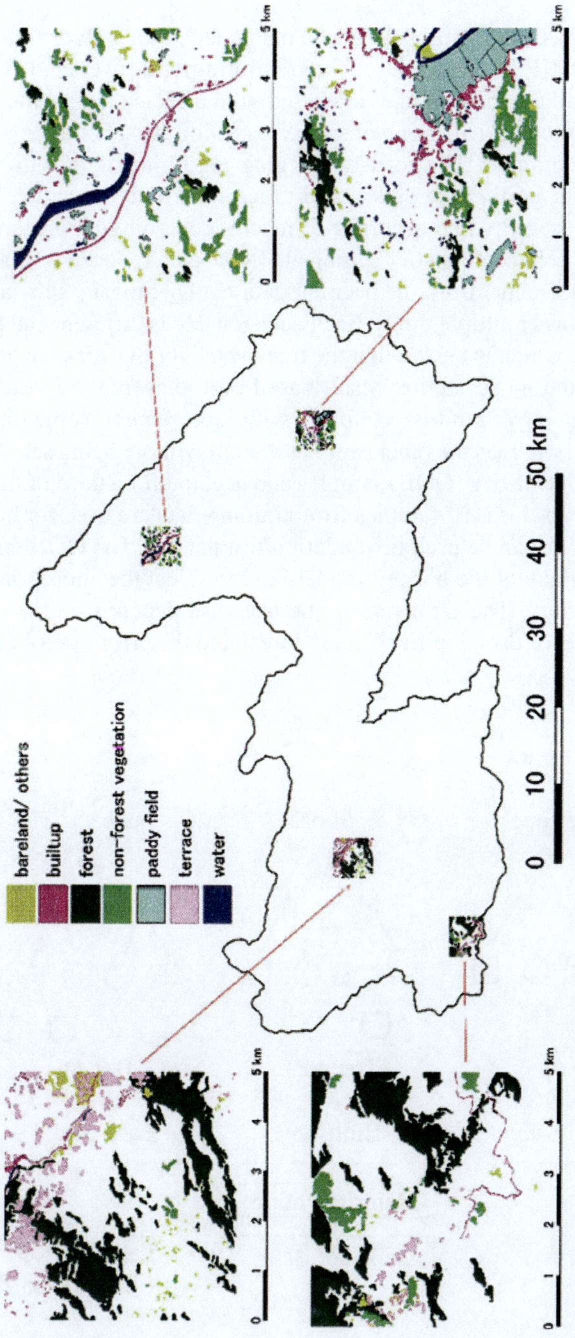

Fig. 2 Research reference data

3.1.2 Random Forest (RF) Algorithm

Random forest (RF) is a powerful supervised machine learning algorithm presented by Leo Breiman in 2001. It is widely used for monitoring land cover land use mapping in many previous studies. RF combines many individual decision tree votes by generating high number of trees [13]. The decision tree is built using bootstrapping aggregating (bagging) technique. The idea of bagging algorithm is that the ensemble predictor or classifier performs a better result than individual classifier. This technique increases the stability and accuracy of machine learning algorithm. Random forest method is the enhancement of bagging [14]. The data for each tree is randomly selected (with replacement) from the training dataset. That means sub-samples can contain the same values multiple times. Each decision tree is different and performs a different prediction, which is known that the tree "vote" for that classification. Then, the forest chooses the most popular result based on the average of votes (Fig. 3). Since RF method employs random sampling with replacement, some observations exist multiple times, whereas the other samples are left without being selected. These samples are called out-of-bag (OOB) samples and account for a third of the samples. That means about two-thirds of samples from training data are used for building the model and the left one can be used in validation/test dataset. The OOB error is used to evaluate the accuracy of the bagged models and to select the appropriate tunning parameter such as mtry. The error rate in the test case depends on the correlation between each couple of decision trees in the forest and the error rate of each tree.

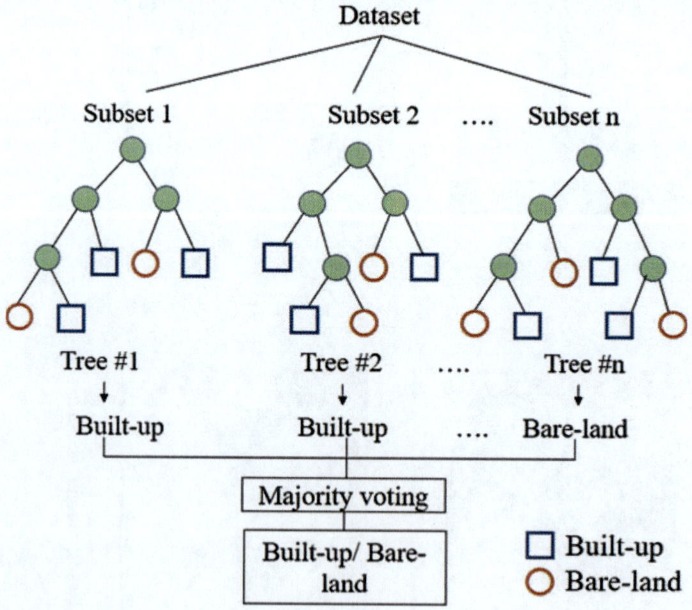

Fig. 3 Random forest algorithm

RF algorithm is an ensemble of decision trees and has improved its limitations. In decision tree algorithm, when the tree is at a particular deep, it perfectly fits the given samples in the training dataset, which results in the inaccurate prediction of the outcome of the untrained data. In this condition, the model is overfitting and has high variance. In the case of RF, each tree has random training data and features, and is uncorrelated or weakly correlated to each other, thus the variance is reduced without increasing the bias and the tendency of overfitting also decreases. RF method is expected to perform better with a higher level of accuracy in predicting final result over the decision trees. RF is useful for both classification as well as regression tasks. RF algorithm can deal with large datasets and handle missing values.

The RF algorithm requires selection of two main parameters: number of decision tree (ntree) and variables in the random subset at each node (mtry). The optimal values for ntree and mtry are determined based on the OOB error estimation in the shortest time. The value of ntree should not be too small so that the input data can be predicted. The selection of ntree and mtry has been analyzed in many previous experiments [8, 15]. In [16], the OOB errors were found smallest for ntree = 500 trees for all seasonal image combinations.

3.2 *LULC Classification and Change Detection*

The research steps for the LULC classification and change detection are demonstrated in Fig. 4. The Sentinel-2 images were collected in two paths for each year (2019, 2020) to cover the whole study area. We sorted the image data from October to November of predetermined years. The two scenes per year were merged together to cover the whole study area.

For the LULC classification process, the study area was classified into seven classes: water, built-up, paddy field, rice terrace, non-forest vegetation, forest, bare-land/others using RF. The process was completed in R-studio, a free platform

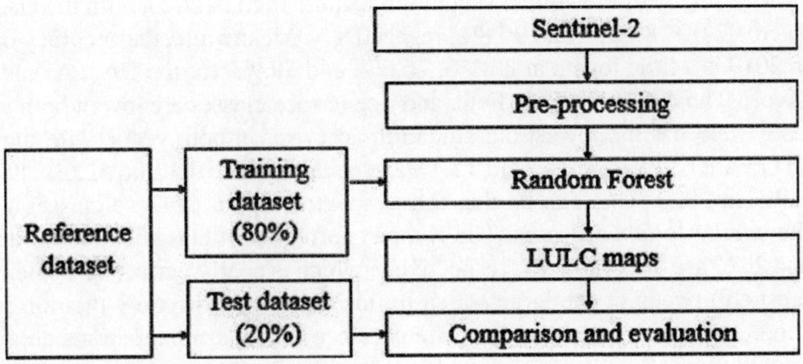

Fig. 4 Research methodology flow chart

for statistical computing, and graphics. The package classification and regression training (**caret**) was used for implementing the machine learning model. Sufficient value of ntree is set to be 500, and mtry values are chosen by taking square root of number of training samples for the classification.

The accuracy of a classification map is evaluated by constructing a confusion matrix between reference and classified pixels including criteria such as the overall accuracy (OA), user's accuracy (UA) (or commission error), producer's accuracy (PA) (or omission error), and Kappa statistics coefficient for each category in the classification [17]. Producer accuracy is the probability that a value in a given class was classified correctly. User accuracy is the probability that a value predicted to be in a certain class really is that class. The overall accuracy and Kappa statistics are assessed were computed by these equations

$$\text{Overall accuracy (OA)} = \frac{\text{Number of correctly classified values}}{\text{Total number of values}} \times 100 \quad (1)$$

$$\text{Kappa coefficient}(\kappa) \equiv \frac{p_0 - p_e}{1 - p_e} \quad (2)$$

where p_0 is observed accuracy and p_e is change agreement.

4 Result

4.1 LULC Classification

LULC map of the years 2019 and 2020 is shown in Figs. 5 and 6. Classification accuracies of the two years are performed in Table 2. The RF algorithm does not provide high accuracy (UA, PA) for all categories; however, the accuracies of water and forest are extremely high, which means these classes are well classified in both years. The accuracy of non-forest vegetation is quite high in 2020, with the UA, PA, and OA are 92.0%, 89.5%, and 90.8%, respectively. Meanwhile, the accuracy of this class in 2019 is a little lower, at 83.1%, 76.6%, and 79.9% for the UA, PA, and OA, respectively. The built-up, paddy field, and rice terrace classes are low in both years. The paddy field has the lowest classification accuracy in both years, with the UA, PA, and OA are 61.8%, 83.0%, and 72.4%, respectively in 2019, and 52.6%, 85.4%, and 68.9% in 2020. This can be due to the spectral signal of these classes which are quite similar to other classes. The Kappa coefficient of classification models in 2019 and 2020 are 0.85 and 0.87, respectively, which is nearly perfect agreement [7]. The forest still occupies the largest area in this region, followed by the non-forest vegetation. The build-up lands are scattered since this is a mountainous area with sparse population.

A traditional classification method namely maximum likelihood (ML) is applied for the Sentinel-2 data in order to compare the effectiveness of the two algorithms.

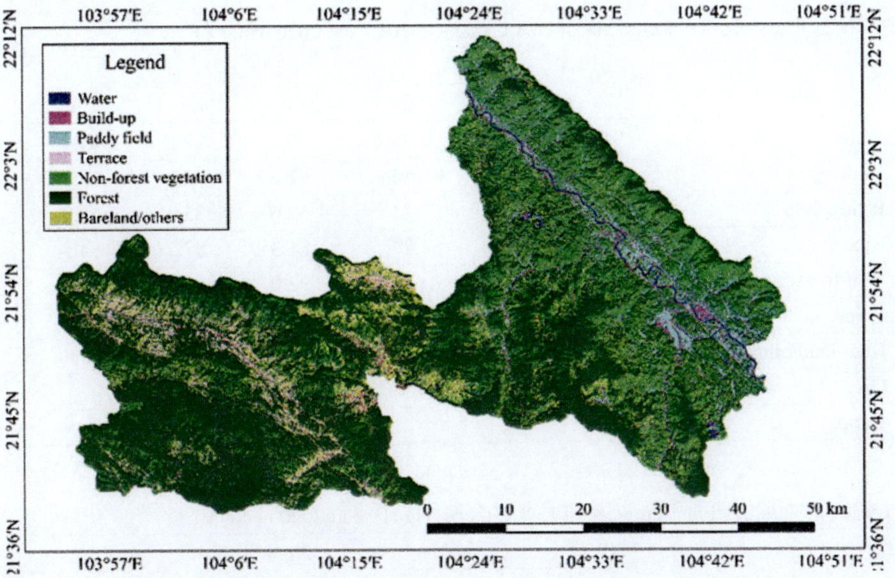

Fig. 5 LULC map 2019

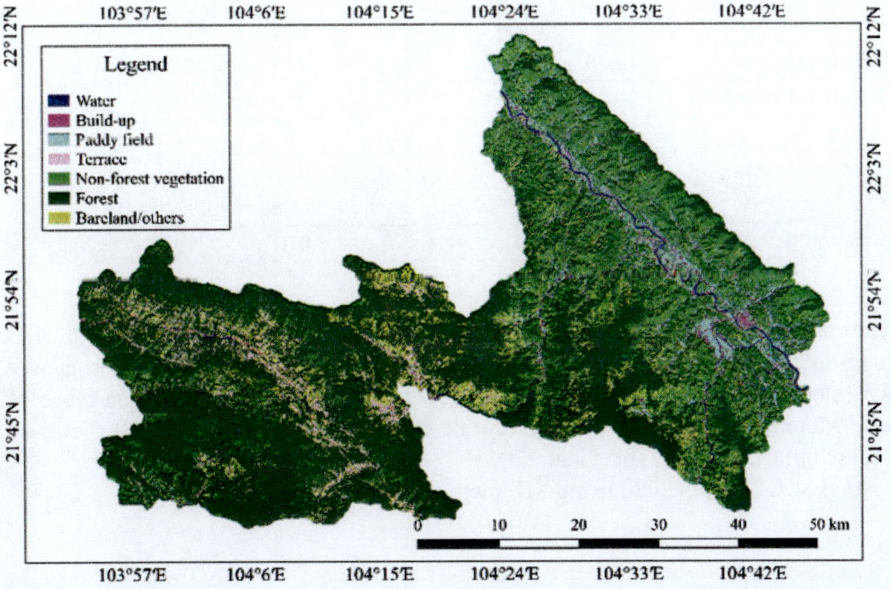

Fig. 6 LULC map 2020

Table 2 Classification accuracies of LULC maps in 2019 and 2020 using RF

	2019			2020		
	UA	PA	OA	UA	PA	OA
Water	96.7	98.1	97.4	95.4	92.2	95.4
Built-up	94.7	78.1	86.4	83.6	73.2	78.4
Paddy field	61.8	83	72.4	52.6	85.4	68.9
Rice terrace	82.4	91.5	87	81.3	87.2	84.3
Non-forest vegetation	83.1	76.6	79.9	92.0	89.5	90.8
Forest	96.1	95.5	95.8	97.1	94.0	95.6
Bare-land/others	53.7	61.8	57.8	65.5	92.2	78.9
OA	90.7			91.1		
Kappa	0.85			0.87		

Table 3 Classification accuracies of LULC maps in 2019 and 2020 using ML

	2019			2020		
	UA	PA	OA	UA	PA	OA
Water	91.8	98.3	95.1	79.6	79.6	79.6
Built-up	88	90	89	85	85	85
Paddy field	40	64.4	52.2	62.7	62.7	62.7
Rice terrace	75.6	92.2	83.9	80.1	80.1	80.1
Non-forest vegetation	82.1	64.6	73.4	90.3	90.3	90.3
Forest	96.4	93.7	95.1	94.5	94.5	94.5
Bare-land/others	48.9	49.2	49.1	61.8	61.8	61.8
OA	88.2			86.8		
Kappa	0.81			0.81		

The classification accuracy of ML is shown in Table 3. Figures 7 and 8 present LULC maps in 2019 and 2020, respectively. It can be clearly seen that RF outperforms over ML method since both OA and Kappa coefficient of RF are higher than those of ML. Moreover, RF performs more accurate classification results of all LULC classes except built-up class (Tables 2 and 3).

Fig. 8LULC map 2020 using ML method

4.2 LULC Change

The LULC change matrix between 2019 and 2020 is presented in Table 4. The diagonal line demonstrates the area of land cover remained unchanged between 2019 and 2020. The other values are the area change from one LULC type in 2019 shift to

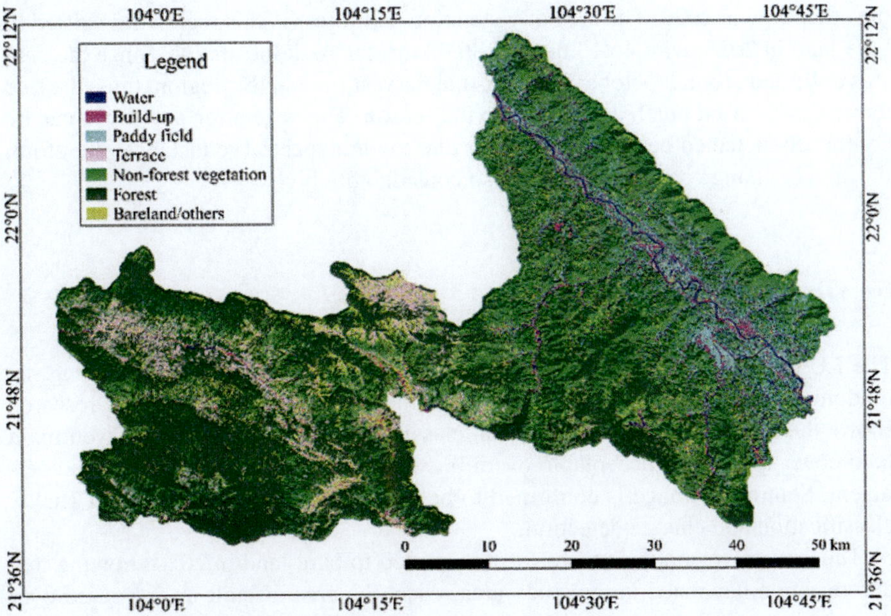

Fig. 7 LULC map 2019 using ML method

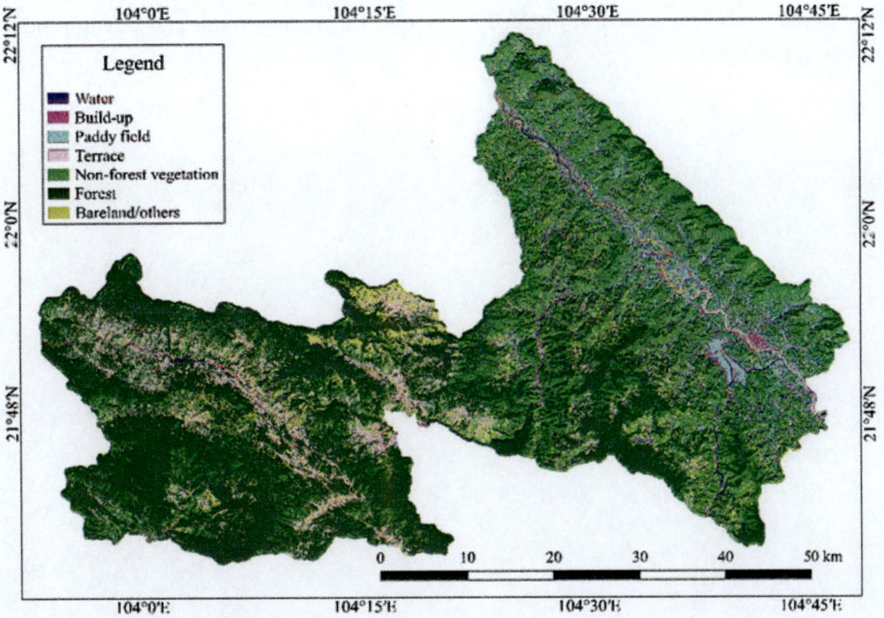

Fig. 8 LULC map 2020 using ML method

the other type in 2020. The area of vegetation and forest in this region has shifted to bare-land in 2020, with 46.1 km^2 and 86.9 km^2 due to deforestation. Since the data was collected around October, which is the harvest time in this region, thus, the rice terrace area varied much depending on the season. The vegetation and forest can be confusedly detected between each other due to their spectral reflectance, therefore, the area of change between them is also considerably high.

5 Discussion and Conclusion

The LULC classification map in the period 2019–2020 was constructed using the random forests classifier based on the Sentinel-2 satellite imagery. The research shows the potential of RF algorithm in classifying the study area into seven main land cover types with acceptable overall accuracy and nearly perfect Kappa coefficient. Sentinel-2 image is confirmed to be a good remote sensing data for LULC classification and change detection.

The area of vegetation and forest was shifted to bare-land in 2020 showing that deforestation happened in the study area. Also, the area of built-up increased over the two years confirming the expansion of residential area.

In the future work, we expect to study LULC mapping and change in a longer period in order to understand the trend of the change in the study area as well as provide useful information on LULC change for management and development planning purposes.

Table 4 The area of LULC change in 2019-2020 (km^2)

2019	LULC Classes	2020 (area in km^2)						
		Water	Built-up	Agriculture	Rice terrace	Vegetation	Forest	Bareland/others
(Area in km^2)	Water	23.9	1.3	2.5	0.3	0.5	4.9	1.4
	Built-up	0.6	15.4	5.5	1	1.4	1.5	3.7
	Agriculture	3.2	6.5	65.2	0.4	13.3	7.4	7.4
	Rice terrace	1.1	4.2	2.4	55.5	4.9	11.3	26.6
	Vegetation	0.4	1.8	17.9	2.2	411.6	77.8	45.4
	Forest	5.2	2.3	4.7	18.9	118.3	1468.6	46.1
	Bareland/others	0.6	6.7	11.4	28.1	61.3	20.8	86.9

Acknowledgements The work is partially funded by the Italian Ministry of Foreign Affairs and International Cooperation within the project "Geoinformatics and Earth Observation for Landslide Monitoring"—CUP D19C21000480001 (Italian side) and partially funded by Ministry of Science and Technology of Vietnam (MOST) (Vietnamese side) by the bilateral scientific research project between Vietnam and Italy, CODE: NĐT/IT/21/14.

References

1. Jansen, L., Di Gregorio, A.: Land cover classification system (LCCS): classification concepts and user manual. Fao, 53 (2000)
2. Houghton, R.A., House, J.I., Pongratz, J., van der Werf, G.R., DeFries, R.S., Hansen, M.C., Le Quéré, C., Ramankutty, N.: Carbon emissions from land use and land-cover change. Biogeosciences **9**, 5125–5142 (2012)
3. Gasser, T., Crepin, L., Quilcaille, Y., Houghton, R.A., Ciais, P., Obersteiner, M.: Historical CO_2 emissions from land use and land cover change and their uncertainty. Biogeosciences **17**, 4075–4101 (2020)
4. Tran, V.A., Le, T.L., Nguyen, N.H., Le, T.N., Tran, H.H.: Monitoring vegetation cover changes by Sentinel-1 radar images using Random Forest classification method. Inzyneria Mineralna **2–46**, 441–452 (2021)
5. Thanh Noi, P., Kappas, M.: Comparison of random forest, k-nearest neighbor, and support vector machine classifiers for land cover classification using sentinel-2 imagery. Sensors **18**(1), 18 (2018)
6. Piao, Y., Jeong, S., Park, S., Lee, D.: Analysis of land use and land cover change using time-series data and random forest in North Korea. Remote Sensing **13**(17), 3501 (2021)
7. Thonfeld, F., Steinbach, S., Muro, J., Kirimi, F.: Long-term land use/land cover change assessment of the Kilombero catchment in Tanzania using random forest classification and robust change vector analysis. Remote Sens. **12**(7), 1057 (2020)
8. Nguyen, H. T. T., Doan, T. M., Radeloff, V.: Applying Random Forest classification to map Land use/Land cover using Landsat 8 OLI. International Archives of the Photogrammetry, Remote Sensing and Spatial Information Sciences - ISPRS Archives, 42-3W4 (2018)
9. Whiteside, T.G., Boggs, G.S., Maier, S.W.: Comparing object-based and pixel-based classifications for mapping savannas. Int. J. Appl. Earth Observation Geoinfo., 13–16 (2011)
10. Shih, H.C, Stow, D.A., Tsai, Y.H.: Guidance on and comparison of machine learning classifiers for Landsat-based land cover and land use mapping. Int. J. Remote Sensing, 40–44 (2019)
11. Maxwell, A.E., Warner, T.A., Fang, F.: Implementation of machine-learning classification in remote sensing: an applied review. Int. J. Remote Sens., 39 (2018)
12. Abdi, A. M.: Land cover and land use classification performance of machine learning algorithms in a boreal landscape using Sentinel-2 data. GIScience Remote Sens., 57 (2020)
13. Breiman, L.: Random forests. Machine Learning, 45 (2001)
14. Altman, N., Krzywinski, M.: Ensemble methods: bagging and random forests. Nature Methods, 14–10 (2017)
15. Abdullah, A.Y.M., Masrur, A., Gani Adnan, M.S., Baky, M.A., Hassan, Q.K., Dewan, A.: Spatio-temporal patterns of land use/land cover change in the heterogeneous coastal region of Bangladesh between 1990 and 2017. Remote Sens., 11–17 (2019)
16. Nguyen, H.T.T., Doan, T.M., Tomppo, E., McRoberts, R.E.: Land use/land cover mapping using multitemporal sentinel-2 imagery and four classification methods-A case study from Dak Nong, Vietnam. Remote Sensing, 12–19 (2020)

17. Congalton, R. G.: A review of assessing the accuracy of classifications of remotely sensed data. Remote Sens. Environ., 37 (1991)
18. Landis, J.R., Koch, G.G.: The measurement of observer agreement for categorical data. Biometrics, 33 (1977)

Engineering Geological Problems of Foundation Pit Construction in Quaternary Strata: Taking Suzhou Area as an Example

Xinyu Luo, Peng Yin, Yongsheng Zheng, Xuedong Li, Yi Zhang, and Qiang Tang

Abstract The Quaternary strata in Suzhou, China, are typical because of its abundant water system and intricate geological conditions. The characteristics of Quaternary sediments in Suzhou area were analyzed and integrated through borehole sampling, field investigation, and data collection. The results show that the thickness of Quaternary sediments in Suzhou area is approximately 200 m. Each aquifer group is divided into water-bearing stratum and water-resisting stratum with a thickness of approximately 15 m and 30 m, respectively. The soil layers are mostly silty clay with approximately 30% water content, which has a loose structure and poor engineering properties. Meanwhile, there are multiple fault zones, mainly East-North and East–West trending, accompanied by North–North-East trending tectonic traces. This common geological situation in coastal and lakeside areas often leads to several problems in foundation pit construction: failure of supporting structure and instability of foundation pit slope, seepage deformation, land subsidence and ground subsidence, and heave and inrush at the bottom of the pit. In this paper, considering the excavation depth, excavation mode, soil properties, and surrounding loads and other key influencing factors, the corresponding prevention and control measures are put forward for practical engineering. Meanwhile, a variety of protective drainage measures are used to enclose and lower the water for its foundation pit project, which effectively reduces the construction risk of the foundation pit project.

X. Luo (✉) · Q. Tang
School of Rail Transportation, Soochow University, Suzhou 215131, China
e-mail: sswdhflxy@outlook.com

Q. Tang
e-mail: tangqiang@suda.edu.cn

P. Yin · Y. Zheng
The Second Engineering Co., Ltd. of CTCE Group, Suzhou 215000, China

X. Li · Y. Zhang
Jiangsu Suzhou Geological Engineering Survey Institute, Suzhou 215011, China

The Fourth Geological Brigade of Jiangsu Geological and Mineral Bureau, Suzhou 215004, China

L. Q. Nguyen et al. (eds.), *Advances in Geospatial Technology in Mining and Earth Sciences*, Environmental Science and Engineering,
https://doi.org/10.1007/978-3-031-20463-0_28

Keywords Quaternary strata · Geological formation · Soil properties · Foundation pit

1 Introduction

With the acceleration of China's urbanization process, the urban population is highly concentrated, and some infrastructure cannot satisfy the demands of people's production and living. In order to improve the pressure of urban life and traffic congestion, many developed areas, especially some coastal cities, have seen a lot of new or renovation projects [1]. However, with the continuous construction of various buildings, it brings the problem of increasing ground load. In order to improve the bearing capacity of foundation and the stability of buildings, the depth of foundation pit excavation becomes larger and larger. At the same time, urban road and railway construction have entered a period of rapid development, many road constructions will adopt deep foundation pit construction method as well. This is the development direction of foundation pit engineering at present [2, 3].

At present, the study of quaternary has become global, and the classification of quaternary strata is mainly based on the confirmation of mammal fossils and paleoclimate analysis [4]. This paper mainly divides the strata by analyzing the lithology, color, bedding structure, sedimentary cycle, mineral combination, animal and plant fossils, and geological age of the borehole sediments [5]. Quaternary sediments are widely distributed in the world. They are loosely formed since 2.6 Ma and basically preserve their original morphology [6, 7]. Fluvial, lacustrine, glacial, eolian, Marine, and cave deposits are the main sedimentary types, while fluvial, residual, slope, pluvial, biological, and volcanic deposits are the secondary sedimentary types. Most of the earth's surface is covered by quaternary sediments, with the exception of steep rocky slopes [8].

As one of the developed cities in the southeast coastal area of China, Suzhou, Jiangsu province, has a very representative Quaternary stratigraphy. The total area of this city is 8657.30 km^2, and the built-up area is 477.63 km^2. Deep layers of Quaternary sediments, generally 150–260 m thick, are widely distributed in Suzhou area, which have been in a state of slow decline since the Quaternary crustal movement. In the process of Quaternary deposition, influenced by many factors, the lithology and genesis are complex, with the basic characteristics of the alternation of multilayer viscous soil layer and loose sand layer, and the deep and large foundation pit engineering just lies in these soft soil layers. And foundation pit engineering is a temporary project, generally speaking, the design safety factor is small, and a little careless will occur engineering accidents [9], for example, support structure failure, slope instability of foundation pit, seepage deformation, ground settlement, ground settlement, bottom uplift water inrush, and other problems [10].

This paper is to introduce the characteristics of the fourth layer of the sediment in Suzhou area, according to these characteristics summed up the problems that may occur in the foundation pit engineering, and according to these problems summed up the corresponding solutions, to provide a theoretical basis for the actual project.

2 Geological and Hydrogeological Characteristics of Suzhou

2.1 Stratigraphic Division of Suzhou

Suzhou city is located in the Yangtze River delta with dense population, developed economy, high level of urbanization and industrialization, and many deep and large foundation pit projects [11]. Due to the continuous improvement of foundation pit depth, the research of Quaternary strata has become a key issue. Quaternary Stratigraphy refers to a recent geological unit in the history of earth development. The lower limit of the stratification and age determination of the Quaternary stratification and age determination in Suzhou city and other Yangtze River delta areas have always been controversial in various regions. Many paleontologists claim 1.8 Ma (Megaannus, means a million years), while paleomagnetic scholars claim 2.48 Ma. However, because 2.6 Ma is the age of loess deposition in China, Chinese geologists tend to regard 2.6 Ma as the bottom boundary of the Quaternary [12, 13]. This paper also adopts this definition, that is, the strata formed after this period become the Quaternary strata [14]. Through borehole sampling and related data collection in some areas, the Quaternary stratigraphy in Suzhou area was observed and described, including soil lithology, color, stratum elevation, and some stratification in the soil layer, as shown in Table 1. Drilling samples were taken from Xiangcheng, Kunshan, Industrial Park, Wuzhong, and Wujiang in Suzhou. The hole diameter of drilling rig is 129 mm. The thickness of the Quaternary strata in Suzhou area is about 180 m, so the drilling depth is 200 m. When the drilling reaches 200 m, the formation is half lithified and no further drilling is required.

Based on the sampling data and relevant data, the aquifer groups of the Quaternary strata in Suzhou area are divided, and the results are shown in Table 1. There are three confined aquifer groups in Suzhou area: The I confined aquifer group has a depth of 5.3–41.2 m, of which 37.0–41.2 m is the first confined aquifer with a thickness of 4.2 m. The II confined aquifer group has a depth of 41.2–87.7 m, of which 74.4–87.7 m is the II confined aquifer with a thickness of 13.4 m. Confined aquifer group III has a depth elevation of 87.7–162.1 m, of which 137.9–151 m is confined aquifer III with a thickness of 13.1 m. It should be noted that the III confined aquifer group contains two weakly permeable layers at depths of 97.1–137.9 m and 151–165.1 m, respectively. It can also be seen from the table that the aquifer is generally fine sand and fine-medium sand, the weakly permeable layer is generally silt and cohesive soil mixed with silt, and the waterproof layer is cohesive soil and silty soil. The strata below the confined aquifer group III are generally low in water content, hard and close to diagenesis [15].

Table 1 Strata distribution and distribution of groundwater layer in Suzhou

Chronostratigraphic		Code name	Depth (m)	Lithology	Confined aquifer group	
Holocene Series		Qh	2	Filled soil		
			5.3	Bluish gray mucky clay		
Pleistocene	Upper Pleistocene	Qp3	19.8	Bluish gray, yellowish-gray silty clay	Group I	Water-resisting layer (I - 1)
			24.55	Bluish gray, brown clay		
			25.95	Bluish gray mucky clay		
			31.53	Grayish yellow, cyan silty clay		
			37	Cyan silty clay		
			41.2	Gray powder sand		Water-bearing stratum (I - 2)
	Middle Pleistocene	Qp2	50.9	Silty clay of bluish gray and taupe	Group II	Water-resisting layer (II - 1)
			57.2	Gray, gray-green silty clay		
			68.5	Polychromatic silty clay, silty sand		
			74.35	Gray-green, gray-brown silty clay		
			87.7	Gray, light gray silt		Water-bearing stratum (II - 2)
	Lower Pleistocene	Qp1	97.1	Gray-green silty clay	Group III	Water-resisting layer (III - 1)
			100.1	Gray-green, gray-brown silty clay, silty sand		Aquitard (III - 2)
			107	Gray-green silty clay, silty sand		
			113.2	Cyan silty clay		
			124.5	Gray-green, yellow–brown silty clay, silty sand		

(continued)

Table 1 (continued)

Chronostratigraphic		Code name	Depth (m)	Lithology	Confined aquifer group	
			137.9	Polychromatic silty clay		
			145.8	Dark gray, gray-green silty clay, silty sand		Water-bearing stratum (III - 3)
Pliocene		N2	151	Dark gray, gray fine sand		
			155.4	Gray-green silty clay, silty sand		Aquitard (III - 4)
			158.3	Gray-green silty clay and fine sand		
			160.3	Tan fine sand		
			165.1	Tawny fine sand, silty sand		
Miocene		N1	169.5	Yellowish brown, cyan silty clay	Semi-solid layer	
			172.7	Yellowish brown, bluish gray fine-medium sand		
			180	Gray yellowish-brown silty clay, fine sand		
			185.1	Cyan gray, tan clay, fine sand		
			186.5	Grayish yellow medium-coarse sand		
			189.7	Yellow silty clay, silty sand		
			193.5	Dark gray, gray-green silty clay, fine sand		
			198.1	Yellowish brown, gray-green silty clay, fine sand		
			200	Bluish gray gravel sand		

2.2 *Characteristics of Quaternary Strata Soil*

Table 2 shows the basic physical and mechanical properties of Quaternary soils in Suzhou area except surface fill. According to the difference of genesis type, age, lithology, and physical and mechanical properties, considering the requirements of foundation engineering, it is divided into 10 engineering geological layers and several sub-layers.

Soil layer 1 is tillage soil. Soil layer 2 is gray, grayish-brown silt clay, silty clay, with high water content, large void ratio, low bearing capacity characteristics. Soil layer 3 is gray, blue gray clay and silty clay, water content and void ratio lower than the previous layer, higher bearing capacity. Soil 4a is yellowish-brown clay with low water content and high standard bearing capacity. Soil layer 4b is yellowish-brown silty clay with high bearing capacity and other properties similar to soil layer 2. Soil layer 4c is yellowish-brown silt and silty sand with low plastic index and high liquid index. Soil 5a is gray-brown silt with low plastic index and high liquid index. Soil layer 5b is gray silt, which has a high standard bearing capacity. Soil 6a is gray-yellow and brown-yellow silty clay with low void ratio and high standard bearing capacity. Soil 6b is composed of gray-brown silty clay, silty soil, clay, and silty sand, with high water content and high liquid index. Soil layer 7 is gray, light gray silty sand, silty soil mixed with silty clay, and the gap is relatively low. Soil layer 8a is dark green, yellow-green, and yellow–brown clay, which is characterized by low water content, large natural weight, low void ratio, and high standard bearing capacity. Soil layer 8b is yellowish-brown and grayish-brown silty clay with low void ratio and high standard bearing capacity. Soil layer 9a is bluish gray and light gray silty clay. 9b is blue gray, gray silt, and silt. Soil layer 10 consists of grayish brown and gray silty clay, clay, and silt, with high water content and high plasticity index [16].

2.3 *Fracture Structure*

Suzhou, Jiangsu Province, is located in the second uplift zone of the Neocaysian system in eastern China, and the eastern section of the southern branch of the east–west Qinling structural belt is in the composite zone of the southern and northern Parts of Jiangsu. The distribution of the Suzhou fault zone is shown in Fig. 1. The EN and EW trending structures are obviously strengthened in the Yanshan area, and the NNE trending structural traces are also formed. The NWW trending (290°–300°) and NNW trending (340°–350°) faults were associated with the development of these structures. Since Quaternary, there have been basalt eruptions along the fault zone with strong activity. A series of NE and NW trending faults are developed in Yanshan granites. The occurrence of fault planes is generally steep, and the fault structure is cataclasite with a width of 0.3–1.5 m [17–19].

Table 2 Physical and mechanical properties of stratums in Suzhou

Layer number	Water content ω (%)	Natural gravity ρ (kN/m^3)	Void ratio €	Saturation Sr (%)	Plasticity index IP	Liquidity index IL	Cohesive force C (kPa)	Internal friction angle φ (rad)	Bearing capacity value fk (kPa)
1	–	–	–	–	–	–	–	–	–
2	37.2–59.4	16.8–18.4	1.02–1.62	96.0–100.0	15.9–24.5	0.76–1.50	11.0–35.0	4.0–12.5	55–90
3	25.5–35.1	18.7–19.9	0.72–0.98	95.0–100.0	12.6–22.8	0.41–0.95	12.0–51.0	7.5–13.7	95–180
4a	23.1–29.1	18.4–20.3	0.68–0.79	94.7–100.0	16.4–23.5	0.18–0.44	38.0–67.0	7.8–14.8	180–260
4b	22.7–34.8	18.4–20.3	0.70–0.96	92.7–99.5	11.2–16.2	0.45–0.88	22.0–57.0	9.5–19.1	150–200
4c	30.9–37.7	18.0–19.2	0.84–1.02	95.2–100.0	6.4–9.5	0.97–1.85	14.0–31.0	17.5–29.0	100–150
5a	31.6–38.9	18.0–19.3	0.82–1.07	96.4–100.0	6.6–12.1	0.80–1.95	12.0–35.0	10.6–24.2	100–140
5b	30.0–35.8	18.5–19.2	0.84–0.96	95.6–100.0	–	–	3.0–14.0	11.5–29.5	120–180
6a	27.1–32.6	19.0–19.8	0.74–0.89	94.7–99.5	10.7–15.4	0.45–0.95	19.0–37.0	9.8–20.0	145–200
6b	29.5–38.7	18.7–19.4	0.83–1.04	96.3–100.0	6.2–17.6	0.65–1.78	6.0–32.0	10.2–28.0	100–160
7	29.5–33.4	18.5–19.1	0.84–0.96	93.6–100.0	–	–	–	–	140–200
8a	21.3–26.6	19.5–20.5	0.62–0.75	94.7–99.6	14.6–20.9	0.17–0.38	36.0–74.0	14.6–20.5	240–300
8b	24.3–31.3	19.2–20.2	0.68–0.86	96.0–99.1	10.5–15.7	0.26–0.64	27.0–61.0	13.0–24.5	160–280
9a	24.7–33.0	19.1–19.8	0.73–0.91	97.1–99.7	9.9–14.3	0.56–0.97	18.0–34.0	13.8–26.5	120–180
9b	25.1–30.0	19.2–19.8	0.69–0.82	95.2–99.7	–	–	–	–	180–220
10	28.9–39.2	18.4–19.3	0.82–1.09	96.0–100.0	12.0–20.8	0.64–1.07	13.0–41.0	11.7–18.0	100–160

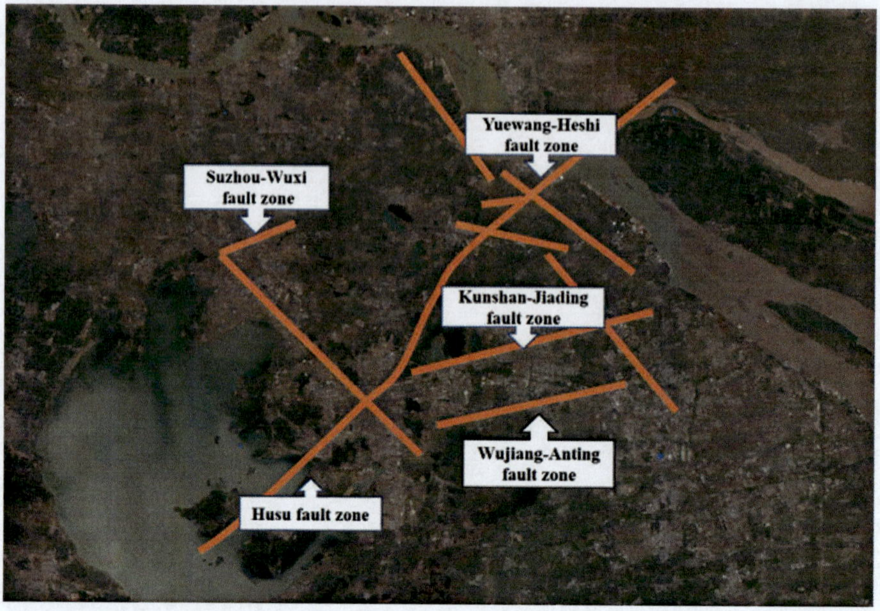

Fig. 1 Distribution of Suzhou fault zones

3 Major Engineering Problems and Solutions

3.1 Main Engineering Geological Problems in Foundation Pit Construction

According to the above analysis, thick soft soil is widely distributed in Suzhou, and its engineering properties are poor. The main manifestations are high content of clay minerals, loose structure, high void ratio and water content, low strength, large compressibility, and easy to produce flow deformation [20, 21]. Therefore, several problems often appear in foundation pit construction: (1) failure of supporting structure and instability of foundation pit slope; (2) seepage deformation; (3) land subsidence and ground subsidence; and (4) heave and inrush at the bottom of the pit.

Failure of supporting structure and instability of foundation pit slope

Due to its remarkable rheological and thixotropic properties, soft soil will creep slowly under the action of various forms of load during construction. Especially, the structure of the water-rich soft soil in Suzhou is more seriously damaged under dynamic load, which is prone to liquefaction, suspension, flow, and settlement. Therefore, the rheological and thixotropy of soft soil have a great influence on the stability of foundation pit and the deformation of supporting structure, and it is toilless to cause the overturning failure of the supporting structure of foundation pit, such as the kicking of the lower part of the supporting structure, the swelling of the middle part, and the overturning of the upper part of the supporting structure. In serious cases, the instability of the foundation pit slope will be caused.

Seepage deformation

The distribution area of soft soil in Suzhou has high groundwater level and high water content, and soft soil often intersects with silt and fine sand aquifers. During the excavation of the foundation pit, there is a large water level difference between the inside and outside of the pit, which is prone to gushing sand and water, resulting in deformation and collapse of the sidewalls of the foundation pit.

Land subsidence and ground subsidence

When the foundation pit is drained, it often causes soil erosion and potential erosion, resulting in uneven ground settlement around the pit, causing cracking and damage to surrounding buildings, municipal roads, and pipelines. At the same time, the thixotropy of soft soil under dynamic load can also cause ground settlement around the foundation pit, and even extensive ground settlement.

Heave and inrush at the bottom of the pit

The excavation of the foundation pit destroys the natural stress state of the soil. During the excavation of the foundation pit, the lateral and bottom of the foundation pit are unloaded, and the lateral soil is prone to extrusion and uplift at the bottom of the foundation pit. Especially when the soil layer at the bottom of the foundation pit is double geological structure, the upper soft soil with poor permeability is the waterproof layer, and the lower fine sand layer is the pressurized water-bearing layer. When the pit is excavated on the soft ground, the damage of the bottom of the pit is very easy to happen because of the low shear strength of the soft ground.

3.2 *Influencing Factors of Engineering Problems*

Foundation pit engineering is a typical geological engineering, and the safety of foundation pit is greatly affected by the nature of geological body. Therefore, the design and construction of foundation pit support in Suzhou soft soil area is more difficult than that in common rock and soil area, and the deformation control is difficult [22, 23]. In order to ensure the safety of foundation pit construction, it is necessary to find out the main factors affecting the stability of foundation pit and put forward the safe and reasonable supporting form with pertinence. Based on the analysis of foundation pit engineering in many water-rich soft soil areas such as sampling area and historical accident data, the main factors affecting the safety of foundation pit are preliminarily identified and summarized as follows [24–26].

Depth of foundation pit

Soft soil has low strength and poor self-stability. Therefore, the deeper the foundation pit is excavated, the more difficult it is to support the foundation pit. At the same time, the larger the deformation of the supporting structure, the larger its risk coefficient is, which is the main risk target.

Physical and mechanical properties of soft soil and thickness of soft soil

Soft soils are susceptible to thixotropy and rheology due to their high water content, loose structure, and high sensitivity. At the same time, the greater the thickness of soft soil, the more obvious its deformation. It is conducive to the safety of the foundation pit.

For the complex foundation pit engineering in soft soil area, the plastic strain of soil is mainly concentrated in the bottom excavation surface attachment, and the influence of soil plastic parameters on the deformation of foundation pit is mainly concentrated in the bottom excavation surface attachment. Therefore, it is beneficial for soil consolidation and soil parameters improvement to take precipitation measures for the excavation surface and the following areas at a certain time before excavation.

Combination of soft soil and non-soft soil

Generally, the silty sand is located below the soft soil. The silty sand layer has good water capacity and is mostly confined water, which is not conducive to the stability of the foundation pit bottom.

Peripheral load of foundation pit

The building loads, vehicle loads, and other piles around the foundation pit will have adverse effects on the safety of the foundation pit. Pit unloading and pit side loading are very common in foundation pit excavation. However, this common phenomenon is the most unfavorable to the stability of the foundation pit, so the pile load should be reduced as much as possible in the foundation pit construction.

The closer the surrounding load is to the pit, the greater the settlement. The settlement near the middle of the pit is more important than the settlement at the ends of the pit. This may be related to the disturbance of the soil around the pit; the greater the disturbance, the greater the relative settlement.

Excavation mode of foundation pit

Excavation of foundation pit is a stress release process, and the soil confining pressure decreases and the shear strength decreases. When excavation of foundation pit, too fast and too large range of soil stress release will lead to sudden deformation of side wall and failure.

3.3 Measures for Prevention and Treatment

Through the analysis of the main problems and influencing factors mentioned above, foundation pit support should be based on site conditions and geological conditions around the pit [27, 28]. The construction should ensure a reasonable number of support forms and choose the correct construction method to minimize the risk factor of the foundation pit [29]. Preventive measures should be considered in the design and construction of engineering geological problems from the following aspects [30–32].

(1) According to the overall consideration of the project, a reasonable supporting form must be adopted according to the conditions of the foundation pit itself to ensure the safety and effectiveness of the supporting structure and prevent the instability of the foundation pit slope. In design and construction, upper unloading and bottom reinforcement are often used. Or the way of back pressure, increase the safety of foundation pit.
(2) Appropriate drainage measures should be selected according to the hydrogeological conditions of the soil layer of the foundation pit. The water cutting curtain is added outside the foundation pit to prevent groundwater seepage and reduce the creep of soft soil. Most foundation pit projects in Suzhou often adopt grouting method, double- and three-axis cement soil mixing pile, high pressure rotary jet grouting pile, and so on as the water curtain. Combined with the surrounding environmental conditions of the foundation pit, if there are important buildings requiring deformation control, recharge Wells should be set up within a certain distance outside the pit for recharge.
(3) The excavation of thc foundation pit soil is carried out in three ways: block, strip, and layer. This can make full use of the bearing capacity of the soil, but also effectively control the deformation of the foundation pit side walls and the bottom of the foundation pit.
(4) Adopt a variety of reinforcement measures to reduce the uplift deformation in the pit and eliminate the uplift damage at the bottom of the pit. The passive area is prone to uplift of foundation pit, and effective reinforcement of passive area is the most direct and effective method to reduce uplift. Slope release support can use the bottom of the slope back pressure, driving piles. Vertical support can be adopted by increasing the embedment ratio of support body and reinforcing the bottom of the pit.
(5) Foundation pit supporting structure design, choose the appropriate supporting structure type, can consider the support structure and permanent structure organic combination. If the underground continuous wall pit wall retaining pile is combined with the basement side wall, the pit bottom anti-uplift pile is joined with anti-floating pile. The supporting structure can be combined with the ring beam, purlin and beam in the basement, and the excavation of the foundation pit is convenient. Truly achieve economic reasonable, convenient construction.
(6) Information construction should be adopted. During construction, the construction progress and construction method should be timely controlled and adjusted

according to the monitoring data, and the whole construction process shall be dynamically controlled.

3.4 Case Analysis

Taking Caohu Road improvement project in Suzhou as an example, the project is located in Taidong Road, Xiangcheng, Suzhou, with a total length of 4689 m. The proposed tunnel engineering foundation pit excavation depth of the project is about 0–17 m. Because the characteristics of soil layers at different depths of foundation pit are different, this project also adopts a variety of different maintenance schemes for foundation pit, respectively. Plate type maintenance system is used when 0– m, steel sheet pile or SMW construction pile + support system is used when 4–10 m, and underground continuous wall or bored cast-in-place pile + support system is used when 10–17 m. It can be seen from Table 2 that the upper layer is mostly soil with low permeability, and the precipitation scheme of open ditch and catchment drainage is mainly adopted here. For the deeper layers, gushing may occur. Therefore, the artesian water level observation well was set up and decompression precipitation was adopted, and the water stop curtain was also installed.

In this paper, 2D numerical analysis software is used to simulate the cross-section structure of a certain foundation pit in the project, and the schematic diagram of the calculation model is shown in Fig. 2. The main loads on beam elements include concentrated load, distributed load, temperature load, and prestressed load. In addition, it also includes the dead weight of the structure, pile side earth pressure, and the lateral overloading of the foundation pit. In this structure, the pile body and crown beam are made of C25 concrete, and the elastic modulus is 3.0×10^4 MPa. The horizontal inner support in the foundation pit adopts steel pipe with outer diameter of 600 mm, wall thickness of 9 mm, and elastic modulus of 2.0×10^5 MPa.

The simulation analysis results show that the maximum horizontal displacement of ground wall is 0.01522 m, the maximum shear force is 226.0 kN/m, and the maximum bending moment is 316.8 kN/m. The maximum axial force of transverse support is 1285.79 kN, the maximum vertical displacement of ground is 0.016 m, and the maximum horizontal displacement is 0.013 m. All the data are less than the characteristic value of the allowable bearing capacity of the structure, indicating that the correct retaining method can effectively ensure the safety of the foundation pit structure in soft stratum.

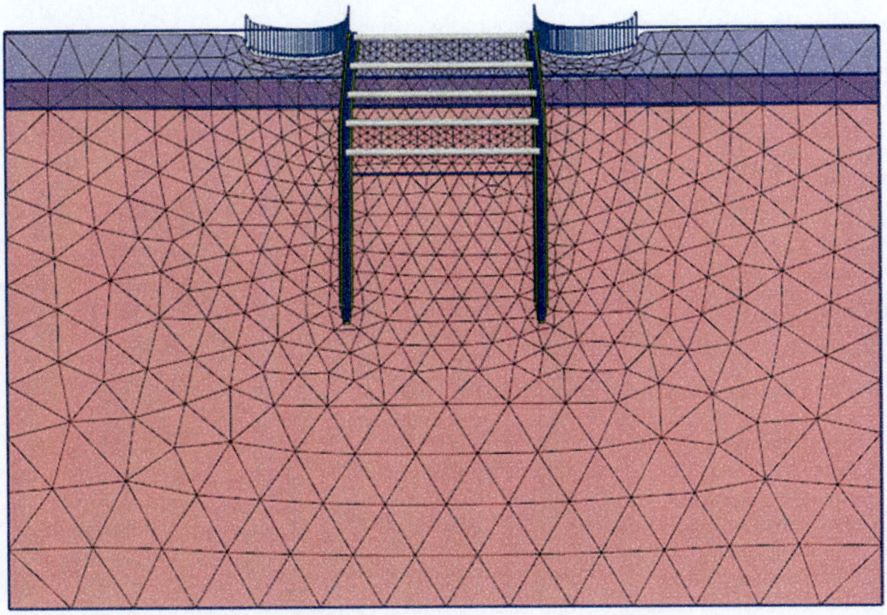

Fig. 2 2D integral structure calculation diagram

4 Conclusions

The Quaternary strata in Suzhou, China, are typical because of its abundant water system and intricate geological conditions. The characteristics of Quaternary sediments in Suzhou area were analyzed and integrated through borehole sampling, field investigation, and data collection. The results are as following.

(1) The thickness of Quaternary sediments in Suzhou is approximately 200 m, which can be divided into three aquifer groups or ten different layers. Each aquifer group is divided into water-bearing stratum and water-resisting stratum with a thickness of approximately 15 m and 30 m, respectively. The soil layers are mostly silty clay with approximately 30% water content, which has a loose structure and poor engineering properties.
(2) There are many fault zones, mainly in the direction of East-North and East–West, accompanied by the structural track of North–North-East.
(3) The geological conditions in Suzhou area often lead to the failure of the supporting structure, the instability of the foundation pit slope, seepage deformation, ground settlement, ground settlement, and water inrush at the bottom of the pit. Therefore, it is necessary to consider the excavation depth, excavation mode, soil properties, and surrounding loads and other key influencing factors and put forward corresponding prevention and control measures for the actual project.

Acknowledgements The research presented here is supported by the National Natural Science Foundation of China (52078317), Natural Science Foundation of Jiangsu Province for Excellent Young Scholars (BK20211597), project from Bureau of Housing and Urban-Rural Development of Suzhou (2021-25; 2021ZD02; 2021ZD30), Bureau of Geology and Mineral Exploration of Jiangsu (2021KY06), China Tiesiju Civil Engineering Group (2021-19), CCCC First Highway Engineering Group Company Limited (KJYF-2021-B-19), and CCCC Tunnel Engineering Company Limited (8gs-2021-04).

References

1. Wang, Q., Li, H., Xia, S., Yang, J., Zhang, W., Liao, G.Z., Wang, D.H.: Geometry of the Quaternary strata along the middle segment of the Longmen Shan and its formation mechanism: Insights from AMT, ERT and borehole data (2022)
2. Wang, L., Li, G., Liu, J., Mei, X., Zhang, Y.: Astronomical dating of quaternary strata in the south yellow sea and its indication for paleoclimatic evolution. Marine Geol, 106557 (2021)
3. Sizov, O., Volvakh, A., Molodkov, A., Vishnevskiy, A., Abakumov, E.: Lithological and geomorphological indicators of glacial genesis in the upper quaternary strata, nadym river basin, western siberia. Solid Earth **11**, 2047–2074 (2020)
4. Yu, X.H., Shen, J., Dai, X.Y., Wang, C.S.: Growth strata reveal the quaternary tectonic feature of gudian fault in songyuan, songliao basin, china. Dizhen Dizhi **40**(6), 1240–1253 (2018)
5. Deng, C.L., Hao, Q.Z., Guo Z.T., Zhu R.: Quaternary integrative stratigraphy and timescale of china. Sci. China (2019)
6. He, X.C., Xu, Y.S., Shen, S.L., Zhou, A.N.: Geological environment problems during metro shield tunnelling in Shenzhen, China. Arabian J. Geosci., **13**(2) (2020)
7. Ren, D.J., Shen, S.L., Cheng, W.C., Zhang, N., Wang, Z.F.: Geological formation and geo-hazards during subway construction in guangzhou. Environmental Earth Sci. **75**(11), 934 (2016)
8. Fang, G., Jiang, G., Yuan, D., Polk, J.S.: Evolution of major environmental geological problems in karst areas of southwestern china. Environ. Earth Sci. **69**(7), 2427–2435 (2013)
9. Cui, Q.L., Shen, S.L., Xu, Y.S., Wu, H.N., Yin, Z.Y.: Mitigation of geohazards during deep excavations in karst regions with caverns: a case study. Eng. Geol. **195**, 16–27 (2015)
10. Yoo, C., Kim, S.B.: Three-dimensional numerical investigations of new Austrian tunnelling method (NATM) twin tunnel interactions. Can. Geotech. J. **45**(10), 1467–1486 (2008)
11. Yilmaz, I.: GIS based susceptibility mapping of karst depression in gypsum: a case study from Sivas basin (Turkey). Eng. Geol. **90**(1–2), 89–103 (2007)
12. Editorial Department of China Journal of Highway and Transport. Review of Academic Research on Subgrade Engineering in China ·2021. China J. Highway Transport, **34**(3), 49 (2021) (in Chinese)
13. Ye, G.L., Hashimoto, T., Shen, S.L., Zhu, H.H., Bai, T.H.: Lessons learnt from unusual ground settlement during Double-O-Tube tunneling in soft ground. Tunn. Undergr. Space Technol. **49**, 79–91 (2015)
14. Ni, J.C., Cheng, W.C.: Characterising the failure pattern of a station box of Taipei rapid transit system (TRTS) and its rehabilitation. Tunn. Undergr. Space Technol. **32**, 260–272 (2012)
15. Shen, S.L., Wu, H.N., Cui, Y.J., Yin, Z.Y.: Long-term settlement behavior of the metro tunnel in Shanghai. Tunn. Undergr. Space Technol. **40**, 309–323 (2014)
16. Gong, Q.M., Yin, L.J., She, Q.R.: TBM tunneling in marble rock masses with high in situ stress and large groundwater inflow: a case study in China. Bull. Eng. Geol. Environ. **72**(2), 163–172 (2013)
17. Zhao, K., Janutolo, M., Barla, G., Chen, G.X.: 3D simulation of TBM excavation in brittle rock associated with fault zones: the Brenner Exploratory Tunnel case. Eng. Geol. **181**, 93–111 (2014)

18. Xu, Y.S., Shen, S.L., Ma, L., Sun, W.J., Yin, Z.Y.: Evaluation of the blocking effect of retaining walls on groundwater seepage in aquifers with different insertion depths. Eng. Geol. **183**, 254–264 (2014)
19. Ji, F., Lu, J.F., Shi, Y.C., Zhou, C.H.: Mechanical response of surrounding rock of tunnels constructed with the TBM and drillblasting method. Nat. Hazards **66**(2), 545–556 (2013)
20. Lai, K., Ren, Y.S., Hao, Y.J., Sun, Q., Liu, J., Jun-Ying, L.I.: Formation age and geological significance of yangbishan bif type iron deposit in jiamusi area, heilongjiang. Global Geol. (2017)
21. Zhu, W.: Geological and geochemical characteristic analyses of lower cambrian hetang formation in jiangshan-tonglu area. Acta Geologica Sinica (English edition) **93**(z2) (2019)
22. Wang, J., Wu, Y., Liu, X., Yang, T., Wang, H.: Zhu, Y,: Areal subsidence under pumping well–curtain interaction in subway foundation pit dewatering: conceptual model and numerical simulations. Environmental Earth Sci. **75**(3), 198 (2016)
23. Zhang, H.Z., Ya-Nan, L.I.: University L N . Research on the influence of the recharge of deep foundation pit dewatering to the surface subsidence. J. Langfang Normal University (Natural Science Edition) (2019)
24. Li, X., Zhou, T., Wang, Y., Han, J., Wen, J.: Response analysis of deep foundation excavation and dewatering on surface settlements. Adv. Civil Eng. **2020**(6), 1–10 (2020)
25. Zhaolin, M. A., Wang, X., Jiao, L., Survey.: Analysis of influence of dewatering of deep foundation pit on ground surface settlement. Low Temperature Architect. Technol. (2017)
26. Liu, J., Huo, J.: Analysis of influence of well point dewatering on surrounding environment in deep foundation excavation. Const. Technol. (2019)
27. Ahmed, M.D., Shiekha, A.A., Al-Obaidy, B.: Damage pattern scope prediction of well point dewatering on building foundation. In: 2nd International Conference on Engineering Sciences, at the University of Karbala-College of Engineering (2018)
28. Ahmed, M.D., Khudair, B.H., Shiekha, A.A.: Damage pattern scope prediction for well point dewatering on building foundations. Iop Conference, 433 (2018)
29. Xue, B.J.: On drainage and dewatering measures in construction of architectural projects. Shanxi Architecture (2013)
30. Zeng, C.F., Zheng, G., Zhou, X.F., Xue, X.L., Zhou, H.Z.: Behaviours of wall and soil during pre-excavation dewatering under different foundation pit widths. Comp. Geotech., **115**(Nov), 103169.1-103169.13 (2019)
31. Wang, Z.: Numerical analysis of deformation control of deep foundation pit in Ulanqab city. Geotech. Geol. Eng., 1–13 (2021)
32. Haibo, L.I.: Selection of deep foundation excavation support and dewatering style. Construction Technol. (2017)

Roof Condition Characteristics Affecting the Stability of Coal Pillars and Retained Roadway

Quang Phuc Le and Van Chi Dao

Abstract The stability of the retained roadway next to the coal pillar in the longwall mining system is mainly affected by displacement and rock pressure of the roof rock layers of the coal seam. With the objective of forecasting negative impacts in underground coal mining, this article analyzes the roof structure of the coal seam to illustrate the mechanism that causes instability of roadway retained in different roof conditions. A computer program is then used to simulate and assess the instability of the coal pillars and roadways in various roof variations according to the geological and technical conditions of the Khe Cham mine #11 seam. The results show that lowering and rotation of the console of main roof seriously affect the stability of the protective coal pillars. Vertical stresses distribution on the coal pillar increases significantly when the thickness of the immediate roof rock is very thin and insufficient to fill the goaf. Therefore, corresponding to different roof conditions, there is an effective solution for coal pillar design and roadway support to ensure production safety. This article can be used as a reference to designate technical recommendations in similar geological conditions.

Keywords Mining system · Temporary coal pillars · Main roof collapse state · Numerical modeling · Hard-caving main roof rocks · Deformation of the roadways · Roadway retained

1 Introduction

Controlling roadway deformation, increasing coal recovery rate, and reducing production costs have always been important strategic goals of underground coal mines [1, 2]. In underground coal mines of Vietnam, the longwall mining system

Q. P. Le (✉) · V. C. Dao
Hanoi University of Mining and Geology, 18 Vien Street, Hanoi 100000, Vietnam
e-mail: lequangphuc@humg.edu.vn

V. C. Dao
e-mail: daovanchi@humg.edu.vn

L. Q. Nguyen et al. (eds.), *Advances in Geospatial Technology in Mining and Earth Sciences*, Environmental Science and Engineering,
https://doi.org/10.1007/978-3-031-20463-0_29

with the retention of coal pillars to protect the roadway (Fig. 1) has long been chosen for wide use because of its advantages, such as (1) coal pillar has the role of isolating goaf with the retained roadway for the adjacent panel, (2) a high level of production concentration and ability to raise output from longwall face. However, its significant disadvantages are the loss of coal in the coal pillars accounting for 20% of the reserve and the high danger of rock bump in the high roof pressure area. In addition, when the width of the coal pillar is not properly designed, it will lead to the risk of roadway deformation problems, and the maintenance and repair costs are high.

Calculating the design of coal pillars and roadway supports is a complex process, as it considers a combination of factors of geology, engineering, and experience. The fact shows that abutment pressure is the main cause of instability of coal pillars and roadways. As a rule, after coal is extracted from the longwall face, the roof rock will crack and collapse into the goaf. Simultaneously, the stress in the rock mass around the goaf is redistributed until the displacement of the rock layers stops [3, 4]. The characteristics of appearance, distribution, and maintenance of stress elevation, abutment pressure on the edge of the coal seam, caused by the displacement of the roof rock layers, depend on their structural characteristics. The negative effects arising from the formation of the raised pressure zone on the retained roadway and coal pillars next to the goaf are roadway deformation, destruction of coal pillars, increased roadway repair costs, increased mining works, and occupational accidents.

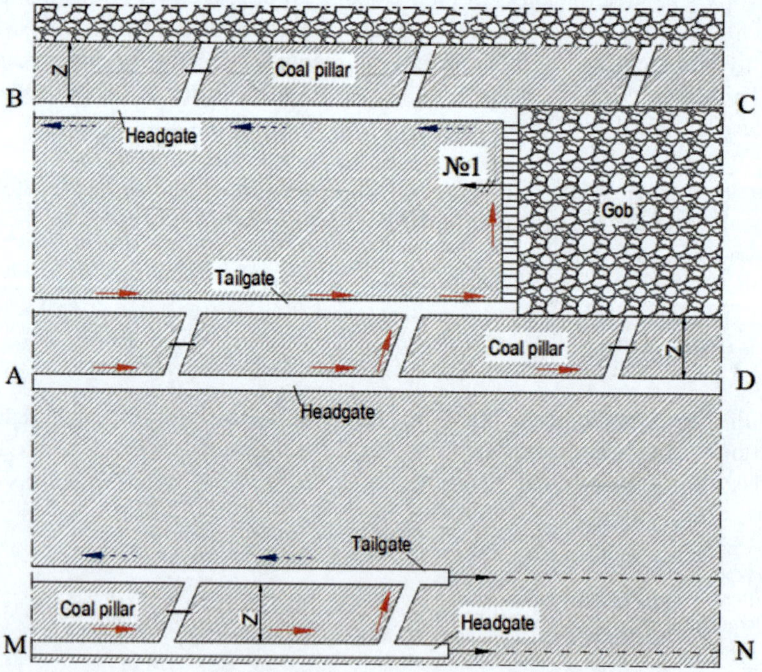

Fig. 1 Plan of longwall mining system applied in underground coal mines in Vietnam

The final state of the abutment pressure around the goaf is determined by the structural condition of the immediate roof and the main roof. As a rule, three collapse zones of roof rock [5, 6] will form above the goaf (from bottom up), which are cave zone, fractured zone, and continuous deformation zone. The value of the abutment pressure will be proportional to the size of the collapse area and depends on the properties of the roof rock layers. When the immediate roof is thick enough, the goaf can be completely backfilled by the cave rock, and the main roof will not crack. The intact main roof can then bear the load of the rock layers above and lead to safer coal seam mining because the retained roadway and coal pillars are less affected by the rock pressure. In contrast, when the main roof is cracked, the height of the roof's failure zone increases and leads to serious effects of abutment pressure caused by displacement of the roof rock layers.

Many studies on roadway stability control have been performed recently, but most of these works focus on the problem of roadway stability under a certain roof condition. For example, Feng et al. [7] studied the size and location of retained roadway under soft rock roofs. Kvardakov et al. [8] proposed a rigid and flexible support for the roadways under the hard rock roof. Zhang et al. [9–11] studied the stability of the roadway under the roof layer containing coal. Wang [12] proposed a method of constructing protective pillar from artificial materials to protect the retained roadway under complex roofs. Ning [13] analyzed the stability of the artificial pillar strip construction on the coal seam with large slope angle.

It can be seen that the study of the difference in roof rock conditions to provide solutions for designing coal pillars and supporting the roadway is absent. In addition, the high stress [14] and elastic–plastic rock mechanics [15–17] in deep mines make it more difficult to control the stability of the roadway and coal pillars. Therefore, the research problem that provides reliable data for finding solutions to design coal pillar and support roadway before impacts from unstable sources is that the roof rock layers will have great significance for the sustainable development of underground coal mining industry.

2 Research Methods

In this paper, the theoretical analysis of the roof rock structures of the coal seam and their collapse characteristics in the goaf is used. Then, a computer program is used to simulate the stability of the coal pillars and the deformation of the roadway during the mining of the panels with different cases of roof rock conditions. In our analysis, we use the typical geological conditions from the Khe Cham coal mine in the Quang Ninh coal basin, which has all the typical roof rock features of Vietnam's underground coal mines.

Building simulation models

A numerical model using the program FLAC3D [18] is developed to investigate the relationship between the stability of a reusable roadway and different conditions of roofs in shallow coal seams. FLAC3D is a software for numerical simulation of the geotechnical state of a continuum for studies of stress–strain state of rock mass. Such an analysis is practiced in the design of structures, strength calculations, research and testing, as well as for solving inverse problems of determining the margin of safety of structures. FLAC3D uses an explicit finite volume specification that captures the complex behavior of models going through several successive stages in their state, experience large displacements and deformations, taking into account nonlinear material behavior or instability (including the possibility of failure over large areas or a general loss of continuity).

The size of the model is 310 m × 200 m × 150 m, which is determined based on a sensitivity analysis of the model with respect to the size and density of the network. Two mining panels #1 and #2 and roadways are included in the model as shown in Fig. 2. In this model, roadways have a rectangular cross-section of 4 m wide and 3.0 m high. Assuming that the coal pillar is built in the model with a width of 20 m. A vertical stress of 7.0 MPa is applied to the upper boundary of the model to simulate natural loading, and the upper specific gravity is assumed to be 0.025 MN/m^3, and gravity is also applied [19]. The horizontal offset of the four vertical planes of the model is constrained to the normal direction, and the vertical offset at the bottom of the model is set to zero. This model uses the Mohr–Coulomb model.

In this model, the immediate roof is mudstone, and the main roof is sandstone. The thickness of the main roof is set to a certain value of 19 m, and the thickness of the immediate roof is changed according to the cases with the maximum value being

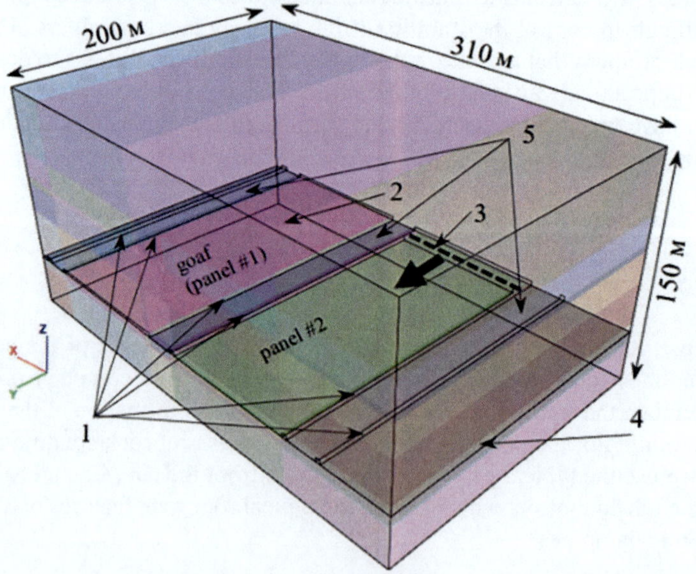

Fig. 2 Model configuration made with the program FLAC3D: 1—roadways, 2—goaf of mining panel #1, 3—mining panel #2, 4—coal seam, 5—coal pillar

able to fill the goaf when a 3-m-thick coal seam is exploited and can be calculated according to Eq. (1).

The model is built with the values of the coefficients taken as $k_p = 1.25$ and $k = 0.05$. Thus, the maximum thickness of the immediate roof determined in the corresponding model will be 12 m (when $\Delta h = 0$).

The article applies an assumption with 5 different types of roof structures to simulate and evaluate their effects on the stability of coal pillar and roadways. The parameters of the 5 types of roof structures are presented in Table 1.

In the model, the rock floor is assumed to be a siltstone with a thickness of 3 m. The rock strata consisting of sandstone, siltstone, mudstone, and coal are referred to as Mohr–Coulomb constitutive material. The mechanical parameters of the rock and coal layers used in this model are collected according to the results of testing rock samples in the laboratory and on-site statistical investigation at the Khe Cham coal mine, Vietnam. The physico-mechanical properties of the rock and coal layers are listed in Table 2.

In this simulation, the numerical model is solved using the following steps: (1) calculation of the initial stress caused by gravity; (2) driving roadways; (3) mining panel #1; and (4) mining panel #2. Five different variants (Table 1) of the roof are implemented in the model. The scheme of the algorithm for performing the simulation is shown in Fig. 3.

Table 1 Thickness of rock strata in different simulation schemes

Type roof seams	Thickness of seams (m)	Thickness of the immediate roof, h_1 (m)	Thickness of the main roof, h_2 (m)	Sink the main roof, Δh (m)
I	3,0	12	19	0
II	3,0	9	19	0.69
III	3,0	6	19	1.44
IV	3,0	3	19	2.19
V	3,0	0	19	2.94

Table 2 Physico-mechanical properties of rocks

Type rock	Tensile strength (MPa)	Bulk modulus (GPa)	Shear modulus (GPa)	Poisson's ratio	Cohesion (MPa)	Friction angle (deg.)	Density (kg/m3)
Sandstone	1.6	7.456	3.244	0,31	3,2	34	2780
Mudstone	0.9	2.333	0.955	0,32	2,1	30	2550
Siltstone	1.2	1.822	0.607	0,35	1,8	26	2250
Coal	0.4	0.748	0.484	0,26	1,5	19	1450

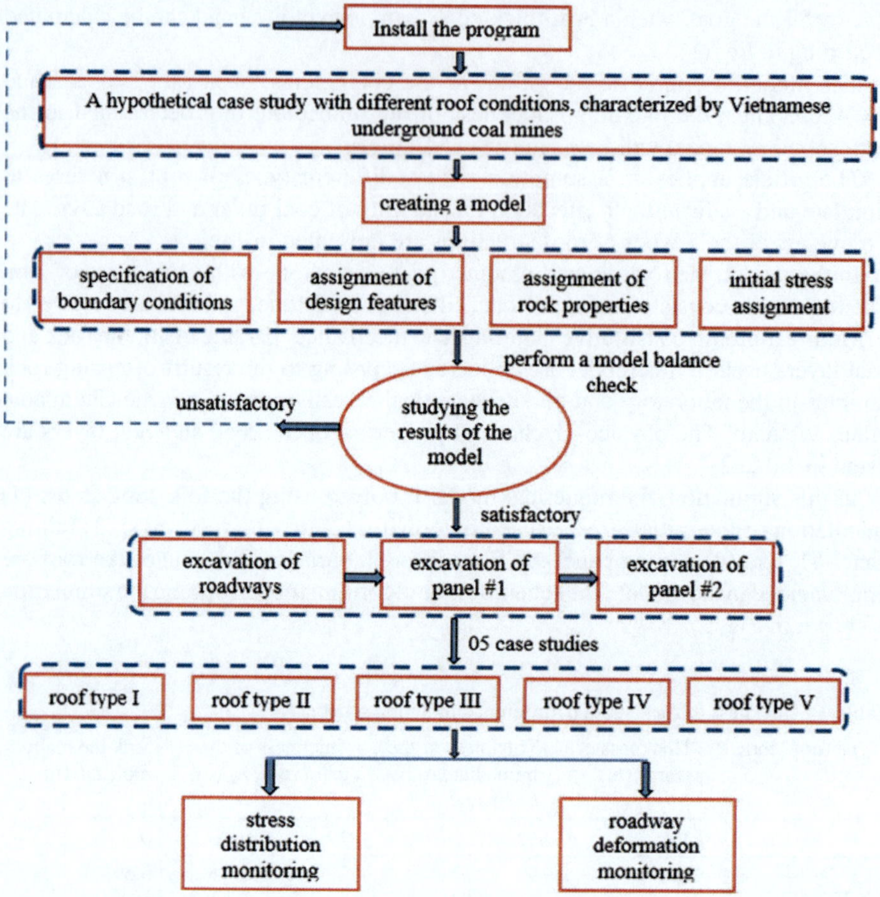

Fig. 3 Block diagram of the algorithm for performing numerical simulations with the FLAC3D program

3 Results and Discussion

3.1 Analysis of the Roof Rock Structures of the Coal Seam and Their Collapse Characteristics

The displacement of the roof rock layers above the goaf forms the console beams of the main roof on the edge of the coal pillar [20, 21]. Over time, the load of the roof rock layers above the console causes the edge of the coal pillar next to the goaf to be compressed and destroyed, facilitating the cracking and fracture process of the console. This static and dynamic load has a strong influence on the stability of the coal pillar and retained roadway [22–24]. As shown in Fig. 4, four deformation models have been constructed to illustrate the collapse of the main roof in different roof structures of the coal seam.

Figure 4 depicts different roof conditions in the following cases.

When the thickness (h_1) of the immediate roof is very thick, it collapses and is capable of filling the goaf (Fig. 4a). In this case, the main roof has almost no cracks, and most of the load of the rock layers is supported by the intact main roof. The influence of the load of the roof rock layers on the coal pillar and the roadway is minimal.

When the immediate roof has a small thickness (h_1), it collapses but cannot fill the goaf (Fig. 4b, c). Therefore, the main roof will collapse on the edge of the coal pillar with the rotation angle φ and subsidence Δh. The displacement of the block of the main roof when fracture can be simulated as a rotation around the crack above the coal pillar. In this situation, the collapsing process of the console beam will compress the underlying rock, including the coal pillar and retained roadway. The estimated impact intensity of rock pressure on the coal pillar is larger than that in the first case.

The subsidence Δh of the main roof console on the edge of the coal pillar can be calculated:

$$\Delta h = m_s \times (1 - k) - \left(k_p - 1\right) \times h_1 \tag{1}$$

where m_s—the thickness of the coal seam m, k—the coefficient of coal loss in the goaf, k_p—the coefficient of loose expansion of caved rocks of immediate roof collapse, h_I—the thickness of the immediate roof.

The rotation angle of the console of the main roof when collapsing can be calculated by:

$$\varphi = \sin^{-1}\left(\frac{\Delta \mathrm{h}}{\mathrm{L}}\right) \tag{2}$$

where L—the length of the collapse block of the main roof.

When the coal seam lacks an immediate roof ($h_1 = 0$) (Fig. 4d), the console of the main roof is located directly on the coal seam. Therefore, the entire load of the roof layers is transmitted directly to the edge of the coal seam, where the coal pillar and roadway are designed. When the coal seam is exploited, a large space ($\Delta h = m$) appears below the console of the main roof. The large deformation energy generated by the console rotation and subsidence is directly absorbed by the coal pillar. As a result, the coal pillar is strongly compressed and weakened significantly and is the cause of rock bump incidents, deformation, and roadway collapse. In this situation, roadway support should be focused on reducing the effect of main roof displacement.

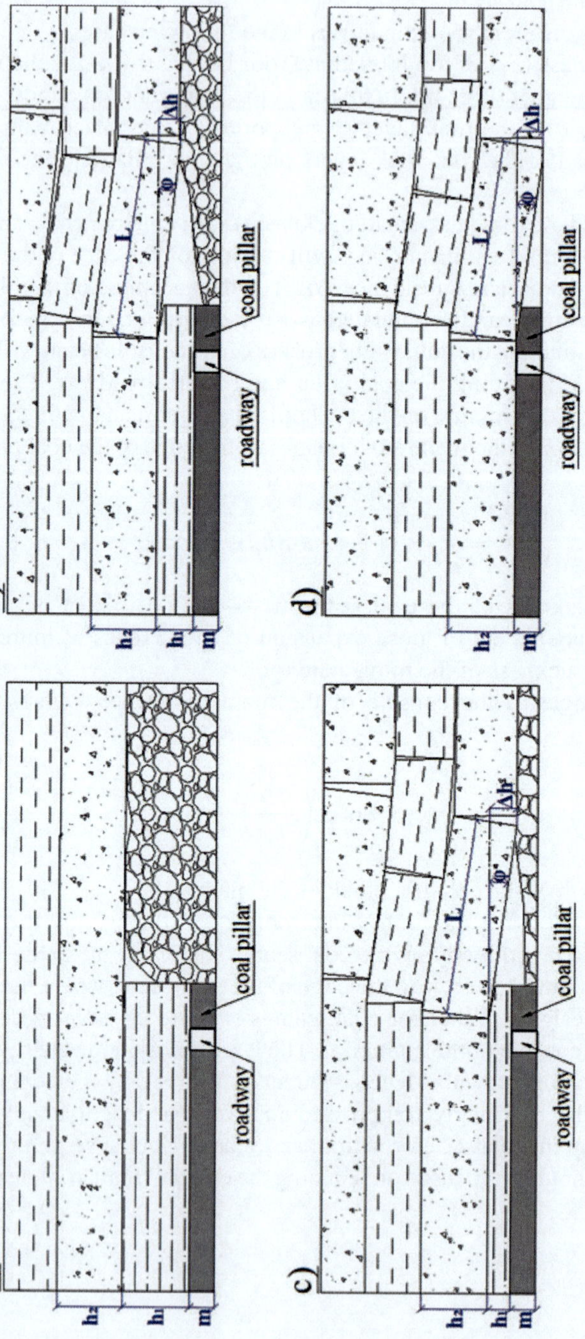

Fig. 4 Sate of collapse of the main roof in various roof structures: **a** very thick immediate roof, **b** immediate roof of medium thickness, **c** thin immediate roof, **d** no immediate roof

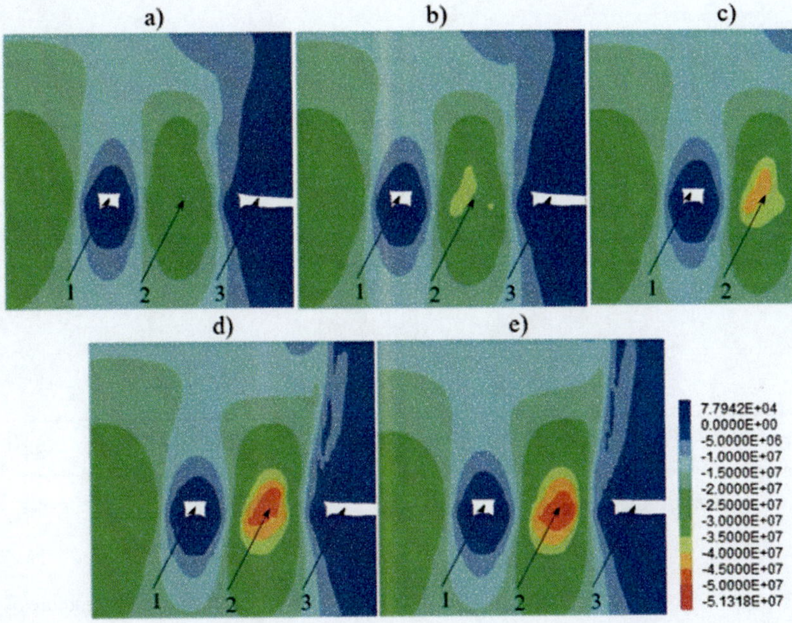

Fig. 5 Vertical stresses distribution around roadway and coal pillar (legend of vertical stress, Pa): **a** roof type I, **b** roof type II, **c** roof type III, **d** roof type IV, **e** roof type V, 1—retained roadway, 2—coal pillar, 3—goaf

3.2 *Numerical Simulation Results*

Results of monitoring stress evolution

The results of stress monitoring on coal pillars in different roof types are shown in Figs. 5 and 6. The position of the stress monitoring cross-section is taken at a distance of 20 m in front of longwall face #2.

Figures 5 and 6 show that there is a change in stress distribution on coal pillars when the roof rock structure changes. When the roof is very thick (types I, II), the stress distribution on the coal pillar is in the form of a double arch with two stress peaks distributed near the edges of the coal pillar. These stress peaks tend to get closer together as the immediate roof thickness decreases. The maximum value of the stress peak also increased, respectively, from 34 MPa (type I) to 39 MPa (type II). However, it is observed that, in the middle of the coal pillar in these two cases, there appears an elastic region (concave part of the stress curve) to ensure that the coal pillar always reaches a stable state.

Next, these two stress peaks merge into a single stress peak, and the integrated maximum stress on it also increases as the thickness of the immediate roof rock is further reduced. The value of stress peaks also increased from 42 MPa (type III) to 49.5 MPa (type IV) and 50.1 MPa (type V). The evolved into a single stress peak

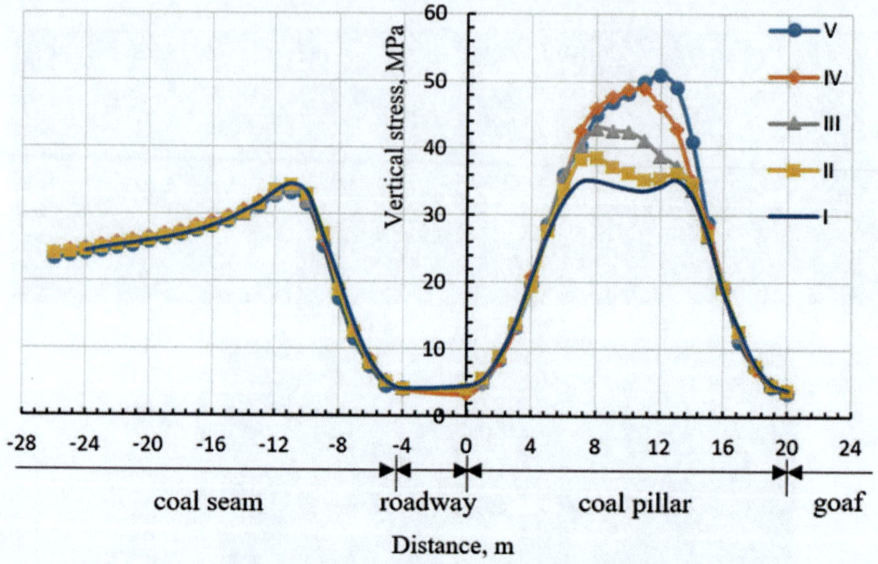

Fig. 6 Distribution of vertical stresses on coal pillar and coal seam when changing the structure of rock roof layers

demonstrated that the elastic region at the center of the coal pillar is lost and that the plastic deformation region has completely invaded it. The increase in stress also proves that when there is no immediate roof or it has a small thickness, the main roof will directly act on the coal pillar with great intensity (from the console formation of the main roof and collapse in the goaf). And so, the coal pillar will have to work in the maximum support mode to resist the pressure of the main roof, as well as the ability to protect the roadway. However, it has been shown that, when the stress on the coal pillar increases beyond the maximum limit, it will decrease suddenly due to the breaking of the cohesion of the coal mass and the formation of many cracks distributed in the coal pillar [10, 11]. Basically, the coal pillar is then damaged and causes serious deformation of the retained roadway. The simulation results show that the amplitude of stress change from type IV to type V is very small, which proves that it has reached the maximum state and will suddenly decrease after that.

Results of monitoring deformations

Figures 5 and 6 show that the coal pillar with a width of 20 m shows no signs of being destroyed, as the maximum stress peak on it is still higher than that on the coal seam. That is, the coal pillar still plays the main bearing role in all cases and eliminates the possibility of being destroyed under the pressure of the roof rock. Therefore, the roadway is maintained in a “mass-pillar” structure with a stable coal pillar, and the convergence of the roadway is assumed to be observed in this case. The evolution of the roadway convergence during the mining of longwall faces 1 and 2 is shown in Figs. 7 and 8. It is seen that the convergence of the roadway is greater when the immediate roof is thinner. This corresponds to the result that increasing vertical stress should increase roadway convergence. The special case is that the roof convergence of the roadway will decrease when there is no immediate roof. It is explained that the immediate roof is mudstone with low strength, and the formation of the crack system in it is greater than the main roof. For example, within the distance of 100 m behind longwall face #1, the roof convergence is 19 cm (type IV) versus 14 cm (type I) (Fig. 7a). However, this is not much different when the coal pillar is assessed as showing no signs of being destroyed.

In all cases, the convergence deformation of the roadways mostly takes place in the stage when they are about 30 m in front of longwall face #1, about 80 m behind longwall face #1, and especially, they change intensely when approaching longwall face #2 (about 40 m in front of longwall face #2). The maximum convergence of the roof increases from 27 cm (type I) to 37 cm (type IV) (Fig. 8a). The maximum convergence value of the floor increases from 49 cm (type I) to 65 cm (type V) (Fig. 8d). The maximum convergence value of the right rib (coal pillar side) has increased from 49 cm (type I) to 58 cm (type V) (Fig. 8c), and that of the left rib (coal seam side) is from 17 cm (type I) to 25 cm (type V) (Fig. 8b). It should be noted that, in the roof and floor of the retained roadway, large convergence occurs near the coal pillar side, and on the ribs of the roadway, the maximum convergence occurs in its center. The deformation on the left rib (coal seam side) is always smaller than that on the right rib (coal pillar side) in all cases. It is explained that the stress distributed on the coal pillar is always higher than that on the coal seam side.

In fact, the final deformation of the surrounding rock with respect to the usable space of the roadway consists of integral motion and volume expansion in the surrounding rock mass, due to the formation of cracks and fissures. For roof layers, the contribution of volume expansion to its final deflection is greatly reduced when the roadway is directly covered by a hard main roof. In addition, the rotation and subsidence of the rock console on the main roof compress the coal pillar more than on the coal seam side (Fig. 4). As a result, the coal pillar and right rib expand in the horizontal direction and are compressed in the vertical direction. Meanwhile, a vertical force is transmitted to the floor through the coal pillar, causing large heaving of the floor.

It should be noted that, when the coal pillar loses its bearing capacity, it means that the retained roadway will be strongly deformed. In the analyzed cases, coal pillar tends to be weaker as the immediate roof thickness decreases, and the deformation of the retained roadway also increases according to this law. Therefore, additional solutions are needed to increase the stability of the coal pillar and ensure the control of retained roadway deformation.

4 Conclusion

In this paper, the stability of the coal pillar and the retained roadway under the different roof conditions have been studied. The roof structure, stress distribution,

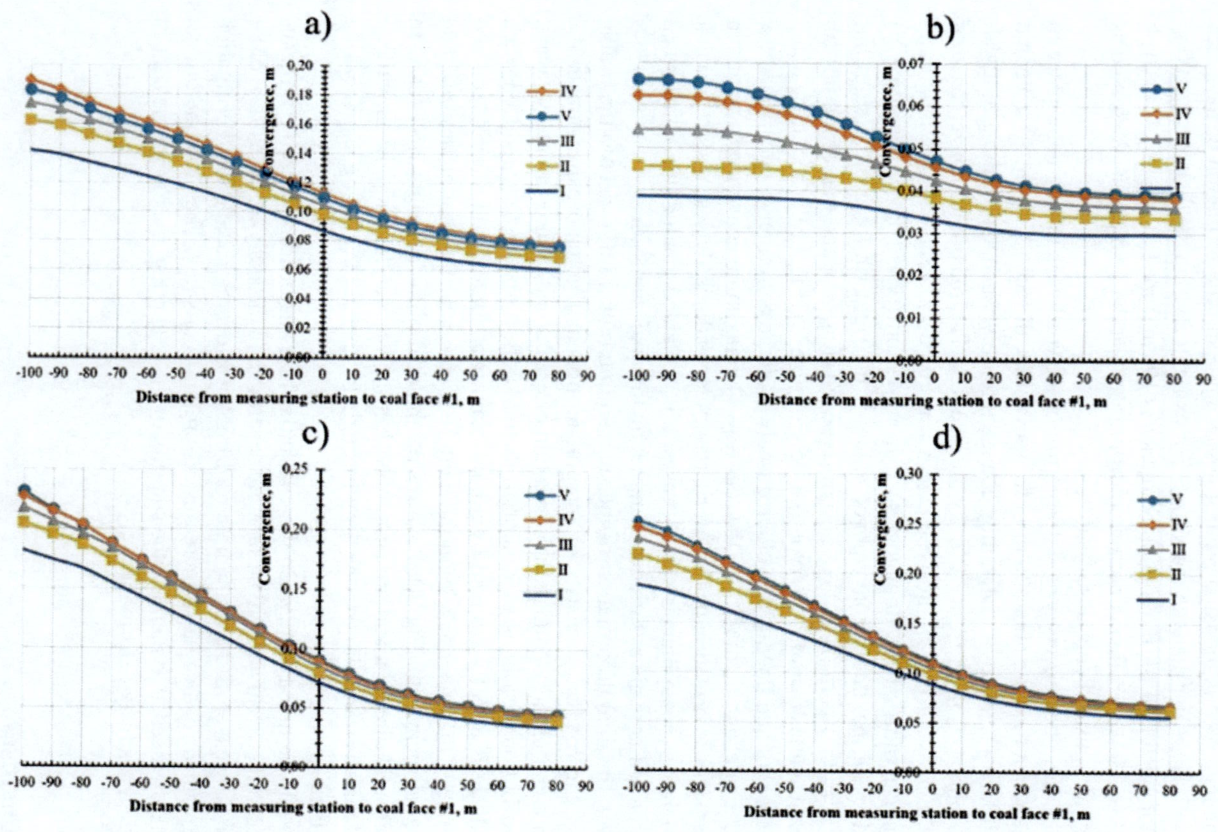

Fig. 7 Convergence of roadway before and after the coal face #1 corresponding to different roof rock structures: **a** roof subsidence, **b** convergence of the left rib (coal seam side), **c** convergence of the right rib (coal pillar side), **d** heaving of the floor

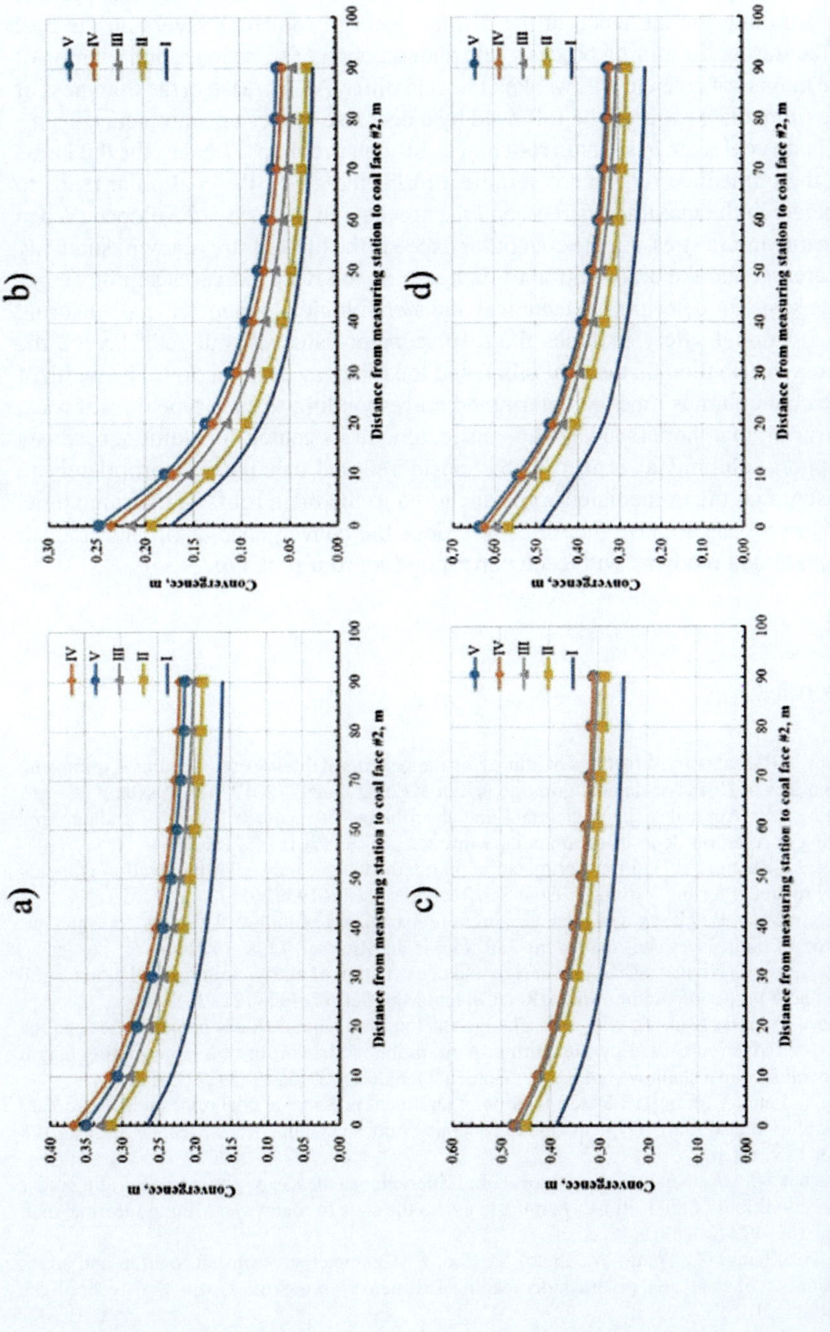

Fig. 8 Convergence of roadway before the coal face #2 corresponding to different roof rock structures: **a** roof subsidence, **b** convergence of the left rib (coal seam side), **c** convergence of the right rib (coal pillar side), **d** heaving of the floor

and the convergence of roadways were considered in this paper. The following conclusions were drawn from this study:

(1) The formation and collapse of the main roof console are the basic principles for determining the influence of the displacement of roof rock layers on the coal pillar during the mining process. The phenomenon of subsidence and rotation of the main roof console will weaken the coal pillar. An increase in the thickness of the immediate roof will be followed by a decrease in these phenomena (Fig. 4).
(2) When a coal seam has a main roof that is difficult to collapse (that is, the thickness of the immediate roof is not sufficient to fill the goaf), the coal pillar tends to operate in the maximum stress, and the cracks will be further developed. When the maximum stress on the coal pillar exceeds the limited stress, it will suddenly decrease, and the deformation of the retained roadway increases significantly.
(3) The basis for determining technical and technological parameters and ensuring occupational safety includes the displacement value of roof rock layers, the stress distribution on the coal pillar, and the roadway deformation. The width of the coal columns must be determined corresponding to each type of roof rock, ensuring that there is always an elastic zone in its center. In addition, roadway support solutions also need to be considered and calculated appropriately to ensure that the immediate roof is anchored to the main roof. At the same time, it is necessary to have measures to reduce the convergence of rib and floor of the retained roadway with each corresponding roof rock type.

References

1. Zubov, V.P.: Status and directions of improving the systems of mining of coal seams at promising coal mines of Kuzbass. Zapiski gornogo instituta **225**, 292–297 (2017). (in Russian)
2. Zubov, V.P.: Applied technologies and actual problems of resource saving in underground mining of reservoir deposits of minerals. Mining J., 77–83 (2018) (in Russian)
3. Feng, X., Zhang, N.: Position-optimization on retained entry and backfilling wall in gob-side entry retaining techniques. Int. J. Coal Sci. Technol. **2**, 186–195 (2015)
4. Yang, W., Lin, B.Q., Yan, Q., Zhai, C.: Stress redistribution of longwall mining stope and gas control of multi-layer coal seams. Int. J. Rock Mech. Min. Sci. **72**, 8–15 (2014)
5. Tan, Y., Yu, F., Ning, J., Zhao, T.: Design and construction of entry retaining wall along a gob side under hard roof stratum. Int. J. Rock Mech. Min. Sci. **77**, 115–121 (2015)
6. Yang, J., Liu, B., Gao, Y., Wang, Y., Cheng, Y., Hou, S.: Dynamic load characteristics and the pressure reduction caused by the cutting seam on the roadside roof of a large mining height longwall face in a shallow coal seam. Geotech. Geol. Eng. **37**, 2949–2962 (2019)
7. Zhu, S., Feng, Y., Jiang, F.: Determination of abutment pressure in coal mines with extremely thick alluvium stratum: a typical kind of rockburst mines in China. Rock Mech. Rock Eng. **49**, 1943–1952 (2016)
8. Kavardakov, A.A., Kuzin, E.G., Pudov, E.J.: Experience in the use of ground-penetrating radar in the conditions of the Kotinskaya mine to assess the state of roadways. Mining Inform. Anal. Bull., 166–173 (2016) (in Russian)
9. Wu, H., Zhang, N., Wang, W., Zhao, Y., Cao, P.: Characteristics of deformation and stress distribution of small coal pillars under leading abutment pressure. Int. J. Min. Sci. Technol. **25**, 921–926 (2015)

10. Zhang, Z., Bai, J., Chen, Y., Yan, S.: An innovative approach for gob-side entry retaining in highly gassy fully-mechanized longwall top-coal caving. Int. J. Rock Mech. Min. Sci. **80**, 1–11 (2015)
11. Zhang, Z., Shimada, H., Sasaoka, T., Hamanaka, A.: Stability control of retained goaf-side gateroad under different roof conditions in deep underground y type longwall mining. Sustainability **9**, 1671 (2017)
12. Wang, H.S., Zhang, D.S., Liu, L., Guo, W.B., Fan, G.W., Song, K.I., Wang, X.F.: Stabilization of gob-side entry with an artificial side for sustaining mining work. Sustainability **8**, 627 (2016)
13. Ning, J., Wang, J., Bu, T., Hu, S., Liu, X.: An innovative support structure for gob-side entry retention in steep coal seam mining. Minerals **7**, 75 (2017)
14. Brown, E.T., Hoek, E.: Trends in relationships between measured in-situ stresses and depth. Int. J. Rock Mech. Mining Sci. Geomech. Abstracts, 211–215. Pergamon
15. Yang, S.Q., Xu, P., Ranjith, P.G., Chen, G.F., Jing, H.W.: Evaluation of creep mechanical behavior of deep-buried marble under triaxial cyclic loading. Arab. J. Geosci. **8**, 6567–6582 (2015)
16. Pankov, I.L., Asanov, V.A., Udarcev, A.A.: Study of the influence of the degree of loading and the shape of samples on the deformation of salt rocks during plastic yield. Mining Inform. Anal. Bulletin, 246–252 (2013) (in Russian)
17. Filimonov, K.A., Renjov, A.A., Kucherenko, A.V., Grechishkin, P.V.: Study of the impact of working faces on the state of roadways retained next to the pillars in the conditions of the mine "IM. VD Yalevskogo". Mining Inform. Anal. Bulletin, 133–148 (2019) (in Russian)
18. Itasca, F.D.: Fast lagrangian analysis of continua in 3 dimensions. Online Manual, 175–180 (2013)
19. Le, Q.P., Zubov, V.P., Phung, M.D.: Improvement of the loading capacity of narrow coal pillars and control roadway deformation in the longwall mining system. a case study at Khe Cham Coal Mine (Vietnam). Inzynieria Mineralna **1**, 115–122 (2020)
20. Chen, M., Yang, S.Q., Zhang, Y.C., Zang, C.W.: Analysis of the failure mechanism and support technology for the Dongtan deep coal roadway. Geomech. Eng. **11**, 401–420 (2016)
21. Le, T.D., Bui, X.N.: Effect of key parameters on top coal first caving and roof first weighting in longwall top coal caving: a case study. Int. J. Geomech. **20**, 04020037 (2020)
22. Gao, Y., Liu, D., Zhang, X., He, M.: Analysis and optimization of entry stability in underground longwall mining. Sustainability **9**, 2079 (2017)
23. Kholmskiy, A.V., Sidorov, D.V.: Arrangements for increase the efficiency of mining operations on the deep ore mines. Scient. Practical Stud. Raw Mat. Issues, 71 (2019)
24. Sun, Y., Li, G., Zhang, J., Xu, J.: Failure mechanisms of rheological coal roadway. Sustainability **12**, 2885 (2020)

On the Flow Assurance for Un-Insulated Subsea Pipeline Systems: Application on the Multiphase Pipeline from Pearl Field to FPSO Ruby II Offshore Vietnam

Van Thinh Nguyen, Thi Hai Yen Nguyen, Sylvain S. Guillou, Thuy Huong Duong, Thi Thanh Thuy Truong, and Thi Thao Nguyen

Abstract The Pearl field is an offshore oil field in Vietnam, located approximately 160 km from Vung Tau. This field signifies high-paraffin crude with wax appearance temperature of approximately 50 °C. With the ambient temperature recorded from 27 °C at the surface to 22 °C at the seabed, the transportation using un-insulated pipeline has to face great challenges such as: the hold-up of liquid, slugging, erosion and corrosion, and the deposition of wax etc.,. Therefore, it is obligatory to investigate the system's performance periodically to predict undesirable mentioned phenomena to assure continuous flow though this pipeline may afford safety and stability to the conveyance of crude oil. The article contributes itself to the build of a dynamic model to characterize this pipeline system at actual working conditions. Flow assurance study is carried out in 3 cases corresponding to 3 different flow rates of the incoming mixture. We then compare these calculations with experimental data to assure its accuracy and reliability. Results of simulation enable us to observe a rapid drop of temperature inside the pipeline after 2 h of shutdown and a risk of slugging if gas appears in the flow and the wax layer of about 0.1 mm after 30 operational days.

V. T. Nguyen (✉) · T. H. Y. Nguyen · T. H. Duong · T. T. T. Truong · T. T. Nguyen
Hanoi University of Mining and Geology (HUMG), Hanoi, Vietnam
e-mail: nguyenvanthinh@humg.edu.vn

T. H. Y. Nguyen
e-mail: nguyenthihaiyen@humg.edu.vn

T. H. Duong
e-mail: duongthuyhuong@humg.edu.vn

T. T. T. Truong
e-mail: truongthithanhthuy@humg.edu.vn

T. T. Nguyen
e-mail: nguyenthithao@humg.edu.vn

T. H. Y. Nguyen · S. S. Guillou
Laboratoire Universitaire de Sciences Appliquées de Cherbourg (LUSAC, EA 4253), Normandy University, UNICAEN, 60 Rue Max Pol Fouchet, CS 20082, 50130 Cherbourg-Octeville, France
e-mail: sylvain.guillou@unicaen.fr

L. Q. Nguyen et al. (eds.), *Advances in Geospatial Technology in Mining and Earth Sciences*, Environmental Science and Engineering,
https://doi.org/10.1007/978-3-031-20463-0_30

This layer is quite acceptable; however, regular maintenance and pigging are required during the transport.

Keywords Offshore oil field · Flow assurance · Subsea pipeline systems

1 Introduction

1.1 Physicochemical Properties of Crude Exploited at Pearl Oil Field

Crude oil transportation in oil fields of Vietnam nowadays faces big challenges when exploited product from depleted reservoirs contain high concentration of paraffin. Pear field, located approximately 160 km away from Vung Tau Port in block 01 and 02 offshore Vietnam, is an operational wellhead platform developed by PetroVietnam Exploration Production Corporation (PVEP), Vietnam. Due to properties of exploited product (Tables 1 and 2) at actual field, crude from Pearl signifies heavy oil with high-paraffin content (19–27%). Rheological properties of crude play a critical role in deciding means of transport [1, 2]. Commonly, high-paraffin content in crude plays an important role in causing dicey problems during the shipment. Analysis results from laboratory inform that at the state of paraffin wax, the tested samples melt at the temperature of about 55–61 °C and crystallizes at about 32–36 °C, which is considered a high freezing point. During the transportation of such a waxy fluid, undesirable situations are likely to happen related to the characteristics of the pipe and the temperature of the surrounding environment. Associated problems to name a few include the accumulation of wax that dwindles the effective diameter of the pipe, an increase of pressure drop and a flow rate below the designed value, etc. [3–5].

Table 1 Bulk properties of pearl dead oil

No.	Test items	Unit	Sample oil-well-Mi-09/10	Sample oil-well-Mi-09/10
1	API gravity	API	28.29	30.90
2	Sulfur content	% mass	0.128	0.098
3	Viscosity at 50 °C	Cst	14.22	11.10
4	Viscosity at 50 °C	Cp	12.26	9.41
5	Pour point	°C	19.00	34.00
6	Water content	% vol	0.17	1.92

Table 2 Rheological properties of Pearl crude oil at shear rate of 20 l/s

No	Shear rate = 201/s			
	Test temperature, °C	Dynamic gel strength, Pa	Plastic viscosity, Pa.s	Effective viscosity, Pa.s
1	45	0	0.008	0.008
2	40	0	0.01	0.01
3	35	0	0.023	0.023
4	30	3.1	0.064	0.08
5	28	7.2	0.106	0.142
6	26	13.9	0.147	0.216
7	24	23.2	0.191	0.307
8	22	38.1	0.229	0.42
9	< = 20	Full solidification		

1.2 *Characteristics of the Transportation Pipeline from Pearl Wellhead to FPSO*

Pearl wellhead platform (WHP) currently produces from 1PS well. The recovery is then diverted to a low-pressure separator (LP separator) where the coming out flow is boosted by transport pump and commingle with gas product from another well (well 2P) to form mixture of gas-saturated oil (Fig. 1). This fluid at this moment is routed to FPSO Ruby II through a subsea pipeline system. The investigated pipeline system consists of two sections: a 10-inch pipeline (0.254 m) routing the full well stream (FWS) from RBDP-B to RBDP-A and a 10-inch flexible pipeline (0.254 m) directing the commingled fluid flow at RBDP-A to the FPSO Ruby II. The system has a total length of 6233 m, mostly soaked in water. In our case, the studied flow enters the gathering line at a temperature of about 30–32 °C, which is under its freezing point, whereas the wellhead RBDP-A is installed at the water depth of 50.91 m with the sea water temperature ranging from 27 °C at the surface to 22 °C at the bottom. Thus, besides facing the problem of transporting the fluid with low flexibility, we are obliged to confront the matter of shipping the crude having pouring point just some few degrees above the ambient temperature. Generally, this, in combination with inconsiderable differential temperature between the inside and the outside of the pipe, causes the mixture to get thicker and consequently increases the flow resistance [4, 6]. Challenges augment in our situation when the transport system is characterized by non-insulated pipeline, which is considered 'abnormal' in the recent epoch where insulation is recommended to almost all pipelines to transport petroleum products across long distances.

Profile of the system is shown in Fig. 2. The pipeline consists of 3 sections with two risers, Fig. 2a and b, assumed to be 20 m and 37 m, respectively, above the mean sea level in the region. Seabed topography, where situated the oil field of interest, is quite plane as measured; thus, the major part of the system on the seabed is taken for

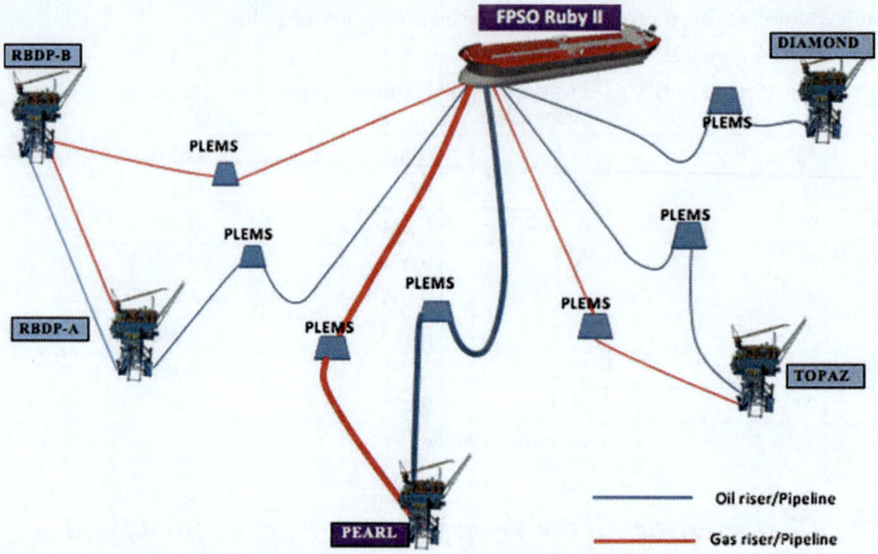

Fig. 1 Transportation pipeline system from Pearl wellhead platform to FPSO Ruby II

granted as at the same altitude. This layout will be fully transferred to simulations at the next section for the sake of reality. The pipeline has an internal diameter of 254.5 mm. It is non-insulated, but the two risers are coated:

- Rigid riser (a) is protected by a multilayer coating 3-layer polypropylene (3LPP)
- Flexible riser (b) is concealed with two coating layers of FBE + PP adhesive + PP topcoat and armor.

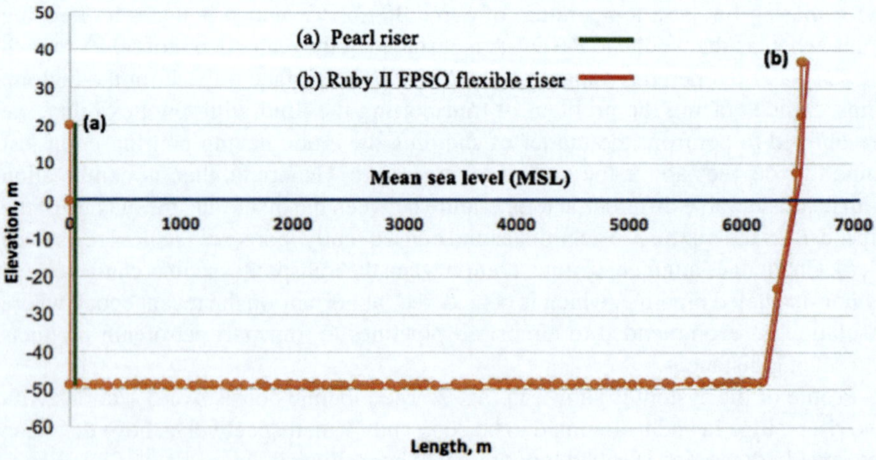

Fig. 2 Pearl—FPSO Ruby II Pipeline profile

Fundamentally, the usage of a submarine pipeline without repetitious thermal insulation treatments in such a condition may easily induce serious predicaments during its working life as a result of higher heat transfer loss [7]. A significant heat loss urges the crystallization of wax in the crude [8, 9]. The over precipitation of wax crystal arises the formation of interlocking structure, which may drive to catastrophic shutdowns [10, 11]. In other words, we predict systematic problem may regularly occur for this pipeline unless legitimate design parameters are utilized and solutions are applied in time.

At this moment, we are not able to deduce whether it is possible or not to deliver recovered product safely and efficiently from Pearl field to FPSO Ruby II by non-insulated submarine pipeline. Therefore, flow assurance is unavoidable for this case.

2 Transportation Equations for Multiphase Flow

If we consider a multiphase (*k* phases) flowing in a finite pipe section, dx, the pipe inclines at θ degree, then the flow in this case is simply described as in the Fig. 3 [12].

The multiphase transporting flow in our case consists mostly of 3 components: crude oil, gas, and water. The equations established for a 3-phase transient flow include mass conservation and momentum conservation. In which, the mass conservation is written in three equations corresponding to 3 phases independently:

$$\text{Mass equation for gas}: \frac{\partial(\alpha_g\rho_g)}{\partial t} + \frac{\partial(\alpha_g\rho_g v_g)}{\partial x} = 0 \tag{1}$$

$$\text{Mass equation for oil}: \frac{\partial(\alpha_o\rho_o)}{\partial t} + \frac{\partial(\alpha_o\rho_o v_o)}{\partial x} = 0 \tag{2}$$

$$\text{Mass equation for water}: \frac{\partial(\alpha_w\rho_w)}{\partial t} + \frac{\partial(\alpha_w\rho_g v_g)}{\partial x} = 0 \tag{3}$$

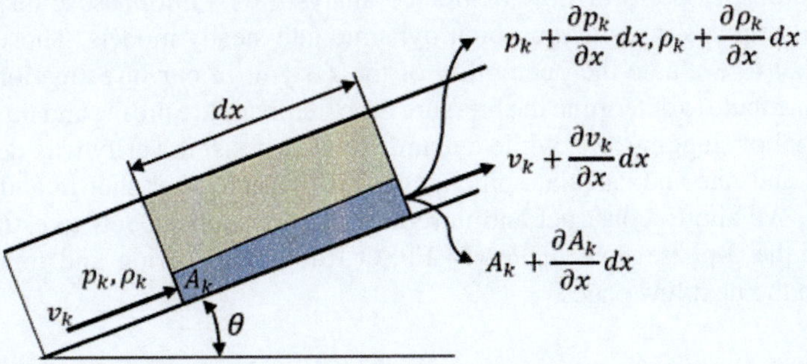

Fig. 3 Multiphase flow in pipeline with the length d*x* [12]

Equations of continuity for each phase are expressed below:

$$\text{Gas phase}: \frac{\partial(\alpha_g \rho_g v_g)}{\partial t} + \frac{\partial\left(\alpha_g \rho_g v_g^2\right)}{\partial x} = -\alpha_g \frac{\partial \rho_g}{\partial x} + \left(R_{gi} + R_{gw}\right) - \alpha_g \rho_g g \sin\theta \quad (4)$$

$$\text{Oil phase}: \frac{\partial(\alpha_o \rho_o v_o)}{\partial t} + \frac{\partial\left(\alpha_o \rho_o v_o^2\right)}{\partial x} = -\alpha_o \frac{\partial \rho_o}{\partial x} + (R_{oi} + R_{ow}) - \alpha_o \rho_o g \sin\theta \quad (5)$$

$$\text{Water phase}: \frac{\partial(\alpha_w \rho_w v_w)}{\partial t} + \frac{\partial\left(\alpha_w \rho_w v_w^2\right)}{\partial x} = -\alpha_w \frac{\partial \rho_w}{\partial x} + (R_{wi} + R_{ww}) - \alpha_w \rho_w g \sin\theta \quad (6)$$

where, R_{gi}, R_{Oi}, and R_{wi} represent friction force on each phase and R_{gw}, R_{Ow}, and R_{ww} describe similarly volume-specific friction force from the pipe's wall on gas, oil, and water, respectively. ρ_g, ρ_o, and ρ_w are functions of density in the relation to pressure and temperature. α_g, α_o, and α_w symbolize ratios of fluid components (phases) which conform to the rule:

$$\alpha_g + \alpha_0 + \alpha_w = 1 \quad (7)$$

3 Model Development

Flow assurance encompasses a wide range of aspects related to the flow stream in order to ensure an economical transmission during the life cycle of a field [13, 14]. Therefore, to perform flow assurance analysis for multiphase transportation pipeline, it requires investigating both dynamic and steady models. Those models are essential to evaluate the operability of the system. In our investigation, steady models contribute to determine the pressure drop, temperature profile and liquid hold-up, potential of slugging, etc. while dynamic models assist to verify wax deposition properties and rate and calculate pipeline restart pressure after shut-in and pigging frequency. We aim to construct and introduce the transport models to estimate the regime of the 3-phase flow in Pearl—FPSO Ruby II gathering and transporting pipeline in the next few pages.

3.1 Modeling Tools

As a well-known tool in petroleum engineering, the software OLGA 2019 is utilized to develop our models by solving the equations mentioned in the Sect. 2. OLGA meets almost all requirements in modeling flow streams with significant phase change and variation of flow structure by time. Theoretically, by reason of accuracy, it is frequently recommended to separate a whole pipeline into smaller sections considering the change of altitude and treat each part independently. It means that calculations get more explicit in our case if we calculate 3 different parts separately. However, OLGA software considers a vast set of pipeline parameters including topography, vertical and horizontal pipe sections, and the closing and opening of valves and the flow fluctuation as well. On the other hand, OLGA empowers us have reliable results to calculate the whole simplified profile of the pipeline as in Fig. 2 without any separation.

3.2 Calculation Data and Study Cases

In our research, the transport models are simulated in 03 cases independently corresponding to 3 different stages of actual shipping mixture's flow rates: last production (denoted as past flow, the oil rate measured in 2021), max oil and liquid (contributes to the rate of production in 2023 which was estimated as the highest value), and min flow rate (which was evaluated as minimum in 2026 due to the rate of depletion of Pearl field) (see Table 3). Exploiting data in 2021 are put into the calculation to bring an overview to the actual production at the field in recent years.

Based on laboratory analysis, results of composition of Pearl commingle oil fluids are characterized and tuned to match the properties of fluid model and laboratory analysis results. Other properties namely wax content and viscosity of Pearl crude oil at shear rate of 20 s^{-1} were used to characterize wax properties of Pearl dead crude oil, as in Tables 1 and 2. The flow assurance models are run with two simulation conditions: steady state and transient state as mentioned previously. The pipeline

Table 3 Flow assurance study cases

Case	Definition	Year	Oil rate, (stb/d)	Water rate, (stb/d)	Liquid rate, (stb/d)	Water cut, %	Gas rate from well 1P, (MMscf/d)
1	Last production	2021	363	298	661	45.1	0.5
2	Max oil and liquid	2023	503	504	1,007	50.0	0.5
3	Min oil and liquid	2026	150	347	497	69.8	0.5

Table 4 Average operating parameters of the transport pipeline PLDP-A-FPSO Ruby II

P, barg	T, °C	$P_{landing}$, barg	$T_{landing}$, °C	ΔP, barg
13.5	59	8.3	29.6	5.2

commences with a pressure of 13.5 barg (gauge pressure) at Pearl riser and a pressure in range of 8.3barg on arrival at FPSO Ruby II. The pressure drop ΔP in the pipeline section approximates 5.2 barg. From the departure to destination, a decrease of temperature ΔT of about 30 °C is recorded (Table 4).

It is not to plainly mention, but other parameter such as wind velocity, seawater current velocity, thickness, or roughness of the pipe, etc. are also employed for the calculations. For example, relying on average operating parameters gained from the Table 3, the models are matched considering an increase of 10% of the roughness of the inner surface of the pipeline due to corrosion and gas association in the flow.

4 Results

4.1 *Steady State Simulation Results*

a. **Flow stability**

The hold-up of liquid along the trajectory of the pipe is plotted in the Fig. 4 where we observe slug in all cases. The hold up of liquid at the inlet, which corresponds to riser a Pearl, is quite the same as the 3 cases of flow rate. As expected, the slug value drops to nearly 0 which means liquid phases are completely moved away at some first few ten meters since the flow is nearly transported vertically down inside the pipeline section a. This harmonizes with the dependency of flow patterns on inclination of pipeline [15–17]. After Ruby II riser, the flow slows down as an effect of distance, gas component separates from liquid and expands to form bubbles of different sizes and builds up [18]. This process continues until attaining its peak and the gas slug is sufficient to move liquid slug forward. This rule repeats over and over until the flow is completely transported to the outlet, FPSO. The figure also indicates a strong accumulation of liquid slug 1 km away from the riser b connecting submarine part of the pipeline with the container in FPSO. This phenomenon could be explained as the sudden change of inclination of the pipe from horizontal to vertical. In addition, energy of the flow gradually depletes by time and distance relating to its deceleration before advancing upward. Thereupon, for all the 3 cases, the huge concentration of liquid is observed at this part of the pipeline system. Noticeably, the hold-up of liquid oscillates strongly for case 1 (production in 2021) and case 3 (minimum production in 2026) while it tends to be more stable with case 2 (production in 2023). It enables us to confirm that the transportation is more effective with case 2 when the production rate is highest.

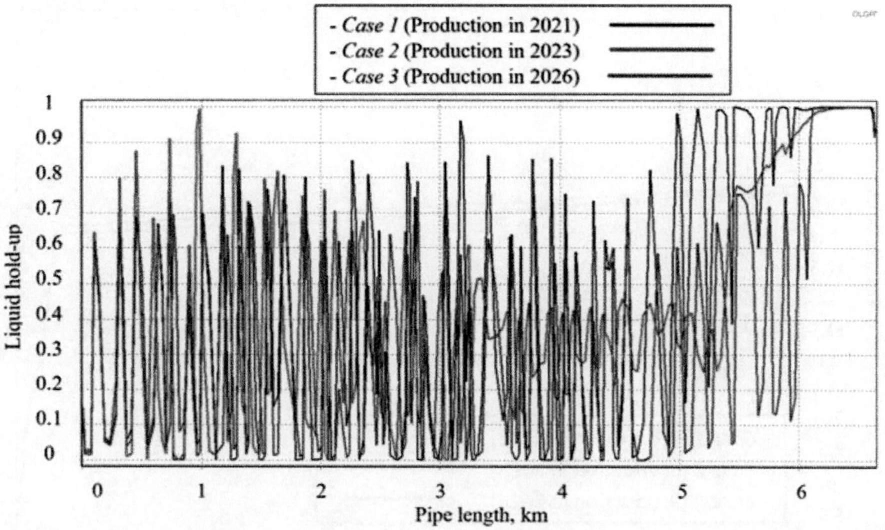

Fig. 4 Liquid hold up along the pipeline for study cases

b. **Pressure**

The dynamic steady state profiles of pressures for 3 study cases are plotted in Fig. 5. The picture clearly displays the variation of pressure corresponding to 3-section profile of the pipeline. With all the 3 cases, the change obeys strictly one same rule. At the riser a, pressure slightly increases due to the tendency to move vertically downward of the flow. The pressure going out of this riser section starts to drop all along the length of the pipeline until the arrival of the stream at riser b FPSO. A differential pressure ΔP ~ 0.4 barg is recorded at this section. After that, the flow changes its direction to go straight upward, we observe a dramatic drop of pressure from 13barg to 8.5 barg at this part of the system.

c. **Thermal issues**

The temperature profiles of the pipeline for 03 study cases are presented in Fig. 6. As it can be seen in the diagram, the flow receives an inlet temperature of nearly 65 °C. A sharp drop is witnessed at the section of riser and some hundred meters away due to the change in the environment. The fluid soon cools down to the temperature of about 22 °C as of seawater at the seabed. Thus, the temperature of transporting fluid of studied cases is 22 °C and is equal to the seabed temperature for along more half-ending part of the uninsulated pipeline. At about 1 km away from riser b, we remark a slight increase of about 5 °C thanks to the gradual elevation of the pipeline from the seabed to the surface and FPSO, where the ambient temperature increases from 22 °C to 27 °C at the time of measure.

Compared to researches of Izwei et al. [7], Dong et al. [19] which presented the drop of approximate 20–30 °C through the burial pipeline, this rate is obviously high.

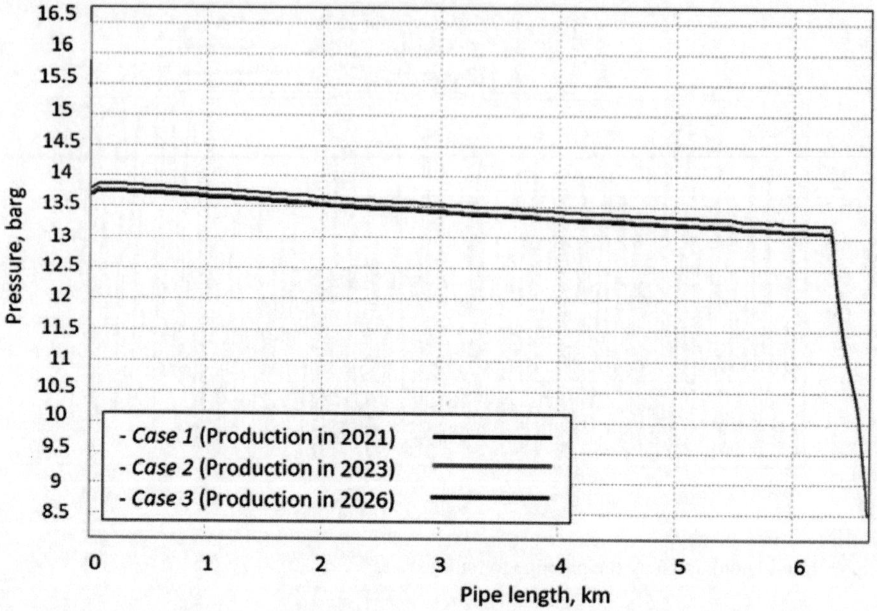

Fig. 5 Pressure profiles along the pipeline during steady state production

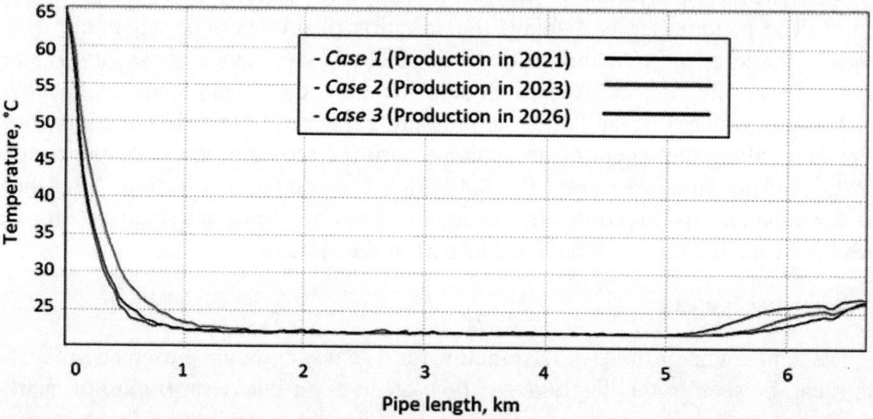

Fig. 6 Temperature profiles along the pipeline during steady state production for the study cases

The matter is that crude extracted from Pearl field is transported at a temperature which is 14 °C below its pour point temperature of 36 °C. Unquestionably, unless precise methods are utilized to change the properties of crude or to apply more heat to the pipeline, we might face disruption of the flow as a result of high thickening possibility [20]. Therefore, we propose using PPD method (pour point depressant) which implements chemical product to lower pour point temperature of the fluid to improve efficiency of the transport and eliminate relevant risk considering heat loss [21, 22].

4.2 *Transient Simulation Result and Analysis*

a. **Pipeline Shut-in**

During the operation of the pipeline, the system may pause due to undesirable or technical reasons. The protracted shut-in, especially with un-insulated pipeline, permits the transported fluid to cool down and provokes the crystallization of its waxy ingredients and the formation of gel-like accumulation along the inner side of the pipe. This may, consequently, increase critical restart pressure to restore the conveyance and require recalculating input parameters necessary for the restart. Therefore, the shut-down analysis needs to be considered. In our case, shut-in simulations are performed where Pearl pipeline is isolated and shut-in for a period of 24 h. A full-bore shutdown valve at the terminal FPSO has assumed to close in 1 min and Pearl wellhead production also reduces linearly from 100% production to zero in 1 min.

The temperature profile of fluid after 24 h of shut-in was plotted in Fig. 7. As it can be seen in the picture, for our uninsulated pipeline, temperature of fluid was quickly cooled to ambient temperature of about 22 °C after only 2 h shutting down the system. This confirms a rapid temperature drop compared to insulated pipeline systems as in study results of Miao [23]. With this rate, we believe to observe the wax formation after 20 h of shutdown as for Yang et al. [24] at sections where temperature declines promptly.

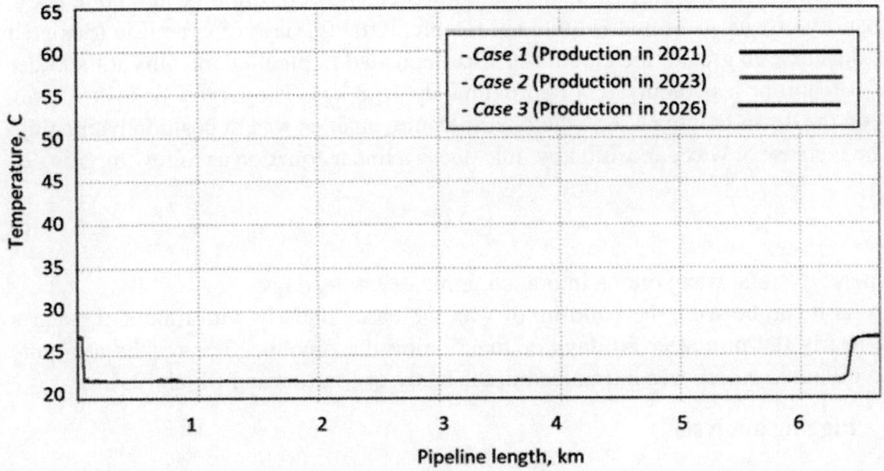

Fig. 7 Temperature of transported fluid after 24 h of shut-in

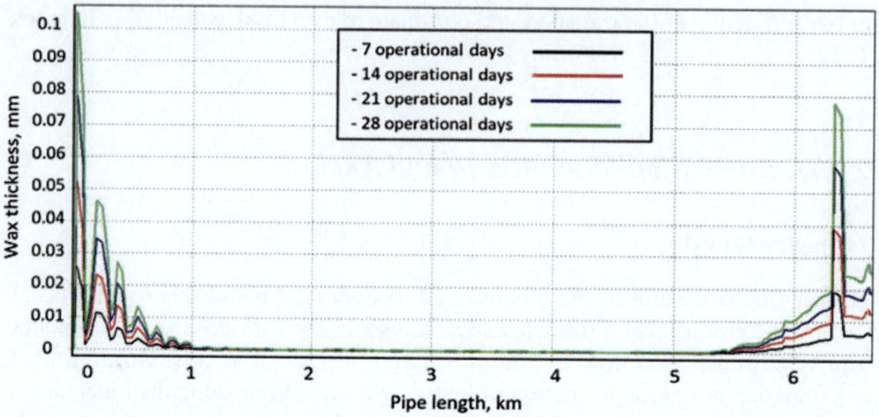

Fig. 8 Wax thickness along the pipeline after 7, 14, 21, and 28 days of operation

b. **Wax deposition**

The analysis of wax deposition is considered for the max flow case, case 2 (production predicted next year, 2023). This case is expected to be the case with highest rate of wax deposition, as it has the highest oil flow rate in the pipeline. Figure 8 displays the wax thickness along the pipeline after 7, 14, 21, and 28 days of operation. It can be seen that the wax only started to deposit at the parts of the pipeline where the system has cooled in temperature range to enable deposition, i.e., at the risers where the dramatic cool down happens. This result coincides with the large temperature differentials illustrated in Figs. 6 and 7 and the high concentrations of liquid at risers are presented in Fig. 4.

The wax thickness diagram also enables us to see that the accumulation gets higher and higher with the time of operation. However, the deposition occurs quite soon, quantity of wax measured is quite acceptable. After 28 days of operation (depicted as the curve in green), the amount of wax deposited in pipeline remains at the order of 0.1 mm or less (equivalent approximately 16.8 kg). The largest deposit of wax is on the down-coming riser, where temperature enables wax deposition happening. The increase of wax deposition by time obeys a linear equation as following (Fig. 9):

$$y = 0.0071x \tag{8}$$

where $y =$ total wax volume in branch, mm; $x =$ time, days.

As it can be seen, the build-up of wax increases linearly with time and reaches to nearly 0.4 mm after 60 days of functioning the pipeline. The results are quite accordant to which was introduced in the study of Martins and Elijah [25].

c. **Pigging analysis**

To face high risk of precipitation and flow redundant, we propose pigging analysis to verify its potency of clearance. In order to achieve this purpose, a pig was launched

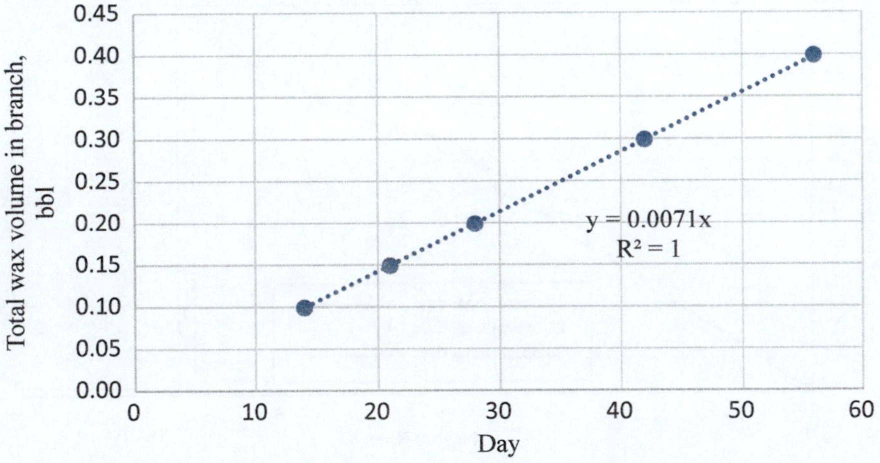

Fig. 9 Dependence of total wax volume in branch on the time

at the top of the Pearl, riser a, and caught at the FPSO, upstream of the flexible riser. The motive source of the pig is the Pearl production at the production rate for case 1 (100% of year 2021). The mass of the pig is assumed to be 4.5 kg and other specifications are assumed as OLGA defaults. Pigging simulations have been performed in this study for 3 situations in which there is one time run after 14, 28, and 42 days of operation correspondingly. The results of the simulation are illustrated in Figs. 10 and 11. It is shown that 100% production from Pearl field in 2021 does not produce a satisfactory velocity profile for the pig (average in range of 1 m/s). Moreover, pig velocity decreases when time interval of pigging increases due to the large of wax deposition in pipe. For case of pigging frequency is once a month, the pig took approximately 4.7 h to traverse the pipeline.

5 Conclusions and Recommendations

For both steady and transition states, crude is transported under its gelling point. After coming out of riser Pearl, the fluid almost immediately cools down to the ambient temperature of 22 °C and remains at this state all along the length of the submarine pipeline. In the case of shutdown, the temperature of the transported fluid drops dramatically about 30–40 °C at the two risers after only 2 h. The rate of wax deposition at normal steady state condition is relatively low. Wax is significantly witnessed in the down-coming risers where fluid cooled temperature range of max wax precipitation. After 30 days of operation, the max thickness of wax is about 0.1 mm. However, due to low temperature of oil transported in pipeline, it is recommended to maintain chemical depressant to lower pour point of Pearl crude to improve crude oil rheological properties and minimize impact of solid wax build-up.

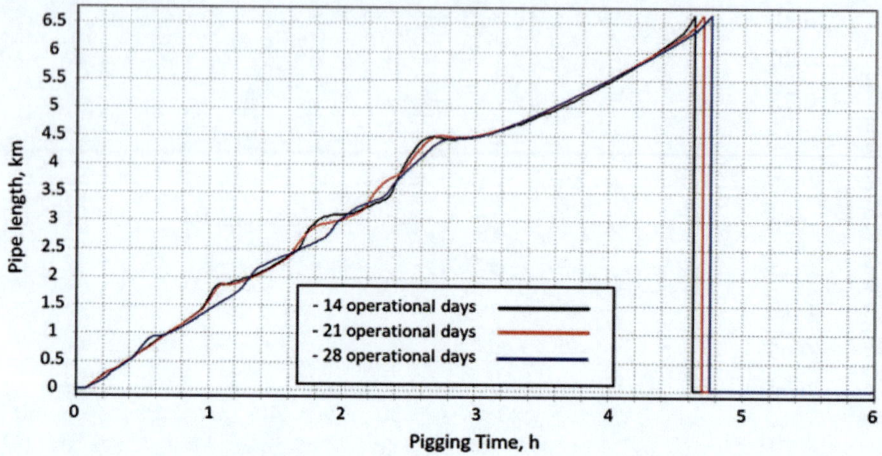

Fig. 10 Position of the pig for case 1 after 14, 28, and 42 days of operation

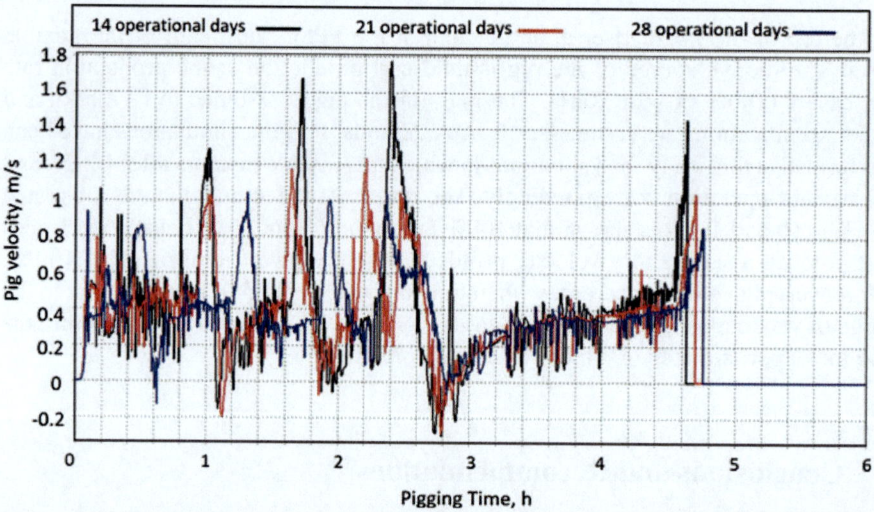

Fig. 11 Pig velocity during pigging at 100% production after 14, 28, 42 and days of operation

In order to ensure the efficiency of the transport, we propose to launch the pig more than one time a month. Apparently, it is possible to transport waxy crude with uninsulated pipeline. However, operational parameters should be precisely considered in our case.

Acknowledgements The authors are grateful for PetroVietnam Exploration Production Corporation (PVEP) supporting to provide invaluable information and appreciate the cooperation and exchange of simulation data.

References

1. Soliman, E.: Flow of heavy oils at low temperatures: potential challenges and solutions. Processing of Heavy Crude Oils - Challenges and Opportunities. IntechOpen (2019)
2. Farid, S., Safri, A., Benmounah, A.: A review on the rheology of heavy crude oil for pipeline transportation. Petroleum Research (2021)
3. Thuc, P.D., Bich, H.V., Son, T.C., Hoe, L.D.: The problem in transportation of high waxy crude oils through submarine pipelines at JV Vietsovpetro oil fields, Offshore Vietnam. J. Canadian Petroleum Technol. (2003)
4. Mohyaldinn, M.E., Husin, H., Hasan, N., Elmubarak, M.M., Genefid, A.M., Dheeb, M.E.: Challenges during operation and shutdown of Waxy crude pipelines. Processing of Heavy Crude Oils, Malaysia (2019)
5. Chala, G.T., Sulaiman, S.A.: Japper-Jaafar: Flow start-up and transportation of waxy crude oil in Pipelines-a review. J. Non Newtonian Fluid Mech. **251**, 69–87 (2018)
6. Lin, T.R., Guo, B., Song, S., Ghalambor, A., Chacko, J.: Offshore pipelines. Elsevier (2005)
7. Izwei, M.A.A., Sanoussi, N.S., Zuwa, S.A.A.: Study of inlet cargo temperature of crude oil in transportation pipeline at a critical section to avoid gel formation using explicit method. J. Eng. Res. Appl., 27–36 (2019)
8. Bern, P.A., Withers, V.R., Cairns, R.J.: Wax deposition in crude oil pipelines. OnePetro (1980)
9. Misra, S., Simanta, B., Kulwant, S.: Paraffin problems in crude oil production and transportation: a review. OnePetro (1995)
10. Bomba, J.G.: Offshore pipeline transport of waxy crude oils. OnePetro (1986)
11. Zhao, J., Dong, H., Wei, L., Zhou, G.: Research on heat transfer characteristic of waxy crude oil after oil pipeline shutdown. J. Thermal Analysis Calor., 487–508 (2017)
12. Bratland, O.: Pipe flow 2: multi-phase flow assurance (2010)
13. Gupta, A., Sircar, A.: Need of flow assurance for crude oil pipelines: a review. Int. J. Comput. Sci. Eng. **6**, 43–49 (2015)
14. Theyab, A.M.: Fluid flow assurance issues: literature review. SciFed J. Petroleum, 1–11 (2018)
15. Okoro, E.E., Rachael, J.E., Sanni, S.E., Emetere, M.E.: Liquid holdup measurement in crude oil transportation using capacitance sensors and electrical capacitance tomography: concept review. Conference Earth Environ., 1–14 (2021)
16. Ganat, T., Hrairi, M.: Effect of flow patterns on multiphase flow measurements in vertical pipes. J. Adv. Res. Fluid Mech. Thermal Sci., 17–26 (2018)
17. Tocci, F.: Assessment of a hybrid VOF two-fluid CFD solver for simulation of gas-liquid flows in vertical pipelines in OpenFOAM. Politecnico di Milano (2016)
18. Shi, S., Wu, X., Han, G., Zhong, Z., li, Z., Sun, K.: Numerical slug flow model of curved pipes with experimental validation. ACS Omega, 14831–14840 (2019)
19. Dong, H., Zhao, J., Zhao, W., Si, M., Li, J.: Study on the thermal characteristics of crude oil pipeline during its consecutive process from shutdown to restart. Case Studies in Thermal Engineering, 14 (2019)
20. Li, S., Huang, Q., Fan, K., Zhao, D.: Transportation technology with pour point depressant and wax deposition in a crude oil pipeline. Petroleum Sci. Tech., 1240–1247 (2016)
21. Yu, H., Sun, Z., Jing, G., Zhen, Z., Liu, Y., Guo, a.K.: Effect of a magnetic nanocomposite pour point depressant on the structural properties of Daqing waxy crude oil. Energy Fuels, 6069–6075 (2019)
22. Cao, D., Li, C., Yang, F., Li, H.: Effect of drag reducer and pour point depressant on the wax deposition of crude oil pipeline. Petroleum Sci. Technol., 1277–1284 (2017)
23. Miao, Q.: Numerical study on the effect of wax deposition on the restart process of a waxy crude oil pipeline. Adv. Mech. Eng. (2012)
24. Yang, M., Jing, J., Lei, L., Tan, J., Chen, J.: Simulation testing on the shutdown and safe restart of crude oil pipelines. ICPTT 2011: Sustainable solutions for water, sewer, gas, and oil pipelines, 1454–1464 (2011)
25. Martins, O., Elijah, P.T.: Dynamic modeling and prediction of wax deposition thickness in crude oil pipelines. J. King Saud University-Engineering Sci., 437–445 (2021)

Detection of Underground Anomalies by Evaluation of Ground Penetrating Radar Attribute Combination

Duy Hoang Dang, Cuong Van Anh Le, and Thuan Van Nguyen

Abstract Detection of underground utilities can be done by application of the ground penetrating radar (GPR) method. Diffraction or reflection phenomena of high-frequency electromagnetic waves provide tools and physics principles for imaging their locations, sizes, and even shapes. In urban zones, GPR machines equipped with shielded antennas are often used in common offset recording settings which provide high quality data after filtering most of artificial noises from the air. Although the data is still contaminated by different noises, the underground works are often recognized through appearances of different hyperbolae. In our paper, we apply different processing techniques for researching underground anomalies to map 3D metal gas pipes in Ba Ria–Vung Tau province, Vietnam. We have collected many 2D GPR profiles in 2019 and conducted a new technique of their 3D representation. In the new 3D representation, different GPR attributes as processed data and its maximum energy difference attribute are applied for defining the two metal gas pipes and revealing their other nearby anomalies as optical fiber cables.

Keywords Ground penetrating radar · Migration · Energy difference · Ba Ria–Vung Tau (Vietnam)

1 Introduction

Ground penetrating radar (GPR) uses diffraction and reflection of high-frequency electromagnetic waves to understand characteristics of underground anomalies and shallow geology structures which are beneficial in many areas from civil engineering, ground water prospecting [1–9]. The electromagnetic waves propagate from the

---

D. H. Dang · C. Van Anh Le (✉) · T. Van Nguyen
University of Science, Ho Chi Minh City, Vietnam
e-mail: lvacuong@hcmus.edu.vn

T. Van Nguyen
e-mail: nvthuan@hcmus.edu.vn

Vietnam National University Ho Chi Minh City, Ho Chi Minh City, Vietnam

L. Q. Nguyen et al. (eds.), *Advances in Geospatial Technology in Mining and Earth Sciences*, Environmental Science and Engineering,
https://doi.org/10.1007/978-3-031-20463-0_31

transmitting antenna, bounce back at boundaries of soil layers having different electric permittivity and go to the receiver antenna. In receiving information, waveforms can have propagation time and amplitude characteristics showing contrasts of the electric permittivity. Analysis of the dynamic information can provide the distribution of environment velocity, locations of underground anomalies (i.e., water pipes, sinkholes), and shallow geology structures. Unlike other low frequency electromagnetic methods, GPR has shallow investigated depth from meters to tens of meters because of the "skin-depth" effect.

Diffraction hyperbole phenomena are commonly well-known helping to detect positions of an underground anomaly in GPR slice in 2D dimensional (distance–time).

Behaving like seismic vibration in mechanic theory, original high-frequency electromagnetic waves can be computed via seismic attributes' equations for better interpretation [10, 11]. GPR attributes can enhance more features than the original processed GPR data do just by using naked eyes. More research has focused on applying of GPR attributes in different areas as civil engineering, archeology, and environment disasters. For example, Zhao et al. [6] have used GPR textural attributes to detect remains of the Bronze Age as well as geology structures. For finding sink holes in old mining site, GPR attributes were suggested [12]. In the civil engineering, research papers showed the application of GPR attributes in mapping underground utilities (i.e., water/electric pipes or tunnels) [4, 13, 14]. Hidden water tunnels are better targeted through the attributes in an underground utilities mapping project, Ho Chi Minh city [13]. Moreover, GPR attributes can provide tools to evaluate propagation wave velocity [4, 14].

In our research, we are tempted to map two metal gas pipes in Ba Ria–Vung Tau province, Vietnam by applying different GPR attributes. Moreover, different nearby objects (i.e., optical fiber wires) can be systematically seen by our innovative 3D visualization approach.

2 Study Area

The study area is Long Hai, Ba Ria–Vung Tau province, Vietnam (Fig. 1). Ba Ria–Vung Tau is one of the most powerful economic Southern Vietnam areas.

The setup of a new gas pipe from offshore to areas of Ba Ria–Vung Tau province needs mapping old gas pipes for preventing any duplicate installed locations. There are specific land spaces for the old pipes, but they have also crossed urban areas. In this case, it is necessary to detect possible underground anomalies that may interfere the pathway of the new pipeline. We have set up sixteen GPR profiles across the pathway of the existed pipes in light of 3D visualization of possible underground utilities. In Fig. 1, there are sixteen GPR lines (Profiles 74–90). The distance between two nearby lines is around 50 m. The GPR profile length can reach 20 m depending on the field trip situation.

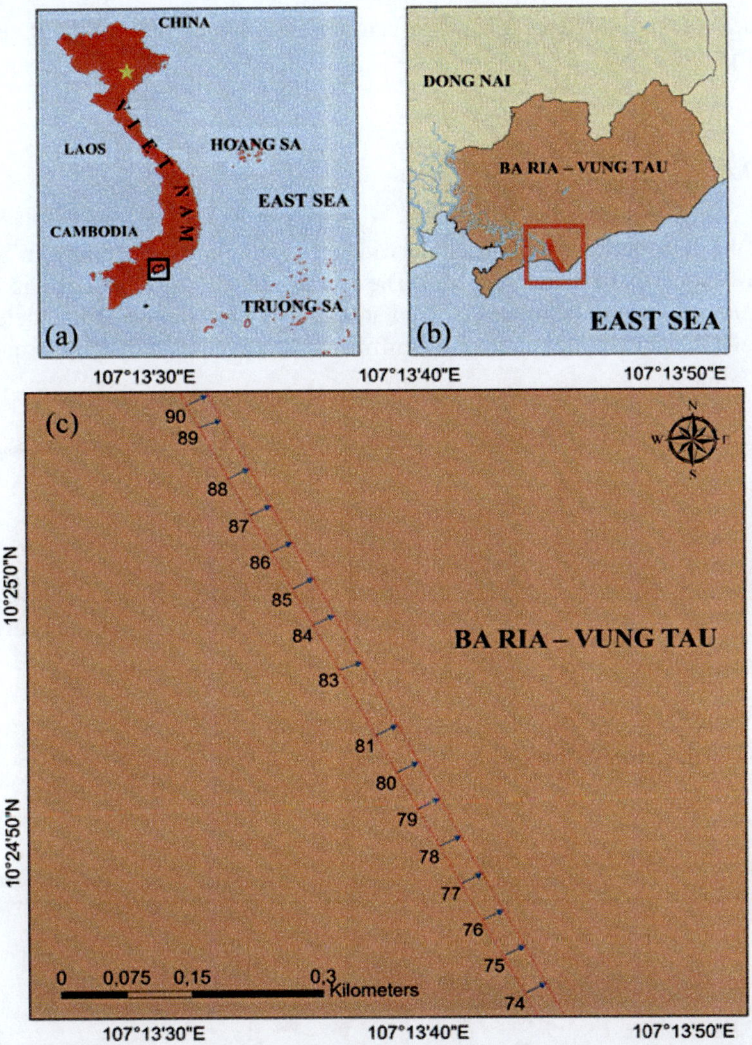

Fig. 1 Location of the survey area in Ba Ria–Vung Tau (**b**), Vietnam map (**a**), and one example of GPR profiles (**c**). The sixteen GPR profiles are set up for detecting the pathway of the existed gas pipes

3 Methodology

Common offset GPR data measurement is used in our research. Interpretable data is conventionally processed from the raw data. Then, we apply different visualization techniques to research the processed data in many aspects. Many different 2D across the pathway of the pipes can assume that their intersection points can play a role as diffractors. Note that each diffractor can present a hyperbola in a GPR section [4, 13–15].

3.1 Data Collection

We use the IDS duo detector machine with the common offset setting in which 700 MHz frequency data are processed. The average 2D GPR profile length is 18 m, and the time sample on its traces is fixed at 0.25 ns. One example is illustrated on how the GPR data are measured and a diffractor is proven as a hyperbola in GPR slice (Fig. 2).

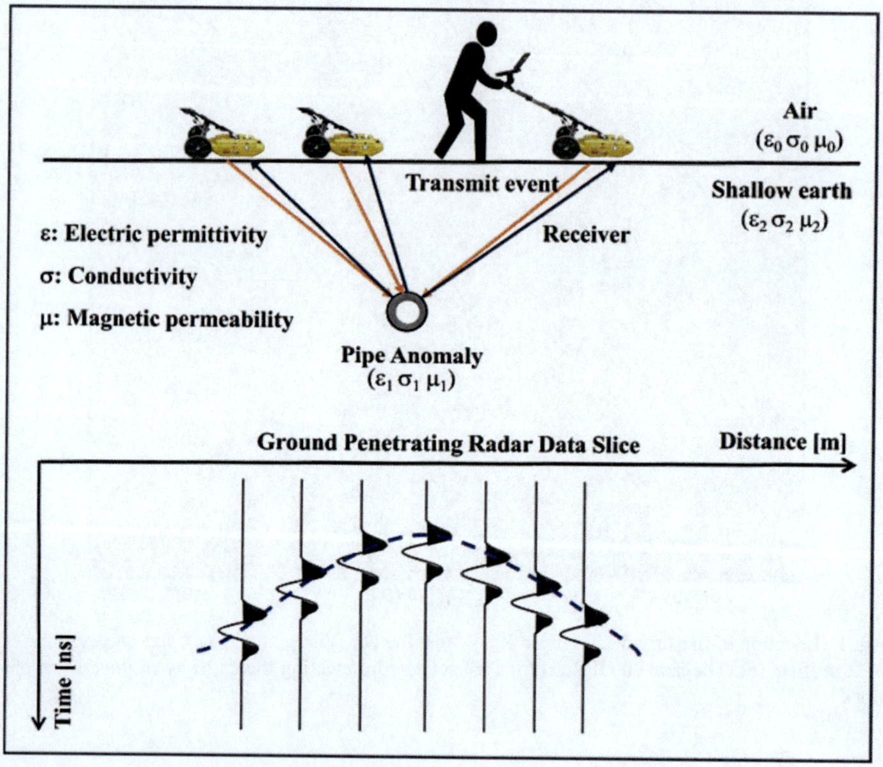

Fig. 2 Graphical figure showing how a GPR machine collected data. The common offset setting is applied to the geophysical measurement in our research

3.2 Data Processing and Analysis

The reasons that make raw data uninterpretable are the energy loss and noise contamination during electromagnetic wave propagation. We have applied three stages for data processing analysis:

- Stage 1: In this stage, the conventional workflow includes (i) importing raw data, (ii) move-start time, (iii) bandpass Butterworth filter, and (iv) gain function that are applied to the raw data.

Four main steps are shown for transforming raw data into interpretable processed data (see Fig. 3). The first step is to import the data with format *.dt (recorded from the GPR machine IDS duo detector) (see Fig. 3a). In the second step, a time-shift constant is applied for all GPR datasets that the travel-time of electromagnetic waves can be correctly defined (see Fig. 3b). Bandpass Butterworth filter can allow available frequency content ranging from 1/2 main frequency to 3/2 main frequency (see Fig. 3c) [4]. Gain function helps to compensate data loss when waves travel down to the ground (see Fig. 3d).

- Stage 2: We use the maximum energy difference technique [4] to derive migrated data with the correct velocity. Cuong Le suggests the computational workflow of the maximum energy difference attribute [4]. Moreover, the maximum energy difference can be a supportive GPR attribute in detecting underground anomaly. Figure 4 illustrates one example of making the maximum energy difference section.

The input of this Stage 2 is the processed data from Stage 1, known as the unmigrated data (see Fig. 4a). We have used the Kirchhoff migration technique for all the GPR data with the velocity band from 0.07 to 0.14 m/ns. The migrated datasets (see Fig. 4b) are input for calculating the maximum energy difference section (see Fig. 4c) sharing the same time-distance space as the unmigrated data. Figure 4d shows energy values in the position $x = 6$ m, $t = 0.4 \times 10^{-7}$ s from all the migrated sections versus velocity. Then, the maximum energy difference value in the position is the absolute value of difference between its energy extremes. The calculation procedure is applied to all the positions in the unmigrated data to form the maximum energy difference section.

Combination of the three GPR sections (i.e., unmigrated data, migrated data, and the max-energy difference) can quickly detect underground anomalies. For example, a hyperbola represents an anomaly through diffraction phenomenon, while migrated data provide a tool to calculate environment velocity and the anomaly size. Moreover, maximum energy difference of anomalies is distinguished around their nearby zones (i.e., the high red values in Fig. 4c).

- Stage 3: We have imported all the GPR data in real coordinates with 3D visualization. In this stage, we have interpreted each GPR slice to have key underground anomalies such as the Nam Con Son and Bach Ho pipes. After having the key

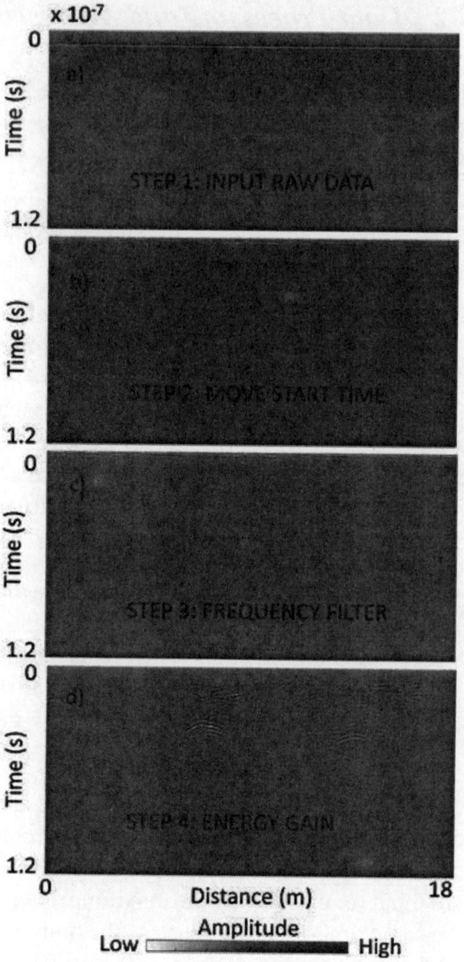

Fig. 3 Processing steps for GPR processing in one example of GPR data profile in Nam Con Son location

coordinates, we set up a new survey setting to build a 3D format survey in which all the GPR profiles are parallel and the locations of each key NCS anomalies have the fixed x coordinate of 0 but different y coordinates (see Fig. 5). The purpose of the new setting is to discover other hidden objects that locate nearby the key object. The hidden objects may have some features systematically connected in 3D view by using GPR attributes.

The tool used for 3D visualization is the OpendTect software [16, 17]. The software is convenient in converting true coordinates into inline, crossline, depth spaces that can depict any 3D object with many flexible color spectrums.

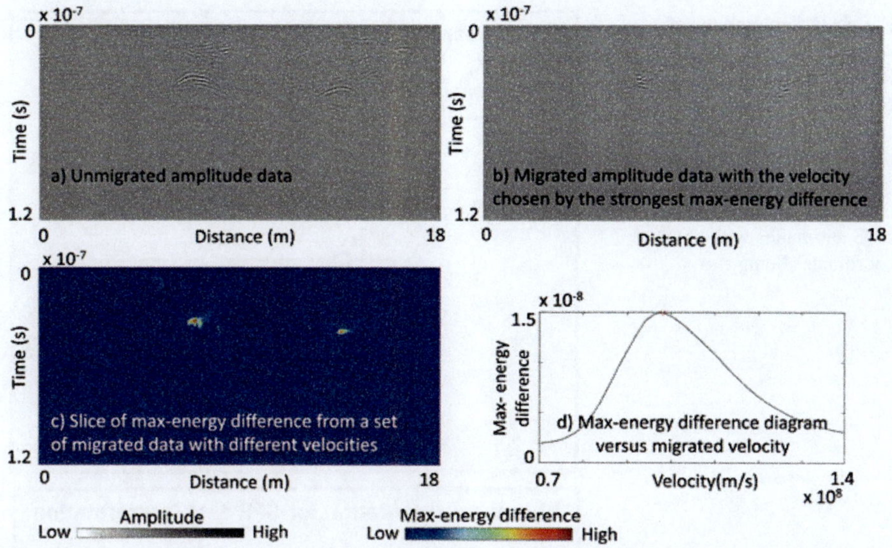

Fig. 4 Maximum energy difference processing in one example of the GPR data profile in Nam Con Son location. **a** The unmigrated processed data from Stage 1 (see Sect. 3.2), **b** migrated data with the velocity chosen by the strongest max-energy difference, **c** slice of maximum energy difference, and **d** the diagram showing max-energy difference versus velocity in the position $x = 6$ m, $t = 0.4 \times 10^{-7}$ s

- Modeling GPR for testing in-phase and out-of-phase case:

We have modeled two examples in which the environment a metal pipe and a sink hole filled by air play a role as diffractors. The main purpose of this modeling task is to investigate how amplitude changes when electromagnetic waves hit a diffractor and bounce back to the antenna receiver. The open source software, gprMax [18], is used to produce synthetic electromagnetic wave data. One model has a metallic pipe and the other has a sink hole filled by air (Fig. 6). Their model sizes are the same with the maximum depth and width of 0.21 m and 0.48 m, respectively. The circular anomalies whose centers locate at the depth 0.15 m and the horizontal position of 0.24 m have big and small radiuses as 0.02 m and 0.015 m, respectively. The relative electric permittivity values (ε) for air, ground, and metal are 1, 6, infinity (perfect conductor [18]), respectively. Note that the two objects contain air inside.

When modeling, the GPR machine with its center frequency of 1.5 GHz moves through all horizontal locations with a step of 0.002 m. In two synthetic GPR data, two diffraction hyperbolae can express the existence of the two anomalies. For the metal case (Fig. 6, left), the out-of-phase characteristics can be seen when comparing waveforms of the first direct event signal and the peak of diffraction hyperbola. Otherwise, the sinkhole case refers to the in-phase characteristics when two waveforms of the first event signal and the peak of diffraction hyperbola are similar (Fig. 6, right). If the electric permittivity of an anomaly is bigger than the surrounding environment one, an out-of-phase phenomenon will occur; otherwise, we have an in-phase

Fig. 5 Different types of survey representation. a) Representation of field trip coordinates and b) New strategy of survey representation. In the new representation, we relocated the new locations of all key NCS anomalies with the x coordinates being zeros

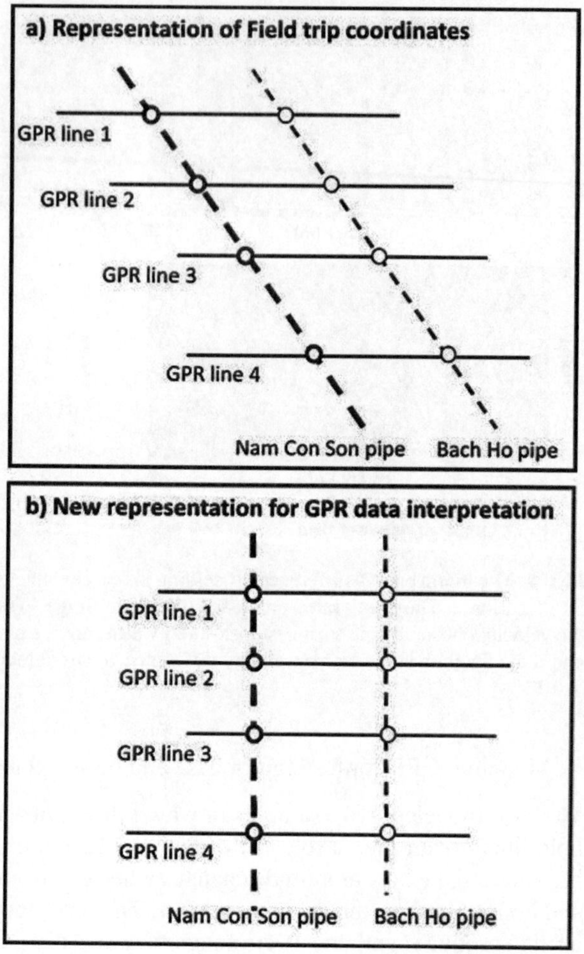

phenomenon [19]. Then, we can apply the phase variation [19] in predicting anomaly material in some extent.

4 Results and Discussion

We can interpret existence of two metal anomalies at the Nam Con Son and Bach Ho metal pipes from many GPR profiles.

For detecting the metal material of the pipes, we investigate the out-of-phase characteristics of two hyperbolae in the GPR slice (Fig. 7). We can see that inversion of amplitudes in the positions of the first arrival time of the hyperbolae shows abrupt

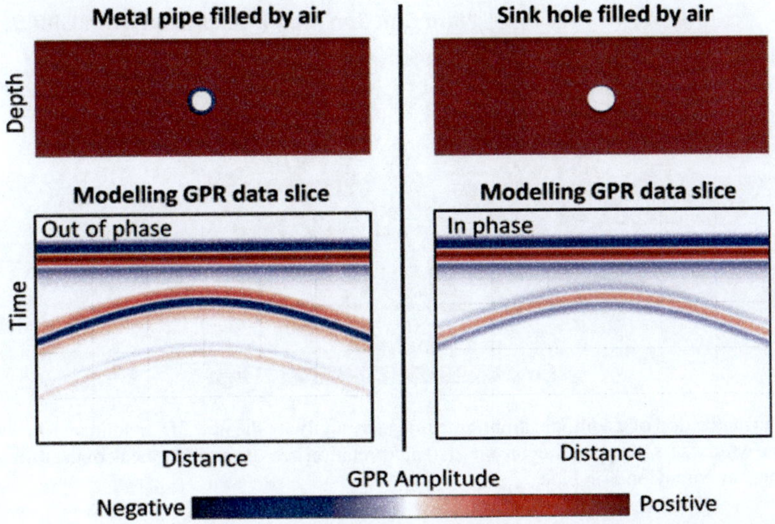

Fig. 6 Modeling GPR for testing in-phase (left image) and out-of-phase cases (right image)

changes of electrical permittivity in the anomaly objects comparing to the environment surroundings. Also, according to the prior information, we can confirm that the objects are metallic.

For revealing existences of potential objects, we have applied the new strategy (see Fig. 5) of performing GPR profiles in new axis where the positions of the Nam Con Son pipe share the same inline index value (Fig. 8). We can find out that all the strongest maximum energy differences in Fig. 9 also refer as the positions of

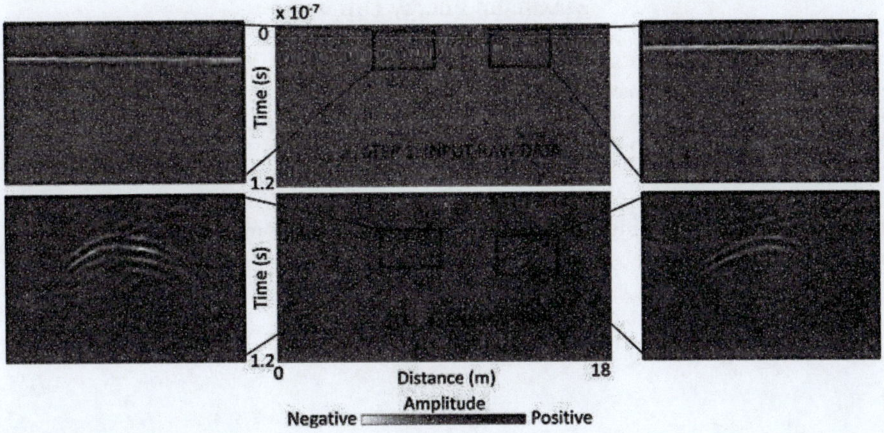

Fig. 7 Amplitude analysis for the detection of metal characteristics of two pipes of Nam Con Son and Bach Ho. The out-of-phase case type is defined in the two zooming images of the diffraction hyperbolae

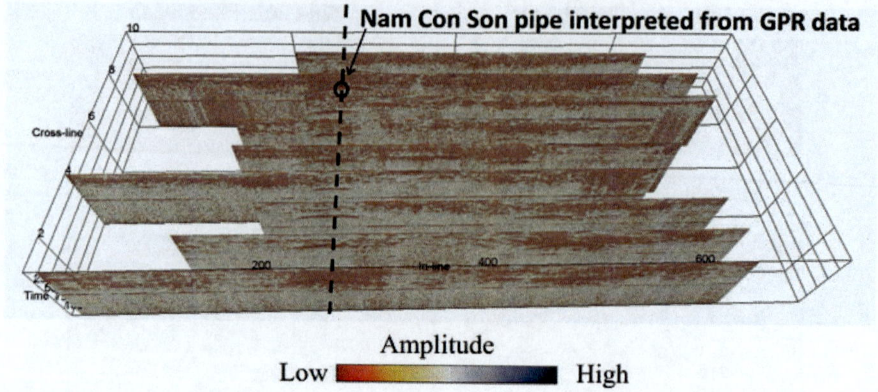

Fig. 8 Visualization of traditional amplitude image by applying the new 3D technique. The dashed black line which is fixed depending on our first interpretation goes through the peak of the diffraction hyperbolae as Nam Con Son pipe

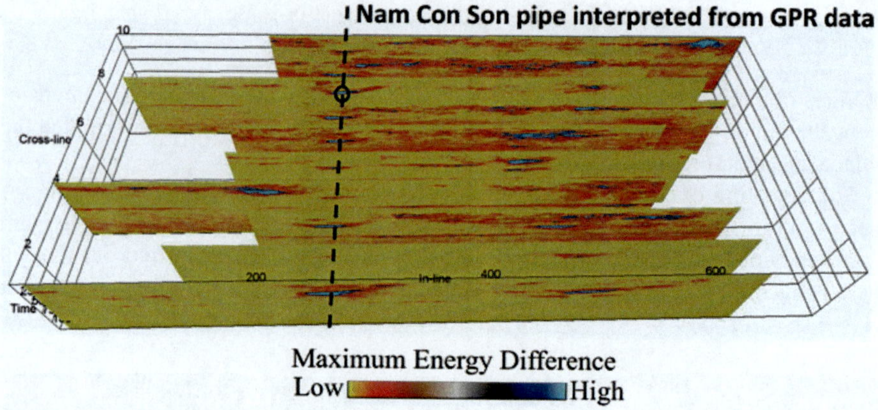

Fig. 9 3D visualization of maximum energy difference attribute image by applying the new 3D technique. All the strongest maximum energy difference values in Fig. 9 also refer as the positions of the Nam Con Son pipe which are in the dashed black line

the Nam Con Son pipe. Figures 9 and 10 also represent the interpretation of the two pipes with the amplitude and maximum energy difference sections.

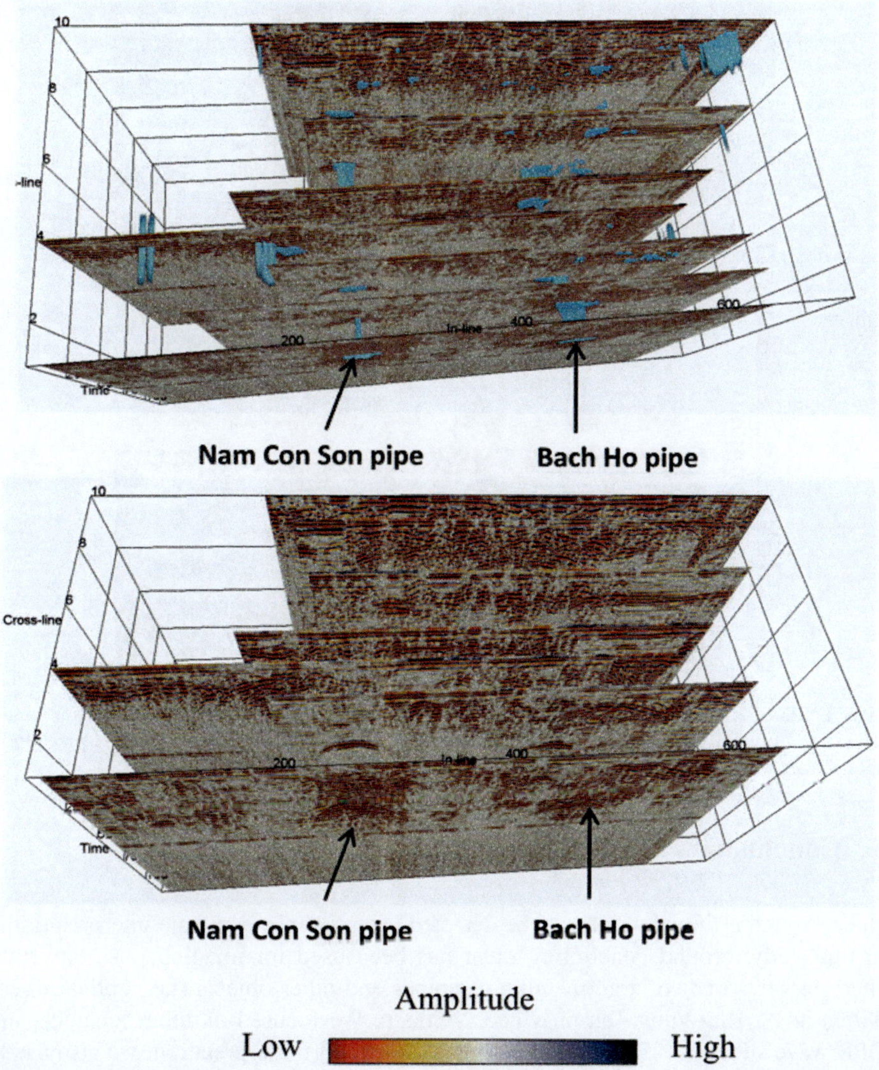

Fig. 10 3D rendering volume of the migration amplitude energy to see the 3D voxel volume of the NCS or other nearby objects. In the top image, strongest maximum energy difference values in cyan zones refer to the anomalies. The below image shows migrated GPR sections

For clear 2D interpretation, we also show some GPR lines to detect potential anomalies. A group or anomalous zones of hyperbolae appear systematically in the GPR sections (Fig. 11). This can be explained as the optical fiber wire system. Its footprint is seen in the field trip record (Figs. 11 and 12).

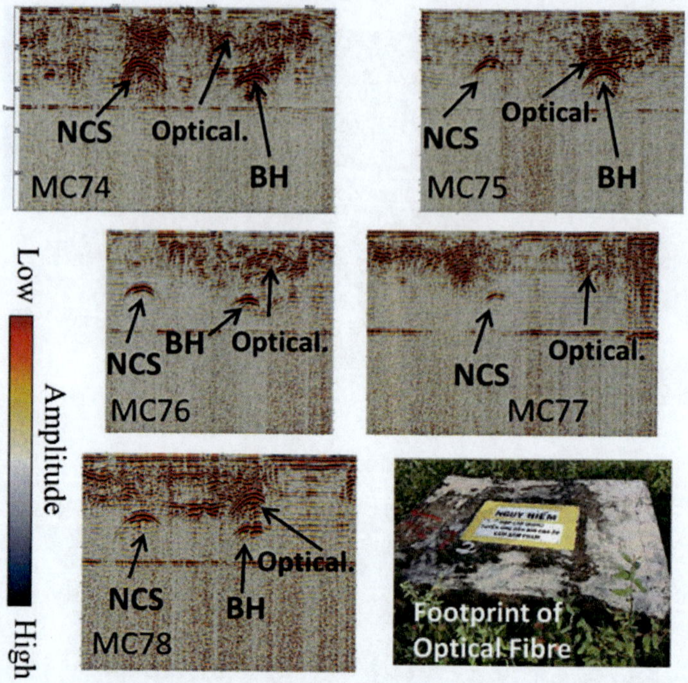

Fig. 11 2D GPR data illustration of underground objects for the 2D profiles MC74 to MC78 and footprint of optical fiber existence. The black arrows show locations of NCS pipe, BH pipe, and optical fiber wire

5 Conclusion

Underground artificial works can be detected using geophysical data interpretation. In this study, ground penetrating radar has been used for imaging positions and characteristics of two large metallic gas pipes and other objects (i.e., optical fiber wires) in Ba Ria–Vung Tau province, Vietnam. We focused on their behaviors in GPR wave slices as diffraction hyperbolae. To support interpretation, we also used their attributes as processed amplitude and maximum energy difference to locate these artificial targets. Moreover, we suggested using new representation of these GPR slices through defining the prior easily seen target as Nam Con Son pipe to investigate other nearby underground objects.

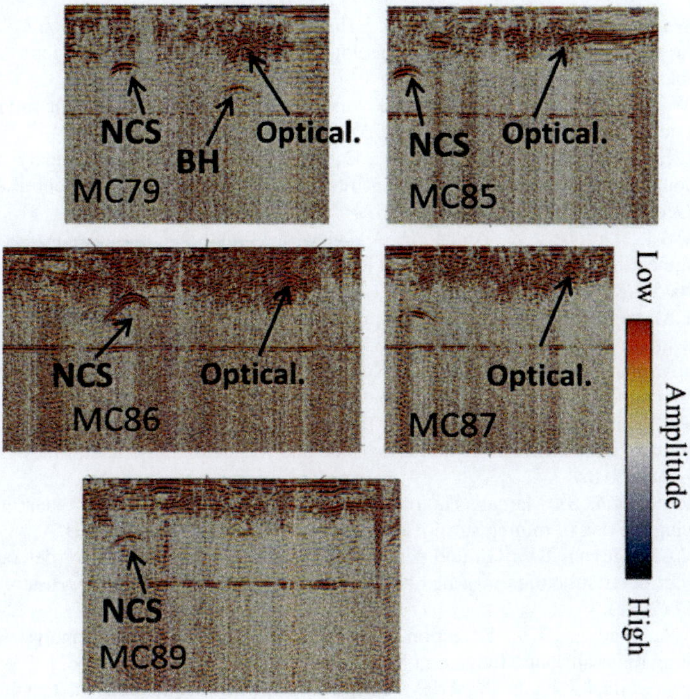

Fig. 12 2D GPR data illustration of underground objects for the 2D profiles MC79, MC85, MC86, MC87, and MC89. The arrows show locations of NCS pipe

Acknowledgements This research is funded by Vietnam National University Ho Chi Minh City (VNU-HCM) under grant number C2022-18-06. We would like to thank Mr. Doan Manh Cuong for his help.

Conflicts of Interest The authors declare no conflict of interest.

References

1. Liner, C.L., Liner, J.L.: Application of GPR to a site investigation involving shallow faults. Lead. Edge **16**, 1649–1651 (1997)
2. Nguyen, V.G., Marquis, G., Le, M.: EM and GPR investigations of contaminant spread around the Hoc Mon waste site, Vietnam. Acta Geophys. **58**, 1040–1055 (2010)
3. Nguyen, V.G., Ziętek, J., Nguyen, B.D., Karczewski, J., Gołębiowski, T.: Study of geological sedimentary structures of Mekong river banks by ground penetrating radar: forecasting avulsion-prone zones. Acta Geophysica Polonica **53**, 167–181 (2005)
4. Nguyen, T.V., Le, C.V.A., Nguyen, V.T., Dang, T.H., Vo, T.M., Vo, L.N.L.: Energy analysis in semiautomatic and automatic velocity estimation for ground penetrating radar data in urban areas: case study in Ho Chi Minh City, Vietnam. In: International Conference on Geo-Spatial Technologies and Earth resources. Springer, Ha Noi, Vietnam (2017)

5. Zhao, W., Forte, E., Levi, S.T., Pipan, M., Tian, G.: Improved high-resolution GPR imaging and characterization of prehistoric archaeological features by means of attribute analysis. J. Archaeol. Sci. **54**, 77–85 (2015)
6. Zhao, W., Forte, E., Pipan, M.: Texture attribute analysis of GPR data for archaeological prospection. Pure Appl. Geophys. **173**, 2737–2751 (2016)
7. Ercoli, M., Pauselli, C., Frigeri, A., Forte, E., Federico, C.: 3-D GPR data analysis for high-resolution imaging of shallow subsurface faults: the Mt Vettore case study (Central Apennines, Italy). Geophys. J. Int. **198**, 609–621 (2014)
8. Lai, W.W.-L., Derobert, X., Annan, P.: A review of ground penetrating radar application in civil engineering: a 30-year journey from locating and testing to imaging and diagnosis. NDT E Int. **96**, 58–78 (2018)
9. Rashed, M.A., Al-Garni, M.A.: On the application of GPR for locating underground utilities in urban areas. Arab. J. Geosci. **6**, 3505–3511 (2013)
10. Le, C.V.A., Harris, B.D., Pethick, A.M.: New perspectives on solid earth geology from seismic texture to cooperative inversion. Sci. Rep. **9**, 14737 (2019)
11. Le, C.V.A., Harris, B.D., Pethick, A.M., Takam Takougang, E.M., Howe, B.: Semiautomatic and automatic cooperative inversion of seismic and magnetotelluric data. Surveys Geophys. **37**, 845–896 (2016)
12. Tomecka-Suchoń, S., Marcak, H.: Interpretation of ground penetrating radar attributes in identifying the risk of mining subsidence. Arch. Min. Sci. **60**, 645–656 (2015)
13. Le, C.V.A., Nguyen, T.V.: Ground penetrating radar attributes analysis for detecting underground artificial structures in urban areas, Vietnam. Lowland Technology International **22**, 249–257 (2020)
14. Le, C.V.A., Nguyen, T.V.: Detection of underground anomalies using analysis of ground penetrating radar attribute. Inżynieria Mineralna J. Polish Mineral Eng. Soc. **1**, 23–34 (2020)
15. Prego, F.J., Solla, M., Puente, I., Arias, P.: Efficient GPR data acquisition to detect underground pipes. NDT E Int. **91**, 22–31 (2017)
16. Huck, H.: The road to open source: sharing a ten years' experience in building OpendTect, the open source seismic interpretation software. In: 74th EAGE Conference and Exhibition, Copenhagen, Denmark (2012)
17. dGB Earth Sciences: OpendTect dGB Plugins User Documentation version 4.6 (2015)
18. Warren, C., Giannopoulos, A., Giannakis, I.: gprMax: open source software to simulate electromagnetic wave propagation for Ground Penetrating Radar. Comput. Phys. Commun. **209**, 163–170 (2016)
19. Park, B., Kim, J., Lee, J., Kang, M.-S., An, Y.-K.: Underground object classification for urban roads using instantaneous phase analysis of ground-penetrating radar (GPR) data. Remote Sens. **10**, 1417 (2018)

Dynamic Failure Process of Soil Particles at the End of Shield Tunnel Based on Discrete Element

Zheyuan Feng, Bin Liu, Sijun Zhang, Fei Kang, Haipeng Hui, and Qiang Tang

Abstract The soil at the end of the shield tunnel has large local deformation and strong nonlinearity; thus, the finite element method based on continuum mechanics cannot simulate the failure process of the soil at the end well. In order to analyze the dynamic failure process of soil particles at the end, a discrete element model of shield tunnel was established; the displacement, stress and settlement of soil particles were tracked, and the failure mechanism was analyzed. The results show that the end soil finally forms a circular arc shear zone with a slip radius of 0.87 m. The contact stress of the soil at the entrance shows a decreasing trend with time. Nonlinear settlement occurs on top of soil, and the final settlement reached 38.20 mm. The soil failure at the end of the shield tunnel presents a gradual development of slip failure characteristics. The soil slip failure surface at the end of the shallow-buried shield tunnel is a combined slip mode.

Keywords Discrete element numerical analysis · Shield tunnel · Soil particles · Subsidence

1 Introduction

At present, human beings are facing four major challenges of population, food, resources, and environment [1, 2]. Under this severe situation, many countries in the world have taken "this century as the century of using underground space" as their

---

Z. Feng · Q. Tang (✉)
School of Rail Transportation, Soochow University, Suzhou 215131, China
e-mail: tangqiang@suda.edu.cn

Z. Feng
e-mail: Fengzheyuan1999@sina.com

B. Liu · S. Zhang · F. Kang · H. Hui
NO. Eight Engineering Co., Ltd. of CCCC First Highway Engineering Co., Ltd, Tianjin 300171, China

CCCC Tunnel Engineering Bureau Co., Ltd, Suzhou 215004, China

L. Q. Nguyen et al. (eds.), *Advances in Geospatial Technology in Mining and Earth Sciences*, Environmental Science and Engineering,
https://doi.org/10.1007/978-3-031-20463-0_32

national policy [2]. With the increasingly serious problems of living space congestion and traffic congestion caused by the explosive growth of urban population, the development and use of underground space has become more and more urgent [3–5]. Shield tunnels account for a considerable proportion of urban underground space development [6, 7]. The use of end reinforcement is an important technical link in the origination and reception of shield machines, and the success or failure of end reinforcement directly affects whether the shield machines can safely originate and receive [8, 9]. Therefore, the reasonable selection of the end reinforcement construction method is a very important technical measure to ensure the smooth construction of the shield [10]. With the development of the industry, the reinforcement forms of shield tunnel end are gradually enriched, including deep mixing method, soil mixing wall method, and grouting method. End reinforcement refers to improving the strength and self-stabilizing ability of the soil at the end by improving the soil at the end. In order to ensure the safety of the tunnel door, prevent the collapse of the tunnel door and stabilize the deformation of the stratum at the end when the enclosure structure is broken. It is necessary to ensure the starting of the shield and receiving security [11–14].

Researchers mainly rely on the empirical method to determine the reinforcement scheme of the end soil [15–18]. There is no systematic theory and quantitative standard to determine the reinforcement scheme in actual engineering. On the one hand, it will affect the economy of the end soil reinforcement. On the other hand, there are also hidden dangers for the security of the reinforcement scheme. Based on this foundation, this paper intends to carry out numerical simulation of end reinforcement, to draw certain conclusions, and bring inspiration to the solution of practical engineering problems of safety and economy.

With the application and development of shield tunneling technology, the above research results have played an important role in analyzing the failure mechanism of the end soil and ensuring the stability of the end soil. However, due to the complexity of stratum and environmental conditions, engineering accidents also occur from time to time. In order to further study the failure process and slip failure mode of the end soil, based on the previous research, this paper analyzes the particle flow of the end soil, in order to provide a reference for the analysis of the soil stability at the end of the shield tunnel.

2 Particle Flow Model

2.1 Theoretical Analysis

It is reasonable for two-dimension disk particles to simulate soil medium particles because the soil is a discontinuous medium. According to the principle of discrete element method, soil particles can be in close contact or detached. Loading will change the position of soil particles and the change of soil particle contact force.

Each calculation step of the program will automatically identify and dynamically update the particle's position and contact force and reach a new equilibrium state. The contact between soil particles adopts a linear contact bonding model, and the contact relationship between particles depends on the contact stiffness, bonding, and slippage [19].

The contact force composition between particles is given by Eq. 1.

$$F_c = F^1 + F^{\mathrm{d}} \tag{1}$$

where F_c represents the contact force between particles, F^1 represents the linear elastic (non-tensile) friction behavior, F^d represents the sticky behavior.

The linear force is generated by the normal and tangential stiffness of the linear spring. The damping force is generated by the damping with viscosity according to the normal and tangential critical damping ratio, and the linear force is parallel to the damping force. The slip between particles is achieved by setting the Coulomb limit shear force, and the friction factor is used after the slip.

The condition for slippage between particles is Eq. 2.

$$F_s^1 > F_{s\,\max}^1 = \mu |F_n^1| \tag{2}$$

where $F1\ s$ represents linear tangential force which is updated by relative shear displacement increments, $F\,s\mathrm{max}$ represents the maximum value of the linear tangential force, μ represents the friction factor, F_n represents linear normal force which is updated based on the absolute amount of surface clearance.

2.2 *Model Dimension*

Based on 25 years' experience in the development and application of discrete element method in manufacturing and materials, civil engineering, mining, oil and gas exploitation and transportation, power generation and other industries, Itasca Consulting Co., Ltd. has re structured and designed the particle flow program PFC developed based on discrete element method and launched PFC5.0. PFC5.0 has undergone innovative changes in modeling convenience, complexity of problem processing, preprocessing and post-processing capacity, computing speed, interface visualization, etc. Using PFC5.0 as a research tool for granular media problems, the modeling is simpler and more efficient; the authenticity is higher, and the post-processing is more powerful. It can deal with more complex large deformation problems (including disengagement) and observe the fracture and crushing process more simply and accurately.

In this paper, taking a section of Suzhou Metro as an example, the particle flow program PFC5.0 is used to establish the numerical model of the end soil. Generates model particles with a uniform distribution of the minimum and maximum radii of the given particles and the specified porosity. In order to clearly and intuitively

Fig. 1 Model size drawing

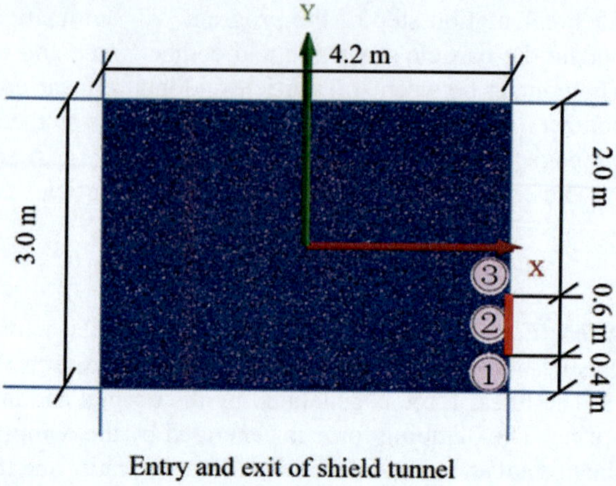

observe the soil displacement at the end and the generated slip shear zone, the soil was dyed to form a mesh after adding its own weight. Finally, the progressive failure model of the soil at the end of the shield tunnel is 4.2 m long, 3.0 m high, and 1.0 m deep. The diameter of the tunnel is 0.3 m, and the total number of particles is 16.709. Three monitoring points are set at different positions of the tunnel door of the shield tunnel to observe the stress change of the soil at the end with time. The layout of the monitoring points is shown in Fig. 1.

In order to further study the effect of end reinforcement, the deformation process of soil at the end of shield tunnel is simulated by particle flow discrete element software. In the simulation, the parameters are calibrated by uniaxial or biaxial compression tests.

In the modeling process, ball-ball contact and ball-facet contact will run through all the time, so the linear model is selected as the contact stiffness model for this simulation. The contact particle is simplified as an elastic beam whose two ends are in the center of the particle. If particle A and particle B contact, the radius of the beam is

$$\bar{R} = \frac{R^{(A)} + R^{(B)}}{2} \tag{3}$$

where $R^{(A)}$ and $R^{(B)}$ are the radius of the contact particles, respectively.

The stress characteristics between particles in contact with each other are equivalent to the pure axial or tangential load on the end of the elastic beam. The cross-sectional area and moment of inertia of the beam are

$$A = 2\bar{R}t \tag{4}$$

$$I = \frac{1}{12}t(2\bar{R})^3 \tag{5}$$

Table 1 Fine view parameters of model specimens

Meso parameters	Value
Particle density (kg·m^{-3})	2000
Minimum particle radius (mm)	0.6
Maximum particle radius (mm)	0.9
Porosity	0.08
Friction coefficient	0.5
Normal contact stiffness (MPa)	28
Tangential contact stiffness (MPa)	28

where t is the assumed the thickness of particle disk.

The normal contact stiffness and tangential contact stiffness are taken as

$$k_n = \frac{AE_C}{L} \tag{6}$$

$$K_s = \frac{12IE_C}{L^3} \tag{7}$$

where E_C is the contact Young's modulus.

2.3 *Parameter Setting*

In the simulation, parameter calibration is performed by uniaxial or biaxial compression tests. In the process of particle generation, non-uniform stress within the system will be generated, and a cyclic calculation must be performed to eliminate the non-uniform stress within the particle. In order to make the initial stress level of the formation close to the actual stress level, according to the model similarity ratio relationship, the acceleration of gravity is set to 100 g (1 g = 9.8 m/s^2). Considering that the research object of this paper is sandy soil, the modeling is carried out according to the document [20], and particle parameters are shown in Table 1. The particle flow discrete element model sample includes the following main mesoscopic parameters, and the specific parameter values are shown in Table 1.

3 Results and Discussion

3.1 *Stress of the Entry and Exit*

It can be seen from Fig. 2 that the sliding and failure of the soil at the end of the shield tunnel is characterized by progressive failure. After the cave door is opened, a small amount of soil particles gushing out through the opening. The particles in the model box undergo a small-scale displacement under the action of gravity due to the change of porosity, and the stress between the particle changes. The stress showed an overall decreasing trend. By monitoring the stress at the measurement point, the stress of the soil below the tunnel door is the largest, and the stress of the soil inside the tunnel door is the smallest. The stress in the x-direction of the soil body in the part of the tunnel door decreases greatly, and the value exceeds 2000 kPa. The opening of the tunnel door causes certain stress changes to the nearby soil particles, but the magnitude of the principal stress in the y-direction is mainly related to the vertical direction of the particles. It is related to the direction and position, and the different values caused by the gravity of the upper soil with different thicknesses are the main reasons.

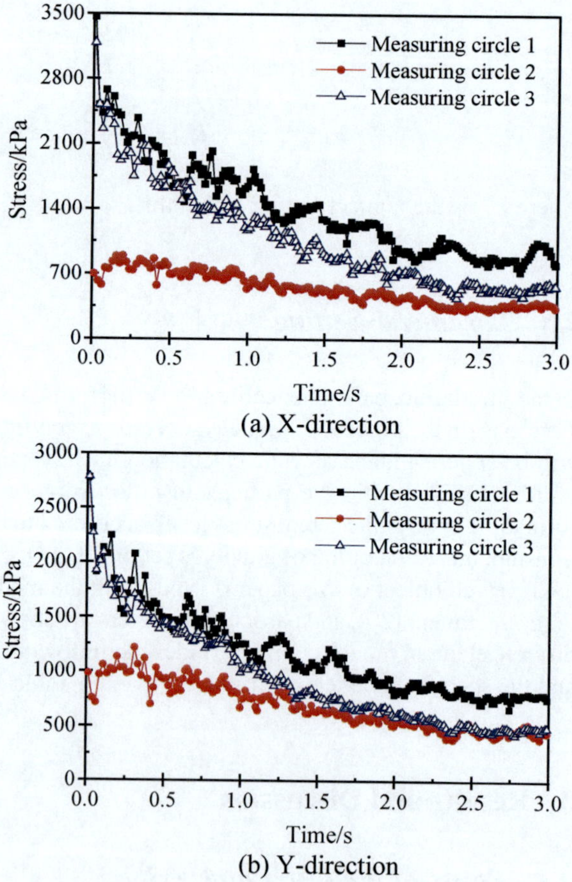

Fig. 2 Stress curves of different measurement circles

3.2 Soil Displacement at the End

The friction and disturbance of the surrounding rock caused by shield propulsion can cause surface subsidence, but it does not cause obvious surface subsidence. Figure 3 is the settlement curve of the top soil. At the end of the model calculation, the top of the soil on the right side of the stirring pile settles, while the top of the soil on the left hardly settles. This is because the hole is on the right side of the model box. When the hole is opened, some soil particles flow out and cause formation loss.

Figure 4 shows the horizontal displacement of particles at different positions of the hole. Over time, the horizontal displacement of soil particles in the hole first increased and then stabilized at 0.24 m. Due to the opening of the tunnel door, some soil particles leave the model box, and the soil particles at the top of the tunnel have a large horizontal displacement within 0.3 s in order to compensate for the loss at the bottom. In addition, the model box lost some soil particles, which also affected the soil particle displacement at the lower part of the tunnel door, resulting in a nonlinear increase in the horizontal displacement of soil particles with time, and the final displacement reached 1.02 m.

Figure 5 shows the particle displacement velocity at different positions of the hole. When the soil particles at the mouth of the hole flow out of the model box, the particles in the hole are basically stable, which is consistent with the conclusion in Fig. 4. However, combined with the analysis in Figs. 3 and 4, the soil particles in the upper part of the hole have a large horizontal displacement before 0.3 s, so the particle displacement velocity is large. With the increase of time, the moving speed of the particles changes irregularly.

Fig. 3 Settlement at different locations on top

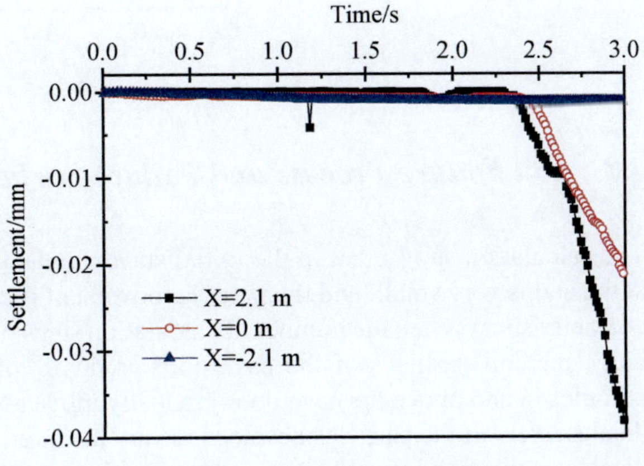

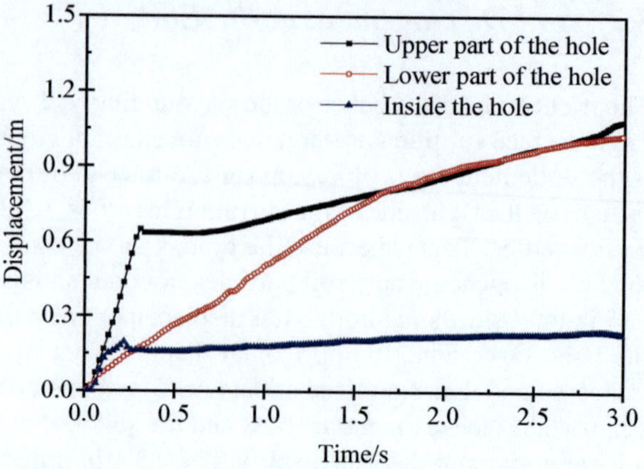

Fig. 4 Horizontal displacement of particles at different positions of the hole

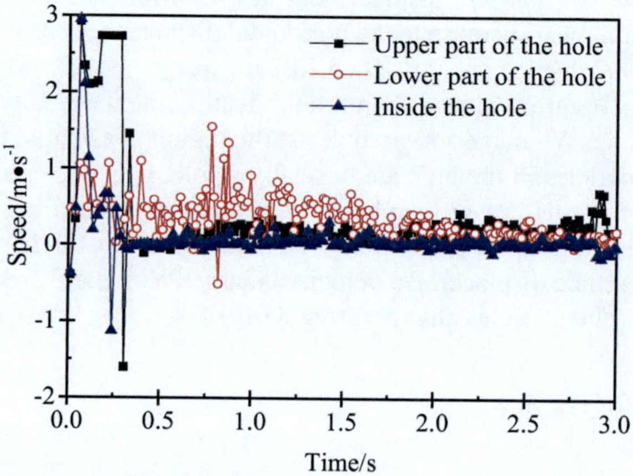

Fig. 5 Particle displacement velocity at different positions of the hole

3.3 *Soil Failure Process and Failure Surface Form*

The calculation shows that in the initial stage the displacement of the soil particles at the end is very small, and the particle movement is chaotic and has no directional characteristics. When the number of calculation steps of the program increases gradually, the soil particles of the cave door gradually detach and slide out; the soil particles in and above the cave door gradually move toward the cave door; the loose displacement of the particles develops toward the interior of the soil, and the particle displacement analysis gradually tends to consistency; the particle slip failure surface gradually formed and extended in the soil. With different excavation calculation steps, the development process and displacement orientation of the model slip surface are shown in Fig. 6. In engineering practice, the final damage of the soil at the end of the shield tunnel is mostly sudden, but the soil must experience the accumulation of displacement and cracks before the instantaneous damage. As a mechanism and process study, the failure process can be summarized into the following two stages, as shown in Fig. 6. The first stage: In the initial stage of breaking the tunnel door, local loosening and displacement of the soil immediately inside the tunnel door occurs. Particle loosening failure gradually expands and extends to the interior of the soil. The second stage: The particle loosening and damage range continues to expand and develops obliquely upward. The slip surface gradually penetrated, and the soil at the end showed obvious slip failure. If buried deep enough, slump arches may form in the upper part.

Figure 7 is an arc-shaped sliding surface. After the tunnel door is broken, the displacement direction of the soil particles at the end has been significantly changed, and the directions are gradually becoming consistent, especially the soil particles inside the diaphragm wall of the tunnel door, forming a relatively clear sliding surface. On both sides of the sliding surface, the displacement of soil particles has obvious discontinuity. On the left side of the sliding surface, the displacement of soil particles is small, and the displacement direction is not uniform. On the right side of the sliding surface, the direction of soil particle displacement is more consistent. Since the soil at the end of the tunnel is constrained by the support of the diaphragm wall, the shear-slip failure of the soil does not form a shear zone, but a shear-slip surface with no thickness.

Fig. 6 Failure process of end soil

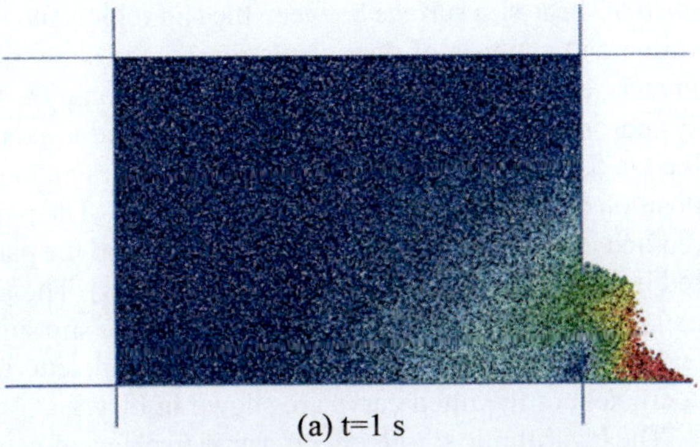

(a) t=1 s

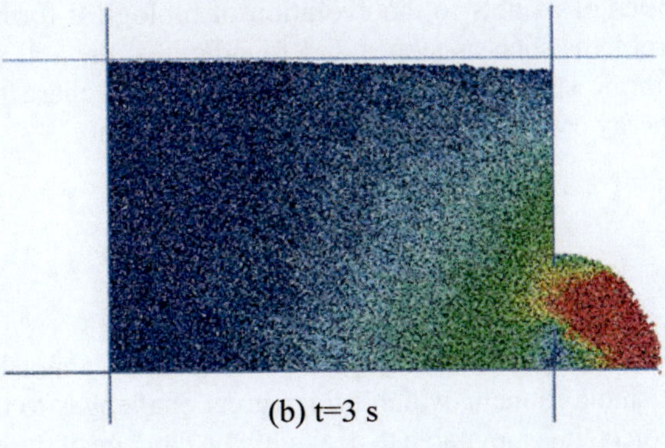

(b) t=3 s

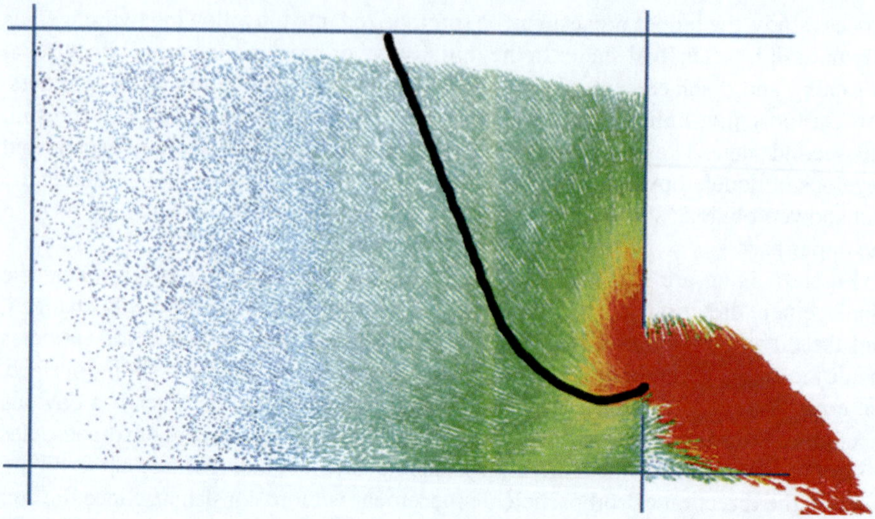

Fig. 7 Arc-shaped sliding surface

It can be seen from the above numerical calculation results that there is a certain form of shear-slip surface between the slip region and the non-slip region. The observation and analysis of the calculation results show that the slip surface is approximately composed of two parts. In order to analyze the slip surface, a coordinate system is established as shown in Fig. 1. The x-axis is horizontal to the left, and the y-axis is the vertical direction of the diaphragm wall. The coordinate system and position of the slip surface are shown in Fig. 8. The part of the slip surface above the red line is close to a straight line (section I), and the part of the slip surface below the red line is similar to a certain curve (section II). The Fish function was compiled to extract the x-coordinate and y-coordinate of the slip surface and 100 nearby particles, and the data were fitted to the slip surface profile curve. The formula and correlation coefficient of the fitted curve are shown in Fig. 8.

The logarithmic spiral reflects a universal law of natural selection, from the trajectories of planets to the evolution of biological forms. In geotechnical engineering problems such as slopes and foundations, the soil slides and fails along the logarithmic spiral, precisely because it has the characteristics of low resistance, low energy consumption, smoothness, and stability.

4 Conclusion

Soil damage at the end is a problem that cannot be ignored in the construction of shield tunnels, which brings great challenges to the timeliness and safety of the project. It is inevitable to deal with the damage of the end soil in engineering practice.

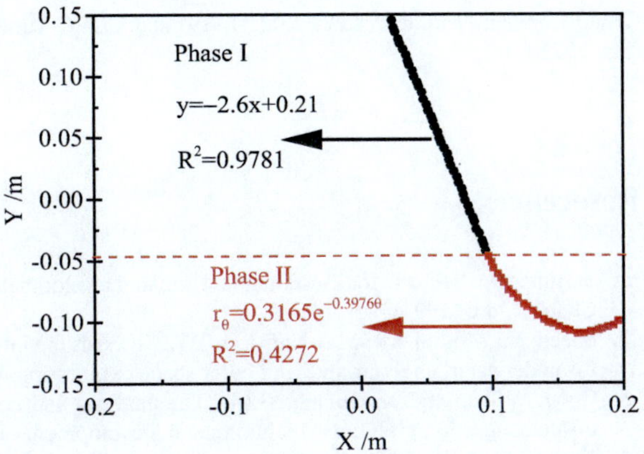

Fig. 8 Fitting curve of soil sliding surface at the end of shield tunnel

Based on the actual engineering of Suzhou subway shield tunnel, this paper conducts numerical simulation research and draws the following conclusions:

1. The soil failure at the end of the shield tunnel presents a gradual development of slip failure characteristics. The particle force chain breaks and the soil loosened and damaged firstly in the soil inside the tunnel door, and then, the particle failure extends inward and upward; the failure zone gradually expands; the slip surface gradually extends, and a dominant shear-slip surface is formed. Due to the support of the shaft diaphragm wall at the end of the shield tunnel, the displacement is small, and the internal shear zone of the soil body degenerates into a shear-slip surface.
2. On the left side of the slip surface, the displacement of soil particles is small, and the displacement direction is not disordered. On the right side of the slip surface, the reorientation of particle displacement is clear. When the soil failure occurs at the end of the tunnel, the soil particles in the slip zone have a large vertical displacement and are accompanied by obvious rotation.
3. The soil slip failure surface at the end of the shallow-buried shield tunnel is a combined slip mode. It is assumed that the soil slip surface in phase I is an oblique line, and the soil slip surface in phase II is a logarithmic spiral. Fitting two slip surfaces based on particle flow discrete metadata.
4. The research and writing of this paper mainly based on the PFC particle flow software, but the current computer capability cannot analyze millions or even hundreds of millions of soil composition particles at the end one by one, so there is still a gap between the model and the reality. When the computer performance improves to a certain level, the method in this paper will be widely used.

Acknowledgements The research presented here is supported by the National Natural Science Foundation of China (52078317), Natural Science Foundation of Jiangsu Province for Excellent Young Scholars (BK20211597), project from Bureau of Housing and Urban-Rural Development of Suzhou (2021–25; 2021ZD02; 2021ZD30), Bureau of Geology and Mineral Exploration of Jiangsu (2021KY06), China Tiesiju Civil Engineering Group (2021–19), CCCC First Highway Engineering

Group Company Limited (KJYF-2021-B-19) and CCCC Tunnel Engineering Company Limited (8gs-2021-04).

References

1. Serrate, O.: Science for Sustainable Human Development, Technologies for Vital Rights, Chap. 7. Idrc (1997)
2. Saeed, N., Alouini, M.S., Al-Naffouri, T.Y.: Towards the internet of X-things: new possibilities for underwater, underground, and outer space exploration, vol. 147, pp. 195–197 (2019)
3. Chen, W.T., Song, X., Huang, G.M.: The analysis and countermeasures of Xiangyang city traffic congestion problems. Technological Development of Enterprise (2014)
4. Kumarage, A.S.: Urban traffic congestion: the problem & solutions. Asian Econ. Rev. (2004)
5. Chen, C., Mao, H.: Research on the traffic congestion problems in Wenzhou City. J. Zhejiang Indus. Trade Voc. College (2012)
6. Yang, W., Cui, G., Xu, Z., et al.: An experimental study of ground-borne vibration from shield tunnels. Tunn. Undergr. Space Technol. **71**, 244–252 (2018)
7. Cui, J., Allan, A., Lin, D.: SWOT analysis and development strategies for underground pedestrian systems. Tunn. Undergr. Space Technol. **87**, 127–133 (2019)
8. Zhu, S., Lin, Z., Gui, C.: Reinforcement technology for and stability analysis on end soil in shield launching and arrival. Tunnel Construction (2012)
9. Pozowski, A., Kowal, M., et al.: The use of basket and cross shell reinforcement in revision acetabulum arthroplasty. Ortopedia Traumatologia Rehabilitacja (2018)
10. Venkatachalam, M.N., Balu, S.: A review on the application of industrial waste as reinforced earth fills in mechanically stabilized earth retaining walls (2022)
11. Ramadhan, M.R., Putra, H.: Evaluation of carbonate precipitation methods for improving the strength of peat soil. In: The First International Seminar on Civil and Environmental Engineering (2020)
12. Mahannopkul, K., Jotisankasa, A.: Influences of root concentration and suction on Chrysopogon zizanioides reinforcement of soil. Soils Found. **59**(2), 500–516 (2019)
13. Ogawa, S., Sato, T., Katoh, M.: Enhancing pyromorphite formation in lead-contaminated soils by improving soil physical parameters using hydroxyapatite treatment. Sci. Total Environ. **747**, 141292 (2020)
14. Singh, M., Trivedi, A., Shukla, S.K.: Strength enhancement of the subgrade soil of unpaved road with geosynthetic reinforcement layers. Transport. Geotech. (2019)
15. Park, J., Kim, I., Kang, J.K.: Root reinforcement effect on cover slopes of solid waste landfill in soil bioengineering. Sustainability **13** (2021)
16. Gotman, N.Z., Evdokimov, A.G.: Determination of additional load on the bridge foundation pile under karst deformation. J. Phys. Conf. Ser. **1928**(1), 012058 (2021)
17. Li, Q., Shi, W., Yang, Q.: Method to determine van der Waals potential energy of particle interactions for soil clay by dynamic light scattering. Eur. J. Soil Sci. (2021)
18. Becknell, R.E., Showalter, K.G., Albrecht, M.A.: Positive soil mutualisms potentially determine the reintroduction outcomes of an endangered legume. Restor. Ecol. (2021)
19. Kezhi, S., Yi, S., Mengshu, W., et al.: Analysis of soil sliding failure mode at the end of sand shield tunnel. Chinese J. Highway **30**(8), 8 (2017)
20. Jian, Z., Jiaquan, W.: Zeng yuan, Zhang Jiao Particle flow simulation of soil slope stability analysis. Geotech. Mech. **30**(01), 86–90 (2009)

Early Triassic Tectonic Evolution of the Northeastern Kontum Massif: New Constraints from the S-type Granite in Ba To Area, Quang Ngai Province, Central Vietnam

Ha Thanh Tran, Bui Vinh Hau, Ngo Xuan Thanh, Nguyen Huu Hiep, and Ngo Thi Kim Chi

Abstract The granitoids exposed in Ba To area, northeastern portion of the Kontum Massif, central Vietnam were previously assigned as part of the Hai Van complex. Field investigation and petrographic study reveal that the magmatic rocks in Ba To area comprise dominantly medium- to coarse-grained, per-aluminous two-mica granite of S-type origin. These rocks are strongly overprinted by several phases of tectonic activities including the emplacement of felsic and mafic dikes and veins, formation of weak ductile foliation, brittle fracturing, and widespread alteration under the influence of several post-magmatic thermotectonic processes. Results of U–Pb dating of zircons collected from granite in Ba To area show a crystallization age of 245.8 ± 1.5 Ma for the magmatic rock. The petrographic study and absolute age dating results conducted by this work are comparable to other recent works on similar granitoid occurred in the northern margin of the Kontum Massif. This indicates that the widespread Ca 245 Ma Hai Van-type S-type granitoid along the north and northeastern portions of the Kontum Massif were derived from a collisional orogenic event, which extended from Late Paleozoic onward during the assembly of Sibumasu Block to Indochina Block to form the Proto-Southeast Asia, which is correspondent to the Indosinian Orogeny. Several post-intrusive thermotectonic events due to the tectonic activities along the newly formed Proto-Indochina Block

---

H. T. Tran (✉) · B. V. Hau (✉) · N. X. Thanh · N. H. Hiep · N. T. K. Chi
Department of Geology, Hanoi University of Mining and Geology, Hanoi, Vietnam
e-mail: tranthanhhai@humg.edu.vn

B. V. Hau
e-mail: vinhhau.humg@gmail.com

B. V. Hau · N. H. Hiep
Centre for Excellence in Analyses and Experiment, Hanoi University of Mining and Geology, Hanoi, Vietnam

H. T. Tran
Research Group on Tectonics and Geodynamics for Geo-resources, Environment and Sustainable Development, Hanoi University of Mining and Geology, Hanoi, Vietnam

B. V. Hau
Hanoi, Vietnam

L. Q. Nguyen et al. (eds.), *Advances in Geospatial Technology in Mining and Earth Sciences*, Environmental Science and Engineering,
https://doi.org/10.1007/978-3-031-20463-0_33

had resulted in subsequent deformation, injection of late magmas and alterations of the Hai Van—Type magmatic rocks.

Keyworks Kontum Massif · Hai Van Complex · Granitoid · U-Pb zircon age

1 Introduction

The Indochina Block, a major tectonic entity of mainland Southeast Asia, comprises of two main components: Truong Son Belt and Kontum Massif [1–3]. The Kontum Massif, which forms the core portion of the Indochina Block (Fig. 1), is mainly composed of high-grade metamorphic rocks and grouped into different units including Kannack, Ngọc Linh, and Kham Duc complexes [1]. These metamorphic rocks were intruded by voluminous magmatic suites of different ages and origins [1, 4] (Fig. 1). Previous studies suggested that magmatic rocks in the Kontum Massif have multi-origins with ages are variable from Archean to Late Proterozoic, Paleozoic to Mesozoic [1, 2, 4, 5], which are tectonically intercalated and formed complex exposures of many types of rocks in the same area. Because of that complex spatial distribution, many magmatic bodies of different ages and origins were doubtfully identified that they had been derived from the same magmatic source and age, which had been grouped into a single magmatic complex in previous mapping without understanding of the absolute age and tectonic environment related to magma emplacement. Such inadequate interpretation has led to unconvincing reconstruction of the regional tectonic evolution of the Kontum Massif.

In order to distinguish magmatic complex of different ages and origins in an area, in addition to field studies to determine the spatial relationship and relative timing of the magmatic suites, an important factor is to perform absolute age dating of the magmatic rocks in a synchronous and systematic manner. For this approach, this study had conducted field mapping in Ba To area, Quang Ngai Province, which occurs in the northeastern part of Kontum Massif (Fig. 1), where magmatic intrusive rocks are widely exposed and is formerly classified as a part of the Hai Van Complex of Late Triassic age [e.g., 1, 4] (Fig. 1). During the mapping, samples had been collected to determine the absolute crystallization age of the observed magmatic rocks in this area using U–Pb isotopes in zircons by LA-ICP-MS technique. This paper presents new results on the crystallization ages of the samples from S-type magmatic rocks and, in conjunction with available geological data, discusses this new finding in the framework of regional tectonic evolution.

2 Geological Background

The Ba To area is located in Quang Ngai Province and forms part of the northeastern portion of the Kontum Massif (Fig. 1). The area is underlain by a lithologically and

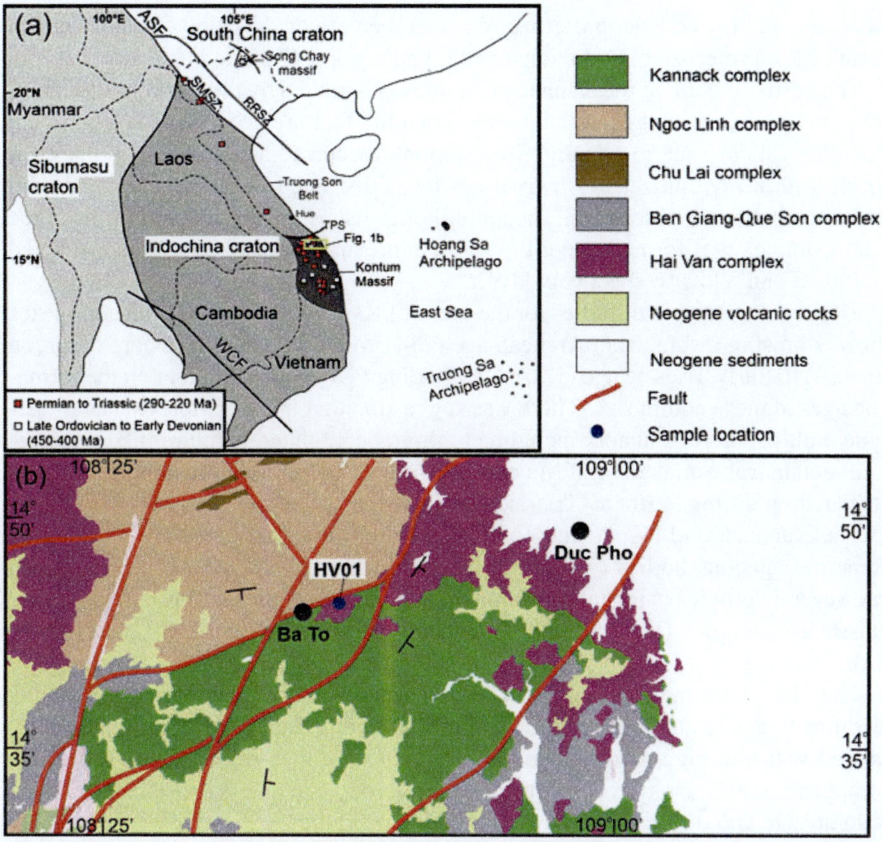

Fig. 1 **a** Location of study area in Vietnam and generalized tectonic map of part of Southeast Asia [6] showing the relative position of Indochina Block, Kontum Massif and the study area; **b** generalized geological map of the study area (compiled from various sources) and location of sampling for this study. Abbreviations in Fig. **a**: *ASF* AilaoShan Fault; *SMSZ* Song Ma Suture Zone; *RRSZ* Red River Shear Zone; *TPS* Tam Ky-Phuoc Son Suture Zone; and *WCF* Wang Chao Fault

structurally complex basement (Fig. 1b), indicated by abundant high-grade metamorphic assemblages, multiple magmatic complexes, and sedimentary cover of differing ages and origins [e.g., see 1, 4].

The high-grade rocks of the Kannack complex exposes in the southern part of the Ba To area (Fig. 1b), mainly comprises of high temperature to ultrahigh temperature (UHT) granulite-facies and minor amphibolite-facies rocks, which are predominant pelite with intercalated minor amounts of mafic and calc-silicate rocks [7, 8]. Intrusive rocks including granitoid and gabbroic bodies that are exposed as stocks and lenses within granulite-facies metamorphic rocks [9]. The peak metamorphic condition of the granulites in the Kannack Complex estimated as 800–1000 °C and 8–11 kbar [10, 11]. The age of this complex was suggested to be Archean to Paleoproterozoic [1, 4, 12, 13]. Recent studies have suggested that the formation age of this complex may be Paleoproterozoic, which then affected by poly-metamorphism during Mesoproterozoic, Ordovician and Late Permian–Triassic [1].

The northern flank of the Kannack Complex is bounded by the amphibolite-facies felsic mylonite, pelitic gneiss, and mafic granulite that are grouped into Ngoc Linh Complex [1]. The mafic granulite is commonly occurred as boudinaged blocks or lenses within mylonitized felsic gneisses or as disrupted bodies intercalated with felsic and pelitic gneisses [10]. The amphibolite-facies metamorphism of the Ngoc Linh Complex was formed under the temperature and pressure condition estimated as 700 °C and 5 kbar respectively [10].

The geochronological studies for the rocks of Kannack and Ngoc Linh complexes show confusing results from Archean age (2541 Ma, U–Pb age from core of zircon crystals) to Early Triassic age [1, 14, 15], leading to a major argument on the formation ages of these complexes. There was suggestion that these complexes had undergone multiple thermal history including high-grade, high temperature-medium pressure metamorphism at ca. 450 Ma and overprinted by regional metamorphism and deformation during Permian–Triassic [1, 16–18].

The Kannack and Ngoc Linh complexes within the study area are intruded by numerous igneous bodies of differing age and origins, which have been grouped into several complexes including Nam Nin and Ta Vi complexes (Proterozoic), Chu Lai-Ba To Complex (Early Middle Paleozoic) or Hai Van Complex (Late Triassic; [1]).

The Nam Nin and Ta Vi complexes consist of mainly plagiogranite, tonalite, medium-coarse grain sized gabbro amphibolite bodies that are scatted and intercalated with the high-grade metamorphic rocks of either Kannack or Ngoc Linh complexes. Although had been considered Proterozoic in age in previous works, the absolute age and origin of these rocks remain uncertain.

The Chu Lai-Ba To Complex comprises of batholith-style and smaller bodies of predominantly coarse-grained granitoid, which were strongly deformed and intercalated with high-grade rocks of the Kannack and Ngoc Linh complexes. Previous studies suggested that the Chu Lai-Ba To-type rocks were intruded in Late Proterozoic [4, 12]. Recent studies, however, have showed that the crystallization age of this complex occurred adjacent to this study area is much younger, at about 440–430 Ma on the basis of U–Pb zircon dating [e.g., 19, 20].

Late Paleozoic intrusive rocks are abundantly exposed in the eastern and southeastern part of the study area (Fig. 1), comprising dominantly mica-rich S-type granitoid containing numerous inclusions of surrounding sedimentary and/or metamorphic rocks that are assigned to the Hai Van Complex (e.g., see [1] for discussion). Previous studies suggested that the age of Hai Van Complex is Late Triassic [1]. Recent studies in adjacent areas show that part of Hai Van Complex have various ages, ranging from 245 to 290 Ma [2, 5, 21]. This indicates that the classification intrusive rocks with different ages and origins into a single complex is unconvincing without detailed studies on geochronology and geochemistry.

3 Sample Collection and Analytical Methods

3.1 Sample Collection

In the study area, magmatic rocks that were mapped as a part of the Hai Van Complex are exposed as a large body in the northeast of the study area, with the biggest body is exposed as several tens of square kilometers in size and many smaller bodies that are intruded into, and scattered within, the high-grade rocks of the Kannack and Ngoc Linh complexes (Fig. 1b).

Several samples had been collected from medium grained granite that exposed near Ba To Town, Quang Ngai Province (14°45′160″ N and 108°42′522″ E (Fig. 1b) for petrographic study and absolute U–Pb zircon dating.

3.2 Analytical Methods

The collected samples were prepared for thin sections for petrographic analysis and separation of zircons for U–Pb dating at the Centre for Excellence in Analyses and Experiments (CEAE) of the Hanoi University of Mining and Geology.

A sample (HV01) was chosen for collection of zircon grains by conventional method [22] including grinding, washing, and separation by using heavy liquid and Franz magnetic separator before hand-picked, mounted in epoxy resin prior to polishing to expose the interior of the zircon crystals.

Cathodoluminescence (CL) images were taken to examine zircon morphology, internal textures and to guide spot selections for the U–Pb analyses. The U–Th–Pb isotopic compositions of zircons from sample HV01 were measured using Nu Plasma II multicollection inductively coupled plasma mass spectrometer equipped with a New Wave Research 193 nm ArF excimer Laser Ablation system (LA-MC-ICP-MS) housed at the Ochang Campus of the Korea Basic Science Institute (KBSI). The Nu Plasma II mass spectrometer contains fixed collectors of sixteen Faraday detectors and five ion-counting electron multipliers. The collectors were set to simultaneously detect for U–Pb age determination in the following array: ^{202}Hg (IC 4), ^{204}(Hg + Pb) (IC 3), ^{206}Pb (IC 2), ^{207}Pb (IC 1), ^{208}Pb (IC 0), ^{232}Th (high 7), and 238U (high 9), respectively. ^{235}U was calculated from the signal at mass ^{238}U using a natural ^{238}U/^{235}U = 137.88. Mass number 204 was used as a monitor for common ^{204}Pb after discarding the ^{204}Hg background. All analyses were carried out with a spot size of 20 μm in diameter, 5 Hz repetition rate, and energy density of 3 J/cm^2. He (940 ml/min) was used as carrier gas. Background intensities, dwell time, and wash out time were measured as 30 s, 30 s, and 20 s, respectively. A Time Resolved Analytical (TRA) procedure was employed to monitor the measured isotope ratio. Signal intensities for each collector were collected every 0.2 s (integration time). Raw data were corrected for the background, laser-induced elemental fractionation, mass discrimination, and drift in ion counter gains. U–Pb isotope ratios were calibrated by concordant reference zircons 91,500 (1065 Ma) [23] and Plešovice zicrcon (337.13 ± 0.37 Ma) [24] that was used at the beginning and end of each analytical session, and at regular intervals during each session, using protocols adapted from [25] a correlation of signal vs. time was also assumed for the reference zircons. The Iolite software running in IgorPro (WaveMetrics, Inc; www.wavemetrics.com) was used for the data reduction of LA–MC–ICP–MS analyses. Once processed with Iolite, the data were exported in EXCEL, and Isoplot [26] was used to calculate weighted average ages and make Tera-Wasserburg plots.

4 Analytical Results

4.1 Petrography

The studied rocks are medium to coarse grained, non to weakly spacely foliated, dark-gray granite with minor late pegmatite, aplite and quartz veins intruded and/or filled the fractures within the rocks (Fig. 2a–c). In part, the pegmatic and quartz dike and veins are flattened and reoriented sub-parallel to the foliation (Fig. 2b), whereas the aplitic veins are either irregular or fracture-filling (Fig. 2c). The mineral composition of the rocks mainly comprises of plagioclase (40%), K-feldspar (35%), quartz (10–15%), muscovite and biotite (~5%), and other minor minerals (Fig. 2d). The mineral composition of the granitic rocks in this area indicate that they are peraluminous S-type granite, possibly derived from syn-orogenic collisional setting. In this area, the rocks had undergone a post-crystallization deformation and alteration, indicated by the fracturing of the mineral grains, undulose extinction and widespread of alteration rims around the K-felspar and plagioclase grains (Fig. 2d).

Both the field investigation and thin section interpretation indicate that the granitic rocks in the study area had undergone a complex post-intrusive history, including an early veining, which led to the intrusion of felsic (pegmatite) dikes/veins followed by a phase of regional deformation to produce overprinting foliation and recrystallization, brittle fracturing, intrusion of late aplitic veins, and hydrothermal alteration.

4.2 U–Pb Dating of Zircons

Zircon crystals from granitic rocks collected in this study are commonly 70–200 μm in the longest dimension, with euhedral to subhedral morphology, and their length-to-width ratios from 2:1 to 3:1 (Fig. 3). Some zircon crystals exhibit core–mantle structure in CL images and all the zircon grains show well-developed oscillatory zoning.

Fig. 2 **a–c** Field exposure of the granitic rocks in the study area, east of Ba To town. In places, the rocks are weakly foliated with sub-parallel alignment of pegmatite/quartz veins (**b**). In places, the main rocks are intruded by several generations of veins with differing composition (felsic or mafic dikes and veins, **c**), or cut by brittle structures (**b**, **c**). **d** is the photograph of the rocks under thin section, which shows post-intrusive recrystallization indicated by undulose extinction of the quartz. Late hydrothermal alteration is indicated by the pale reaction rims around the K-feldspar and plagioclase crystals. The observed petrological and structural relationship indicate a complex intrusive and post-intrusive tectonic history (see text for discussion). Abbreviations: *Plg* plagioclase, *KF* K-feldspar, *Ms* muscovite, *Bt* biotite, *Qtz* quartz

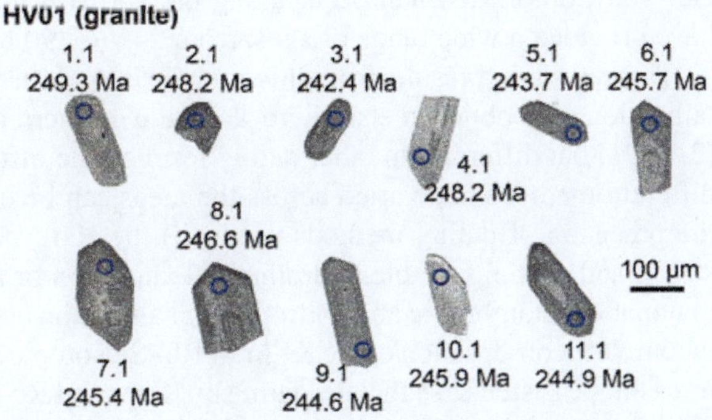

Fig. 3 Cathodoluminescence (CL) image of the representative zircon grains used for U–Pb analysis with positions of analytical points (in circles) and their determined ages from sample HV01

The U-Th-Pb isotopic compositions of zircon were presented in the Table 1 and plotted in the Tera-Wasserburg concordia and Weighted mean age diagrams (Fig. 4). The U-Th-Pb isotopic analyses result show that U and Th contents range from 289 to 4590 ppm and 229 to 4049 ppm, respectively. Whereas the Th/U ratios of the analyzed zircon crystals range from 0.62 to 1.08 (Table 1), which strongly demonstrate a direct crystallization of the zircon from a magmatic source [27].

The most suitable $^{238}U/^{206}Pb$ age results were corrected using the ^{207}Pb correction by Isoplot 3.6 software (Ludwig 2008) with an error of 2-sigma show the $^{238}U/^{206}Pb$ age varies from 242.4 to 249.3 Ma (Table 1). In the Tera-Wasserburg Concordia diagram, eleven analyses from eleven zircon crystals have well-concordant $^{238}U/^{206}Pb$ age (Fig. 4a). Eleven analyses yield a weighted mean $^{238}U/^{206}Pb$ age of 245.8 ± 1.5 Ma (MSDW = 0.62, probability = 0.80) (Fig. 4b). This age is interpreted as the crystallization age for the granitic rocks in the study area.

5 Discussions

The U–Th–Pb zircon ages obtained from this work shows that the crystallization age of the granitoid in the study area is 245.8 ± 1.5 Ma (ca. 246 Ma). If these rocks are part of the Hai Van Complex, it can be confirmed that the age of this Complex in the study area is ca. 246 Ma.

According to previous geological studies, the granitoid assigned to Hai Van Complex is considered to be Late Triassic in age [1, 28]. However, the previous isotopic ages dated by different methods gives confusing ages of the Hai Van Complex, such as 138 ± 4 Ma [29], 220 ± 4 Ma by K/Ar method [30], 236 ± 4.6 Ma [31] by Rb–Sr method. More recently, new results by U–Pb dating (conventional, SHRIMP or LA-ICP-MS) performed on granitoid considered as parts of the Hai Van Complex, which occur along the northern margin of the Kontum Massif, have provided a wide range of ages, from 245 to 290 Ma [2, 5, 21].

The new dating results from this study is in accordance with the ca. 245 Ma crystallization ages obtained elsewhere along the northern margin of the Kontum Massif [2, 5, 21] but differs from other dating results. The different age ranges obtained by different methods and varied across the areas can be explained by several facts: (i) the precision of dating methods used; (ii) the Hai Van Complex was subjected to strong and spatial variable hydrothermal alteration or recrystallization during post-magmatic metamorphic and hydrothermal alteration events. These would lead to the incomplete conservation of the K–Ar or Rb–Sr isotopic system (low closure temperature isotope systems) so that the dating by K–Ar and Rb–Sr or U–Pb methods produce different age values for the same rocks that have experimented different degree of post-intrusive deformation and associated metamorphic and/or alteration; or iii) the rocks assigned to the Hai Van Complex at different localities within the Kontum Massif are in fact derived from differing magmatic sources that were generated at different time frames and tectonic setting during the regional tectonic evolution.

Table 1 Results of U-Th-Pb isotopic analyses by LA-MC-ICP-MS performed on zircons of sample HV01 collected from granitic rocks in Ba To area

No.	U (ppm)	Th (ppm)	Th/U	Isotopic ratios						Age (Ma)			
				$^{238}U/^{206}Pb$	2σ	$^{207}U/^{206}U$	2σ	$^{207}U/^{235}U$	2σ	$^{207}U/^{235}U$	2σ	$^{238}U/^{206}Pb$	2σ
1.1	2730	2535	0.99	25.35497	0.295722	0.05136	0.00084	0.2801	0.0052	250.7	4.1	249.3	2.9
2.1	2478	2031	0.89	25.47771	0.22719	0.05124	0.00093	0.275	0.0052	246.6	4.1	248.2	2.1
3.1	2150	2119	1.08	26.10285	0.252103	0.0514	0.0011	0.2718	0.0059	244.1	4.7	242.4	2.3
4.1	383	308.1	0.89	25.47771	0.40245	0.0525	0.0026	0.283	0.013	252	11	248.2	3.8
5.1	2460	1980	0.90	25.9538	0.350272	0.05139	0.00086	0.2756	0.005	247.1	3.9	243.7	3.2
6.1	2860	2062	0.82	25.74003	0.245143	0.05107	0.00083	0.2765	0.0055	247.8	4.3	245.7	2.3
7.1	1946	1023	0.62	25.76656	0.165979	0.05086	0.00083	0.2733	0.0046	245.3	3.6	245.4	1.5
8.1	1639	1170	0.84	25.64103	0.282709	0.05171	0.00084	0.2793	0.0047	250	3.7	246.6	2.7
9.1	3799	2467	0.78	25.85315	0.360928	0.05138	0.00089	0.2756	0.0067	247.1	5.3	244.6	3.4
10.1	289.4	229.5	0.95	25.71355	0.337205	0.0512	0.0019	0.273	0.01	244.9	8.1	245.9	3.2
11.1	4590	4049	1.06	25.82645	0.273472	0.05118	0.00065	0.2739	0.0048	246.6	4	244.9	2.6

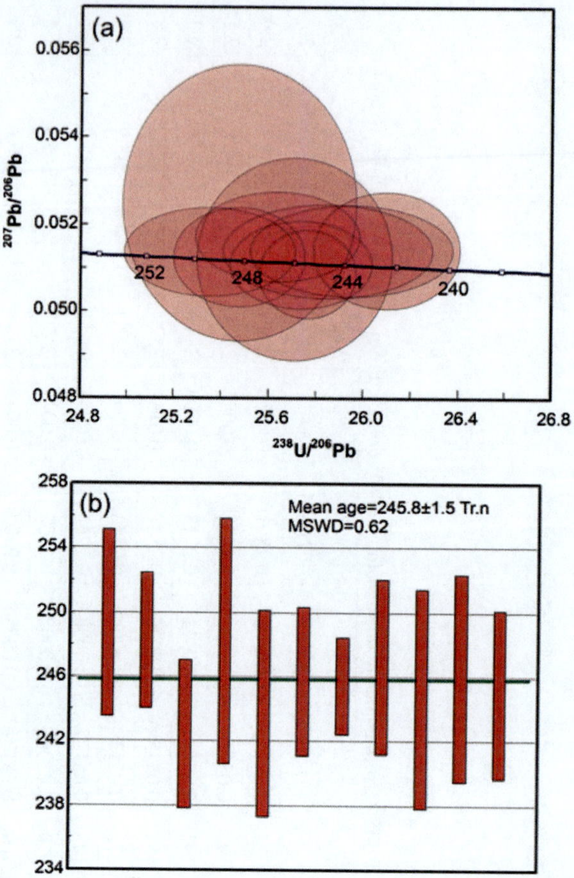

Fig. 4 **a** The Tera-Wasserburg Concordia diagram showing $^{207}Pb/^{206}Pb$ and $^{238}U/^{206}Pb$ plots for the analytical results for zircon dating show in Table 1, in which most of the ages are concordant and fall in the 242.4–249.3 Ma. **b** The weighted mean age diagram shows absolute ages and SWD of dating results as well as mean age of the zircon grains of sample HV01

They are temporarily grouped in to a magmatic complex during the mapping and other studies based solely on the tentative comparison of petrographic similarity, field occurrence, or relative age dating (e.g., see [5] for detail discussion).

The results of this work and comparable results of similar works conducted recently clearly demonstrate the existence of a regional widespread of ca. 245 Ma magmatic event. This also shows that during Late Paleozoic to Early Mesozoic, the Kontum Massif had experienced a complex thermotectonic history, which led to the formation of multiple magmatic suites of different ages and origins that intruded successively in the same area.

The mineral composition of the granitic rocks in this area shows that they are peraluminous S-type granite, possibly derived from syn-orogenic collisional setting (see also [32]). The widespread of Hai Van-type, ca. 245 Ma S-type granitoid complexes along the northern margin of the Kontum Massif may indicate that the collisional activities may have taken place since at least Late Permian, leading to the deep burial and melt of the supracrustal materials to form S-type granitic magma that was widely intruded into the upper crust in this area. This orogenic activity may correspond to the Indosinian orogenic phase, which is directly related to the collision between the Sibumasu and Indochina blocks to create a Proto-Southeast Asia Block, which has been discussed by many recent studies [2, 6, 33–36]. Subsequent tectonic overprinting due to terrane collision and assembly around the Proto-Southeast Asia would have led to the overprinting deformation and alteration of the ca. 245 Ma magmatic rocks in this areas and other parts of the Kontum Massif [2, 6, 35].

6 Conclusion

Detailed field investigation, petrographic study, and U–Pb age dating of zircons from the rocks by LA-ICP-MS method for the granitic rocks in the Ba To area, northeastern Kontum Massif has revealed that the Hai Van-type magmatic rocks in this area consists of per-aluminous, S-type granite that was crystallized at ca. 246 Ma and then overprinted by several phases of tectonic activity, including the intrusions of felsic magmas, sub-ductile deformation, brittle fracturing, mafic dikes and veins injection, and hydrothermal alteration. The occurrence of ca. 246 Ma S-type granite in the area as well as widespread exposure of similar Hai Van-type magmatic rocks is correspondent to a phase of Late Paleozoic to Early Mesozoic tectonic collision and regional orogeny that are likely linked to the assembly of Sibumasu Plate to Indochina Block to form the Proto-Southeast Asia mainland, which is equivalent to the Indosinian Orogeny.

The results of this study also show that there is an urgent need for more quantitative and systematic studies on the field occurrence, composition, and age of magmatic rocks, especially in the areas that are multiply and complexly intruded by numerous types of magmas of differing ages and origins that cannot be distinguished based solely on petrographic composition or field appearance. The combination of field studies, chemical and isotopic analysis should be carried out systematically in any mapping and/or regional geological study to precisely determine the age, environment, and nature of the magma evolution as well as its roles in the regional tectonic evolution.

Acknowledgements This work was funded by the Ministral Project No. B2019-MDA-562-15ĐT to Tran Thanh Hai. We thank the Center for Excellence in Analyses and Experiments (CEAE)—Hanoi University of Mining and Geology for providing experimental facilities to prepare the thin sections and zircons for U–Pb dating. The assistance of the Korea Basic Science Institute (KBSI) in Ochang Campus for U–Pb dating is highly appreciated.

References

1. Tran, V.T., Vu, K.: Geology and Resources of Vietnam. Natural Sciences and Technology Publisher, Hanoi, Vietnam (2009)
2. Tran, T.H., Zaw, K., Halpin, J.A., Manaka, T., Meffre, S., Lai, C.K., Lee, Y., Le, H.V., Dinh, S.: The Tam Ky-Phuoc Son shear zone in central Vietnam: tectonic and metallogenic implications. Gondwana Res (2014). https://doi.org/10.1016/j.gr.2013.04.008
3. Nguyen, M.Q., Feng, Q., Zi, J.W., Zhao, T., Tran, T.H., Ngo, X.T., Tran, M.D., Nguyen, Q.H.: Cambrian intra-oceanic arc trondhjemite and tonalite in the Tam Ky–Phuoc Son Suture Zone, central Vietnam: implications for the early Paleozoic assembly of the Indochina Block. Gondwana Res. **70**(2019), 151–170 (2019)
4. Nguyen, V.T.: Geological map of Huế-Quảng Ngãi sheets of 1:200 000 scale. Geological Archives, Hà Nội, Vietnam (1986) (In Vietnamese)
5. Nguyen, T.G., Tran, T.H.: Multiple magmatic intrusion along the magrin of Mesozoic Nông Sơn Basin, central Vietnam, and their tectonic significance. J. Geol. Vietnam **336**, 37–49 (2016). (In Vietnamese with English abstract)
6. Matcalfe, I.: Gondwana dispersion and Asian accrition: tectonic and paleogeographic evolution of eastern Tethys. J. Asian Earth Sci. **66**, 1–63 (2013)
7. Osanai, Y., Owada, M., Tsunogae, T., Toyoshima, T., Hokada, T., Long, T.V., Sajeev, K., Nakano, N.: Ultrahigh-temperature pelitic granulites from Kontum Massif, central Vietnam: evidence for East Asian juxtaposition at ca. 250 Ma. Gondwana Res. **4**(4), 720–723 (2001)
8. Osanai, Y., Nakano, N., Owada, M., Nam, T.N., Toyoshima, T., Tsunogae, T., Binh, P.: Permo-Triassic ultrahigh-temperature metamorphism in the Kontum Massif, central Vietnam. J. Mineral. Petrol. Sci. **99**(4), 225–241 (2004)
9. Owada, M., Osanai, Y., Nakano, N., Adachi, T., Kitano, I., Tri, T.V., Kagami, H.: Late Permian plume-related magmatism and tectonothermal events in the Kontum Massif, central Vietnam. J. Mineral. Petrol. Sci. **111**(3), 181–195 (2016)
10. Nakano, N., Osanai, Y., Owada, M., Nam, T.N., Toyoshima, T., Binh, P., Tsunogae, T., Kagami, H.: Geologic and metamorphic evolution of the basement complexes in the Kontum Massif, central Vietnam. Gondwana Res. **12**, 438–453 (2007)
11. Osanai, Y., Nakano, N., Owada, M., Nam, T.N., Miyamoto, T., Minh, N.T., Nam, N.V., Tri, T.V.: Collision zone metamorphism in Vietnam and adjacent South-eastern Asia: proposition for Trans Vietnam Orogenic Belt. J. Mineral. Petrol. Sci. **103**(4), 226–241 (2008)
12. Tran, D.L., Nguyen, X.B.: Geological map of Vietnam at 1:500 000 scale. Geological Archives, Hanoi, Vietnam (1982) (In Vietnamese)
13. Tong, D.T., Vu, K. (eds.): Stratigraphic units of Vietnam. Vietnam National University Publisher, pp. 554 (2011) (In Vietnamese)
14. Nakano, N., Osanai, Y., Owada, M., Tran, N.N., Charusiri, P., Khamphavong, K.: Tectonic evolution of high-grade metamorphic terranes in central Vietnam: constraints from large-scale monazite geochronology. J. Asian Earth Sci. **73**, 520–539 (2013)
15. Nakano, N., Osanai, Y., Owada, M., Binh, P., Hokada, T., Kaiden, H., Vuong, T.S., Bui, K.: Evolution of the Indochina block from its formation to amalgamation with Asia: constraints from protoliths in the Kontum Massif, Vietnam. Gondwana Res. **90**, 47–62 (2021)
16. Usuki, T., Lan, C.Y., Yeh, M.W., Anh, T.T., Iizuka, Y.: Metamorphic evolution in the northern Kontum Massif, central Vietnam. EOS Trans. Am. Geophys. Union **85**(47) (2004); Fall Meeting Supplement, Abstract T11C-1273
17. Usuki, T., Lan, C.Y., Yui, T.F., Iizuka, Y., Vu, V.T., Tran, T.A., Okamoto, K., Wooden, J.L., Liou, J.G.: Early Paleozoic medium-pressure metamorphism in central Vietnam: evidence from SHRIMP U-Pb zircon ages. Geosci. J. **13**, 245–256 (2009)
18. Usuki, T., Lan, C.Y., Wang, K.L., Chiu, H.Y.: Linking the Indochina block and Gondwana during the early Paleozoic: evidence from U-Pb ages and Hf isotopes of detrital zircons. Tectonophysics **586**, 145–159 (2013). https://doi.org/10.1016/j.tecto

19. Dinh, Q.S.: Petrographic characteristics and zircon U-Pb geochronology of granitogneiss rocks in the Chu Lai—Kham Duc area (Quang Nam province). Sci. Technol. Devel. J. Natural Sci. **1**, 258–272 (2017)
20. Luc, T.T., Tran, T.H., Nguyen, H.H., Bui, H.B., Carter, A.: New discovery on the absolute age of the granodiorite of Chu Lai complex in northeast Quang Ngai. J. Mining Earth Sci. **60**, 7–14 (2019). (In Vietnamese with English abstracts)
21. Pham, T.H., Yi, Z.Y., Nguyen, T.B.T., Le, T.D., Chen, F.: Late Permian to early Triassic crustal evolution of the Kontum massif, central Vietnam: zircon U-Pb ages and geochemical and Nd–Hf isotopic composition of the Hai Van granitoid complex. Int. Rev. (2015). https://doi.org/10.1080/00206814.2015.1031194
22. Cheong, W.S., Cho, M., Kim, Y.: An efficient method for zircon separation using gold pen. J. Petrol. Soc. Korea **22**, 63–70 (2013). (In Korean with English abstract)
23. Wiedenbeck, M., Allé, P., Corfu, F., Griffin, W.L., Meier, M., Oberli, F., von Quadt, A., Roddick, J.C., Spiegel, W.: Three natural zircon standards for U–Th–Pb, Lu–Hf, trace element and REE analyses. Geostand. Newsl. **19**, 1–23 (1995)
24. Sláma, J., Košler, J., Condon, D.J., Crowley, J.L., Gerdes, A., Hanchar, J.M., Horstwood, M.S.A., Morris, G.A., Nasdala, L., Norberg, N., Schaltegger, U., Schoene, B., Tubrett, M.N., Whitehouse, M.J.: Plešovice zircon—a new natural reference material for U-Pb and Hf isotopic microanalysis. Chem. Geol. **1–2**, 1–35 (2008)
25. Andersen, T.: Correction of common lead in U-Pb analyses that do not report 204Pb. Chem. Geol. **192**, 59–79 (2002)
26. Ludwig, K.R.: User's manual for Isoplot 3.6: a geo-chronological toolkit for Microsoft Excel. Berkeley Geochronology Center, Special Publication, vol. 4, 77 p. (2008)
27. Wu, Y.B., Zheng, Y.F.: Genesis of zircon and its constraints in interpretation of U–Pb. Chin. Sci. Bull. **49**, 1554–1569 (2004)
28. Dao, D.T., Huynh, T.: Geology of Vietnam—Volume 2: Intrusive Rocks. Scientific and Technical Publisher, Hanoi, Vietnam (1995)
29. Huynh, T., Tran, P.H., Le, D.P.: Principles of Petrology and Metallogeny. Ho Chi inh City National University Publisher, Ho Chi Minh city, Vietnam (2006).(In Vietnamese)
30. Nguyen, X.B.: Report on the study of tectonics and metallogeny of Southern Vietnam, Geological Archives, Hanoi, Vietnam (2000) (In Vietnamese)
31. Phan, L.A., Tran, T.H.: Condition of formation of Hải Vân, Ba Na—type granitoid, on the basics of new data on rare elements and isotopes. Vietnam Jiurnal of Earth Sciences (1995) (In Vietnamese)
32. Tran, T.H., Ansdell, K.M., Bethune, K.M., Ashton, K., Hamilton, M.A.: Provenance and tectonic setting of Paleoproterozoic metasedimentary rocks along the eastern margin of Hearne craton: constraints from SHRIMP geochronology, Wollaston Group, Saskatchewan, Canada. Precambr. Res. **167**, 171–185 (2008)
33. Lan, C.Y., Chung, S.L., Van Long, T., Lo, C.H., Lee, T.Y., Mertzman, S.A., Shen, J.J.S.: Geochemical and Sr–Nd isotopic constraints from the Kontum massif, central Vietnam on the crustal evolution of the Indochina block. Precambr. Res. **22**, 7–27 (2003)
34. Carter, A., Clift, P.D.: Was the Indosinian orogeny a Triassic mountain building or a thermotectonic reactivation event? C.R. Geosci. **340**, 83–89 (2008)
35. Hall, R.: Late Jurassic-Cenozoic reconstruction of the Indonesian region and the Indian Ocean. Tectonophysics **570–570**, 1–41 (2012)
36. Tran, V.T., Faure, M., Nguyen, V.V., Hoang, H.B, Fyhn, M.B.W., Nguyen, T.Q., Lepvrier, C., Thomsen, T.B, Tani., K., Charusiri, P.: Neoproterozoic to early Triassic tectono-stratigraphic evolution of Indochina and adjacent areas: a review with new data. J. Asian Earth Sci. **191**, 104231 (2020)

conducted. Therefore, a workflow in the part of the discussion in this paper is introduced that shows how subsidence data can be interpreted with the help of coupled hydro-mechanical simulations conducted with numerical multi-physics software. We present the data basis and model setup for the planned modeling study with the open-source software platform OpenGeoSys.

Keywords Land subsidence · InSAR · Hydro-mechanical modeling · Hanoi · OpenGeoSys

1 Introduction

Land subsidence is a global problem due to natural and human activities with gradually increasing consequences, e.g., infrastructure damage, land surface cracks, increasing the risk of flooding, reducing the capacity of aquifers to store water, and eventually posing a risk to society and the economy [1]. Severe problems caused by land subsidence were observed in megacities in the world, e.g., Bangkok (Thailand), Houston (USA), Mexico City (Mexico), Tokyo (Japan), Shanghai (China), and Venice (Italy) [2]. Over extraction of groundwater, which lowered the water table, was the main cause of land subsidence in these cities [3].

Hanoi, located in the northern part, is the capital city and a cultural, political, and societal center of Vietnam. Since the expansion of the city in 2008 covering parts or the entire adjacent provinces, the newly formed Hanoi with more inhabitants has put a lot more pressure on groundwater extraction and high demand on infrastructure for fast urbanization. Given that more water was pumped out of the ground, the surface soil became more compacted, and increasing severe land subsidence in the populated areas was observed. Some areas had especially high subsidence rates, including Ngo Sy Lien District (32 mm/year), Thanh Cong District (41 − 42 mm/year), and Phap Van District (22 mm/year) [4]. Due to the scarcely distributed number of ground monitoring stations in Hanoi, it is challenging to monitor the land deformation in this city both in terms of time and space.

There have been several studies aiming at monitoring surface deformation over Hanoi City based on different types of data (e.g., geotechnical, geological, or hydrological data) [5–7]. These studies used high-quality in situ data but with poor spatial resolutions, restricted coverage, and time discontinuity. These restricted their capacity to monitor surface deformation and patterns in space, particularly to detect deformation hot spots [8]. Interferometric Synthetic Aperture Radar (InSAR) provides the capacity to measure the time series of surface displacements induced by earthquakes, volcanoes, subsidence, and uplift processes with high spatial resolution on a large scale [9]. The analysis of SAR radar data is based on the amplitude and phase. While the amplitude indicates the strength of the signal response, the phase is used to measure the distance between the satellite and the target, and thus the deformation. Differential InSAR (DInSAR) is one of the most widely used phase-related methods for measuring surface displacement [10].

Radar remote sensing data with InSAR have been successfully deployed to monitor the displacement in Hanoi in the 1995–2020 period [4, 8, 11–18]. The authors have successfully proved, from different perspectives, the use of various radar remote sensing techniques for land subsidence monitoring. However, these studies only showed the detection of land subsidence, and there is still a lack of assessment of the connection between land subsidence and triggering factors. Additionally, all of the studies have assumed that the horizontal movement over the city of Hanoi is insignificant and is thus neglected. Based on the literature review of InSAR applications in Hanoi City, including some aspects of the study area, InSAR methods, and SAR data, this study further extends the land subsidence monitoring by introducing the use of radar remote sensing combined with hydro-mechanical modeling to quantify the relationship between hydrogeological factors and land subsidence in the urban setting of Hanoi over the 2016–2019 period. The results presented in this article are the initial phase before the comprehensive assessment using the subsequent hydro-mechanical modeling.

The rest of the study is organized as follows: Sect. 2 introduces the city of Hanoi. InSAR time series analysis methods and data are provided in Sect. 3. Section 4 reviews literary studies on InSAR application over Hanoi City. Section 5 discusses the combination of InSAR and hydro-mechanical modeling, and Sect. 6 concludes the study.

2 The City of Hanoi

Hanoi is located in the northeast of Vietnam between $20^\circ 53' - 21^\circ 23'$ latitudes and $105^\circ 44'$–$106^0 02'$ longitudes, covering an area of 3.358.9 km^2 (Fig. 1). Hanoi is one of the most populous cities in Asia, with an average population density of 2.087 people per km^2 (population statistics made in 2019 [19]). It is the core city in the Red River Delta with crucial political, socio-economic, cultural, and scientific importance besides Ho Chi Minh City in the Mekong River Delta in the South.

With the average height ranging from 5 to 20 m above sea level, the terrain gradually lowers from north to south and from west to east. Quaternary sediments cover the entire area; specifically, there are different formations divided by the origin and age in Hanoi from top to bottom as follows: Late Holocene alluvial; Early-Middle Holocene deposits, marine and bogged sediments; Late Pleistocene alluvial, lacustrine; Middle-Late Pleistocene alluvial; alluvial-proluvial deposits; and Early Pleistocene alluvial deposits [14]. Hanoi's aquifer system consists of four primary units from the surface down to 80–100 m depth, including the Holocene aquifer, Holocene and Pleistocene aquitards, the Pleistocene aquifer, and Neogene sandstone bedrock [6].

The consequences of ground subsidence in the center of Hanoi (i.e., urban districts) have been recorded since the late 1980s, especially in old apartment buildings. Recently, some new urban areas such as Dong Tau and Van Phu are also affected by high-rate subsidence in the 2017–2019 period. Land subsidence in the setting of

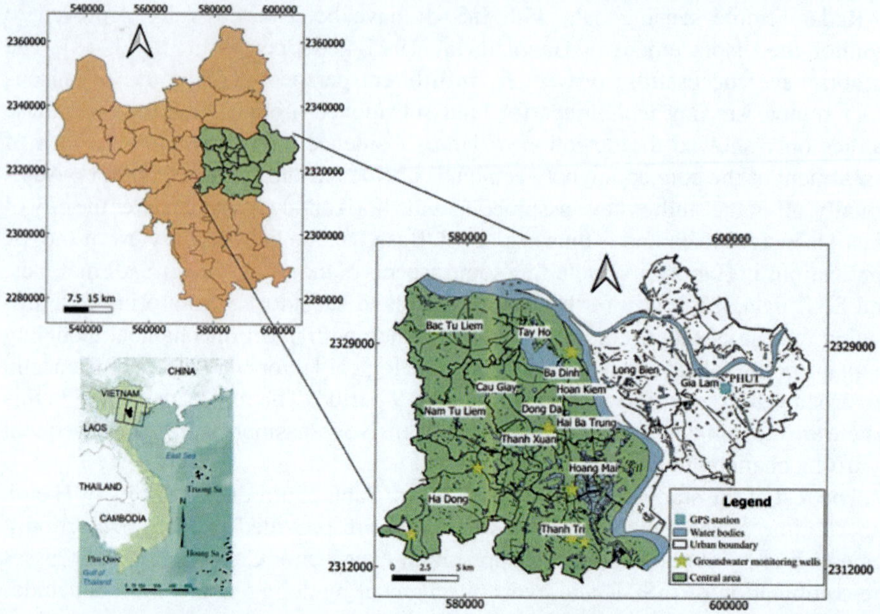

Fig. 1 The city of Hanoi (upper figure) with its urban districts (blue area) and spatial distribution of groundwater and GNSS monitoring stations (right figure)

Hanoi, Vietnam, is mostly considered to be the consequence of human activities such as excessive groundwater extraction, rapid urbanization, and rapid urban population increase [11].

3 Interferometric Synthetic Aperture Radar

3.1 InSAR Time Series Analysis Methods

DInSAR is a remote sensing method for precisely computing the deformation time series for a large area. The phase difference between two complex radar SAR observations over the same area indicates the Earth's surface deformation and other error and noise terms [20]:

$$\Delta\phi = \Delta\phi_{\text{disp}} + \Delta\phi_{\text{flat}} + \Delta\phi_{\text{elev}} + \Delta\phi_{\text{atm}} + \Delta\phi_{\text{noise}} + \Delta\phi_{\text{err}} \quad (1)$$

where $\Delta\phi_{\text{disp}}$ refers to the differential deformation pattern along the line-of-sight (LOS) direction, $\Delta\phi_{\text{flat}}$ is the phase associated with the assumption of an ideally flat earth terrain, $\Delta\phi_{\text{elev}}$ is the interferometric phase caused by terrain topography, $\Delta\phi_{\text{atm}}$ denotes the atmospheric effect, $\Delta\phi_{\text{noise}}$ refers to the noise contribution from the radar instrument and temporal deceleration, and $\Delta\phi_{\text{err}}$ due to orbital error and topographic height information. In DInSAR, accurate identification of ground deformation ($\Delta\phi_{\text{disp}}$) requires eliminating all the other contributions [20].

Persistent Scatterer InSAR (PSInSAR) [21, 22] and Small BAseline Subset (SBAS) [23, 24] are among the widely used InSAR time series analysis methods working with the DInSAR technique. SBAS is advantageous that multiple interferograms are used in an interferogram network [25], thereby resulting in redundant interferograms adopted to reduce the noise in the InSAR time series, particularly in the case of a low signal to noise ratio [26]. In contrast, PSInSAR exploits a network of interferograms with a single primary image to work with highly coherent scatterers [21, 22]. As a result, it allows for a reduction in the primary image noise contribution, which is present in all interferograms. PSInSAR and SBAS improve the ability to identify ground deformation with millimetric precision, which helps to overcome the limitations of DInSAR [21]. Both PSInSAR and SBAS have recently been applied over the city of Hanoi, which showed consistent results [8]. The main steps applied in SBAS and PSInSAR data processing with the InSAR Scientific Computing Environment (ISCE), Generic InSAR Analysis Toolbox (GIAnT), and Stanford Method for Persistent Scatterer (StaMPS) software packages are shown in Fig. 2.

If the horizontal movement is insignificant, which was an assumption in many literary studies, the deformation in the LOS direction can be converted to that in the vertical direction by [9]:

$$d_U = \frac{d_{\text{LOS}}}{\cos(\theta_{\text{inc}})} \tag{2}$$

where d_U and d_{LOS} are the deformations in the vertical and LOS directions, and θ_{inc} is the radar incidence angle.

3.2 SAR Data

On a global scale, the use of radar satellites for remote sensing has a long history. Starting with Seasat in the late 1970s, numerous missions have contributed to an enormous range of data in the main three radar frequencies (X, C, and L bands) that have improved environmental and Earth system sciences. Figure 3 shows the progress of several SAR platforms in time which are summarized in Table 1.

At present, some active SAR satellites include Sentinel-1 (C-band), Radar SAT-2 (C-band), ALOS-2 (L-band), SAOCOM (L-band), TerraSAR-X (X-band), TanDEM-X (X-band), and COSMO-SkyMed (X-band) (see Fig. 3 and Table 1). However, the rule of the use of data acquisitions of these missions is highly inconsistent in the literature, depending on various factors, e.g., the background missions (regular acquisition over a defined area) or data availability. Typically, a C-band PSI study usually necessitates a minimum of 15–20 images [27]. In contrast, because of the better resolution and shorter wavelength of the X-band PSI, it is possible to employ

4 Review on the Studies Using InSAR in Land Subsidence Monitoring in Hanoi

During the last decades, the large amount of data from the abovementioned satellites and the development of various techniques enable utilizing the radar images over Hanoi. One of the earliest studies applying radar remote sensing to detect land subsidence in Hanoi was by Tran et al. [17]. In this study, three JERS-1 SAR L-band images from 1995 to 1998 were used with the three-pass differential interferometry (one of the DInSAR methods) to show some areas with high deformation rates in the urban area of Hanoi, such as Yen Phu, Long Bien, Ba Dinh, and Ngo Sy Lien area. However, some high rates of subsidence in the southern part of Hanoi City (Phap Van, Van Dien, or Ba La) could not be detected by DInSAR analysis because of the low coherence from pairs of scenes.

Dang et al. [11] used L-band ALOS-PALSAR images to quantify the spatial distribution of land subsidence in the whole Hanoi urban region using a multi-temporal InSAR technique from 2007 to 2011. Some regions with high deformation rates were found including Hoang Mai, Ha Dong, and Hoai Duc–Tu Liem districts. The unsaturated layer's nature, the reduction of groundwater levels in aquifers due to pumping extraction capacity (Qp aquifer), the expansion of built-up surfaces, and the kind of building's foundation are all factors that were considered to influence the spatial distribution of surface deformation during this study time.

Over the period from 2011 to 2014, some studies combining TerraSAR and COSMO-SkyMed data demonstrated the feasibility of ground subsidence by X-band SAR data in the condition impacted by a strong atmosphere in Vietnam [12, 13, 15, 18, 30]. These studies also concentrated on the urban area of Hanoi using different InSAR time series analysis methods, such as PSInSAR and PS/DS techniques. In specific, Ho et al. [12, 13] showed that subsidence is particularly powerful in the HaiHung silt loam regions in the city's south and one of the primary causes of ground sinking is groundwater overexploitation. Le et al. [14] applied TerraSAR data with the SBAS method to extract the displacement in the historical center of Hanoi. This research denoted that the image oversampling implemented in the SB InSAR processing chain helps to eliminate some of the noisiest locations and allows for the assessment of the status of a single structure or monument, which is required for the monitoring of cultural heritage buildings, monuments, and sites for subsidence.

With the free C-band Sentinel-1 data from the European Space Agency (ESA), researchers have more opportunities to access easily the SAR data in the 2016-present period. Therefore, Sentinel-1 SAR data are used mainly for the ground deformation in the Hanoi area recently [8, 16]. Bui et al. [8] showed the high consistency between the results derived from SBAS, PSInSAR, and stacking methods in the spatio-temporal displacement rate in the urban area in Hanoi from 2016 to 2020.

By applying both PSInSAR and SBAS methods, Nguyen et al. [31] used different SAR sensors including ALOS, COSMO-SkyMed, and Sentinel-1 data to produce a long InSAR deformation time series in the urban area in Hanoi from 2008 to 2018. With InSAR deformations validated by leveling and GNSS data for a time-series ten years, this study denoted that subsidence in downtown Hanoi is waning, whereas the southern and western suburbs (Ha Dong and Hoai Duc districts) have emerged as the new subsidence hotspots in recent years. Besides, Hanoi's subsidence problem is relatively mild in comparison with other Southeast Asian cities.

Table 2 summarizes literary studies of InSAR application to monitor surface subsidence in Hanoi City.

As can be seen from the previous InSAR studies, SAR data of X, C, and L bands were applied successfully for extracting surface deformation mainly in the urban setting of Hanoi from 1995 to the present. It helped to detect the areas of high subsiding rates (e.g., higher than −20 mm/year) in the southern part of Hanoi, including Ha Dong, Hoang Mai, and Thanh Tri. Besides, the common philosophy of

Table 2 Literary studies of InSAR application to monitor surface subsidence in Hanoi City

Reference	SAR data	Time span	Number of images	Time series analysis method
Tran et al. [17]	JERS-1 (L-band)	1995–1998	3	Three-pass DInSAR
Dang et al. [11]	ALOS PALSAR (L-band)	2007–2011	22	SBAS, PSInSAR
Tran et al. [30]	ALOS PALSAR (L-band) TerraSAR (X-band) COSMO-SkyMed (X-band)	2007–2011 2012–2014 2011–2013	22 18 17	PS/DS (SqueeSAR)
Ho et al. [12]	ALOS PALSAR (L-band) TerraSAR (X-band) COSMO-SkyMed (X-band)	2007–2011 2012–2014 2011–2013	22 18 17	PS/DS (SqueeSAR)
Ho et al. [13]	TerraSAR (X-band) COSMO-SkyMed (X-band)	2012–2014 2011–2014	19 27	PS/DS (SqueeSAR)
Le et al. [14]	TerraSAR (X-band)	2012–2013	23	SBAS
Nguyen et al. [15]	TerraSAR (X-band)	2012–2015	23	PSInSAR
Bui et al. [8]	Sentinel-1A (C-band)	2016 – 2020	114	SBAS, PSInSAR
Nguyen et al. [31]	ALOS PALSAR (L-band) COSMO-SkyMed (X-band) Sentinel-1	2007–2011 2011–2015 2015–2018	22 17 98	PS/DS (StaMPS) PS/DS (StaMPS) SBAS

the abovementioned research was to derive the deformation in the vertical direction only (converted from line-of-sight deformation) with the assumption that the horizontal movement in the urban setting of Hanoi is insignificant. Due to a lack of data on ground monitoring stations in Hanoi, some available sources as an alternative for validation of InSAR deformation are Global Positioning System (GPS) deformation, a comparison of deformation hot spots [11].

5 Proposal of a Further Study on the Combination Between InSAR Results and Hydro-mechanical Modeling

The vertical deformation rates in the urban area in Hanoi measured by Sentinel-1 data from 2016 will be used as a basis for subsequent hydro-mechanical modeling. The results [8] are shown in Fig. 4 with the negative displacements shown by yellow to red colors representing the movements away from the satellite (i.e., land subsidence), while the positive values (green) represent the movements toward the satellite (i.e., land uplift). The velocity of ±2 mm/year is characterized by light-green color and is considered stable.

The vertical deformation time series derived from InSAR will be validated with GNSS data at the PHUT station [8] (Fig. 5). The InSAR deformation is calculated

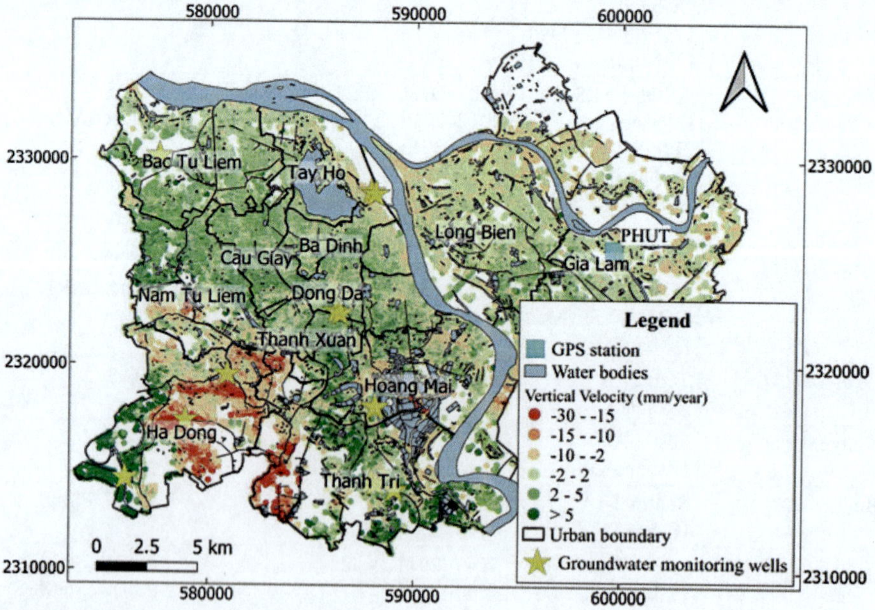

Fig. 4 Displacement in the vertical direction in the urban area of Hanoi from April 03, 2016, to December 20, 2019

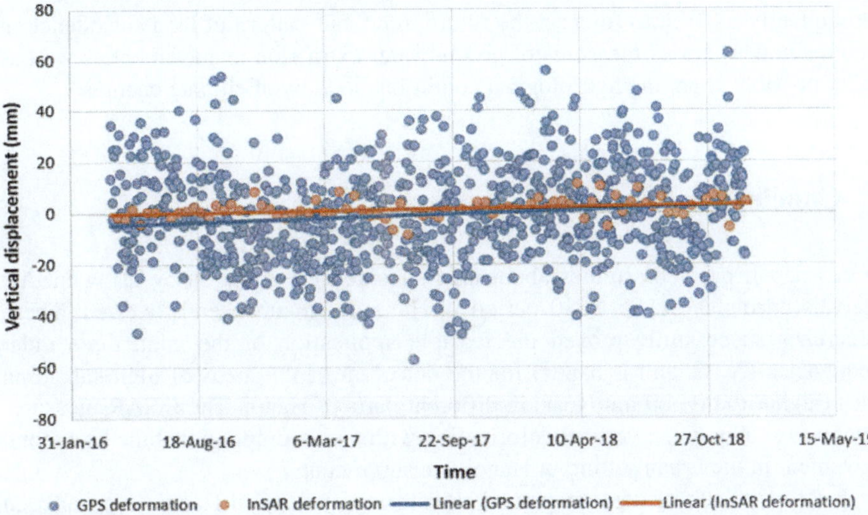

Fig. 5 Comparison between PSInSAR and GNSS-derived deformations in the vertical direction at the PHUT station

by averaging the time series within a 100-m radius centered at the GNSS station. This comparison is to test whether the InSAR and GNSS deformation time series are consistent (i.e., an agreement in the subsidence/uplift trend).

Besides, a large number of data are recently available, including boreholes with geological data, DEM, rainfall distribution, groundwater levels, and pumping rates, which will be used as inputs for a planned study with the hydro-mechanical model. The model will be generated using the open-access software OpenGeoSys [32], a finite element code developed for thermo-hydro-mechanical-chemical coupled modeling of geo-environmental processes. For the hydro-mechanical simulation of land subsidence, a poroelastic constitutive model using the Biot formulation can be used to calculate displacements.

The model will be calibrated using the InSAR-derived deformations mentioned above. The hydraulic conductivities and geotechnical parameters are specifically calibrated in OpenGeoSys. This proposed workflow includes multiple steps, including data standardization, the development of a conceptual and geological model using borehole data, numerical model setup (including boundary conditions, and assignment of hydraulic and mechanical properties to the geological layers), model calibration using InSAR-derived deformations, and analysis and scenario simulations.

The calibrated model will help assess the triggering factors that have impacts on the spatio-temporal distribution of land subsidence observed from the remotely sensed data. Specifically, we will analyze the impact of both pumping rates and the distribution of clay layers (aquitards) in the subsurface to better understand observed subsidence patterns. Scenario simulations with the calibrated model will subsequently be used to forecast the future trend and pattern of land subsidence in the area under various measures of groundwater extraction management as well as under possible changing hydrological conditions in view of climate change.

6 Conclusion

In this paper, previous studies about land subsidence estimation by using InSAR analysis during the 1995–2020 period in Hanoi, Vietnam, were reviewed. These researches successfully proved the feasible application of the main three radar frequencies (X, C, and L bands) for the detection of the areas of high subsiding rates (higher than −20 mm/year) in different parts of Hanoi. The InSAR measurement was shown in the vertical deformation with the assumption that the horizontal movement in the urban setting of Hanoi is insignificant.

Due to the lack of data on ground monitoring stations in Hanoi (especially in the period from 2012 to the present), some available sources as an alternative for validation of InSAR deformation are GPS deformation, a comparison of deformation hot spots [11].

The InSAR deformations derived by Bui et al. [8] that have high consistency with GNSS deformations will be an important input for the calibration of future numerical hydro-mechanical models. Using the open-source software platform OpenGeoSys, subsidence in the entire aquifer system of the area will be simulated. The data needed for the next phase of the numerical analysis include InSAR measurements and other input data. The general modeling workflow was also proposed in this paper.

References

1. Holzer, T.L.: Galloway, impacts of land subsidence caused by withdrawal of underground fluids in the United States. In: Humans as Geologic Agents (2005)
2. Holzer, T.L., Ivan, J.A.: Land subsidence caused by ground water withdrawal in urban areas. Geojournal **11**(3), 245–255 (1985)
3. Richa, B., et al.: Detection of land subsidence in Kathmandu Valley, Nepal, using DInSAR technique. Land **6**(2) (2017)
4. Tran, Q.C., et al.: Research on scientific basis and proposed solution of land subsidence prediction in Hanoi by using Radar interferometry technique. Institute of Geology at Vietnamese Academy of Science and Technology: Hanoi (2016)
5. Nhu, V.H: Spatio-temporal prediction of land subsidence using GIS-based 3D geological, geotechnical and groundwater flow models in Hanoi city (Vietnam). In: Department of Hydrology and Hydraulic Engineering, p. 178. Vrije Universiteit Brussel: Belgium (2015)
6. Pham, H.G., et al.: Land subsidence prediction for a new urban mass rapid transit line in Hanoi. Underground Space (2018)
7. Phi, T.H., Strokova, L.A.: Prediction maps of land subsidence caused by groundwater exploitation in Hanoi, Vietnam. Resource-Eff. Technol. **1**(2), 80–89 (2015)

8. Bui, K.L., et al.: Recent land deformation detected by Sentinel-1A InSAR data (2016–2020) over Hanoi, Vietnam, and the relationship with groundwater level change. GIScience Remote Sens., 1–19 (2021)
9. Kampes, B.M.: Radar Interferometry—Persistent Scatterer Technique. Springer, Germany (2006)
10. Fárová, K., et al.: Comparing DInSAR and PSI techniques employed to sentinel-1 data to monitor highway stability: a case study of a massive Dobkovičky landslide, Czech Republic. Remote Sens. **11**(22) (2019)
11. Dang, V.K., et al.: Recent land subsidence caused by the rapid urban development in the Hanoi region (Vietnam) using ALOS InSAR data. Nat. Hazard. **14**(3), 657–674 (2014)
12. Ho, T.M.D., et al.: Measuring ground subsidence in Hanoi city by radar interferometry. Sci. Technol. Devel. **19**(K4), 8 (2016)
13. Ho, T.M.D., et al.: Measuring ground subsidence in Ha Noi through the radar interferometry technique using TerraSAR-X and cosmos SkyMed data. IEEE J. Sel. Topics Appl. Earth Observ. Remote Sens. 1–11 (2019)
14. Le, T., et al.: TerraSAR-X data for high-precision land subsidence monitoring: a case study in the historical centre of Hanoi, Vietnam. Remote Sens. **8**(4) (2016)
15. Nguyen, D.A., et al.: Application Terrasar-X data for studying land subsidence in Hanoi city. Vietnam J. Earth Sci. **41**(4), 339–357 (2019)
16. Nguyen, M., Chang, C.-P., Tseng, K.-H.: Using sentinel-1 TOPS SAR and SBAS for land subsidence monitoring in Hanoi, Vietnam. In: Conference Paper: International Symposium on Remote Sensing (2018)
17. Tran, V.A., et al.: Spatial distribution of subsidence in Hanoi detected by JER-1 SAR Interferometry. Geoinformatics **18**(1), 3–13 (2007)
18. Tran, V.A., et al.: Subsidence assessment of building blocks in Hanoi urban area from 2011 to 2014 using TerraSAR-X and COSMO-SkyMed images and PSInSAR. In: Remote Sensing and GIScience, pp. 127–150 (2021)
19. General Statistics Office of Vietnam in Average population in term of regions (2019)
20. Reigber, A., et al.: I.D.I.O.T.: a free and easy-to-use software tool for DInSAR analysis. In: Proceedings of the Envisat Symposium 2007. Montreux, Switzerland (2007)
21. Ferretti, A., Prati, C., Rocca, F.: Permanent scatterers in SAR interferometry. IEEE Trans. Geosci. Remote Sens. **39**(1), 8–20 (2001)
22. Hooper, A., et al.: A new method for measuring deformation on volcanoes and other natural terrains using InSAR persistent scatterers. Geophys. Res. Lett. **31**(23), L23611 (2004)
23. Berardino, P., et al.: A new algorithm for surface deformation monitoring based on small baseline differential SAR interferograms. IEEE Trans. Geosci. Remote Sens. **40**(11), 2375–2383 (2002)
24. Lanari, R., et al.: Satellite radar interferometry time series analysis of surface deformation for Los Angeles, California. Geophys. Res. Lett. **31**(23), 1–5 (2004)
25. Castellazzi, P., et al.: Assessing groundwater depletion and dynamics using GRACE and InSAR: potential and limitations. Groundwater **54**(6), 768–780 (2016)
26. Bui, L.K., Featherstone, W.E., Filmer, M.S.: Disruptive influences of residual noise, network configuration and data gaps on InSAR-derived land motion rates using the SBAS technique. Remote Sens. Environ. **247**, 111941 (2020)
27. Crosetto, M., et al.: Persistent Scatterer Interferometry: a review. ISPRS J. Photogramm. Remote. Sens. **115**, 78–89 (2016)
28. Bovenga, F., et al.: Using COSMO/SkyMed X-band and ENVISAT C-band SAR interferometry for landslides analysis. Remote Sens. Environ. **119**, 272–285 (2012)
29. Bui, L.K.: Small baseline subset InSAR data processing: design of interferogram networks and noise analysis in InSAR-derived deformation time series. Curtin University (2021)
30. Tran, Q.C., et al.: Ground subsidence monitoring in Vietnam by multi-temporal InSAR technique. In: IEEE International Geoscience and Remote Sensing Symposium (IGARSS), pp. 3540–3543 (2015)

31. Nguyen, M., et al.: Assessment of long-term ground subsidence and groundwater depletion in Hanoi, Vietnam. Eng. Geol. **299** (2022)
32. Kolditz, O., et al.: OpenGeoSys: an open-source initiative for numerical simulation of thermo-hydro-mechanical/chemical (THM/C) processes in porous media. Environ. Earth Sci. **67**(2), 589–599 (2012)